Smart Process Plants

About the Author

Miguel J. Bagajewicz, Ph.D., is the Sam Wilson Professor of Chemical Engineering at the University of Oklahoma. His research is in the fields of design, operation, simulation, and optimization of process plants and product design. Dr. Bagajewicz specializes in financial risk, environmentally benign processes, and microeconomics, as applied to product design.

Smart Process Plants

Software and Hardware Solutions for Accurate Data and Profitable Operations

Miguel J. Bagajewicz, Ph.D.
University of Oklahoma

Donald J. Chmielewski Contributor
DuyQuang Nguyen Tanth Contributor

New York Chicago San Francisco
Lisbon London Madrid Mexico City
Milan New Delhi San Juan
Seoul Singapore Sydney Toronto

The McGraw·Hill Companies

Cataloging-in-Publication Data is on file with the Library of Congress

1 2 3 4 5 6 7 8 9 0 DOC/DOC 0 1 5 4 3 2 1 0 9

ISBN 978-0-07-160471-0
MHID 0-07-160471-5

Sponsoring Editor
Taisuke Soda

Editing Supervisor
Stephen M. Smith

Production Supervisor
Pamela A. Pelton

Acquisitions Coordinator
Michael Mulcahy

Project Manager
Ekta Dixit,
Glyph International

Copy Editor
Megha Roy Choudhary

Proofreader
Ragini Pandey,
Glyph International

Indexer
Robert Swanson

Art Director, Cover
Jeff Weeks

Composition
Glyph International

Printed and bound by RR Donnelley.

McGraw-Hill books are available at special quantity discounts to use as premiums and sales promotions, or for use in corporate training programs. To contact a representative, please e-mail us at bulksales@mcgraw-hill.com.

This book is printed on acid-free paper.

A Mita, el tigre; a mis inquietos y creativos hijos, a mis perspicaces y curiosos nietos y a todos los amigos lejanos y cercanos eternamente pendientes de nuestros progresos y éxitos.

"Do or do not. There is no try."

Yoda

"The possession of facts is knowledge; the use of them is wisdom."

Thomas Jefferson

Contents

Preface

Book prefaces are boring, so I will get to the point.

A few years ago, a new name was coined for process plants that are run efficiently. The name "smart plants" suggests that somehow there is some intelligence and self-awareness added to existing plants that will let them run more efficiently. The key then is to define efficiency.

Aside from preventing accidents that would cost lives or cause injuries, all the rest of the objectives of efficiency can be summarized in only one short statement: plants are efficient when their profitability is maximized, or in some other economic contexts, operating costs are minimized. While the driving force is cost, the long-term underlying concept that would accomplish this new improved efficiency is that problems (disturbances, wear and tear, accidents, faults, etc.) can be prevented, in addition to the current philosophy of having means to react to them, bringing back the plant to the required state. In Chap. 1, I elaborate further on the many existing definitions of smart plants.

Many of us in the process systems engineering community believe that, ultimately, the design of control, fault detection, and alarm systems; production accounting and quality control; and maintenance are driven by the relationship between the economic value they generate and their cost. In addition, at the root of all these activities are the estimators of key variables, which are obtained from the instrumentation network. Thus, the more accurate (precise and bias free) the measurement and the estimators, the more value one obtains. However, accurate measurements require redundancy and increase costs.

Thus, maximizing the difference between the added value and its cost is the main theme of this book. We therefore depart conceptually from what we call substitute objectives: accuracy for plant monitoring, stability and other considerations for control, and fault observability. We simply reduce everything to plant economic welfare.

In the production accounting and quality control area, I started to discuss the value of precision first and the value of accuracy later. Having worked for years in the design of instrumentation networks with cost as an objective, I proposed the value minus cost paradigm a few years ago. We are now embarked on the efficient solution of this instrumentation network design problem. Thus, Chaps. 2 through 12 cover methods to obtain accurate estimators of key variables (data reconciliation, Kalman filtering, etc.) and discuss means to obtain the economic value of the information produced. The control community is already moving toward this paradigm of value minus cost, by tying the design of control systems to profit. This is covered in Chap. 13 by Donald Chmielewski, a guest contributor. Nonstructural faults and the value obtained from their identification is covered in Chap. 14. In turn, because the bulk of the cost of implementing these methodologies stems from picking a redundant instrumentation network, in Chaps. 15 and 16 we present methods to obtain the optimal instrumentation type, quality, precision, and location. Finally, Chaps. 17 through 19 cover methodologies, some of them very recent, that allow economically optimal preventive maintenance of all equipment and especially instrumentation.

The book is not a textbook and it does not pretend to be thorough and complete. In fact, understandably so, there is a tendency to refer to my own work more than to others'. I made the honest attempt to be inclusive and refer to everybody's work. The pressures of delivering the book on time prevented me from making a thorough analysis of all existing work in detail. In addition, I can already make a list of shortcomings and things I regret not having the time to include or present better. However, I will leave that judgment to the readers and eventual critics.

I chose to use smaller type for examples, detailed calculations, and historical remarks. In this way, a reader can focus on large-type material and examples only and ignore the rest without losing continuity in the presentation.

Chapters 17, 18, and 19 are coauthored by my graduate student, Quang, who has done an excellent job of researching maintenance methods. A good portion of these chapters is adaptation from the maintenance/reliability literature, but there is also some novel material we produced. He also was instrumental in editing many other chapters.

Chapters 13 and 16 were written entirely by Donald Chmielewski. I am glad I asked him to make the contribution. I nagged a lot for the chapters to blend with the rest of the book and Don complied. As a result, I must say that these are among the best chapters in the book, perhaps one of the best introductions to contemporary value-based control pieces I have seen. They are clear, didactic, and not lacking insights on more sophisticated material. Finally, I must acknowledge Prof. Ragunathan Rengaswamy and Sridharakumar Narasimhan for providing valuable suggestions for Chap. 14.

Because of the intention of presenting the tip of the iceberg of complex and well-developed areas, the book is not comprehensive; that is, it does not intend to be a comprehensive and exhaustive review of the field.

The great staff of the School of Chemical, Biological and Materials Engineering at the University of Oklahoma has facilitated for me so many things that I cannot enumerate them. My family has contributed with a great deal of patience. There is no way to repay them.

Miguel J. Bagajewicz, Ph.D.

CHAPTER 1

Smart Plants

We start with the following analogy:

We drive cars with a certain level of mastery and skillfulness, although we cannot completely avoid accidents. A lot of functionalities run automatically: cruise control, engine temperature control, fuel injection, etc. There are also fault-detection and isolation systems connected to alarms. The driver, however, only reacts to all disturbances, and is unable to alter the driving proactively to avoid the disturbances or better deal with them.

The car of the future is a smart car: it will take your orders about where you want to go and will take you there while you read a paper or relax; it will keep track of where the other cars are so that collisions are minimized or even almost suppressed. In other words, *the smart car will drive itself better than humans can drive them*. The technology for all this exists, but it needs to be put together. The incentive here is comfort and economics.

Process plants are like cars. Current plants operate in two levels. At the process control level:

- They are operated using control loops to keep certain key variables within specified limits under the effect of normally expected disturbances.

At the process management level:

- There is a hierarchy of business decision-making tools designed to maximize profit. They set production rates, sometimes independently of processing constraints, establishing the set points.
- They react to unusual disturbances through a set of alarms and associated actions designed to prevent losses or accidents.
- Planning schemes determine what safety inventory is kept to overcome production shortcomings.
- Production accounting and quality assurance perform their duties using data gathered by measurements.

Christofides et al. (2007) discuss the vision and challenges of smart plant operations that are in part a reflection of their participation in an NSF-sponsored Workshop on Cyberinfrastructure (Davis, 2007). Plant monitoring is based on sensor data transmitted to the plant management system, and plant-wide optimization is based on economic considerations. They also point out that "it is becoming increasingly important to develop plant management strategies that balance the need for maximizing operating profit with sustainable Environmental, Health and Safety (EH&S) performance and corporate social responsibility. Specific EH&S targets may include:

- no damage to facilities and processes or injury to personnel,
- no releases of toxic materials to the environment,
- no negative impact to communities adjacent to operating facilities,
- no physical or cyber-security breaches, and
- no unplanned outages."

The key issues here are that existing plants *react* to disturbances, but are still "driven" by humans for the most part. However, most of the technology to make them fully automated, able to act proactively to avoid problems, exists. The incentives are

- Profit maximization and/or cost reduction.
- The incorporation of the aforementioned EH&S and social responsibility targets, some of which also have a cost reduction component.

The idea that plants should seek aid from advance control and other data-processing technologies for fault detection and therefore that these technologies should migrate into prevention was pointed out first by industrial contributors. As usual, the new paradigm emerges in a form of a new innovative use of existing tools and thus creating the need for and suggesting new ones.

It is unclear who first coined the term "smart plants," but this vision has many early proposers, although they may not have used the term (Koolen, 1994; Pelham and Pharris, 1996). White (2003) for example pointed out the advantages of incorporating increasing levels of measurement and modeling to prevent failures. He was among the first to emphasize "prediction" as opposed to "reaction" and he cited three major incentives: financial returns, safety and environmental issues, and workforce demographics. Gipson (2005) and Humphrey (2005) were among the first to use the name explicitly and started an effort to define the goals and activities that a "smart plant" would be based on.

In an attempt to define the term more precisley, Christofides et al. (2007) stated that

"The objective of a 'Smart Plant' is to make optimal use of the plant assets first and foremost to approach zero-incident and sustained EH&S targets, and maximize the economic operating value of the plant. Ideally, in a 'Smart Plant,' each asset—from the smallest pipe or pump, to a single process unit and to collections of processes—not only executes its basic process function, but also provides feedback and predictive information, through real-time communication networks, on the current and expected performance of that asset to the plant management system. This, in turn, will allow the plant management system, together with human decision-making, to maximize, based on the current and future collective performance of all assets, the use (and, thus, the value) of each asset as business conditions change and warrant."

Christofides et al. (2007) identify three distinct levels:

- The fault-tolerant process control level, where real-time feedback control via wired or wireless communication takes place and where real-time information also feeds the process management levels and receives the set points.
- The plant management level, where operating constraints and conditions are set based on the business strategies.
- The corporate office level, which sets the business strategies.

To accomplish this, they point out the recent advances in key areas as follows:

- *Process control methods* dealing with nonlinearities, model predictive control, and Lyapunov-based/backed-off control to deal simultaneously with nonlinearities, model uncertainty, and actuator constraints. They also point out the advantages of new feedback control systems that include differential equations or differential algebraic systems, as well as nonlinear distributed parameter systems. Finally, another emerging area they mention is that of fault-tolerant robust process control capable of functioning under faults.
- *Plant-wide optimization and supervisory control*: Typically, these are large steady-state optimization models solved less frequently than the control calculations and include all the environmental, health, and safety (EH&S) issues, the logistics, and the plant economics, as well as other limitations. This type of optimization has received a name: real-time optimization (RTO), and when done online, is called "online optimization."
- *Process monitoring*: Fault detection and isolation through the use of dynamic filters and computing residuals to

specific faults in addition to other specialized algorithms for early detection as well as multiresolution smart networks for anomaly detection.

They also point out one important element of this vision: the paradigm "is not hierarchical with respect to information flows. Rather, through extensive interconnected communications and networks individual assets can self-evaluate their individual and collective performances, both current and expected, toward the business goals of the plant." In other words, all flow of information is bidirectional; that is, the corporate level does its profit optimization using information from the plant to incorporate the appropriate constraints. If there were a hierarchy, then the plan produced by the business office might not be feasible and adjustments would have to be made.

Our vision of the plant management and the control sections is depicted in Fig 1.1.

Humphrey (2005) also proposed the smart plant concept in 2005 and has been chairing sessions in the AIChE annual conference ever since. He focuses more on the "proactive" aspect of preventing faults. His definition is the following:

"The focus of the Smart Plant is on combining people-processes-technologies to prevent plant problems before they occur. Examples of problems Smart Plants prevent include unplanned shutdowns, excessive energy use, and unwanted safety events. Accurate real time measurements for individual processes are the foundation of the Smart Plant (e.g., flow rates, compositions, etc.)."

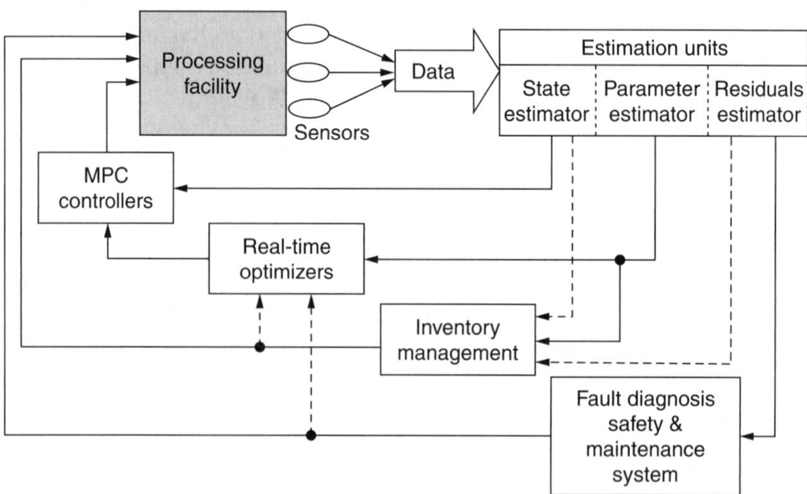

FIGURE 1.1 Smart plant management and control levels.

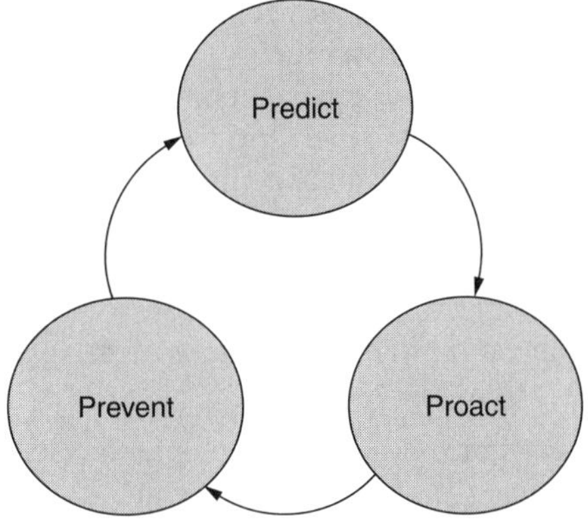

FIGURE 1.2 Smart plant "predict-proact-prevent cycle."

He proposes the use of concept shown in Fig. 1.2 and in the 2008 AIChE annual conference he presented an abbreviated version of his definition of smart plants:

"The Smart Plant utilizes PEOPLE, PROCESSES, and TECHNOLOGIES to (1) prevent plant problems and (2) optimize performance for the good of the enterprise as a whole."

Examples of predictions that allow proactive actions are

- Reduce excessive energy consumption: It could save $20 billion a year.
- Reduce unplanned shutdowns: Calculated savings are also in the order of $20 billion a year.
- Plan maintenance: It can save $5 billion a year.

He also identifies smart managers and accurate data as key enablers.

In turn, in 2008, in another NSF roadmap development workshop (Davis, 2008), smart plants were identified more specifically:

"Smart process manufacturing is the realization of a continuous or batch manufacturing environment wherein all critical elements are maintained within acceptable operating conditions and the product is qualified for sale and delivery based on the automatic adaptation and assurance of in-control operation. In addition to product quality, smart process manufacturing embraces all operational aspects. Smart process manufacturing is the enterprise-wide application of 'smart' technologies, tools, and systems coupled with knowledge-enabled personnel to plan, design, build, operate, maintain, and manage process manufacturing facilities in full concert with the business and manufacturing missions of the enterprise.

"Smart process manufacturing refers to a design and operational paradigm involving the integration of measurement and actuation, ES&H protection, regulatory control, real-time optimization and monitoring, and planning and scheduling. This integrated processes approach provides the basis for a strong predictive and preventive mode of operation with a much swifter incident-response capability."

The report adds that smart process manufacturing "drives towards zero emissions and zero incidents through proactive, predictive enterprise optimization and management." Later, they emphasize the following characteristics of smart process manufacturing systems. They

- "Are capable of intelligent actions and responses."
- "Can adapt to new situations or perturbations (i.e., abnormal situations)."
- Have information that is available and timely appropriate.
- Are proactive.
- Have assets that are integrated and self-aware (via sensors) of their state.
- Use field devices that have intelligent processing capability with the sensors needed for self-awareness.
- Use human resources (people) who are knowledgeable, well trained, empowered, connected (via cyber tools), and able to adapt/improve the system's performance.
- Are sustainable. Sustainable manufacturing includes reuse, with a life-cycle view of products and processes. A minimum environmental footprint (energy, water, emissions) is a high-priority goal of a smart manufacturing environment.

The smart manufacturing environment proposed in Davis et al's report is one that has the following attributes:

- Use multiscale dynamic modeling and simulation. Perform large-scale optimization to address large-scale problems at the strategic, tactical, and operational levels.
- Data interoperability.
- Sensor networks.
- Scalable requirements-driven security at all levels.

The concept of "smart plants" has already permeated into the industrial activity. Lenz and Chester (2008) illustrated activities in power systems that are based on the incorporation of "wisdom" as they put it, through the use of plant models to help "preventing problems and optimizing operations and costs, making plants smarter, safer and more profitable." Huynh and Haake (2008) reviewed the use of advanced data analytics, pattern recognition, semantic data modeling, and knowledge management to "capture, predict, prevent and correct plant problems even before they occur." Sharpe (2009) highlights the use of advanced process control to "drive the operation toward its optimum point." His approach "leverages pre-engineered, multi-use applications across multiple installations."

Our Vision

Our long-term vision is that smart plants, like cars

- "Drive" themselves, with minimal human intervention. In other words, full automation that is fault tolerant and capable of decision making.
- Are capable of gathering information from the environment as well as from markets, to predict problems (faults) and act proactively to prevent them.
- Reduce emissions and meet EH&S factors.
- Operate in such a way that maximum profit is achieved.

This vision is far from being materialized. There are several challenges in control, fault detection, modeling, online optimization, operations planning and scheduling, maintenance planning, etc. While there has been progress in all these areas, one fully integrated smart system including all these functions and running automatically is yet to be developed.

Some of these activities cannot be performed successfully without accurate and consistent data. It is transparent that, aside from state-of-the-art procedures, quality of data is a key aspect for smart plants. However, increased quality of data requires increased quality of instrumentation, which has an attached cost. Bagajewicz (2000) overviews several methodologies for the design and retrofit of sensor networks for this purpose. One chapter of the present book covers some more recent advances.

Value of Information

A key aspect of our vision for smart plants is then one in which

- Information has an economic value.
- Information has a cost that needs to be smaller than the economic value. Otherwise, it is not worthy of gathering.

This value of information paradigm includes

- Software that enhances the quality of measurement data by filtering biases, identifying good estimators, etc.
- Hardware, specifically sensors that are appropriately positioned, and have the adequate precision, to produce information of economic value larger than their cost.

We conclude that software (control methods, fault detection and isolation algorithms, maintenance planning, data reconciliation) and hardware (appropriate sensors located strategically) operate together to generate economic value. *Without considering their interactions, smart plants are not possible. This book focuses on these interactions.*

Industry is starting to be very keen on key performance indicators (KPI), which are measures that indicate how well a system is performing relative to a certain objective, or how well it complies with regulations or imposed limits. This is especially popular in "lean" manufacturing.

As far as smart plants are concerned, there are, in our view, only two KPIs: profit (or cost) and EH&S. If one is able to connect the choice of technology to be used for parameter estimation, real-time optimization, and the associated instrumentation design or retrofit to those KPIs, then there is no need for other indicators. However, because such a connection was not made in the past, other substitute indicators have been used. In data reconciliation, one would use accuracy of estimators; in control, backed-off operating point; in maintenance, availability; in fault diagnosis, observability and resolvability. Because of our insistence on tying everything to profit (or cost), we have no need to use other KPIs that are popular in manufacturing like overall equipment effectiveness, total effective equipment performance, which rely on other metrics like availability, loading, performance, and quality. To us, setting goals for those types of metrics is more difficult than finding the impact of the decision making on economics.

Book Focus and Contents

In the first part of the book, we focus on measurements and the data they produce. We later discuss data reconciliation as a technique that can help identify biases and provide good estimators of process variables.

Recognizing that biases are unavoidable and that some may not be identified, we discuss the concept of accuracy and show means to calculate it for each variable of interest. We later identify the value of such accuracy for production accounting, quality assurance, and control purposes. We then discuss new emerging techniques for fault isolation, the sensors that enable it, and the value of the associated information.

Next, we discuss methodologies that are capable of designing the sensor networks that will function aided by the methods presented in previous chapters. These techniques are those that compare value of information with its cost and determine where sensors should be placed and what is their precision.

Finally, we discuss maintenance optimization, by showing that it should be performed using minimization of expected losses plus costs.

Their importance notwithstanding, we leave out of this book planning and scheduling and other corporate functions that also contribute to smart manufacturing in favor of focusing more on the data processing and the associated decision making.

References

Bagajewicz, M., *Process Plant Instrumentation. Design and Upgrade,* (ISBN:1-56676-998-1), Technomic Publishing Company, available at: http://www.techpub.com. Now CRC Press (http://www.crcpress.com) (2000).

Christofides, P. D., J. F. Davis, N. H. El-Farra, D. Clark, K. D. Harris, J. N. Gipson, "Smart Plant Operations: Vision, Progress and Challenges," *AIChE J.* **53** (11) (2007).

Davis, J. F., "Report from NSF Workshop on Cyberinfrastructure in Chemical and Biological Systems: Impact and Directions," available at: http://www.oit.ucla.edu/nsfci/NSFCIFullReport.pdf (2007).

Davis, J. F., "Report from NSF Workshop on Smart Process Manufacturing," available at: http://www.oit.ucla.edu/nsf-evo-2008/ (2008).

Gipson, J. N. Jr., "Integrated Engineering Solutions Technology Center," *AIChE Annual Meeting*, October 31, Cincinnati, Ohio (2005).

Humphrey, J. L., "The Smart Plant: Opportunities in Operations, Security, and the Environment," *AIChE Annual Meeting*, October 31, Cincinnati, Ohio (2005).

Humphrey, J. L., A. F. Seibert, J. C. Lewis, J. P. Farone, "Smart Manufacturing Plants: Advances and Priorities," *AIChE Annual Meeting*, November 18, Philadelphia, Pa. (2008).

Huynh, S. X. and D. L. Haake, "Early Warning, Cloud Computing & Semantic Modeling in Smart Manufacturing," *AIChE Annual Meeting*, November 18, Philadelphia, Pa. (2008).

Koolen, J. L. A., "Plant Operation in the Future," *Comp. Chem. Eng.*, **18**, S477–S481 (1994).

Lenz, D. H. and D. L. Chester, "Preventing Plant Problems before They Occur," *AIChE Annual Meeting*, November 18, Philadelphia, Pa. (2008).

Pelham, R. and C. Pharris, "Refinery Operations and Control: A Future Vision," *Hydrocarbon Processing* July (1996).

Sharpe, P., "A Standardized Approach to Smart Manufacturing," *AIChE Annual Meeting*, November 8–13, Nashville, Tenn. (2009).

White, D. C., "Creating the Smart Plant," *Hydrocarbon Processing* **82**(10):41–50, October (2003).

CHAPTER 2

Measurement Errors

Measurements are subject to errors, no matter how much the conditions and the apparatus used are improved. To understand the quality of the data that each instrument may provide, there are four basic features that need to be considered: precision, accuracy, sensitivity, and speed of response. The first three are relevant to all applications, process monitoring, and control. Speed of response is more related to control. A second set of instrument properties is also important: conformity, drift, hysteresis, dead band, etc. This chapter briefly overviews these properties. A more complete coverage can be found in the handbooks on instrumentation edited by Webster (1999), Liptak (2003), and Wilson (2005).

Range and Span

We call *range* the region within which a certain variable is measured or transmitted. The range is expressed by stating the lower- and upper-range values. The *span* is simply the difference between the upper- and lower-range values.

Precision

Precision of an instrument is defined as the closeness of agreement among a number of consecutive measurements of a variable that maintains its value static. Therefore, *precision is not related to the true value of the variable being measured.* Figure 2.1 illustrates the concept with a set of measurements of a temperature of water boiling (100°C). Figure 2.1*a* corresponds to one thermometer and Fig. 2.1*b* another, which has similar average value but larger deviations from the average.

Statistical theory is used to define precision. Then the variance of this distribution is estimated using the standard deviation of a sample

$$s = \sqrt{\frac{\sum_{i=1}^{n}(x_i - \bar{x})^2}{n}} \tag{2.1}$$

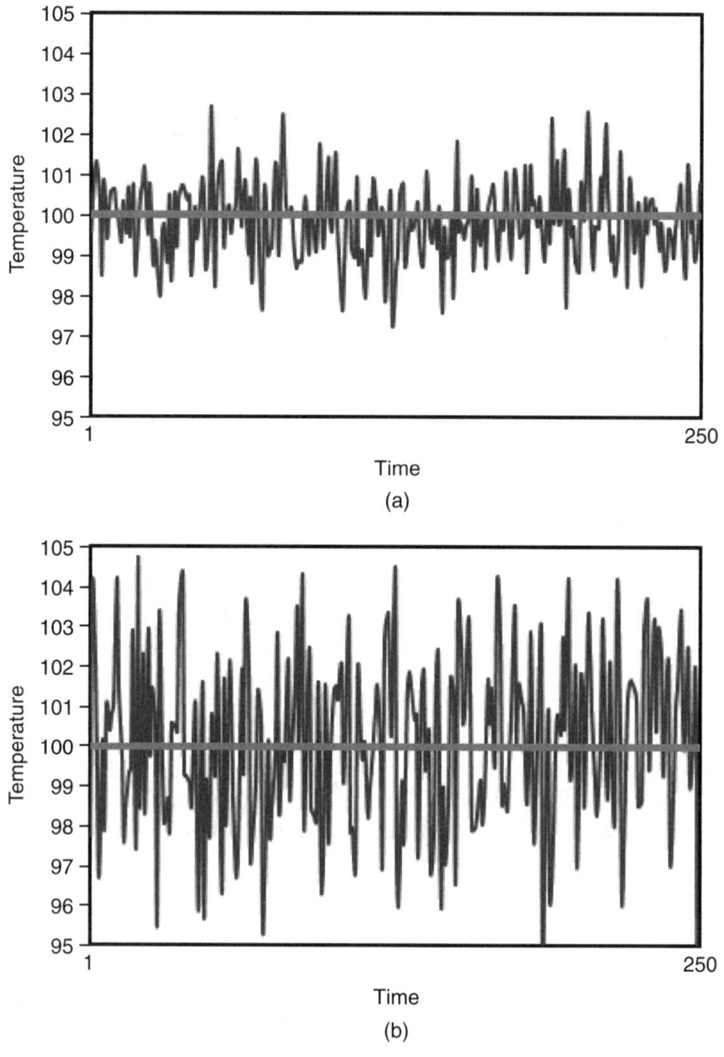

FIGURE 2.1 Precision.

A statistically unbiased estimate of the variance σ^2 is the modified variance $\hat{s}^2$ where

$$\hat{s} = s\sqrt{\frac{n}{n-1}} \tag{2.2}$$

As n measurements are taken from a population that is normally distributed, statistical estimation theory states that the population

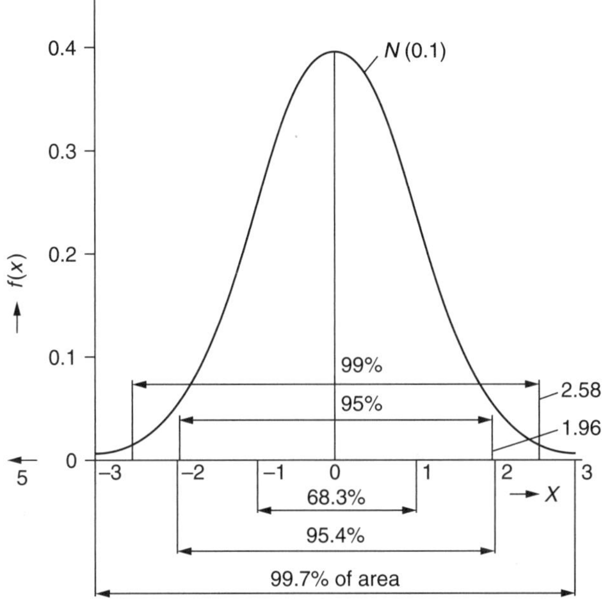

FIGURE 2.2 Normal distribution.

mean μ has a p% probability to be in the interval $\bar{x} \pm z_p \hat{s} / \sqrt{n}$. The parameter z_p is called confidence coefficient, whereas p is the confidence level, or confidence limit. For example, for $p = 95\%$, $z_p = 1.96$, as illustrated in Fig. 2.2.

Since in practice the number of measurements is finite, the t-student distribution is used, and therefore t_p is used instead of z_p, where t_p is the p% confidence coefficient of a t-student distribution of $(n - 1)$ degrees of freedom. When $n \geq 30$, the t-student and normal distribution are practically the same. Precision is thus defined as half of the confidence interval when one measurement is taken, that is,

$$\sigma_p = t_p \hat{s} \tag{2.3}$$

Precision is called more formally *repeatability or reproducibility* by International Society of Automation (ISA) standards and other literature. *Repeatability* is defined by ISA standards as the closeness of agreement among a number of consecutive measurements of the output for the same value of the input under the same operating conditions (ambient temperature, ambient pressure, voltage, etc.), approaching from the same direction. *Reproducibility* is defined the same way, but approaching from both directions. Thus, repeatability does not include hysteresis, dead band, and drift effects whereas reproducibility does. Hysteresis, dead band, and drift are discussed later. Precision is thus a loose term coined by practice that does not have a formal definition set by a standard. Therefore, its usage is also loose and can either refer to repeatability or reproducibility. From now on, unless indication to the contrary, it will substitute reproducibility.

Origin of Fluctuations

Fluctuation of measurements has varied sources. For example, in the case of pressure, its value is affected by small fluctuations originated in pumps and/or compressors' vibrations, as well as other factors. Process temperatures are affected by ambient temperature fluctuations, etc. Turbulent flow is modeled in this way and even laminar flows are subject to such variations, because flow is driven by pressure differences and depends on density.

Remark: The normal distribution of random errors is assumed to arise from the assumption that errors are the product of innumerable sources and consequently the central limit theorem of statistical theory applies. This theorem states that a sum of a large number of disturbances, each having its own distribution, tends to give a disturbance with a normal distribution. Measurements are related to the state variables they measure through a series of signal transformations. These transformations involve the use of measurement devices, transducers, electronic amplifiers, and final reading instruments, among others. Each of these devices adds a random error, sometimes called "noise" in signal processing. This noise is usually assumed to have a normal distribution. This is sometimes not true. For example, when noise has a narrow frequency bandwidth it has a Rayleigh distribution (Brown and Glazier, 1964). The consequences of this type of noise error probability distribution are not explored here.

Remark: One important signal of great importance is the so called white noise. This type of signal is characterized by the fact that it is completely uncorrelated. We cover the definition of white noise in detail in Chap. 13. It is often assumed that gaussian noise, that is noise with a gaussian amplitude distribution, is white noise. A gaussian signal refers to the probability distribution; the term "white" refers to the way the signal power is distributed over time or among frequencies. Thus, one can have gaussian white noise, Poisson, Cauchy, etc., white noises.

Systematic Error (Bias)

The *systematic error* of an instrument is defined as the closeness of agreement of the mean value of a number of consecutive measurements of a variable that maintains its value static. It is also called bias. Figure 2.3 illustrates the concept with a set of measurements of the temperature of boiling water. The measurements in this figure have a +2°C systematic error.

When the true value is known, the amount of bias can be estimated by subtracting the mean value of all measurements from the true value.

$$\delta = \hat{x} - \overline{x} \tag{2.4}$$

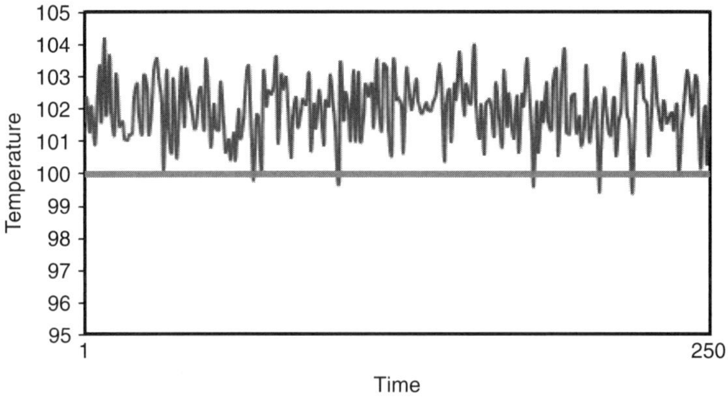

FIGURE 2.3 Systematic error.

Thus when the measurements are higher than the true value, the bias is positive and the reading is said to be high. Conversely, a negative bias corresponds to a low reading.

When true values are not known, other instruments are needed to determine them. These are in fact better estimates of the variable. This process is called calibration. For example, temperature meters can be calibrated using systems with well-known values of temperature, like the boiling or melting point of pure substances. Flowmeters, in turn, need to be calibrated with the aid of some other more precise instrumentation.

Biases can be classified in two major categories: *constant* and *variable bias.*

Sources for constant bias are

- Use of incorrect assumptions in the calibration procedure
- Human faults: improper installation such as violation of upstream and downstream straight run requirement; improper calibration
- Corrections not performed in the calibration procedure
- Unknown errors in reference standards
- Zero shift
- Departure of operating conditions (temperature, pressure, density) from the standard conditions upon which meters are calibrated, or worse, operating outside the operation range of meters (e.g., operating below the minimum Reynolds number requirement)
- Distortion of the flow profile at the measuring point due to entrainment of gas bubbles in the liquid stream

Sources of variable bias are

- Drift in the voltage supply to the instrument
- Span shift
- Wear of the instrument; for example, in flowmeters, erosion, corrosion, and particle deposits that change the roughness of the inside pipe surface

Periodic sensor recalibration and training skillful personnel so that sensors are properly installed and commissioned is imperative if human errors are to be avoided. On the other hand, sensor failures are usually invisible to operation personnel and the biases caused by usually go on undetected unless a gross error detection/data reconciliation system is in place.

Drift is defined as a change in output over a specified period of time for a constant input. *Shifts* are, in turn, independent of time and correspond to errors in the measurement range. Span and zero shifts are illustrated in Fig. 2.4.

We now focus our attention to biases that are caused by sensor failure. Three types of biases are considered:

(a) *Sudden fixed value bias* (Fig. 2.5*a*). This can be described by a step function (positive or negative). The uncertainties are in the time at which the step takes place and the size of the bias. These are biases that typically emerge because of failures in electronic components (the readout system or the

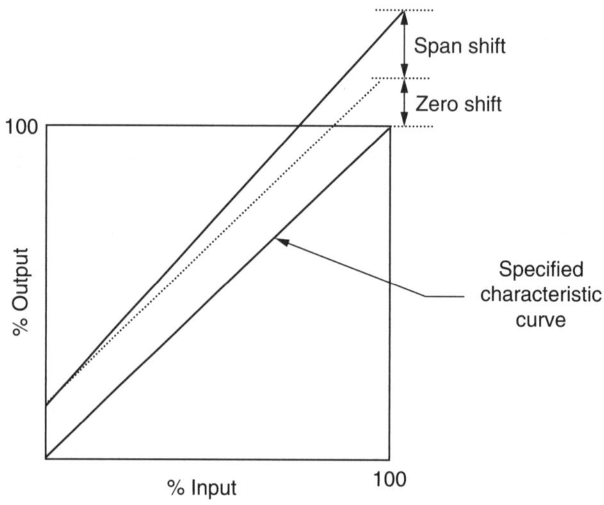

Figure 2.4 Span and zero shift. (*Following Liptak, 2003.*)

signal-processing system) of the sensor. Generally, a sensor is prone to failure when fluid environment and/or ambient environment are harsh. For example, high ambient temperature or high humidity can cause damage to the electronic components. The local presence of power surges or the appearance of sudden electrical effects (e.g., lightening) may interfere with the readings and outputs of some meters. Other events like the change in resistor impedance can cause a shift in output of the electric circuit and cause a bias.

(b) *Randomly emerging drifts*. These are characterized by ramps (up or down and not necessarily linear). The uncertainty is in the time at which the ramp starts and its slope or some parameter describing a nonlinear drift. These are drifts attributable to wear and tear, deterioration of mechanical parts due to cavitations, and sometimes to electronic failure. Some examples are wear and tear of bearing or turbine rotor in turbine meters and erosion of orifice plate in orifice meters. Once they appear, the wear and tear of sensor mechanical parts may continue until the part is damaged and the sensor stops working properly. The shape and slope of the drift are not known. We assume that the shape of randomly emerging drifts is either concave or convex, as shown in Fig. 2.5*b*. We assume that the concave shape corresponds to biases caused by failures where the severity of failure initially increases rapidly with time and then slows down. An example of this type of failure is the corrosion of sensor parts that is gradually slowed down by a protective layer (which is in turn formed by the products of the corrosion process). We assume that the convex shape corresponds to biases caused by failures where the situation is the opposite: the failure starts slowly and then speeds up.

(c) *Deterministic drifts*. These are not random events, but rather predictable continuous processes that develop progressively with time, such as corrosion, coating, deposit of particles on sensor parts that are in contact with the fluid (the primary element of the sensor). If the fluid is corrosive and/or dirty, then one knows with high confidence that this type of bias exists, only parameters like the shape and slope of the drift are eventually unknown. Corrosive/dirty fluids can damage or distort the shape and size of the sensor's primary elements which causes bias in measurement. The most common factor that affects the measurement accuracy is the coating or deposit of particles. Any amount of deposit may cause measurements to be in error; fortunately, these are usually low for orifice plate meters (Upp and LaNasa, 2002). Generally, industrial fluids are not clean, and the fluids may get contaminated with lubricant oil while passing through valves, compressors, and pumps. They can also get contaminated with particles generated from erosion of metal pipes and fittings. The effects of this factor include: (i) changing the roughness of the pipe, which distorts flow profile; (ii) reduction in cross-section of flow; (iii) reduction of the mechanical clearances of moving parts, for example, in turbine meter rotors; (iv) changing the original dimension of orifice plates, vortex bluff bodies; (v) insulation of electromagnetic meter electrodes; (vi) blocking of pressure taps. Therefore, the dependence of bias size with time is usually a nonlinear function because one single factor can cause multiple effects that induce bias in the measurement. The shape and the slope of the drifts depend on the type of sensor and the nature of the problem that causes bias (drift). Particle deposits may reach an equilibrium where the number of particles deposited on the surface is equal to the number of particles on the surface layer swept away by the flowing fluid. In this case, the drifts have an asymptotic shape, as shown in Fig. 2.5*c*. Three cases of errors in measurement of orifice meters due to deposition of particles were discussed by Upp and LaNasa (2002), in which biases go on undetected for a number of years. This implies that the drifts have asymptotic shape. Another cause of deterministic drift is corrosion, in which certain part of sensor that contacts with fluid may be continuously corroded until the sensor part is completely consumed and sensor stops working (drifts have opposite shape: concave shape) or the product of corrosion process (e.g., anhydrous iron oxides) may act as protective layer that slows down the corrosion process (drifts have asymptotic shape).

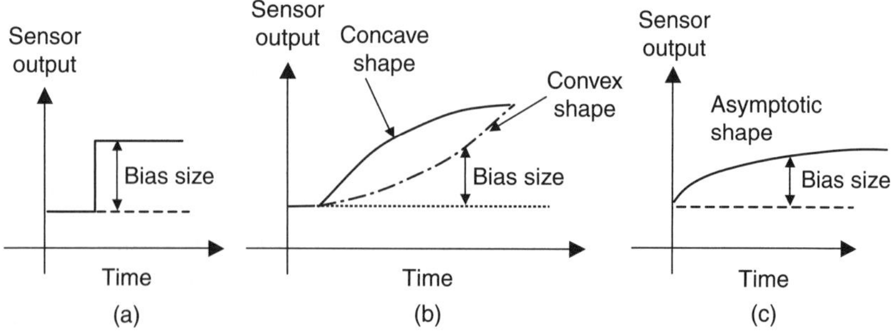

FIGURE 2.5 Types of biases: (a) Sudden fixed-value bias; (b) randomly emerging drift; (c) deterministic drift.

Remark: We now discuss some specific flowmeters.

- *Differential pressure flowmeters:* Differential pressure meters are the most commonly used flowmeters. Their operation is based on the premise that the pressure drop across the meters is proportional to square of the volumetric flowrate. Like most flowmeters, differential pressure flowmeters have a primary and secondary element. The primary element causes a constriction in the flow cross-section area to create change in pressure. The secondary element measures the differential pressure and provides the signal or readout that is converted to the actual flow value (Miller, 1996).

  Different kinds of differential pressure flowmeters are characterized by how the flow cross-section area is constricted. In the orifice flowmeter, the most popular liquid flowmeter in use today, the flow cross-section area is suddenly restricted by making use of an orifice plate. In a Ventury flowmeter, the pipe diameter is gradually constricted. Other differential pressure flowmeters are flow nozzles, pitot tubes, flow tubes, and elbow meters.

  Failure modes involving mechanical parts of the meters include erosion, corrosion, deposit of particles due to dirty and/or corrosive fluids, leak through the flanges at the orifice plates, cavitation damage, and plugging of pressure taps (Taylor, 1994; Padmanabhan, 2000). Failure modes involving the electronic parts or the transmitters (for signal amplification, linearization, etc.) include complete stoppage or step change in the output signal which is usually caused by harsh ambient conditions or significant change in ambient environment; the failure rate of electronic parts, however, is smaller than that of the mechanical parts. Failures involving electronic parts occur suddenly and the measurement bias develops in a stepwise fashion. Conversely, failures involving mechanical parts occur gradually and, consequently, the bias size increases gradually with time. The failure rates of differential pressure flowmeters for measuring liquid flow are given in Table 2.1 (Taylor, 1994).

	Failure Mode	Failure Rate
Mechanical parts faults (for liquid flow)	Drift greater than 5% from calibration	5 per 10^6 h
	No output signal	4 per 10^6 h
	Output signal high	1 per 10^6 h
Transmitter fault	Short circuit/no signal	0.5 per 10^6 h
	Fail high signal/short circuit	0.2 per 10^6 h
	Drift greater than 5%	1 per 10^6 h

Table 2.1 Failure Rate of Differential Pressure Flowmeters (Taylor, 1994)

Different kinds of (simulated) failure modes and their associated measurement errors of orifice flowmeters were studied by the Florida Gas Transmission Company (McMillan and Considine, 1999). Some results of that study are given in Table 2.2.

Condition	% Error
Orifice edge beveled 45° circumference	
1.01 bevel width	–2.2
1.02 bevel width	–4.5
0.05 bevel width	–13.1
Turbulent gas stream (distortion of flow profile) due to	
Upstream valve partially closed	–6.7
Grease and dirt deposits in meter tube	–11.1
Leak around orifice plate	
1. One clean cut through plate sealing unit	
a. Cut on top side of plate	–3.3
b. Cut next to tap holes	–6.1
2. Orifice plate carrier raised approximately 3/8 in from bottom (plate not centered)	–8.2
Valve lubrication on upstream side of plate	
Bottom half of plate coated 1/16 in thick	–9.7
Three gob-type random deposits	0.0
Nine gob-type random deposits	–0.6
Orifice plate uniformly coated 1/16 in over full face	–15.8
Valve lubrication on both sides of plate	
Plate coated 1/8 in both sides of full face	–17.9
Plate coated 1/4 in both sides full face	–24.4

Table 2.2 Failure Modes and Their Associated Measurement Errors of Orifice Flowmeters

- *Mechanical flowmeters:* Mechanical flowmeters measure flow making use of moving parts (rotors), either by delivering isolated, known volumes of fluid through champers (positive displacement meters), or by recording the rotational speed of rotor as a function of fluid velocity (turbine flowmeters). Positive displacement meters operate by isolating and counting a known volume of fluid while moving. The turbine flowmeter, a velocity meter, consists of a multiple-bladed rotor perpendicular to the fluid flow. The rotor spins as the liquid passes through the blades. The rotational speed is a direct function of flow velocity and can be sensed by a magnetic pickup or a photoelectric cell.

  It is well known that the moving parts are the weakest link in the structure of the meters. The rotor and the bearing will be eventually damaged due to abrasion, corrosion, wear and tear, especially when exposed to dirty, corrosive fluid, and this problem causes bias in measurement. Under certain conditions, the pressure drop across the turbine meters can cause flashing, which in turn causes the meters to read high or cavitation, which results in rotor damage (Omega

Engineering Inc., 2005; Padmanabhan, 2000). The failure rate of these mechanical meters is comparable to that of positive displacement pumps, which is 30–60 per 10^6 hour (Taylor, 1994). As in the case of orifice plates, we also consider the possibility of sudden emergence of a steady bias, or drifts due to slow mechanical wear and tear.

- *Electromagnetic meters* operate on Faraday's law of electromagnetic induction, which states that a voltage will be induced when a conductor moves through a magnetic field. The liquid serves as the conductor; the magnetic field is created by energized coils outside the flow tube. A pair of electrodes penetrates through the pipe and its lining to measure the amount of voltage produced, which is directly proportional to the velocity of fluid.
- *Vortex meters* make use of a natural phenomenon that occurs when a liquid flows around a bluff object. Eddies or vortices are shed alternately downstream of the object. Piezoelectric or capacitance-type sensors (located either inside or outside the meter body) are used to detect the frequency of the vortex shedding which is directly proportional to the liquid velocity.
- *Ultrasonic meters* operate on the principle that the speed at which the sound propagates in the liquid is dependent on the fluid's density. If the fluid's density is constant, one can use the time of ultrasonic passage (or reflection) to determine the velocity of the flowing fluid. The ultrasonic meters consist of two transducers: one to transmit the ultrasonic signal and one to receive the ultrasonic signal that passes through the fluid or is reflected when contacted with discontinuities (e.g., particles) in the fluid.
- *Coriolis meters* are based on diverting the flow through a small tubing and measuring the Coriolis force exerted as the fluid turns inside the tube.

Having no moving parts, and being relatively nonintrusive (i.e., they do not directly contact or obstruct fluid flow), electromagnetic, vortex, ultrasonic, and coriolis meters are considered to be reliable, that is, with lower failure rates than orifice and turbine meters. The common failure mode for these advanced meters is the erosion of the coating of the electrodes in electromagnetic meters or the erosion in the coating of the inside surface of the pipe. This erosion leads to a loss in stimulus force (e.g., magnetic field in electromagnetic meters) and/or the output signal and consequently, it affects the accuracy of meters. Finally, the erosion of the coating of the bluff body in vortex meters changes its dimensions (Omega Engineering Inc., 2005; Padmanabhan, 2000) and therefore affects accuracy. Moreover, because these meters make use of sophisticated electronics technology, failure of signal conditioning electronic parts also needs to be considered.

Outliers

An *outlier* is defined as a measurement that can never be explained, calculated, estimated, or anticipated. Human errors are typical sources of outliers. In Fig. 2.6, a signal without outliers is presented. In Fig. 2.7, a signal with small variance and outliers is presented.

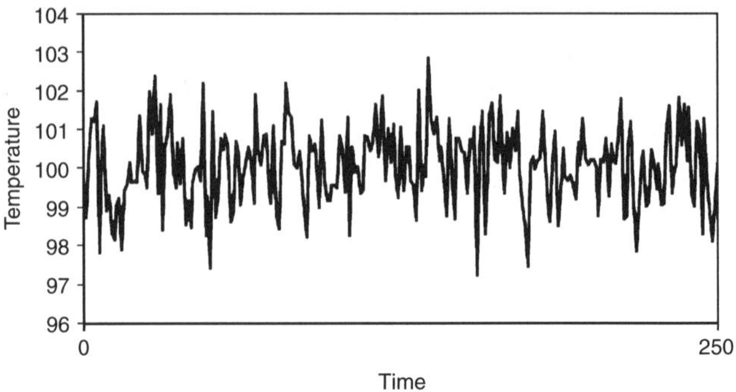

Figure 2.6 Gaussian signal.

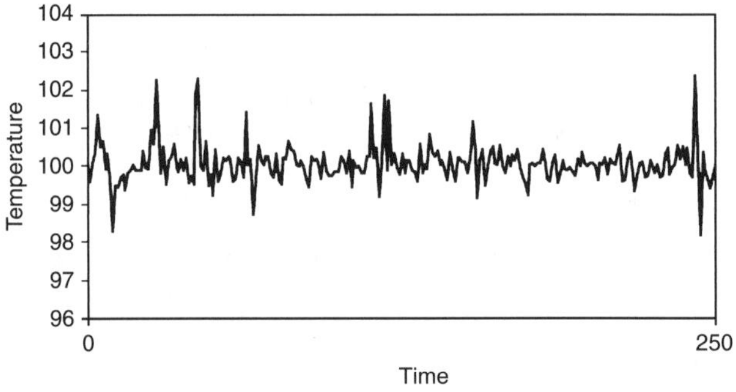

Figure 2.7 Gaussian signal with a few outliers.

In signals that exhibit so clear stability of the mean value, that is, no drift or large process fluctuations, outliers can be detected by using very simple procedures. One procedure consists of constructing the histogram of frequencies. For example, Figs. 2.8 and 2.9 show the histograms corresponding to the signals in Figs. 2.6 and 2.7. Quite clearly, one can observe that the histogram of Fig. 2.7 resembles a normal distribution. However, the histogram of Fig. 2.9 resembles a normal distribution in the center, but exhibits some abnormal values far away from the mean, indicating the presence of negative and positive spikes.

Sometimes, one can observe a set of only positive spikes exhibiting also a normal distribution. Such is the case of the signal shown in Fig. 2.10, whose corresponding histogram is shown in Fig. 2.11.

Another procedure is to determine the mean and the standard deviation of the sample and use that as means of testing a null hypothesis that each individual spike is affected by a systematic error. For example, for the signal of Fig. 2.6, the mean is 100.05 and the standard deviation is 1.026. For such values and assuming a confidence

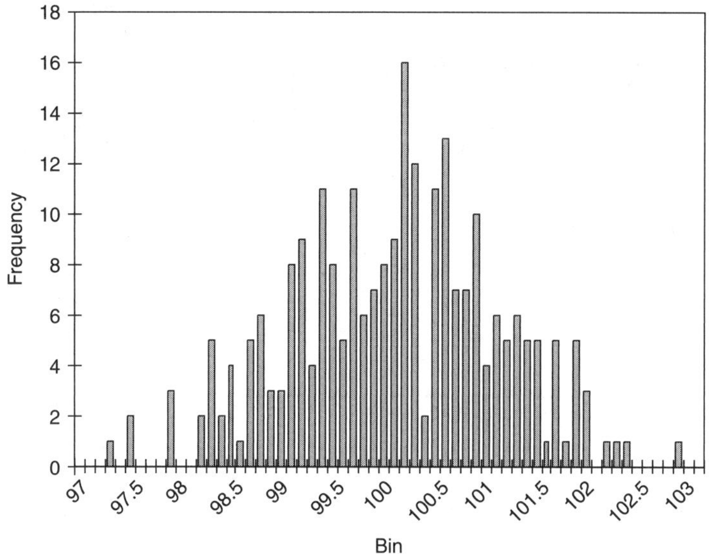

FIGURE 2.8 Histogram corresponding to the signal of Fig. 2.6.

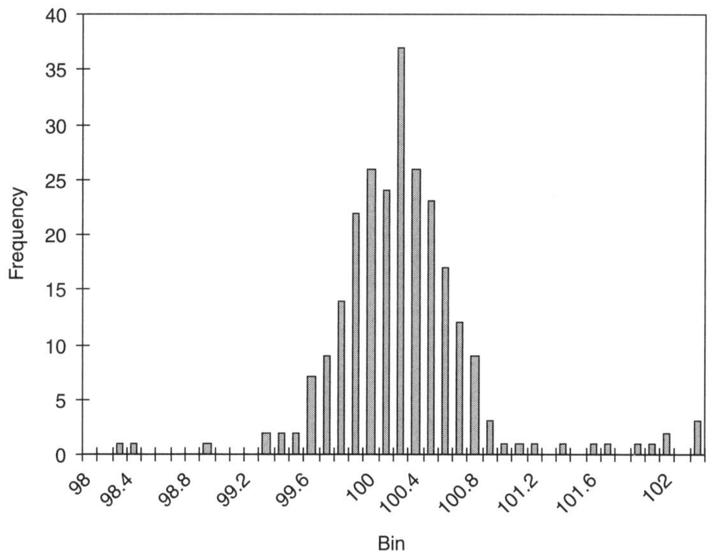

FIGURE 2.9 Histogram corresponding to the signal of Fig. 2.7.

level of 95%, $R = (-u_{1-\alpha/2}, u_{1-\alpha/2})$ = (98.36,101.74). Thus, if 95% or more measurements are inside this interval, we do not reject the hypothesis that the distribution is normal. In fact, the signal of Fig. 2.6 has only 4% outside this range. The test for other confidence

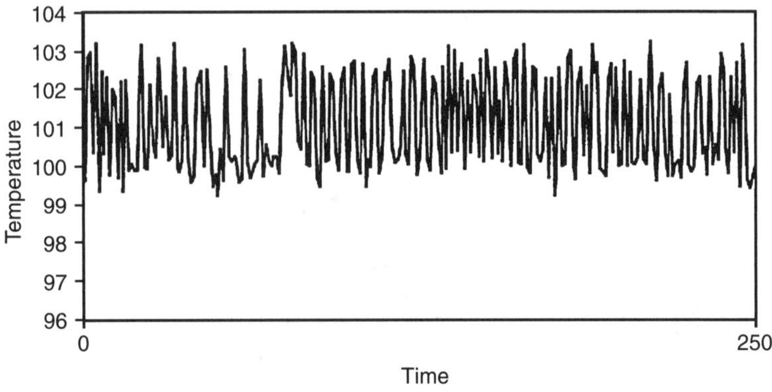

FIGURE 2.10 Signal with outliers normally distributed.

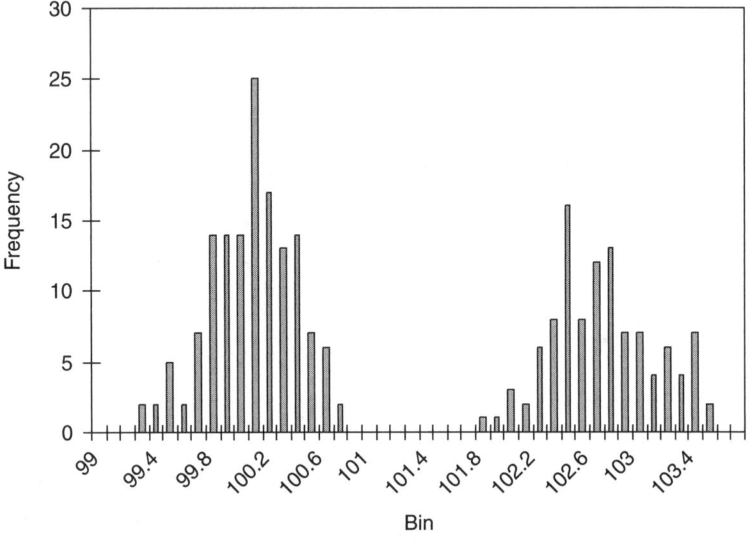

FIGURE 2.11 Histogram corresponding to the signal of Fig. 2.10.

levels (90 and 99%) also suggests the distribution is normal. However, the signal of Fig. 2.7 has 5.2% of the measurements outside the range, making this test rather inconclusive. Moreover for 90% confidence the test gives 7.2% oustide the range, and therefore passing the test and for 99% confidence, 4.2%, failing in this case. Such behavior is typical of these type of signals, where a few spikes are present. Finally, for the signal of Fig. 2.10 at 95% confidence we obtain 39.2% outside the range. Not surprisingly, for 90% confidence the percentage outside the range is 42.4%, which is consistent with the bimodal distribution.

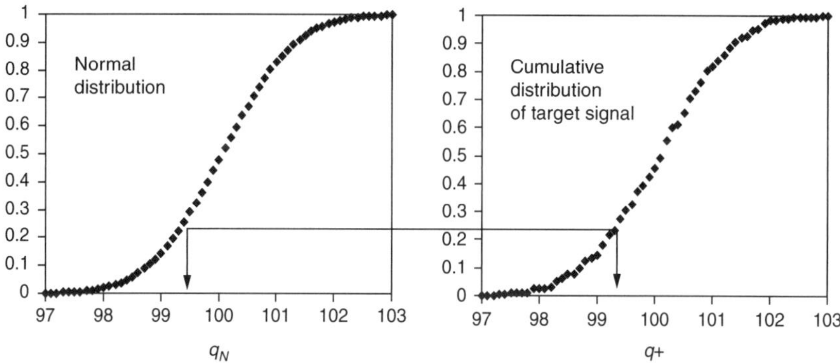

FIGURE 2.12 Construction procedure of a *q-q* plot for the signal of Fig. 2.5.

Another method to test the existence of these outliers is the use of *q-q* plots (Chen et al., 1998). These plots consist of contrasting the cumulative distribution of the data with the standard cumulative distribution. The procedure is as follows:

1. Construct the cumulative frequency plot for the data set and one with the same bin intervals for a normal distribution.
2. For all interval centers, determine the cumulative distribution of the data set (q_+). Do the same for the normal distribution (q_N). Plot all points (q_+, q_N) as indicated in Fig. 2.12.
3. If the data is normally distributed, then the plot should result in a straight line. Deviations from a straight line indicate the existence of outliers.

Thus, *q-q* plots for the signals of Figs. 2.6, 2.7, and 2.10 are shown in Figs. 2.13, 2.14, and 2.15, respectively. Clearly, the signal from

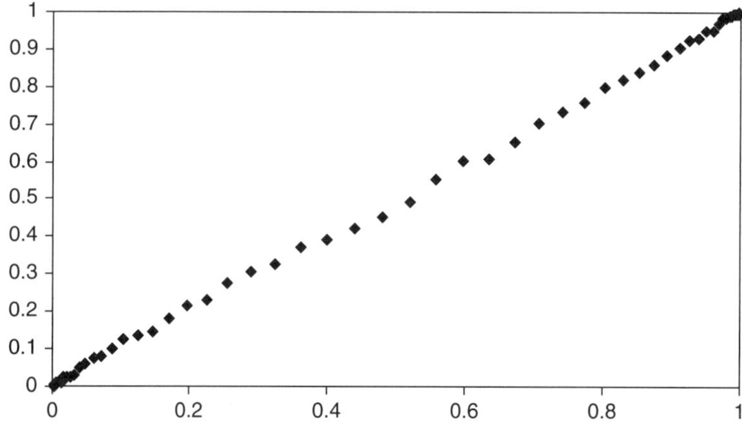

FIGURE 2.13 *q-q* plot corresponding to the signal of Fig. 2.5.

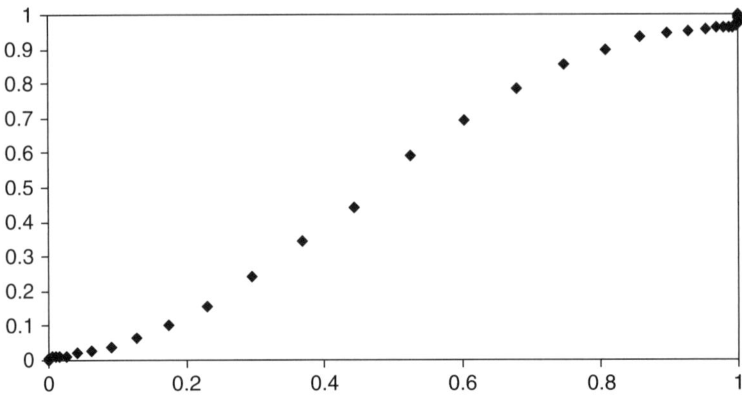

FIGURE 2.14 *q-q* plot corresponding to the signal of Fig. 2.6.

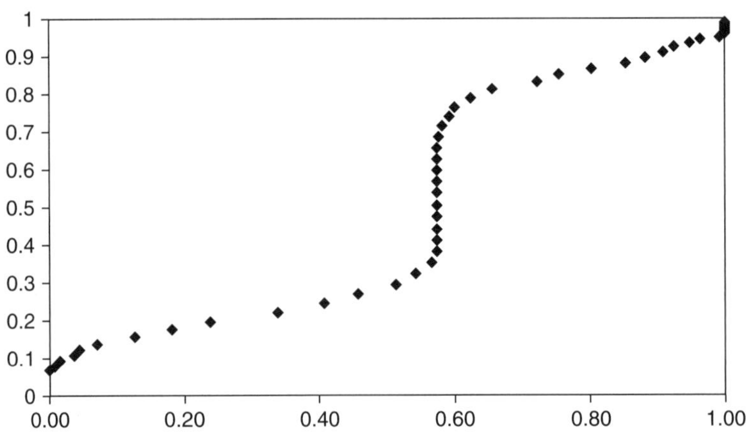

FIGURE 2.15 *q-q* plot corresponding to the signal of Fig. 2.9.

Fig. 2.6 can be considered gaussian, while the one from Fig. 2.7 exhibits small deviations, likely because only a few spikes are present, while the one corresponding to Fig. 2.10 exhibits a clear bimodal distribution pattern, both distributions being normal.

Remark: Chen et al. (1998) use influence functions and propose an optimization procedure that helps in finding the outliers automatically. More recent work includes online outlier detection and data cleaning (Liu et al., 2004) who proposed a data filter-cleaner includes an online outlier-resistant estimate of the process model and combines it with a modified Kalman filter (see Chap. 9) to detect and "clean" outliers. The proposed method does not require a priori knowledge of the process model, and it is applicable to autocorrelated data and preserves all other information in the data. One has to be careful in using method for filtering data to remove so-called "disturbances" (see Cao and Rhinehart, 1997), because these methods may actually remove legitimate random errors, not only outliers. Finally, we cite the elaborate work of Zhao et al. (2004), who proposed an outlier detection method based on radial basis functions–partial least squares (RBF–PLS) approach and the Prescott test, all methods and tests outside the scope of this book.

Accuracy

The *accuracy* of a measurement is defined as the degree of conformity with a standard or true value. Therefore, precision and systematic errors contribute to the accuracy of an instrument. Thus, a device is said to be accurate if it is unbiased and precise. Accuracy (σ_a) is many times reported numerically as follows:

$$\sigma_a = \delta + \sigma_P \tag{2.5}$$

where δ is the bias and σ_p is the precision.

In many other textbooks accuracy is defined similarly to precision, ignoring (or assuming) thus that bias is not present. Finally, precision, bias, and accuracy are also often reported in relative terms, that is, as a percentage of the value measured. For example, a typical expression would be (all ignoring bias)

- In terms of the measured variable (ignoring bias): ±1°C
- In terms of percentage of span: ±0.5%
- In terms of percentage of the upper-range value: ±0.5%
- In terms of percentage of actual output reading: ±0.3%

The accuracy expressed in terms of the measured variable as well as the definitions in terms of the span or upper-range values are independent of the value measured.

Modern nomenclature, not reflected, for example, in ISA standards, uses the term "inaccuracy" instead of "accuracy" to indicate the degree of departure from standard or ideal values. Although we prefer the former term, "accuracy" is so deeply rooted in the sensor location and data reconciliation literature that it will be used throughout this book nonetheless.

Calibration Curves

Instruments are used over a certain range. Therefore, one might expect the precision and bias, and therefore the accuracy, to vary in the range of measurement. A graph that shows the ideal values of an output variable as a function of an input variable is called a *characteristic curve*. For example, an orifice flowmeter measures pressure drop as a

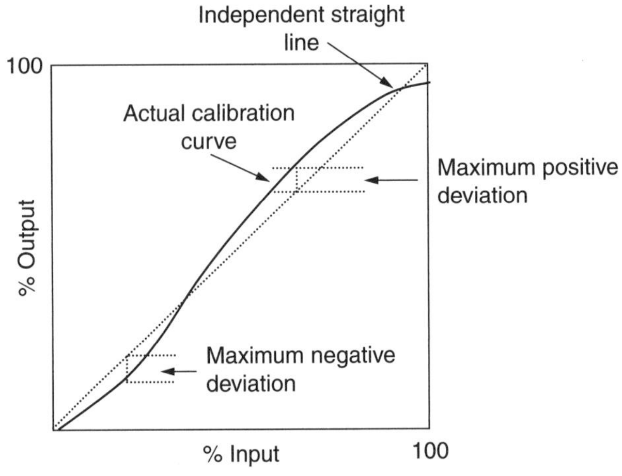

Figure 2.16 Independent conformity. (*Following Liptak, 2003.*)

function of volumetric flowrate. Thus, we expect the characteristic curve of such device to be a square root function. The calibration procedure consists of first obtaining the actual value of the output variable by averaging the upscale and downscale readings. The comparison between the actual and characteristic curve is called *independent conformity*. This is illustrated in Fig. 2.16. To construct such curves, another instrument is needed to determine the value of the input. In practice, the device used for this purpose has at least five times better accuracy and is free from systematic errors (Miller, 1996).

In many cases the characteristic curve cannot be constructed, as many parameters are not known, although the shape is still known. For example, in the case of the orifice flowmeter, one might not be able to calculate the actual value, but it is still anticipated that the ideal behavior is proportional to a square root function. To determine conformity, it is therefore needed to adjust this characteristic curve. When the characteristic curve is adjusted using the upper-range value, the conformity is called *terminal-based conformity*, whereas when the initial value is used, the conformity is called *zero-based conformity* (Fig. 2.17).

When the expected measured output value is the same variable as the input value, then the characteristic curve is a straight 45° line. This is, for example, the case where the flow measured by a flowmeter, after all compensation and conversions have been performed, is plotted against the real flow. Such diagram is shown in Fig. 2.18.

The conformity between the actual measured value and the characteristic line is called *independent linearity*. Terminal-based and zero-based linearity are defined in the same way as terminal-based and zero-based conformity. The deviations from linearity are called *static errors*.

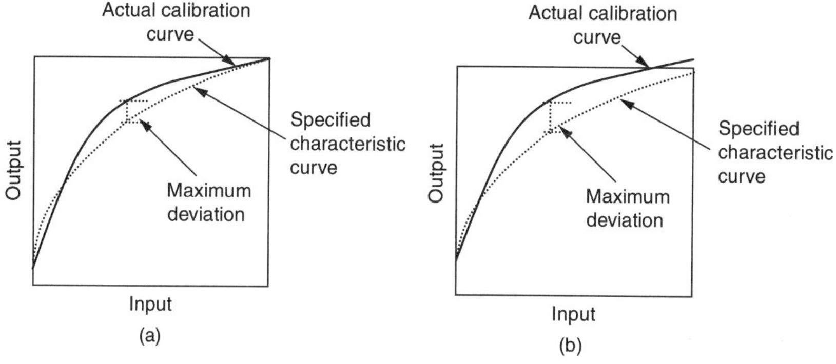

FIGURE 2.17 (a) Terminal-based conformity. (b) Zero-based conformity. (*Following Liptak, 2003.*)

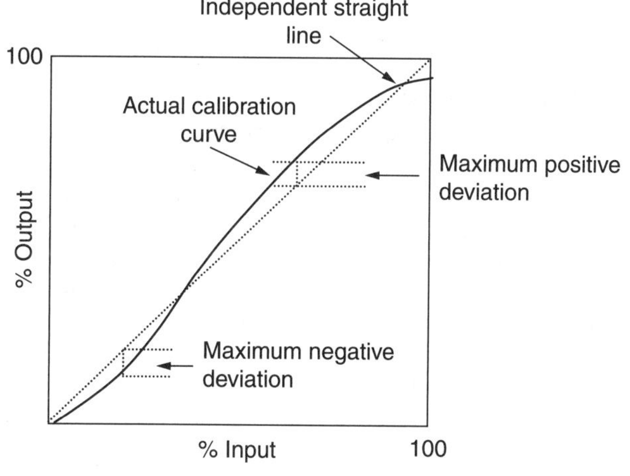

FIGURE 2.18 Independent linearity. (*Following Liptak, 2003.*)

While deviations are to be expected, and can certainly be accounted for if an independent linearity plot is available, there are cases in which these systematic deviations are too many. Indeed, when liquid flows are measured, then temperature and/or pressure variations have little impact on the fluid density, and therefore the independent linearity curve can be eventually used to make corrections. Unfortunately, these curves are not used that often. When in turn a gas is being measured, temperature and pressure and even small variations in composition have a significant effect on the independent linearity. At this point these are too many variations to track, and therefore the compensation that one could make is not done.

In addition to the above, outside ambient temperature and humidity, for example, affect the readings and introduce bias and/or alter the

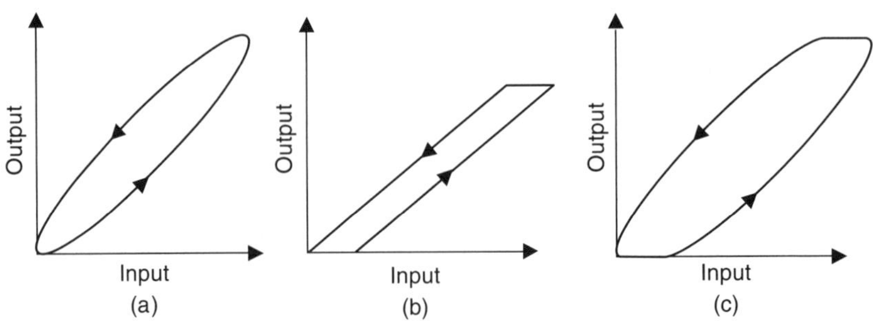

FIGURE 2.19 (a) Hysteresis. (b) Dead band. (c) Hysteresis and dead band.

precision. These are called *influence quantities* (Miller, 1996), and their effects are not linearly additive.

Hysteresis and Dead Band

Hysteresis, illustrated in Fig. 2.19, is a phenomenon where an output corresponding to an increasing input varies through a certain path, which differs from the path the output follows. This occurs when the input is decreased from the maximum reached until the original value. The discrepancy between the paths is typically larger if the interval of change is larger. Dead band, illustrated in the same figure (and in combination with hysteresis), is a range within which the input can be changed without observing a change in the output. This is typically observed when the input direction of change is suddenly reversed. Most often, these effects are due to electronics associated with the instrument or when mechanical parts are present.

Guide to the Expression of Uncertainty in Measurement

The International Organization for Standardization (ISO) published the *Guide to the Expression of Uncertainty in Measurement* (GUM, 1995), which is widely used as an international standard. Later, in 2004, the Bureau International des Poids et Mesures (BIPM) produced a draft Supplement 1 to the ISO-GUM (draft GUMS1, 2004).

ISO-GUM uses as a starting point the measurement equation

$$Y = f(X_1, X_2, \ldots, X_N) \tag{2.6}$$

where Y represents the output variable and $X_1, X_2, \ldots, X_N$ represent the input. Both input and output are assumed to have a certain distribution.

Thus, the estimators of X_i, namely x_i, are used to obtain the estimator of Y, namely y, as follows:

$$y = f(x_1, x_2, \ldots, x_N) \tag{2.7}$$

ISO-GUM proposes that Eq. (2.7) is linearized as follows:

$$Y \approx Y_{\text{linear}} = y + \sum_i c_i (X_i - x_i) \tag{2.8}$$

where c_i are the corresponding partial derivatives.

Now, σ_y^2, the variance of Y is then given by

$$\sigma_y^2 = \sum_i c_i^2 \sigma_{x_i}^2 + 2\sum_{i<j} c_i c_j \operatorname{cov}(x_i, x_j) \tag{2.9}$$

where $\operatorname{cov}(x_i, x_j)$ is the covariance of the two variables. We note that an equivalent expression is

$$\sigma_y^2 = \sum_i c_i^2 \sigma_{x_i}^2 + 2\sum_{i<j} c_i c_j \sigma_{x_i} \sigma_{x_j} \rho_{x_i, x_j} \tag{2.10}$$

where ρ_{x_i, x_j} is the correlation coefficient.

Details: The covariance of a linear combination of variables $y = \sum_i c_i x_i$, is defined as follows:

$$\sigma_y^2 = \int_{-\infty}^{\infty} \cdots \int_{-\infty}^{\infty} \left(\sum_i c_i x_i - \bar{y} \right) \left(\sum_j c_j x_j - \bar{y} \right) p(x) dx_1 dx_2 \cdots dx_n \tag{2.11}$$

where $\bar{y} = \sum_i c_i \bar{x}_i$ and $\bar{x}_i$ is the mean value of the measurements of variable x_i. After some small bookkeeping one can write

$$\sigma_y^2 = \sum_i \sum_j c_i c_j \int_{-\infty}^{\infty} \int_{-\infty}^{\infty} (x_i - \bar{x}_i)(x_j - \bar{x}_j) p(x) dx_i dx_j = \sum_i \sum_j c_i c_j \operatorname{cov}(x_i, x_j) \tag{2.12}$$

where we used the definition of covariance $\operatorname{cov}(x_i, x_j)$. We also recognize that for $i = j$ we have $c_i c_j \operatorname{cov}(x_i, x_j) = c_i^2 \sigma_i^2$, which renders Eq. (2.9). Now since the correlation coefficient is given by $\rho_{x_i, x_j} = \operatorname{cov}(x_i, x_j)/(\sigma_i \sigma_j)$, we have Eq. (2.10).

ISO-GUM calls the random variables "uncertainties," and their evaluation consists of determining mean values, standard deviations, variances, covariances, correlation coefficients, etc.; it also defines two types:

- Type A "uncertainties": Those evaluated using statistical methods.
- Type B "uncertainties": Those evaluated using other methods.

The draft GUM Supplement does not make this distinction and deals with the specifics of different distributions for the input variables, including the case of independent as well as multivariate normal distribution. For the most part, these two documents spend a considerable amount of time to define the uncertainties using intervals.

Thus, the GUM approach considers a linear approach to propagate "uncertainties," as explained above. The draft supplement propagates probability distributions of inputs through a numerical simulation of the measurement equation to determine a probability distribution for the value of the measurand.

This new nomenclature of "uncertainty" has created some confusion because it dumps different errors (random disturbances, biases, hysteresis, nonconformity) into one term and attempts to quantify them collectively with one variance (or interval) instead of attempting to quantify each of them.

An alternative to the GUM approach is bayesian statistics (see Kacker et al., 2006). Bayesian statistics regards all variables as having state-of-knowledge probability distributions, called prior distributions and uses them to obtain posterior distributions representing the state-of-knowledge after the measurement is made.

References

1995 Guide to the Expression of Uncertainty in Measurement, 2d ed. (ISBN 92-67-10188-9), Geneva: International Organization for Standardization) (1995).

"2004 Draft GUM Supplement 1: Numerical Methods for the Propagation of Distributions," BIPM Joint Committee on Guides in Metrology (2004).

Brown, J. and E. V. D. Glazier, *Signal Analysis,* Reinhold, New York (1964).

Cao, S. and R. R. Rhinehart, "A Self-Tuning Filter," *J. Proc. Control* **7**(2):139–148 (1997).

Chen, J., A. Bandoni, and J. A. Romagnoli, "Outlier Detection in Process Plant Data," *Comp. Chem. Eng.* **22**(4/5):641–646 (1998).

Kacker, R., B. Toman, and D. Huang, "Comparison of ISO-GUM, Draft GUM, Supplement 1 and Bayesian Statistics Using Simple Linear Calibration," *Metrologia* **43:**S167–S177 (2006).

Lipták, B. G. (ed.), *Instrument Engineer's Handbook,* 4th ed., vol. 1, CRC Press, Boca Raton, Fla. (2003).

Liu, H., S. Shah, and W. Jiang, "On-line Outlier Detection and Data Cleaning," *Comp. Chem. Eng.* **28:**1635–1647 (2004).

McMillan, G. K. and D. M. Considine, *Process Industrial Instruments and Controls Handbook,* 5th ed., McGraw-Hill, New York (1999).

Miller, R. W. *Flow Measurement Engineering Handbook,* McGraw-Hill, New York (1996).

Omega Engineering Inc., "Transactions in Measurement and Control," vol. 4, OEI (available at www.omega.com, accessed in 2005).

Padmanabhan, T. R., *Industrial Instrumentation: Principles and Design,* Springer-Verlag, London (2000).

Taylor, J. R., *Risk Analysis for Process Plant, Pipelines and Transport,* E & FN Spon, London (1994).

Upp, E. L. and P. J. LaNasa, *Fluid Flow Measurement,* 2d ed., Gulf Professional Publishing, Boston (2002).

Webster, J. G. (ed.), *Measurement, Instrumentation and Sensors Handbook,* CRC Press, LLC (1999).

Wilson, J. S. (ed.), *Sensor Technology Handbook,* Elsevier, Amsterdam, Boston (2005).

Zhao, W., D. Chen, and S. Hu, "Detection of Outlier and a Robust BP Algorithm against Outlier," *Comp. Chem. Eng.* **28**(8):1403–1408 (2004).

CHAPTER 3

Variable Classification

This chapter presents the classification of variables in their different categories. These categories are related to the ability of observing them, that is, to obtain an estimate of their value, especially when they are not measured.

Linear Model

The model for the material balances, written in matrix form, is

$$C\,x = 0 \tag{3.1}$$

where x is the vector of flows of the different streams connecting the units.

Example 3.1: Consider the system shown in Fig. 3.1. It consists of 5 units and 11 streams. Measured streams are indicated by a small star (☆). Also consider that the hold-up of units V_1 through V_5 is negligible). Thus, all balances in the system do not contain derivatives and are represented by equations of the form shown by Eq. (3.1).

The balances are

$$\left.\begin{aligned} F_1 - F_2 - F_3 &= 0 \\ F_2 - F_4 &= 0 \\ F_3 - F_5 &= 0 \\ F_4 + F_5 - F_6 &= 0 \\ F_6 - F_7 &= 0 \\ F_7 - F_8 - F_9 &= 0 \\ F_8 - F_{10} &= 0 \\ F_9 - F_{11} &= 0 \end{aligned}\right\} \tag{3.2}$$

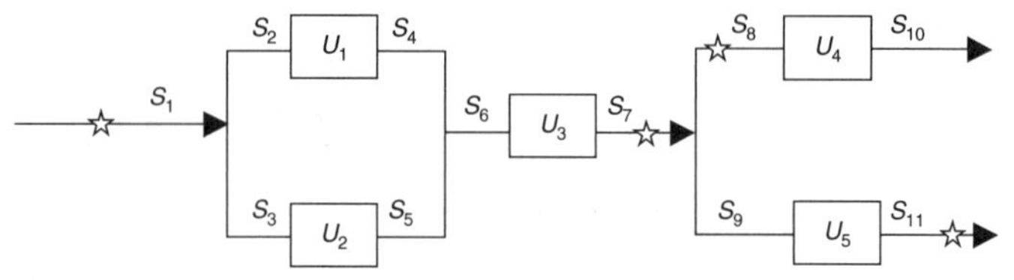

FIGURE 3.1 Flowsheet and flow measurements for Example 3.1.

The matrix C corresponding to these balances is

$$C = \begin{array}{c} \begin{matrix} S_1 & S_2 & S_3 & S_4 & S_5 & S_6 & S_7 & S_8 & S_9 & S_{10} & S_{11} \end{matrix} \\ \begin{bmatrix} 1 & -1 & -1 & & & & & & & & \\ & 1 & & -1 & & & & & & & \\ & & 1 & & -1 & & & & & & \\ & & & 1 & 1 & -1 & & & & & \\ & & & & & 1 & -1 & & & & \\ & & & & & & 1 & -1 & -1 & & \\ & & & & & & & 1 & & -1 & \\ & & & & & & & & 1 & & -1 \end{bmatrix} \end{array} \tag{3.3}$$

where the columns correspond to the variables indicated on top of the matrix.

For convenience, the set of state variables is divided into measured variables (x_M) and unmeasured variables (x_U). In our example

$$x_M = \begin{bmatrix} F_1 \\ F_7 \\ F_8 \\ F_{11} \end{bmatrix} \qquad x_U = \begin{bmatrix} F_2 \\ F_3 \\ F_4 \\ F_5 \\ F_6 \\ F_9 \\ F_{10} \end{bmatrix} \tag{3.4}$$

Observability

Observability of a variable in a system can be defined in a broad sense as the ability to obtain an estimate of the variable using certain measurements performed in the system. We will see next that this term will be restricted to unmeasured variables. We now concentrate on a variable classification for linear steady-state systems.

Notice first that all the flowrates of Fig. 3.1 after the first split (F_2, F_3, F_4, F_5) cannot be calculated, as there is no material balance that can be used to obtain them. These variables are called *unobservable variables*. The rest of the unmeasured variables (F_6, F_9, F_{10}) can be obtained from material balances using the measured values. We call these variables *observable variables*. This leads to our first classification

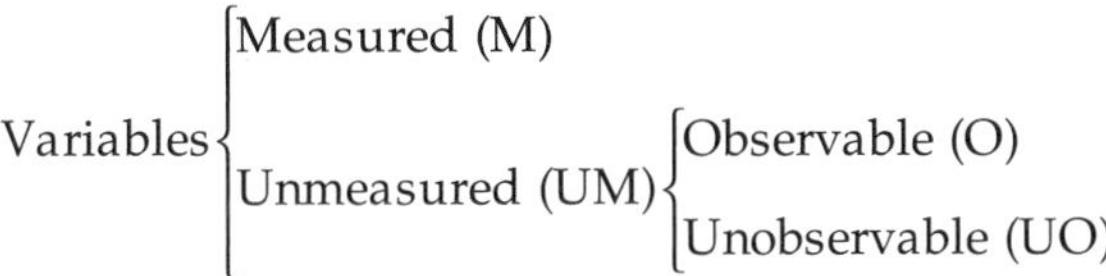

Thus, we formally define observability as follows:

Definition *A nonmeasured variable is observable if it can be calculated in at least one way from the measurements.*

Historical Note: Kalman (1960) introduced the concept of observability for linear dynamic systems. Griffith and Kumar (1971), Kou et al. (1973), and Singh (1975) presented seminal papers discussing observability in nonlinear systems. Václavek (1968, 1969) was one of the first to present the classification problem in process plants connected to steady state data reconciliation and Romagnoli and Stephanopoulos (1980) worked the problem analytically. Earlier, Mah et al. (1976) presented a similar strategy whereas Stanley and Mah (1981a, b) discussed this issue of observability of systems described by steady-state models in depth. In particular, they discuss conditions under which observability can be attained.

Redundancy

Let us turn our attention to the measured streams in Fig. 3.1. If the flowrate of stream S_1 is not measured, it can be estimated by using the value of F_7. However, there is also one additional way: by adding F_8 and F_{11}. In this last case, we say that the system $\{F_1, F_7, F_8, F_{11}\}$ is redundant.

Assume now that the flowrates of streams S_7 and S_8 are not measured. Then F_1 could not be estimated by any balance equation and its removal makes it unobservable: we call it nonredundant. Thus, the following definitions follow:

Definition *A measurement is redundant if it can also be calculated in at least one way from the remaining measurements. That is, the measurement can be deleted and the rest of the measurements can be used to calculate the value of that variable.*

Definition *A set of measurements of a system is redundant if all measurements are redundant.*

Definition *A measurement of a variable is nonredundant if after removing its measurement, the variable becomes unobservable, that is, it cannot be calculated using a balance equation involving the other variables of the system.*

We therefore complete our classification of variables as follows:

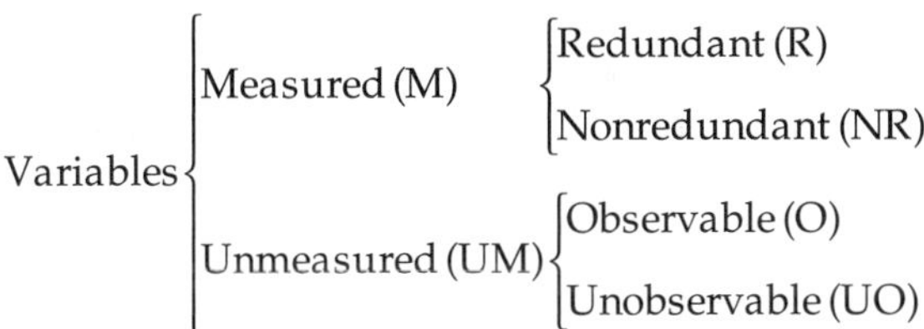

Redundancy is a desirable property of a system because in the case when an instrument fails, its variable can be estimated through balances. Moreover, if the number of different balances that can be used increases, there will be additional ways to calculate the variable. We say that the reliability of such system increases. There is, therefore, a need to distinguish these different levels of redundancy.

Hardware Redundancy

We consider now the case where more than one instrument is used to measure the same variable. This has been traditionally the way redundancy has been understood. However, hardware redundancy has no effect on observability, but has an effect on the accuracy and the reliability with which the estimates of the variables are obtained. These issues will be discussed later in the book.

Quantification of Observability and Redundancy

Observability and redundancy can be based on a few or on many balance equations. The amount of balance equations upon which observability and redundancy are based is important because instruments fail and a variable being estimated using many sources is likely to remain observable when some measurements are no longer available. In addition, the use of many sources improves the quality of the estimates because, as we shall see, the standard deviation of these estimates is smaller for a larger number of contributions. In this section, we concentrate on obtaining a number that will help in determining and comparing observable and redundant variables. A unification of both concepts under a single notion of estimability is first presented.

Estimability

A generalized definition of observability was attempted by Ali and Narasimhan (1993) to denote as observable any variable, measured or unmeasured, for which an estimate can be produced. Estimability is a term preferred to avoid confusion and reserve the name observable to unmeasured variables, as it has become popular in the literature. The definition of estimability (Bagajewicz and Sánchez, 1999) is formally presented next.

Definition *A variable S_i is estimable if it is measured or unmeasured, but observable.*

Degree of Observability of Variables

For convenience, we denote $\Theta(p)$ as the set of all possible combinations of p measurements. We call $\Theta_j(p)$, the jth element (combination) of this set. We are now ready for the following definition:

Definition *An unmeasured variable S_i has degree of observability O_i if*

(a) It remains observable after the elimination of <u>any</u> combination $\Theta_j(O_i - 1) \in \Theta(O_i - 1)$, AND

(b) It becomes unobservable when <u>at least</u> one set $\Theta_j(O_i) \in \Theta(O_i)$ is eliminated.

Notice the underlined "any" and "at least" in these definitions. They are crucial for the understanding of the concepts. In addition, also notice that the two conditions need to hold simultaneously.

Example 3.2: Consider the system of Fig. 3.2 and assume that $x_M = \{F_1, F_2\}$. Variable S_6 has degree of observability $O_6 = 1$, because just the elimination of the measurement in S_1 makes it unobservable. Part (*a*) of the definition does not apply, as $O_i - 1 = 0$.

If, for example, F_3 is also measured, that is, $x_M = \{F_1, F_2, F_3\}$, then F_6 would have degree of observability $O_6 = 2$, because elimination of one measurement at a time (in streams S_1, S_2, or S_3) would not make it unobservable. However, a deletion of any of the following two sets: (F_1, F_2), (F_1, F_3), would render it unobservable. Note that the elimination of the set (F_2, F_3) would not make F_6 unobservable.

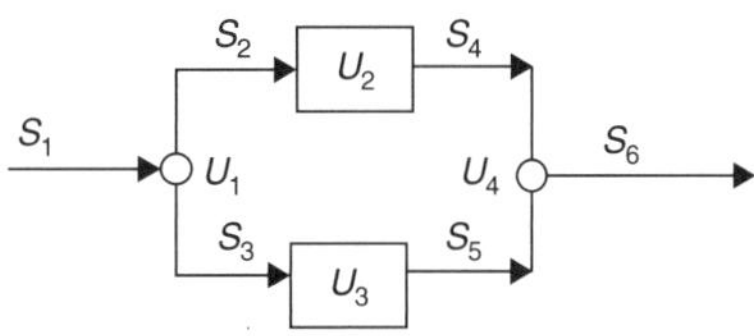

Figure 3.2 Illustration of the concept of degree of observability.

Degree of Redundancy of Variables

For convenience we denote $\vartheta(p, S_i)$ as the set of all possible combinations of p measured variables, not including the measured stream S_i. We call $\vartheta_j(p, S_i)$ the jth element (combination) of this set. The following definition is then presented.

Definition *A redundant measured variable S_i has degree of redundancy R_i if*

(a) It remains redundant after the elimination of <u>any</u> combination $\vartheta_j(R_i - 1, S_i) \in \vartheta(R_i - 1, S_i)$, AND

(b) It becomes nonredundant when <u>at least</u> one set $\vartheta_j(R_i, S_i) \in \vartheta(R_i, S_i)$ is eliminated.

Remark: Accordingly, the degree of redundancy of a nonredundant measurement is zero.

Example 3.3: In the system of Fig. 3.2, for $x_M = \{F_1, F_2\}$, variable F_1 has degree of redundancy $R_1 = 0$ because it is already nonredundant. If F_3 is measured, that is, if $x_M = \{F_1, F_2, F_3\}$, then F_1 has degree of redundancy $R_1 = 1$, because it is sufficient to eliminate the measurements in S_2 or S_3 to make it nonredundant.

A redundant measurement is such that the variable becomes observable when the measurement is eliminated. A nonredundant variable, in turn, becomes unobservable when its measurement is eliminated. Thus, if a variable has degree of redundancy R_i, the elimination of its measurement will make it a variable of degree of observability $O_i = R_i$.

Example 3.4: If for Fig. 3.2, $x_M = \{F_1, F_2, F_3\}$, then the elimination of the measurement in S_1, makes it a nonmeasured variable with degree of observability $O_1 = 1$, because it is enough to eliminate the measurements in S_2 or S_3 to make it unobservable.

Historical Note: Definitions of degree of observability and redundancy were first introduced by Maquin et al. (1991, 1995). The version presented above is the result of slight modifications made by Bagajewicz and Sánchez (1999). These modifications are, however, not conceptual. In addition, Maquin et al. (1991, 1995) introduced the concept of degree of redundancy of unmeasured variables. This concept is not used in this book, because it is preferred to leave the concept of redundancy of variables confined to measured ones.

The concept of degree of estimability is presented next. This concept unifies the concepts of degree of observability and degree of redundancy in a single one. This is, for example, important for the design of sensor networks, as requirements can be easier established mathematically.

Degree of Estimability of Variables

Definition *A variable S_i (measured or not) has degree of estimability E_i if*

(a) *It remains estimable after the elimination of <u>any</u> combination $\Theta_j(E_i - 1) \in \Theta(E_i - 1)$, AND*

(b) *It becomes unobservable when <u>at least</u> one set $\Theta_j(E_i) \in \Theta(E_i)$ is eliminated.*

Example 3.5: To illustrate the above definition, consider the process graph in Fig. 3.3, which includes the environment as a node, and assumes all flowrates are measured, that is $x_M = \{F_1, F_2, F_3, F_4\}$.

In this case, $\vartheta(1,S_2) = \{(S_1), (S_3), (S_4)\}$. Therefore, the measurement in stream S_2 has degree of redundancy $R_2 = 1$ because just the elimination of the measurement

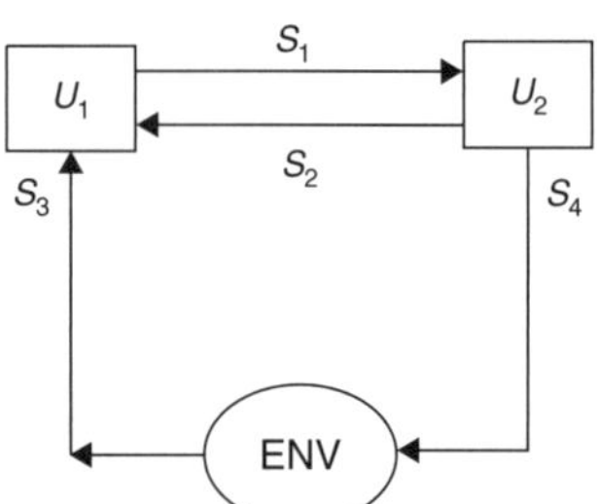

FIGURE 3.3 Illustration of degree of estimability.

in S_1 makes F_2 nonredundant. Since the elimination of the measurement in S_1 makes F_2 nonredundant, and the elimination of the measurement in either S_3 or S_4 does not alter its redundancy status, its degree of estimability is larger than one. Consequently, it is necessary to evaluate the elements of the set $\Theta(2) = \{(S_1, S_2), (S_1, S_3), (S_1, S_4), (S_2, S_3), (S_2, S_4), (S_3, S_4)\}$. From this analysis, it can be easily seen that the elimination of the measurements in (S_1, S_2) makes S_2 unobservable; thus $E_2 = 2$.

In a similar way, it can be shown that $R_3 = 2$ because if all the elements of $\vartheta(1,S_3) = \{(S_1), (S_2), (S_4)\}$ are individually eliminated, variable F_3 stays redundant, but becomes nonredundant if the measurements in (S_1, S_4) or (S_2, S_4) are deleted from $\vartheta(2,S_3) = \{(S_1, S_2), (S_1, S_4), (S_2, S_4)\}$. The inspection of $\Theta(2)$ and $\Theta(3) = \{S_1, S_2, S_3), (S_1, S_2, S_4), (S_2, S_3, S_4)\}$ helps to conclude that $E_3 = 3$, because the elimination of the measurements in (S_2, S_3, S_4) renders S_3 unobservable.

The following properties are natural consequences of the definition.

Property *The degree of estimability of a nonmeasured variable is equal to the degree of observability.*

Property *The degree of estimability of a measured redundant variable is its degree of redundancy plus one.*

Property *A measured nonredundant variable has degree of redundancy zero and degree of estimability one.*

The degree of estimability is an important property, because it does not distinguish whether a variable is measured or not. Systems with all variables with degree of estimability one are nonredundant, and as we shall see later, feature a minimum number of sensors.

Lemma 3.1 *A system where all variables have degree of estimability equal to one is a system of nonredundant measured variables and observable unmeasured ones.*

Proof If a variable is measured and has degree of estimability one, then it is nonredundant. If in turn, it is unmeasured, it is observable (because its estimability is one) and its observability is given by nonredundant variables.

Q.E.D.

Property *When the degree of estimability is larger than one for any variable, then the number of sensors that need to fail before the variable in question becomes unobservable is equal to its degree of estimability.*

Canonical Representation

We now resort to the mathematical representation to develop an analytical way to determine observable, unobservable, and redundant variables. We first consider the partition of the system of equations into measured and unmeasured variables.

By using simple rearrangement of columns in matrix C, the system can be rewritten in the following way:

$$\begin{bmatrix} C_U & C_M \end{bmatrix} \begin{bmatrix} x_U \\ x_M \end{bmatrix} = 0 \tag{3.5}$$

Example 3.6: In the case of Fig. 3.1, we identify matrices C_U and C_M as being

$$C_U = \begin{array}{c} \begin{array}{ccccccc} S_2 & S_3 & S_4 & S_5 & S_6 & S_9 & S_{10} \end{array} \\ \begin{bmatrix} -1 & -1 & & & & & \\ 1 & & -1 & & & & \\ & 1 & & -1 & & & \\ & & 1 & 1 & -1 & & \\ & & & & 1 & & \\ & & & & & -1 & \\ & & & & & & -1 \\ & & & & & 1 & \end{bmatrix} \end{array} \quad C_M = \begin{array}{c} \begin{array}{cccc} S_1 & S_7 & S_8 & S_{11} \end{array} \\ \begin{bmatrix} 1 & & & \\ & & & \\ & & & \\ & & & \\ & -1 & & \\ & 1 & -1 & \\ & & 1 & \\ & & & -1 \end{bmatrix} \end{array} \tag{3.6}$$

The basic elementary rearrangements are

1. The multiplication of two rows of the matrix by a number different from zero
2. The interchange of position of two rows (or columns) of the matrix
3. The addition of a row to another row

By elementary rearrangements, any rectangular matrix can be transformed to the so-called canonical form. In the canonical form, in the upper-left corner of the matrix one has the unit matrix while the other rows, if present, equal zero. This theory was first developed for data reconciliation by Madron (1992).

Example 3.7: We now discuss the details of how to obtain the canonical form based on our example. We start with the rearranged incidence matrix.

$$C = \begin{array}{c} \begin{array}{ccccccc|cccc} S_2 & S_3 & S_4 & S_5 & S_6 & S_9 & S_{10} & S_1 & S_7 & S_8 & S_{11} \end{array} \\ \left[\begin{array}{ccccccc|cccc} -1 & -1 & & & & & & 1 & & & \\ 1 & & -1 & & & & & & & & \\ & 1 & & -1 & & & & & & & \\ & & 1 & 1 & -1 & & & & & & \\ & & & & 1 & & & & -1 & & \\ & & & & & -1 & & & 1 & -1 & \\ & & & & & & -1 & & & 1 & \\ & & & & & 1 & & & & & -1 \end{array}\right] \end{array} \tag{3.7}$$

Applying the procedure of elementary rearrangements, we obtain the following:

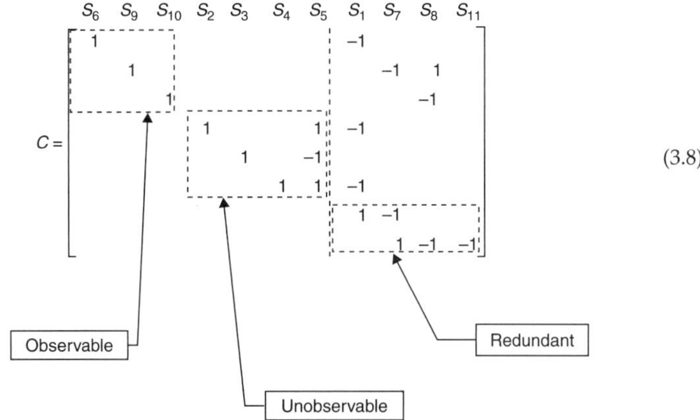

(3.8)

If in addition, variable S_3 is measured, then the result is

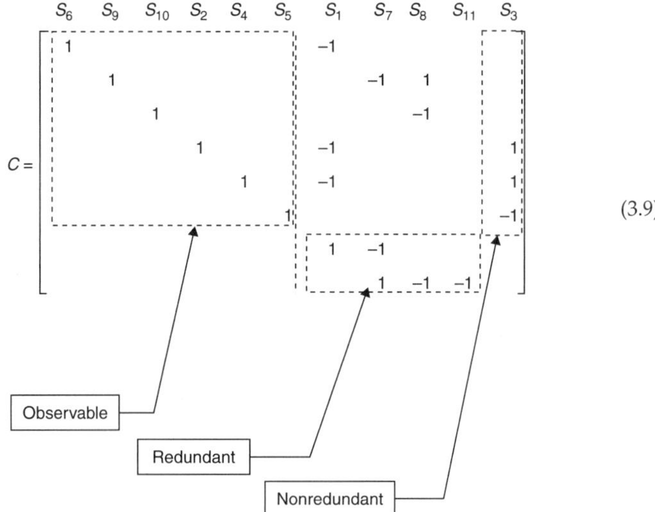

(3.9)

Details of the Procedure: Start with Eq. (3.7) and add row 1 to row 2 and multiply both rows (1 and 2) by (–1). The objective of adding is to make zero all the elements of column 1, except the one in the first row. The change of sign is mostly for convenience, as one would like to see positive numbers in the diagonal.

$$
\begin{array}{c}
\begin{array}{ccccccc:cccc} S_2 & S_3 & S_4 & S_5 & S_6 & S_9 & S_{10} & S_1 & S_7 & S_8 & S_{11} \end{array} \\
C = \left[\begin{array}{ccccccc:cccc}
1 & 1 & & & & & & -1 & & & \\
 & 1 & 1 & & & & & -1 & & & \\
 & 1 & & -1 & & & & & & & \\
 & & 1 & 1 & -1 & & & & & & \\
 & & & & 1 & & & & -1 & & \\
 & & & & & -1 & & & 1 & -1 & \\
 & & & & & & -1 & & & 1 & \\
 & & & & & 1 & & & & & -1
\end{array}\right]
\end{array}
\tag{3.10}
$$

Subtract row 2 from row 1, and from row 3. The objective is to make zero all the elements of column 2, except the one in the second row.

$$
\begin{array}{cccccccccccc}
 & S_2 & S_3 & S_4 & S_5 & S_6 & S_9 & S_{10} & S_1 & S_7 & S_8 & S_{11} \\
\end{array}
$$
$$
C = \left[\begin{array}{ccccccc|cccc}
1 & & -1 & & & & & & & & \\
 & 1 & 1 & & & & & -1 & & & \\
 & & -1 & -1 & & & & 1 & & & \\
 & & 1 & 1 & -1 & & & & & & \\
 & & & & 1 & & & & -1 & & \\
 & & & & & -1 & & & 1 & -1 & \\
 & & & & & & -1 & & & 1 & \\
 & & & & & 1 & & & & & -1
\end{array}\right] \quad (3.11)
$$

Add row 3 to rows 4 and 2 and subtract row 3 from row 1. Finally, multiply row 3 by (–1).

$$
\begin{array}{cccccccccccc}
 & S_2 & S_3 & S_4 & S_5 & S_6 & S_9 & S_{10} & S_1 & S_7 & S_8 & S_{11} \\
\end{array}
$$
$$
C = \left[\begin{array}{ccccccc|cccc}
1 & & & 1 & & & & -1 & & & \\
 & 1 & & -1 & & & & & & & \\
 & & 1 & 1 & & & & -1 & & & \\
 & & & & -1 & & & 1 & & & \\
 & & & & 1 & & & & -1 & & \\
 & & & & & -1 & & & 1 & -1 & \\
 & & & & & & -1 & & & 1 & \\
 & & & & & 1 & & & & & -1
\end{array}\right] \quad (3.12)
$$

As one can observe, we are slowly obtaining a diagonal matrix in the first rows of the unmeasured variables. One may need to also rearrange the position of rows or columns to achieve this. Notice, for example, that we do not have an element on the diagonal in row 4. Thus, we look for a column that has one and exchange columns. Thus, we exchange columns 4 and 5 (S_5 and S_6) and multiply row 4 by (–1).

$$
\begin{array}{cccccccccccc}
 & S_2 & S_3 & S_4 & S_6 & S_5 & S_9 & S_{10} & S_1 & S_7 & S_8 & S_{11} \\
\end{array}
$$
$$
C = \left[\begin{array}{ccccccc|cccc}
1 & & & & 1 & & & -1 & & & \\
 & 1 & & & -1 & & & & & & \\
 & & 1 & & 1 & & & -1 & & & \\
 & & & 1 & & & & -1 & & & \\
 & & & 1 & & & & & -1 & & \\
 & & & & & -1 & & & 1 & -1 & \\
 & & & & & & -1 & & & 1 & \\
 & & & & & 1 & & & & & -1
\end{array}\right] \quad (3.13)
$$

Subtract row 4 from row 5 to make the element in row 5 and column 4, zero.

$$\begin{array}{ccccccc|cccc}
S_2 & S_3 & S_4 & S_6 & S_5 & S_9 & S_{10} & S_1 & S_7 & S_8 & S_{11}
\end{array}$$

$$C = \left[\begin{array}{ccccccc|cccc}
1 & & & & 1 & & & -1 & & & \\
 & 1 & & & -1 & & & & & & \\
 & & 1 & & 1 & & & -1 & & & \\
 & & & 1 & & & & -1 & & & \\
 & & & & & & & 1 & -1 & & \\
 & & & & & -1 & & & 1 & -1 & \\
 & & & & & & -1 & & & 1 & \\
 & & & & & 1 & & & & & -1
\end{array}\right] \tag{3.14}$$

At this point we realize that S_2, S_3, S_4, and S_5 are unobservable because they are unmeasured and it is impossible to obtain a unit matrix by row-column manipulation. Indeed, there is no row after row 4 that would have an entry in the column corresponding to S_5. Thus, rows 1, 2, and 3 contain two unknowns each, and therefore these equations cannot be solved. In other words, it is impossible to calculate these variables. Therefore, we put rows 1, 2, and 3 at the bottom.

$$\begin{array}{ccccccc|cccc}
S_2 & S_3 & S_4 & S_6 & S_5 & S_9 & S_{10} & S_1 & S_7 & S_8 & S_{11}
\end{array}$$

$$C = \left[\begin{array}{ccccccc|cccc}
 & & & 1 & & & & -1 & & & \\
 & & & & & & & 1 & -1 & & \\
 & & & & & -1 & & & 1 & -1 & \\
 & & & & & & -1 & & & 1 & \\
 & & & & & 1 & & & & & -1 \\
1 & & & & 1 & & & -1 & & & \\
 & 1 & & & -1 & & & & & & \\
 & & 1 & & 1 & & & -1 & & &
\end{array}\right] \tag{3.15}$$

We now put columns 1, 2, 3, and 5 at the end of C_U because they are unobservable.

$$\begin{array}{ccccccc|cccc}
S_6 & S_9 & S_{10} & S_2 & S_3 & S_4 & S_5 & S_1 & S_7 & S_8 & S_{11}
\end{array}$$

$$C = \left[\begin{array}{ccccccc|cccc}
1 & & & & & & & -1 & & & \\
 & & & & & & & 1 & -1 & & \\
 & -1 & & & & & & & 1 & -1 & \\
 & & -1 & & & & & & & 1 & \\
 & 1 & & & & & & & & & -1 \\
 & & & 1 & & & 1 & -1 & & & \\
 & & & & 1 & & -1 & & & & \\
 & & & & & 1 & 1 & -1 & & &
\end{array}\right] \tag{3.16}$$

We now start the canonization process again: Multiply row 3 by (–1) and exchange rows 2 and 3.

$$\begin{array}{ccccccc|cccc}
S_6 & S_9 & S_{10} & S_2 & S_3 & S_4 & S_5 & S_1 & S_7 & S_8 & S_{11}
\end{array}$$

$$C = \left[\begin{array}{ccccccc|cccc}
1 & & & & & & & -1 & & & \\
 & 1 & & & & & & & -1 & 1 & \\
 & & & & & & & 1 & -1 & & \\
 & & -1 & & & & & & & 1 & \\
 & 1 & & & & & & & & & -1 \\
 & & & 1 & & & 1 & -1 & & & \\
 & & & & 1 & & -1 & & & & \\
 & & & & & 1 & 1 & -1 & & &
\end{array}\right] \tag{3.17}$$

Subtract row 2 from row 4 and exchange rows 3 and 4. Finally, change the sign of row 3. We thus get:

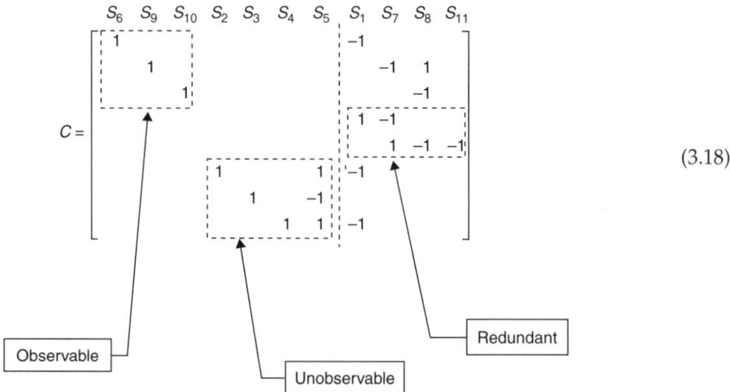

(3.18)

which is the canonical form [Eq. (3.8)].

We now concentrate on obtaining the result of Eq. (3.9). We could start rearranging Eq. (3.7), but we will start from Eq. (3.18). If S_3 is measured, the system in Eq. (3.18) becomes

$$
\begin{array}{c} \\ C = \end{array}
\begin{array}{ccccccc|cccc}
S_6 & S_9 & S_{10} & S_2 & S_4 & S_5 & S_3 & S_1 & S_7 & S_8 & S_{11} \\
1 & & & & & & & -1 & & & \\
& 1 & & & & & & & -1 & 1 & \\
& & 1 & & & & & & & -1 & \\
& & & & & & & 1 & -1 & & \\
& & & & & & & & 1 & -1 & -1 \\
& & & 1 & & 1 & & -1 & & & \\
& & & & & -1 & 1 & & & & \\
& & & & 1 & 1 & & -1 & & &
\end{array}
\tag{3.19}
$$

where we only exchanged columns. We now have row 4 empty in the part of the matrix corresponding to unmeasured variables. To change this, we exchange rows 4 and 6.

$$
\begin{array}{c} \\ C = \end{array}
\begin{array}{ccccccc|cccc}
S_6 & S_9 & S_{10} & S_2 & S_4 & S_5 & S_3 & S_1 & S_7 & S_8 & S_{11} \\
1 & & & & & & & -1 & & & \\
& 1 & & & & & & & -1 & 1 & \\
& & 1 & & & & & & & -1 & \\
& & & 1 & & 1 & & -1 & & & \\
& & & & & & & & 1 & -1 & -1 \\
& & & & & & & 1 & -1 & & \\
& & & & & -1 & 1 & & & & \\
& & & & 1 & 1 & & -1 & & &
\end{array}
\tag{3.20}
$$

We now proceed to obtain a diagonal nonzero element in row 5 by exchanging rows 8 and 5.

$$
\begin{array}{c}
\begin{array}{cccccc:ccccc}
S_6 & S_9 & S_{10} & S_2 & S_4 & S_5 & S_3 & S_1 & S_7 & S_8 & S_{11}
\end{array}\\
C=\left[\begin{array}{cccccc:ccccc}
1 & & & & & & & -1 & & & \\
 & 1 & & & & & & & -1 & 1 & \\
 & & 1 & & & & & & & -1 & \\
 & & & 1 & & 1 & & -1 & & & \\
 & & & & 1 & 1 & & -1 & & & \\
 & & & & & & & 1 & -1 & & \\
 & & & & & -1 & 1 & & & & \\
 & & & & & & & & 1 & -1 & -1
\end{array}\right]
\end{array}
\tag{3.21}
$$

We also put an element in the diagonal of row 6 by multiplying row 7 by (–1) and exchanging rows 6 and 7.

$$
\begin{array}{c}
\begin{array}{cccccc:ccccc}
S_6 & S_9 & S_{10} & S_2 & S_4 & S_5 & S_3 & S_1 & S_7 & S_8 & S_{11}
\end{array}\\
C=\left[\begin{array}{cccccc:ccccc}
1 & & & & & & & -1 & & & \\
 & 1 & & & & & & & -1 & 1 & \\
 & & 1 & & & & & & & -1 & \\
 & & & 1 & & 1 & & -1 & & & \\
 & & & & 1 & 1 & & -1 & & & \\
 & & & & & 1 & -1 & & & & \\
 & & & & & & & 1 & -1 & & \\
 & & & & & & & & 1 & -1 & -1
\end{array}\right]
\end{array}
\tag{3.22}
$$

Next, the objective is to make zero all the nondiagonal elements of column 6. Subtract row 6 from rows 4 and 5 and obtain the following:

$$
\begin{array}{c}
\begin{array}{cccccc:ccccc}
S_6 & S_9 & S_{10} & S_2 & S_4 & S_5 & S_3 & S_1 & S_7 & S_8 & S_{11}
\end{array}\\
C=\left[\begin{array}{cccccc:ccccc}
1 & & & & & & & -1 & & & \\
 & 1 & & & & & & & -1 & 1 & \\
 & & 1 & & & & & & & -1 & \\
 & & & 1 & & & 1 & -1 & & & \\
 & & & & 1 & & 1 & -1 & & & \\
 & & & & & 1 & -1 & & & & \\
 & & & & & & & 1 & -1 & & \\
 & & & & & & & & 1 & -1 & -1
\end{array}\right]
\end{array}
\tag{3.23}
$$

At this point, we realize that only rows 7 and 8 can contain redundant variables because the part of the matrix corresponding to unmeasured variables is empty in those rows. We also recognize that the column corresponding to S_3 contains elements that are in rows corresponding to unmeasured observable variables but not in rows corresponding to redundant variables. Therefore, S_3 is nonredundant. We thus, move column 7 to the end to get:

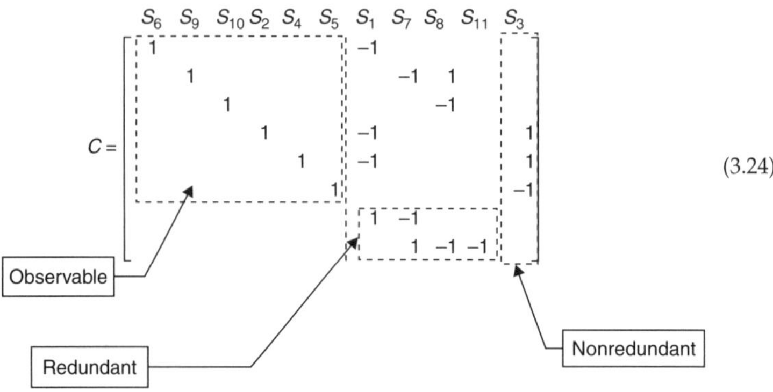

(3.24)

The General Case

In general we obtain the following:

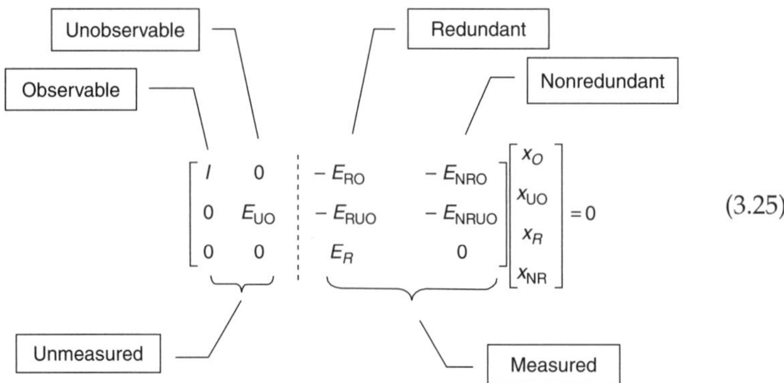

(3.25)

This is called the canonical form of *C*. The last set of variables is called nonredundant, as their values cannot be improved and their measurement has to be taken at face value.

Rewriting the system one obtains

$$x_O = E_{\mathrm{RO}}\, x_R + E_{\mathrm{NRO}}\, x_{\mathrm{NR}} \tag{3.26}$$

$$E_{\mathrm{UO}} x_{\mathrm{UO}} = -E_{\mathrm{RUO}} x_R - E_{\mathrm{NRUO}} x_{\mathrm{NR}} \tag{3.27}$$

$$E_R x_R = 0 \tag{3.28}$$

Equation (3.28) cannot be satisfied by the measurements and will be used later to adjust these measured variables through data reconciliation. Equation (3.26) allows the calculation of the observable variables. Finally, Eq. (3.27) represents a system of equations that cannot be solved. Finally, we realize that knowledge of x_R and x_{NR} allows the calculation of the observable variables uniquely.

Remark: There are several alternatives to the above procedure. Among the most well known are the matrix projection (Crowe et al., 1983; Crowe, 1989) and the QR decomposition (Swartz, 1989; Sánchez and Romagnoli, 1996). They are completely equivalent. A good review of all work can be found in Sánchez and Romagnoli (2000) and Narasimhan and Jordache (2000). Also, Ragot et al (1996) proposed a method based on looking at the circuits of a graph and Kelly (1998,1999) proposed alternative methods that allows easy updating when measurements are removed.

Remark: In the case of nonlinear systems, that is, systems where aside from material balances, component and energy balances as well as phase equilibria and other relations are included, observability analysis is rather more complicated. One alternative is to linearize the system and obtain an incidence matrix, which in this case will not be full of ones and negative ones. Rather the coefficients of the matrix will be result of the linearization around a nominal operating point (Joris and Kalitventzeff, 1987). We explore this in more detail in Chap. 8.

Remark: Work on classification of nonlinear systems without linearization was presented by Ponzoni et al. (1999), which consists of making a structural rearrangement of the occurrence matrix, a matrix where every row corresponds to an equation and contains a value different from zero in the column corresponding to a variable that participates in that equation. A symbolic derivation of the nonlinear equations is used by Ferraro et al. (2002) to perform redundancy analysis. Building up on this work, Ponzoni et al. (2004) proposed a robust direct method for classification that improves on previous work. Finally, Carballidoa et al. (2009) used an evolutionary approach.

Remark: In Chapter 4, we show a direct methodology for classification that makes use of data reconciliation.

References

Ali, Y. and S. Narasimhan, "Sensor Network Design for Maximizing Reliability of Linear Processes," *AIChE J.* **39**(5):2237–2249 (1993).

Bagajewicz, M. and M. Sánchez, "Design and Upgrade of Non-Redundant and Redundant Linear Sensor Networks," *AIChE J.* **45**(9):1927–1939 (1999).

Carballidoa, J. A., I. Ponzoni, and N. B. Brignole, "SID-GA: An Evolutionary Approach for Improving Observability and Redundancy Analysis in Structural Instrumentation Design," *Comp. and Ind. Eng.* **56**:1419–1428 (2009).

Crowe, C. M., "Observability and Redundancy of Process Data for Steady State Reconciliation," *Chem. Eng. Sci.* **44**:2909–2917 (1989).

Crowe, C. M., Y. A. García Campos, and A. Hrymak, "Reconciliation of Process Flow Rates by Matrix Projection. Part I: Linear Case," *AIChE J.* **29**:881–888 (1983).

Ferraro, S. J., I. Ponzoni, M. C. Sánchez, and N. B. Brignole, "A Symbolic Derivation Approach for Redundancy Analysis," *Ind. Eng. Chem. Res.* **41**(23):5692–5701 (2002).

Griffith, E. W. and K. S. P. Kumar, "On the Observability of Nonlinear Systems," *J. Math. Anal. Appl.* **35:**135 (1971).
Joris, P. and B. Kalitventzeff, "Process Measurement Analysis and Validation," Proceedings of CEF'87: *The Use of Computers in Chemical Engineering.* Taormina, Italy: 41–46 (1987).
Kalman, R. E., "New Approach to Linear Filtering and Prediction Problems," *J. Basic Eng. ASME* **82D**(35) (1960).
Kelly, J. D., "On Finding the Matrix Projection in the Data Reconciliation Solution," *Comp. Chem. Eng.* **22:**1553–1557 (1998).
Kelly, J. D., "Reconciliation of Process Data Using Other Projection Matrices," *Comp. Chem. Eng.* **23:**785–789 (1999).
Kou, S. R., D. L. Elliot, and T. J. Tarn, "Observability of Nonlinear Systems," *Inf. Contr.* **22:**89 (1973).
Madron, F., *Process Plant Performance,* Ellis Horwood, Chichester, England (1992).
Mah, R. S. H., *Chemical Process Structures and Information Flows,* Butterworths, Boston (1990).
Mah, R. S. H., G. Stanley, and D. Downing, "Reconciliation and Rectification of Process Flow and Inventory Data," *Ind. Eng. Chem Res. Proc. Des. Dev.* **15:** 175–183 (1976).
Maquin, D., G. Bloch, and J. Ragot, "Data Reconciliation of Measurements," *Revue Diagnostic et Surete de Fonctionnement* **1**(2):145–181 (1991).
Maquin, D., C. T. Huynh, M. Luong, and J. Ragot, "Observability, Redundancy, Reliability and Integrated Design of Measurement Systems," Proceedings of the *2nd IFAC Symposium on Intelligent Components and Instruments for Control Applications, SICICA'94,* Budapest, Hungary (1994).
Maquin, D., M. Luong, and J. Paris, "Dependability and Analytical Redundancy," *IFAC Symposium on On-Line Fault Detection in the Chemical Process Industries,* Newcastle, UK (1995).
Narasimhan, S. and C. Jordache, *Data Reconciliation and Gross Error Detection. An Intelligent Use of Process Data,* Gulf Publishing Company, Houston, Tex. (2000).
Ponzoni, I., M. C. Sánchez, and N. B. Brignole, "A New Structural Algorithm for Observability Classification," *Ind. Eng. Chem. Res.* **38**(8):3027–3035 (1999).
Ponzoni, I., M. C. Sánchez, and N. B. Brignole, "Direct Method for Structural Observability Analysis," *Ind. Eng. Chem. Res.* **43**(2):577–588 (2004).
Ragot, J., M. Luong, and D. Maquin, "Observability of Systems Involving Flow Circulation," *Int. J. Miner. Process.* **47:**125–140 (1996).
Romagnoli, J. and G. Stephanopoulos, "On the rectification of Measurement Errors for Complex Chemical Plants," *Chem. Eng. Sci.* **35:**1067–1081 (1980).
Sánchez, M. and J. Romagnoli, "Use of Orthogonal Transformations in Data Classification—Reconciliation," *Comput. Chem. Eng.* **20:**483–493 (1996).
Sánchez, M. and J. Romagnoli, "Data Processing and Reconciliation for Chemical Process Operations," Academic Press, San Diego, Calif. (2000).
Singh, S. N., "Observability in Non-Linear Systems with Immeasurable Inputs," *Int. J. Syst. Sci.* **6:**723–732 (1975).
Stanley, G. M. and R. H. S. Mah, "Observability and Redundancy in Process Data Estimation," *CES* **36:**259–272 (1981).
Swartz, C. L. E., "Data Reconciliation for Generalized Flowsheet Applications," *American Chemical Society—National Meeting,* Dallas, Tex. (1989).
Václavek, V., "Studies on System Engineering. I. On the Application of the Calculus of Observations in Calculations of Chemical Engineering Balances," *Coll. Czeck. Chem. Commun.* 34:3653 (1968).

CHAPTER 4

Material Balance Data Reconciliation

In Chap. 3, it was shown that once we distinguish measured from unmeasured variables, the model equation

$$CF = 0 \tag{4.1}$$

can be rewritten as follows:

$$E_R F_R = 0 \tag{4.2}$$

$$F_O = E_{RO} F_R + E_{NRO} F_{NR} \tag{4.3}$$

We have no problem with estimating the observable flowrates using Eq. (4.3) and using the measurements. However, the measurements usually (never) satisfy Eq. (4.2), that is,

$$r = E_R F_R^+ \neq 0 \tag{4.4}$$

where r is the vector of residuals.

Therefore, there is a conflict between these measurements. We need to adjust them so that they satisfy Eq. (4.2). When a statistical approach is taken, the problem of reconciliation consists of minimizing the weighed square of the difference between the measurements F_R^+ and the estimators $\tilde{F}_R$, using as weight the variance of the measurements S_R. This is done by solving the following optimization problem:

$$\left.\begin{array}{l} \text{Min } [\tilde{F}_R - F_R^+]^T S_R^{-1} [\tilde{F}_R - F_R^+] \\ \text{s.t} \\ \quad E_R \tilde{F}_R = 0 \end{array}\right\} \tag{4.5}$$

This optimization problem is a least square minimization of the difference between the estimate to be calculated and the measurement, weighted by the inverse of the variance value. In such case, when the variance is large, the adjustment ($\tilde{F}_R - F_R^+$) may also be large. Finally, note that the variance matrix, which in general is symmetric, is usually considered diagonal.

Example 4.1: Consider the system of Fig. 3.1, whose redundant streams are S_1, S_7, S_8, and S_{11}, with, E_R given by

$$E_R = \begin{array}{c} \begin{matrix} S_1 & S_7 & S_8 & S_{11} \end{matrix} \\ \begin{bmatrix} 1 & -1 & & \\ & 1 & -1 & -1 \end{bmatrix} \end{array} \tag{4.6}$$

which corresponds to the following set of equations:

$$\left.\begin{aligned} F_1 - F_7 &= 0 \\ F_7 - F_8 - F_{11} &= 0 \end{aligned}\right\} \tag{4.7}$$

Thus, if we consider that the variance matrix is diagonal, that is,

$$S_R = \begin{array}{c} \begin{matrix} S_1 & S_7 & S_8 & S_{11} \end{matrix} \\ \begin{bmatrix} \sigma_1^2 & & & \\ & \sigma_7^2 & & \\ & & \sigma_8^2 & \\ & & & \sigma_{11}^2 \end{bmatrix} \end{array} \tag{4.8}$$

Then, Eq. (4.5) is expressed as follows:

$$\left.\begin{aligned} &\text{Min } \frac{[\tilde{F}_1 - F_1^+]^2}{\sigma_1^2} + \frac{[\tilde{F}_7 - F_7^+]^2}{\sigma_7^2} + \frac{[\tilde{F}_8 - F_8^+]^2}{\sigma_8^2} + \frac{[\tilde{F}_{11} - F_{11}^+]^2}{\sigma_{11}^2} \\ &\text{s.t} \\ &\qquad \tilde{F}_1 - \tilde{F}_7 = 0 \\ &\qquad \tilde{F}_7 - \tilde{F}_8 - \tilde{F}_{11} = 0 \end{aligned}\right\} \tag{4.9}$$

From calculus, the minimum is obtained by forming the Lagrangian function, equating its derivatives with respect to the variables and the multipliers, and solving the algebraic system obtained. The solution of Eq. (4.5) is

$$\tilde{F}_R = \left[I - S_R E_R^T \left(E_R S_R E_R^T \right)^{-1} E_R \right] F_R^+ \tag{4.10}$$

Derivation of Eq. (4.10): The Lagrangian function is

$$L = [\tilde{F}_R - F_R^+]^T S_R^{-1} [\tilde{F}_R - F_R^+] - \lambda^T (E_R \tilde{F}_R) \tag{4.11}$$

where λ is the vector of Lagrange multipliers. The necessary conditions of optimality are

$$\frac{\partial L}{\partial \tilde{F}_R} = \frac{\partial [\tilde{F}_R - F_R^+]^T}{\partial \tilde{F}_R} S_R^{-1}[\tilde{F}_R - F_R^+] + \frac{\partial [\tilde{F}_R - F_R^+]^T}{\partial \tilde{F}_R} S_R^{-1}[\tilde{F}_R - F_R^+] - E_R^T \lambda = 0 \tag{4.12}$$

$$\frac{\partial L}{\partial \lambda} = E_R \tilde{F}_R = 0 \tag{4.13}$$

Since $\dfrac{\partial [\tilde{F}_R - F_R^+]^T}{\partial \tilde{F}_R} = I$, Eq. (4.12) renders

$$\tilde{F}_R = F_R^+ + \frac{1}{2} S_R E_R^T \lambda \tag{4.14}$$

Now, substituting Eq. (4.14) into Eq. (4.13) renders

$$E_R F_R^+ = -\frac{1}{2} E_R S_R E_R^T \lambda \tag{4.15}$$

and, since $E_R S_R E_R^T$ is square and nonsingular, then

$$\lambda = -2(E_R S_R E_R^T)^{-1} E_R F_R^+ \tag{4.16}$$

which substituted in Eq. (4.14) renders Eq. (4.10).

Q.E.D.

Historical Note: The above procedure for data reconciliation was first proposed by Kuehn and Davidson (1961). Later Mah et al. (1976) formalized it in more detail and Madron et al. (1977) extended it to consider including chemical reactors. This least square problem can be derived from Bayesian estimation theory using the assumption that the distribution of errors is normal. This was first discussed by Reilly and Patino-Leal, (1981) and later rigorously proven by Johnston and Kramer (1995) where they proposed a maximum likelihood derivation of the steady state linear reconciliation model. Narasimhan and Jordache (2000) also discussed the Bayesian approach, and Bakshi et al. (2001) used Bayesian theory in conjunction with wavelet filtering. Dovi and Del Borghi (2001) also used Bayesian theory presenting a modified likelihood function.

Remark: Crowe (1996) showed that the same result can be derived using information theory. Reis and Saraiva (2005) put this technique in the context of regression theory and Kelly (1999) showed that other projection matrices to remove unobservable variables can be formulated. Finally, Maquin et al. (2000) presented a methodology where the data reconciliation is performed using uncertain models, that is, when matrix E_R is not well known.

Remark: Bagajewicz (1996) considered that even when the original measurements are normal, this distribution can be distorted by signal transformations, giving raise to a different formulation. Maquin et al. (1999) presented a method where the data reconciliation can be performed in the absence of knowledge of the variance-covariance matrix. In turn, Narasimhan and Hari Kumar (1993) as well as Ragot and Maquin (2004) considered the case in which the errors of the measurements are bounded. Finally, Ragot et al. (2005) considered the errors as intervals without particular knowledge or assumption about the density probability function. By using inequality constraints the technique renders a set of reconciled data in a form of intervals. Mandel et al. (1998) had used a similar approach relying on linear matrix inequalities (LMI).

Remark: Several reviews of the field were written by Hlavacek (1977), Tamhane and Mah (1985), Mah (1990), Madron (1992), Crowe (1996), and Veverka and Madron (1997). Finally the books by Sánchez and Romagnoli (2000) and Narasimhan and Jordache (2000) present the material in a formal way.

Determination of the Measurement Vector

In practice one is faced with a chart with the signal containing a lot of noise, or a set of measured values at different time intervals. However, to obtain reconciled values one needs only one number per stream. The solution in practice has been to obtain this number by making an average of all data.

Example 4.2: We return to our system of Fig. 3.1 with an added measurement in stream S_3 (Fig. 4.1)

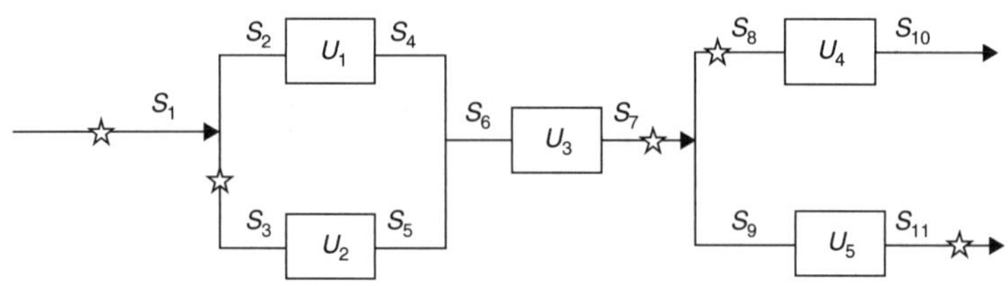

FIGURE 4.1 Flowsheet of Example 4.1.

Consider the "true" values shown in Table 4.1 and the measurements to be those of Table 4.2.

Stream	True Flowrate
S_1	10
S_3	3
S_7	10
S_8	6
S_{11}	4

TABLE 4.1 True Flowrates of Example 4.1

Stream	Measured Values										Average
S_1	9.90	10.15	10.26	9.82	10.15	9.79	10.21	9.93	10.17	9.95	10.03
S_3	3.02	2.95	3.01	2.86	3.06	2.91	2.93	2.94	2.88	3.09	2.97
S_7	9.72	10.17	10.51	9.86	9.85	10.25	9.93	9.88	10.03	10.36	10.06
S_8	5.86	6.13	5.92	6.09	6.06	6.03	5.98	6.06	5.95	5.86	5.99
S_{11}	4.03	3.97	4.04	3.95	4.04	4.03	3.87	3.90	4.02	4.09	3.99

Table 4.2 Measurements for Example 4.1

If they are fairly stable and the distribution around the average is Gaussian (both conditions that may not be known), then we can obtain the standard deviation. To determine the variance, a 95% confidence was used (t_p= 1.83, see Chap. 2). The result is given in Table 4.3.

Stream	Standard Deviation	Variance
S_1	0.17	0.1
S_3	0.08	0.02
S_7	0.26	0.22
S_8	0.10	0.03
S_{11}	0.07	0.16

TABLE 4.3 Resulting Variances for Example 4.1

We now proceed to obtain the reconciled values. Matrix E_R is given by Eq. (4.6) and the averages of the measurements, when plugged into Eq. (4.7) provide the following:

$$\left.\begin{aligned} F_1^+ - F_7^+ &= 0.03 \\ F_7^+ - F_8^+ - F_{11}^+ &= 0.08 \end{aligned}\right\} \tag{4.17}$$

We now want to reconcile these measurements. Given the calculated standard deviations, we have

$$S_R = \begin{array}{c} \begin{matrix} S_1 & S_7 & S_8 & S_{11} \end{matrix} \\ \begin{bmatrix} 0.1 & & & \\ & 0.22 & & \\ & & 0.03 & \\ & & & 0.16 \end{bmatrix} \end{array} \tag{4.18}$$

Then

$$\tilde{F}_R = \left[I - S_R E_R^T \left(E_R S_R E_R^T\right)^{-1} E_R\right] F_R^+ = \begin{bmatrix} 0.505 & 0.230 & 0.266 & 0.266 \\ 0.505 & 0.230 & 0.266 & 0.266 \\ 0.080 & 0.036 & 0.884 & -0.116 \\ 0.425 & 0.193 & -0.618 & 0.382 \end{bmatrix} \begin{bmatrix} 10.03 \\ 10.06 \\ 5.99 \\ 3.99 \end{bmatrix} = \begin{bmatrix} 10.02 \\ 10.02 \\ 6.00 \\ 4.03 \end{bmatrix} \tag{4.19}$$

We now obtain the observable variables using Eq. (4.3). In our case [(Chap. 3, Eqs. (3.24) and (3.25)]

$$E_{RO} = \begin{array}{c} \begin{matrix} S_1 & S_7 & S_8 & S_{11} \end{matrix} \\ \begin{bmatrix} 1 & & & \\ & 1 & -1 & \\ & & 1 & \\ 1 & & & \\ 1 & & & \\ & & & \end{bmatrix} \end{array} \qquad E_{NRO} = \begin{array}{c} S_3 \\ \begin{bmatrix} \\ \\ \\ -1 \\ -1 \\ 1 \end{bmatrix} \end{array} \tag{4.20}$$

Therefore,

$$F_O = \begin{bmatrix} F_6 \\ F_9 \\ F_{10} \\ F_2 \\ F_4 \\ F_5 \end{bmatrix} = \begin{bmatrix} 1 & & & \\ & 1 & -1 & \\ & & 1 & \\ 1 & & & \\ 1 & & & \\ & & & \end{bmatrix} \begin{bmatrix} 10.03 \\ 10.03 \\ 6.00 \\ 4.03 \end{bmatrix} + \begin{bmatrix} \\ \\ \\ -1 \\ -1 \\ 1 \end{bmatrix} 2.97 = \begin{bmatrix} 10.03 \\ 4.03 \\ 6.00 \\ 7.06 \\ 7.06 \\ 2.97 \end{bmatrix} \tag{4.21}$$

Details of the Procedure: We now proceed to obtain the result shown in Eq. (4.19).

$$S_R E_R^T = \begin{bmatrix} \sigma_1^2 & & & \\ & \sigma_7^2 & & \\ & & \sigma_8^2 & \\ & & & \sigma_{11}^2 \end{bmatrix} \begin{bmatrix} 1 & \\ -1 & 1 \\ & -1 \\ & -1 \end{bmatrix} = \begin{bmatrix} \sigma_1^2 & \\ -\sigma_7^2 & \sigma_7^2 \\ & -\sigma_8^2 \\ & -\sigma_{11}^2 \end{bmatrix} = \begin{bmatrix} 0.1 & \\ -0.22 & 0.22 \\ & -0.03 \\ & -0.16 \end{bmatrix} \tag{4.22}$$

Then,

$$E_R S_R E_R^T = \begin{bmatrix} 1 & -1 & & \\ & 1 & -1 & -1 \end{bmatrix} \begin{bmatrix} 0.1 & \\ -0.22 & 0.22 \\ & -0.03 \\ & -0.16 \end{bmatrix} = \begin{bmatrix} 0.32 & -0.22 \\ -0.22 & 0.41 \end{bmatrix} \tag{4.23}$$

The inverse is

$$\left[E_R S_R E_R^T\right]^{-1} = \begin{bmatrix} 4.952 & 2.657 \\ 2.657 & 3.865 \end{bmatrix} \tag{4.24}$$

But,

$$\left.\begin{aligned} \left[E_R S_R E_R^T\right]^{-1} E_R &= \begin{bmatrix} 4.952 & 2.657 \\ 2.657 & 3.865 \end{bmatrix} \begin{bmatrix} 1 & -1 & & \\ & 1 & -1 & -1 \end{bmatrix} = \\ &= \begin{bmatrix} 4.952 & -2.295 & -2.657 & -2.657 \\ 2.657 & 1.208 & -3.865 & -3.865 \end{bmatrix} \end{aligned}\right\} \tag{4.25}$$

Thus,

$$\left.\begin{aligned} S_R E_R^T \left(E_R S_R E_R^T\right)^{-1} E_R &= \begin{bmatrix} 0.1 & \\ -0.22 & 0.22 \\ & -0.03 \\ & -0.16 \end{bmatrix} \begin{bmatrix} 4.952 & -2.295 & -2.657 & -2.657 \\ 2.657 & 1.208 & -3.865 & -3.865 \end{bmatrix} = \\ &= \begin{bmatrix} 0.495 & -0.230 & -0.266 & -0.266 \\ -0.505 & 0.770 & -0.266 & -0.266 \\ -0.080 & -0.036 & 0.116 & 0.116 \\ -0.425 & -0.193 & 0.618 & 0.618 \end{bmatrix} \end{aligned}\right\} \tag{4.26}$$

and therefore,

$$I - S_R E_R^T \left(E_R S_R E_R^T\right)^{-1} E_R = \begin{bmatrix} 0.505 & 0.230 & 0.266 & 0.266 \\ 0.505 & 0.230 & 0.266 & 0.266 \\ 0.080 & 0.036 & 0.884 & -0.116 \\ 0.425 & 0.193 & -0.618 & 0.382 \end{bmatrix} \tag{4.27}$$

from which Eq. (4.19) follows immediately.

Remark: The above procedure for obtaining the variance of the measurements (diagonals of S_R) (Table 4.3) is called the *direct method*. This is a correct procedure if the measurements are independent, the system is truly at steady state, and in addition, there are no outliers. Since the system is never at a true steady state, if one uses the above formulas of the direct method, one will incorporate the process variations, that is, the variance of the natural process oscillations into the calculation. To take into account this problem and to assess the existence of variable interdependence (nondiagonal variance matrix), the indirect method was proposed. This method was originally proposed by Almasy and Mah (1984), later modified by Darouach et al. (1989), Keller et al. (1992), and Chen et al. (1997). Data reconciliation software notoriously lacks a module to perform such estimation, and there are no published results regarding the efficiency of these methods in practice. In theory, this matrix does not have to be diagonal. In addition, we assume that the system is at steady state. We discuss this issue further in Chap. 12.

Precision of the Estimates

Once the steady-state data reconciliation problem is solved, it is desired to know the precision of the estimates obtained. In general, if $y = \Gamma x$, then the variance of y, S_y, is given by

$$S_y = \Gamma S_x \Gamma^T \tag{4.28}$$

where S_x is the variance of the variable x.

Details: In Chap. 2, we discussed the covariance of a linear combination of variables. We generalize now the demonstration for the vector linear relation shown above.

$$\operatorname{cov}(y_i, y_j) = \int_{-\infty}^{\infty} \cdots \int_{-\infty}^{\infty} \left(\sum_k \gamma_{ik} x_k - \overline{y}_i\right)\left(\sum_s \gamma_{js} x_s - \overline{y}_j\right) p(x) dx_1 dx_2 \ldots dx_n \tag{4.29}$$

where $\bar{y}_i = \sum_k \gamma_{ik}\bar{x}_k$ and $\bar{x}_k$ is the mean value of the measurements of variable x_k. After some small bookkeeping one can write

$$\operatorname{cov}(y_i, y_j) = \sum_k \sum_s \gamma_{ik}\gamma_{js} \int_{-\infty}^{\infty}\int_{-\infty}^{\infty} (x_k - \bar{x}_k)(x_s - \bar{x}_s)\, p(x)\, dx_k\, dx_s \tag{4.30}$$

But,

$$\int_{-\infty}^{\infty} \cdots \int_{-\infty}^{\infty} (x_k - \bar{y}_i)(x_s - \bar{y}_j)\, p(x)\, dx_i\, dx_j = \operatorname{cov}(x_i, x_j) \tag{4.31}$$

Then,

$$\operatorname{cov}(y_i, y_j) = \sum_k \sum_s \gamma_{ik}\gamma_{js} \operatorname{cov}(x_k, x_s) \tag{4.32}$$

which is the same as Eq. (4.28).

Thus, the variances of the flowrate estimates obtained from reconciliation are given by

$$\hat{S}_R = \left[I - S_R E_R^T \left(E_R S_R E_R^T\right)^{-1} E_R\right] S_R \tag{4.33}$$

which the reader can verify by simply plugging in the corresponding vectors. However, we are only interested in the diagonal elements of this matrix.

Variance of Observable Quantities

According to Eq. (4.29),

$$\hat{S}_O = \left[E_{\mathrm{RO}}\ E_{\mathrm{NRO}}\right] \begin{bmatrix} \hat{S}_R & \\ & S_{\mathrm{NR}} \end{bmatrix} \left[E_{\mathrm{RO}}\ E_{\mathrm{NRO}}\right]^T \tag{4.34}$$

or

$$\hat{S}_O = E_{\mathrm{RO}} \hat{S}_R E_{\mathrm{RO}}^T + E_{\mathrm{NRO}} S_{\mathrm{NR}} E_{\mathrm{NRO}}^T \tag{4.35}$$

Example 4.3: For our example,

$$\hat{S}_R = \begin{bmatrix} 0.05 & 0.05 & 0.008 & 0.043 \\ 0.05 & 0.05 & 0.008 & 0.043 \\ 0.008 & 0.008 & 0.027 & -0.019 \\ 0.043 & 0.043 & -0.019 & 0.061 \end{bmatrix} \tag{4.36}$$

Thus,

$$\left.\begin{aligned} \hat{\sigma}_{F_1}^2 &= 0.05 < 0.1 \\ \hat{\sigma}_{F_7}^2 &= 0.05 < 0.22 \\ \hat{\sigma}_{F_8}^2 &= 0.027 < 0.03 \\ \hat{\sigma}_{F_{11}}^2 &= 0.061 < 0.16 \end{aligned}\right\} \tag{4.37}$$

Notice how some variables improved substantially their variance while others did not. In addition

$$\hat{S}_O = \begin{bmatrix} 0.050 & 0.042 & 0.008 & 0.050 & 0.050 & 0 \\ 0.042 & 0.061 & -0.019 & 0.043 & 0.043 & 0 \\ 0.008 & -0.019 & 0.026 & 0.008 & 0.008 & 0 \\ 0.050 & 0.043 & 0.008 & 0.071 & 0.071 & -0.02 \\ 0.050 & 0.043 & 0.008 & 0.071 & 0.071 & -0.02 \\ 0 & 0 & 0 & -0.02 & -0.02 & 0.02 \end{bmatrix} \tag{4.38}$$

Therefore,

$$\left.\begin{aligned} \hat{\sigma}^2_{F_6} &= 0.05 \\ \hat{\sigma}^2_{F_9} &= 0.061 \\ \hat{\sigma}^2_{F_{10}} &= 0.026 \\ \hat{\sigma}^2_{F_2} &= 0.071 \\ \hat{\sigma}^2_{F_4} &= 0.071 \\ \hat{\sigma}^2_{F_5} &= 0.02 \end{aligned}\right\} \tag{4.39}$$

Remark: Kretsovalis and Mah (1987) were the first to establish the connections between redundancy and the precision of the estimators. They used this relationship for instrumentation upgrade studies, which we cover in Chap. 15.

Remark: Al-Arfaj (2006) proposed a shortcut data reconciliation technique that eliminates the use of variable classification steps by making the flowsheet simpler through unit merging and other approximations. We believe that although these simplifications are useful as first approximation and when commercial software is not available. Even then, the classification procedure of Chap. 3 and the solution procedure of this chapter are not that difficult to implement.

Method to Avoid the Classification Step

To avoid the classification step, one could use very large variances for unmeasured steams. The unmeasured streams will have to be given some initial value, however. Because large variances are chosen, the observable variables will adjust freely away from these initial values. The unobservable, in turn, may remain unchanged, or "reconciled" rendering values compatible with balances. Finally, the variance-covariance matrix of the estimators will render a finite value for the variance of the observable variables, whereas it will still keep a very large value for the unobservables. This approach was suggested by Kelly (1998) in the context of non-linear cases.

Example 4.4: Consider again Example 4.2 (Fig. 4.1 and Tables 4.2 and 4.3). We know from above that there are no unobservable variables. If we assume large variances in all unmeasured variables, we get

$$S_R = \begin{bmatrix} 0.1 & & & & & & & & & & \\ & 10^6 & & & & & & & & & \\ & & 0.02 & & & & & & & & \\ & & & 10^6 & & & & & & & \\ & & & & 10^6 & & & & & & \\ & & & & & 10^6 & & & & & \\ & & & & & & 0.22 & & & & \\ & & & & & & & 0.1 & & & \\ & & & & & & & & 10^6 & & \\ & & & & & & & & & 10^6 & \\ & & & & & & & & & & 0.16 \end{bmatrix} \tag{4.40}$$

(columns S_1 through S_{11})

$$E_R = \begin{bmatrix} 1 & -1 & -1 & & & & & & & & \\ & 1 & & -1 & & & & & & & \\ & & 1 & & -1 & & & & & & \\ & & & 1 & 1 & -1 & & & & & \\ & & & & & 1 & -1 & & & & \\ & & & & & & 1 & -1 & -1 & & \\ & & & & & & & 1 & & -1 & \\ & & & & & & & & 1 & & -1 \end{bmatrix} \tag{4.41}$$

(columns S_1 through S_{11})

Then, using an arbitrary number (5 in our case) for the unmeasured variables, we get $F_R^{+T} = [10.03 \;\; 5 \;\; 2.97 \;\; 5 \;\; 5 \;\; 5 \;\; 10.06 \;\; 5.99 \;\; 5 \;\; 5 \;\; 3.99]$. Equation (4.10) renders the same as Eq. (4.19). In addition, using Eq. (4.33) renders the same variances of the estimators, and the step of variable classification was omitted altogether.

We now eliminate the measurement in S_3. Thus, the corresponding variance in the diagonal of S_R (0.02) is changed to 10^6. Now using $F_R^{+T} = [10.03 \;\; 5 \;\; 5 \;\; 5 \;\; 5 \;\; 5 \;\; 10.06 \;\; 5.99 \;\; 5 \;\; 5 \;\; 3.99]$. we obtain $\tilde{F}_R^{+T} = [10.02 \;\; 5.01 \;\; 5.01 \;\; 5.01 \;\; 5.01 \;\; 10.02 \;\; 10.02 \;\; 6.00 \;\; 4.03 \;\; 6.00 \;\; 4.03]$. Clearly, the unobservable streams accommodate to values that are compatible with the balances. In addition the variances of the estimators are

$$\left.\begin{aligned} \tilde{\sigma}_1^2 &= 0.0505 \\ \tilde{\sigma}_2^2 &= 250{,}000 \\ \tilde{\sigma}_3^2 &= 250{,}000 \\ \tilde{\sigma}_4^2 &= 250{,}000 \\ \tilde{\sigma}_5^2 &= 250{,}000 \\ \tilde{\sigma}_6^2 &= 0.0505 \\ \tilde{\sigma}_7^2 &= 0.0505 \\ \tilde{\sigma}_8^2 &= 0.0265 \\ \tilde{\sigma}_9^2 &= 0.0611 \\ \tilde{\sigma}_{10}^2 &= 0.0265 \\ \tilde{\sigma}_{11}^2 &= 0.0611 \end{aligned}\right\} \tag{4.42}$$

Clearly the variances of S_2, S_3, S_4, and S_5 are very large and therefore, they should be assumed unobservable.

There is only one warning that needs to be made regarding this method. As matrices are larger, the number to use for unmeasured variables (10^6 in our example) may create numerical problems. Aside from this detail that needs to be monitored carefully, the method is sound.

References

Al-Arfaj, M. A., "Shortcut Data Reconciliation Technique: Development and Industrial Application," *AIChE J.* **52**(1):414–417 (2006).

Almasy, G. A., and R. S. H. Mah, "Estimation of Measurement Error Variances from Process Data." *Ind. Eng. Chem. Process Des. Dev.* **23**: 779 (1984).

Bagajewicz, M. "On the Probability Distribution and Reconciliation of Process Data," *Comp. Chem. Eng.* **20**, 6/7:813–819 (1996).

Bakshi, B. R., M. N. Nounou, P. K. Goel and X. Shen, "Multiscale Bayesian Rectification of Data from Linear Steady-State and Dynamic Systems without Accurate Models." *Ind. Eng. Chem. Res.* **40**:261–274 (2001).

Chen, J., A. Bandoni, and J. A. Romagnoli, "Robust Estimation of Measurement Error Variance/Covariance from Process Sampling Data," *Comp. Chem. Eng.* **21**:593–600 (1997).

Crowe, C. M., "Data Reconciliation—Progress and Challenges," *J. Proc. Control* 89–98 (1996).

Crowe, C. M., "Formulation of Linear Data Reconciliation using Information Theory," *Comp. Chem. Eng.* **51**(12):3359–3366 (1996).

Darouach, M., R. Ragot, M. Zasadzinski, and G. Krzakala, "Maximum Likelihood Estimator of Measurement Error Variances in Data Reconciliation," *IFAC. AIPAC Symp.* 2 135–139 (1989).

Dovi, V. G., A. P. Reverberi, L. Maga, "Reconciliation of Process Measurements When Data Are Subject to Detection Limits," *Chem. Eng. Sci.* **52**(17):3047–3050 (1997).

Dovi, V. G., A. Del Borghi, "Rectification of Flow Measurements in Continuous Processes Subject to Fluctuations," *Chem. Eng. Sci.* **56**:2851–2857 (2001).

Hlavacek, V., "Analysis of a Complex Plant-Steady State and Transient Behavior. I. Plant Data Estimation and Adjustment," *Comp. Chem. Eng.* **1**:75–100 (1977).

Johnston, L. P. M. and M. A. Kramer, "Maximum Likelihood Data Rectification. Steady State Systems," *AIChE J.* **41**(11) (1995).

Kelly, J. D., "A Regularization Approach to the Reconciliation of Constrained Data Sets," *Comp. Chem. Eng.* **22**:1771–1788 (1998).

Kelly, J. D., "Reconciliation of Process Data Using Other Projection Matrices," *Comp. Chem. Eng.* **23**:785–789 (1999).

Keller, J. Y., M. Zasadzinski, and M. Darouach, "Analytical Estimator of Measurement Error Variances in Data Reconciliation," *Comp. Chem. Eng.* **16**:185 (1992).

Kuehn, D. R. and H. Davidson, "Computer Control. II. Mathematics of Control," *Chem. Eng. Prog.* **57**(44) (1961).

Kretsovalis and R. H. S. Mah, "Effect of Redundancy on Estimation Accuracy Process Data Reconciliation," *Chem. Eng. Sci.* **42**(9):2115–2121 (1987).

Madron, F., "Process Plant Performance," Ellis Horwood, Chichester, England (1992).

Madron, F., V. Veverka, and V. Vanecek, "Statistical Analysis of Material Balance of Chemical Reactor," *AIChE J.* **23**(4):482–486 (1977).

Mah, R. S. H., "Chemical Process Structures and Information Flows," Butterworths (1990).

Mah, R. S. H., G. Stanley, and D. Downing, "Reconciliation and Rectification of Process Flow and Inventory Data," *Ind. Eng. Chem Res. Proc. Des. Dev.* **15**: 175–183 (1976).

Mandel, D. A. Abdollahzadeh, D. Maquin, and J. Ragot, "Data Reconciliation by Inequality Balance Equilibrium: A LMI Approach," *Int. J. Miner. Process.* **53**:157–169 (1998).

Maquin, D., S. Narasimhan, J. Ragot, "Data Validation with Unknown Variance Matrix," *Comp. Chem. Eng. Supplement* S609–S612 (1999).

Maquin, D., O. Adrot, J. Ragot, "Data Reconciliation with Uncertain Models," *ISA Transactions* **39**(1):35–45 (2000).

Narasimhan, S. and C. Jordache, "Data Reconciliation and Gross Error Detection. An Intelligent Use of Process Data," Gulf Publishing Company (2000).

Narasimhan, S. and P. Harikumar, "A Method to Incorporate Bounds in Data Reconciliation and Gross Error Detection-I. The Bounded Data Reconciliation Problem," *Comp. Chem. Eng.* **17**:1115–1120 (1993).

Ragot, J. and D. Maquin, "Reformulation of Data Reconciliation Problem with Unknown-but-Bounded Errors," *Ind. Eng. Chem. Res.* **43**:1530–1536 (2004).

Ragot, J., D. Maquin, M. Alhaj-Dibo, "Linear Mass Balance Equilibration: A New Approach for an Old Problem," *ISA Transactions* **44**(1):23–34 (2005).

Reilly, P. M. and H. Patino Leal, "A Bayesian Study of the Error-in-Variables Model," *Technometrics* **23**(3):221–231 (1981).

Resi and Saraiva, "Integration of Data Uncertainty in Linear Regression and Process Optimization," *AIChE J.* **51**(11):3007–3019 (2005).

Sánchez, M. and J. Romagnoli, "Data Processing and Reconciliation for Chemical Process Operations," Academic Press (2000).

Tamhane, A. C. and R. S. H. Mah, "Data Reconciliation and Gross Error Detection Chemical Process Networks," *Technometrics* **27**:409–422 (1985).

Tjoa and Biegler, "Simultaneous Strategies for Data Reconciliation and Gross Error Detection of Nonlinear Systems," *Comp. Chem. Eng.* **15**:679–690 (1991).

Veverka, V. and F. Madron, "Material and Energy Balancing in Process Industries: From Microscopic Balances to Large Plants," Elsevier, Amsterdam (1996).

CHAPTER 5

Gross Error Detection

Data reconciliation performed using the assumption of steady state is implemented in practice by taking averages of single instrument measurements. Thus, only one value per instrument is used to perform data reconciliation. Gross errors from different sources are in this way lumped in a single value.

The sources of gross errors are

1. ***Instrument biases:*** These are consistent fixed-value departures from the averaged values of the signals. Drift is also a bias.
2. ***Leaks:*** Leaks can be classified as predictable (tank evaporation, e.g.) or unpredictable. Both are typically not considered in the plant models, although the predictable ones can be incorporated.
3. ***True outliers:*** These are occasional measurements that depart significantly from all other measurements. If the number and size of these outliers is large and one-sided, the averaged values may get distorted and confused with biases. Figure 2.7 (Chap. 2) presented a case where outliers were present. This type of signals has typically bivariate distributions. If the spikes are eliminated, the remaining signal has a (nearly perfect) Gaussian distribution. Gross errors like these are likely to distort any data reconciliation.
4. ***Departure from steady state:*** When plants present small drifts and oscillations, averaged values, treated as steady-state values, do not reflect real plant behavior. This phenomenon has multiple effects, not only on gross error detection but also on reconciliation itself, as well as on variance estimation.

Gross Error Handling

The challenging task in data reconciliation is to

- Identify the existence of gross errors
- Identify the gross errors location
- Identify the gross error type (bias or leak)
- Determine the size of the gross error

There exists at least one method that allows the detection of the existence of gross errors. Some of the methods for gross error identification are to a certain extent capable of discerning the location and type. Very little work has been performed to address departures from steady state, which will be covered in a later chapter.

After the gross errors are identified, two alternative responses are possible and/or desired.

- Eliminate the measurement containing the gross error, or
- Correct the measurements or the model and run the reconciliation again

The first alternative is the one implemented in commercial software, which in general only considers biases and is not designed to detect leaks. This leaves the system with a smaller degree of redundancy and deterioration of the quality of the reconciliation (redundancy becomes smaller). If one is able to identify the gross errors and obtain a reliable estimate of their values, then the second alternative becomes appealing because redundancy is not lost. Some methods perform such task with reasonable success.

Smearing

We will now consider the case where there is a bias in only one stream. We will show how this gross error affects the reconciled data in all other streams.

Let

$$F_R^+ = F_R + \varepsilon_R + \delta\, e_i \tag{5.1}$$

where e_i is a unit vector (all zeros except in position i) and δ is the size of the gross error.

Then the result of the data reconciliation is (see Eq. 4.10)

$$\hat{F}_R(i) = \left[I - S_R E_R^T \left(E_R S_R E_R^T\right)^{-1} E_R\right](F_R + \varepsilon_R + \delta e_i) \tag{5.2}$$

where $\hat{F}_R(i)$ is the vector of reconciled values in the presence of the gross error in flowrate S_i.

It is easy to see that

$$\hat{F}_R(i) = \tilde{F}_R + \delta\left[I - S_R E_R^T \left(E_R S_R E_R^T\right)^{-1} E_R\right] e_i \tag{5.3}$$

Thus, the result of data reconciliation is equal to the result obtained when there is no gross error plus an additional term. However, because e_i is a vector of zeros with a one in the ith position, the product $[I - S_R E_R^T (E_R S_R E_R^T)^{-1} E_R]e_i$ is a vector exactly equal to the ith column of $[I - S_R E_R^T (E_R S_R E_R^T)^{-1} E_R]$. Since this last matrix is dense, we expect an effect in all the rest of the variables. This phenomenon is called *smearing*.

Leaks have a more devastating effect. We illustrate this next.

Example 5.1: In example 4.2 of Chap. 4, we obtained

$$\left[I - S_R E_R^T \left(E_R S_R E_R^T\right)^{-1} E_R\right] = \begin{bmatrix} 0.505 & 0.230 & 0.266 & 0.266 \\ 0.505 & 0.230 & 0.266 & 0.266 \\ 0.080 & 0.036 & 0.884 & -0.116 \\ 0.425 & 0.193 & -0.618 & 0.618 \end{bmatrix} \tag{5.4}$$

Therefore, if, an error of size δ is introduced in S_1, then the smearing is given by

$$\left[I - S_R E_R^T \left(E_R S_R E_R^T\right)^{-1} E_R\right] = \begin{bmatrix} 0.505 \\ 0.505 \\ 0.080 \\ 0.425 \end{bmatrix} \delta \tag{5.5}$$

As we can see the smearing is serious if δ is significant.

We immediately notice that when there is one biased variable, the smearing effect of data reconciliation reduces the absolute size of the induced error in the variable. Indeed, assume that the only bias is δ_i. Then $\hat{\delta}_i = [I - S_R E_R^T (E_R S_R E_R^T)^{-1} E_R]_{ii}\, \delta_i$, which is smaller than δ_i, because $S_R E_R^T (E_R S_R E_R^T)^{-1} E_R$ has a nonnegative diagonal smaller than 1. We will cover this smearing phenomenon in more detail in Chap. 10.

Example 5.2: Consider the flowsheet of Fig. 3.1, with measurements in $x_M = \{S_1, S_3, S_7, S_8, S_{11}\}$. The redundant measurements are S_1, S_7, S_8, S_{11} and the nonredundant measurement is S_3. Assume now that there is a leak in unit U_1.

Consider now the data of Table 5.1, which is similar to the values given in Table 4.1, but affected by a leak of 1 in unit U_1.

Stream	True Flowrate
S_1	10
S_3	3
S_7	9
S_8	5.4
S_{11}	3.6

Table 5.1 True Flowrates for Example 5.1

If one does not know that there is a leak, one will have a serious smearing effect. Indeed, assume the measurements are those given in Table 5.2, which are very close to the real values.

Stream	Measurement
S_1	10.03
S_3	3.01
S_7	9.01
S_8	5.41
S_{11}	3.58

Table 5.2 Measurements for Example 5.1

Assume they have the same standard deviation as in Table 4.3. If we apply the data reconciliation formula without knowing that there is a leak, the following result is obtained:

$$\tilde{F}_R = \left[I - S_R E_R^T \left(E_R S_R E_R^T\right)^{-1} E_R\right] F_R^+ = \begin{bmatrix} 0.505 & 0.230 & 0.266 & 0.266 \\ 0.505 & 0.230 & 0.266 & 0.266 \\ 0.080 & 0.036 & 0.884 & -0.116 \\ 0.425 & 0.193 & -0.618 & 0.618 \end{bmatrix} \begin{bmatrix} 10.03 \\ 9.01 \\ 5.41 \\ 3.58 \end{bmatrix} = \begin{bmatrix} 9.52 \\ 9.52 \\ 5.49 \\ 4.03 \end{bmatrix} \tag{5.6}$$

which has larger deviations from the true values than the standard deviation of the measurement!!

Tests for Gross Errors

We make use of the theory of hypothesis testing to develop a test for the presence of gross errors.

Hypothesis Testing

The procedure of deciding on the validity of a postulated hypothesis is called hypothesis testing. The hypothesis whose validity is being tested is called the *null hypothesis*. The test of the null hypothesis H_0 against the alternate hypothesis H_1 leads, on the basis of examining a random sample, either to rejecting H_0 or to not rejecting H_0 (i.e., to reject H_1).

In our case, the null hypothesis is that there is no gross error in the system, or that there is no gross error in a certain measurement, or even that there is no leak in the system or in a particular unit.

In testing a hypothesis in our case, one proceeds as follows: For a given set of measurements $x = (X_1, \ldots, X_n)$, one chooses an appropriate statistics $T = T(X_1, \ldots, X_n)$, which in this case is called the testing criterion (we will present a few below). Knowing what is the distribution of the random variable T, one finds the interval where the value of the statistic can be included to make the null hypothesis declared true.

For example, $(-\infty; +\infty)$, is divided into two intervals, the interval R and its complement R'. R is chosen in the manner that under the validity of H_0 the statistics T assumes a value within this interval with probability $(1 - \alpha)$. The value of α is chosen sufficiently small (e.g., 0.05) and it is called the level of significance. The region R' is called the critical region. Finally, from the random sample one now computes the statistic T. If T assumes a value from the critical region R', the hypothesis H_0 is rejected.

Example 5.3: A company produces a chemical product that has *a very well-known* mean composition (μ_1) of 99.98% with a standard deviation of 0.01%. A new process is introduced which claims that the purity has been increased. To test the claim 50 samples are taken and it is found that the mean composition (μ_2) is 99.985%. Can the claim be supported at a 0.01 significance level?

Here H_0:μ_2 = 99.98% (no change in the composition) and H_1:μ_2 >99.98% (the new composition is larger). With the significance level $\alpha = 0.01$ we choose the corresponding interval R. Because these are normal distributions, the difference is also a normal distribution. Then, we take the interval $R = (-\infty, u_{1-\alpha})$, where $u_{1-\alpha}$ is the value of x such that the area under the normal distribution from $-\infty$ to x is equal to $(1 - \alpha)$.

The appropriate statistics T is the following normalized score:

$$T = \frac{\mu_2 - \mu_1}{\sigma_T/\sqrt{N}} = \frac{99.985 - 99.98}{0.01/\sqrt{50}} = 3.5355 \tag{5.7}$$

which is larger than $u_{1-\alpha} = 2.33$. Thus, the null hypothesis is rejected and the claim is accepted as true.

Example 5.4: A particular set of temperature measurements (x_i) is said to follow a normal distribution with a mean value $\bar{x}$. Sometimes, the measurement is affected by a sudden spike of voltage, which results in a biased value. Consider one sampling x^+, that is, one temperature measurement. We will construct a test to determine if this value is biased or not.

Assume the bias is given by δ. Then, the test statistics is given by

$$T = x^+ - \bar{x} \tag{5.8}$$

Notice that $T = e - \delta$, where e is normally distributed. The null hypothesis is H_0:$\delta = 0$ (no bias) and H_1:$\delta \neq 0$ (there is a bias). We take the interval $R = (-u_{1-\alpha/2}, u_{1-\alpha/2})$, where we used $\alpha/2$ instead of α because the bias could be positive or negative. Thus if T is outside R, then we cannot accept H_0.

Example 5.5: A liquid flows through a pipe with a volume flowrate F. Two flowmeters are installed: A and B, measuring the corresponding flows (F_A and F_B). Assume that the errors are normally distributed with variances σ_A^2 and σ_B^2, respectively. The flowmeter A is prone to give a systematic error. Our aim is to judge, on the basis of the measured values, whether such an error actually arises in flowmeter A.

Thus, the test statistics is

$$T = F_A - F_B \tag{5.9}$$

Thus, $T = e_A + \delta_A - e_B$, where e_A and e_B are normally distributed. Thus $H_0{:}\delta_A = 0$, and $H_1{:}\delta_A \neq 0$ (there is a bias). T equals the sum of two random variables with zero mean with normal distribution and it also has a normal distribution $N(0, \sigma_A^2 + \sigma_B^2)$. The interval R is symmetric with respect to zero

$$R = (-\sigma_T u_{1-\alpha/2}, \sigma_T u_{1-\alpha/2}) \tag{5.10}$$

where $\sigma_T = (\sigma_A^2 + \sigma_B^2)^{1/2}$ and $u_{1-\alpha/2}$ is the value of x such that the area under the normal distribution from $-x$ to x is equal to $(1 - \alpha)$.

Thus, if T falls within this interval we accept the null hypothesis. A test hypothesis for the case of gross errors in both flowmeters carries some problems because the two gross errors might exist, but be of opposite sign and therefore cancel each other.

Type I and Type II Errors and the Power of a Test

In testing hypotheses one can commit basically two kinds of errors.

- Type I error: Rejecting the hypothesis when the hypothesis, in fact, is true.
- Type II error: Not rejecting the hypothesis when the hypothesis, in fact, is false.

The power of a particular test is given by the number of type II errors it makes. If this number is small, the power (γ) is said to be large. More precisely, the power is related to the probability of a type II error and is given by

$$\gamma = 1 - \Pr\{\text{type II error}\} \tag{5.11}$$

Example 5.6: The power of the test constructed in Example 5.3 can be obtained as follows: Assume that the new mean composition μ_2 is larger than μ_1. Figure 5.1 illustrates the two probability distributions.

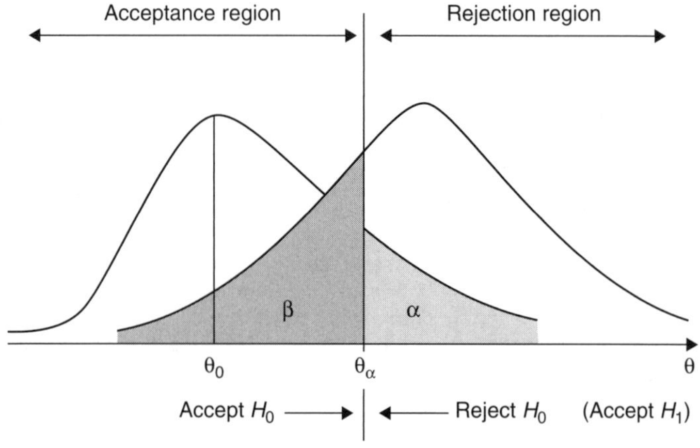

Figure 5.1 Acceptance and rejection regions.

The distribution on the left is the original one, and the one on the right is the new one with $\mu_2 > \mu_1$. If the new distribution is the one on the left, one would reject H_0, if the new calculated mean were larger than $u_{1-\alpha}$, and in the case of the figure, this would have been correct. However, if the new calculated mean is smaller than $u_{1-\alpha}$, accepting H_0 is incorrect. The probability of doing this is equal to the shaded area β. This is the probability of type II error. Thus, one can make a graph of β as a function of the new value of μ_2. Such graph is called the operating characteristic (OC) curve. A typical shape of such curve is shown in Fig. 5.2.

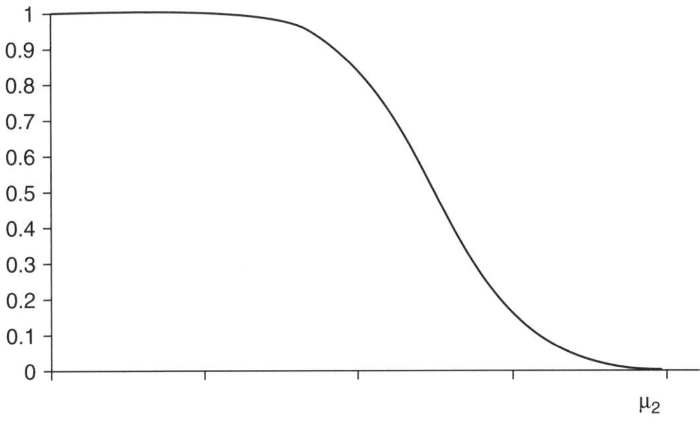

FIGURE 5.2 Power of the test.

Global Test

The global test (GT) was first published by Reilly and Carpani (1963), and also proposed by Ripps (1965), Almasy and Sztano (1975), and Madron et al. (1977); it was further discussed by Madron (1985).

The null hypothesis H_0 is that there is no gross error. Thus, if r is the vector or residuals of the expected value of r is zero, that is,

$$E(r) = 0 \tag{5.12}$$

The variance-covariance matrix of r is given by

$$\mathrm{Cov}(r) = CS_RC^T = J \tag{5.13}$$

where S_R is the variance-covariance matrix of random measurement errors. In the absence of gross errors, it can be proved that the following variable

$$\chi_m^2 = rJ^{-1}r \tag{5.14}$$

follows a chi-squared distribution with m degrees of freedom where m is the number of rows of C. Notice that even before performing the data reconciliation, one instance of this number can be obtained. If this number falls within the interval of confidence, that is, it is lower than the certain critical value (obtained using the confidence level chosen), then the null hypothesis is accepted, that is, no gross error is suspected. On the other hand, if it is larger than the critical value, it is said that a gross error has been detected.

Therefore, the GT is performed as follows:

(a) Compute χ_m^2 using Eq. (5.14).

(b) Determine the threshold value $\chi_{m,\alpha}^2$ of the chi-squared distribution corresponding to a desired level of confidence α (usually 90–95%).

(c) If $\chi_m^2 > \chi_{m,\alpha}^2$, then there is at least one gross error in the data set (but there could be more than one).

Remark: The GT cannot determine where the gross errors are located or how many are there. It only points globally at the existence of inconsistencies.

Nodal Test

Reilly and Carpani (1963) and Mah et al. (1976) proposed to use all streams associated to a unit. In the absence of gross errors, the constraint residuals r follow a m-variate normal distribution (m is the rank of C_R). Therefore the following test statistics

$$|Z_i| = \frac{|r_i|}{\sqrt{J_{ii}}} \tag{5.15}$$

follow a standard normal distribution, $N(0,1)$, under the null hypothesis H_0.

Therefore, if $|Z_i|$ is larger than the critical value ($Z_{\alpha/2}$) based on the level of confidence α, then one concludes that there is at least one gross error in the set of measurement that participates in the corresponding node balance. Notice that $\alpha/2$ is used instead of α because the test is a two-tail test, that is, tests for a positive or a negative deviation.

There are m tests that one can make on a single distribution. Multiple tests are then being made using the same critical value, and for this reason, the likelihood of a type I error (wrongly reject the null hypothesis H_0, i.e., declare the existence of a gross error when there is none) increases. Mah and Tamhane (1982) proposed the use of a new smaller level of significance (α') derived using the Sidak inequality (Sidak, 1967). This new level of significance is given by

$$\alpha' = 1 - (1 - \alpha)^{1/m} \tag{5.16}$$

That is, to use $Z_{\alpha'/2}$ instead of $Z_{\alpha/2}$ as a threshold value (see also Mah, 1990). Rollins and Davis (1992) proposed to use

$$\alpha' = \alpha/m \tag{5.17}$$

based on Bonferroni confidence intervals. Note that when m is large, Eq. (5.16) approaches Eq. (5.17). A word of caution regarding confidence intervals was suggested by Buzzi Ferraris (2000).

The nodal test is summarized as follows: At least one measurement participating in the ith constraint contains a gross error if

$$|Z_i| > Z_{\alpha/2} \tag{5.18}$$

Maximum Power Nodal Test

Crowe (1989, 1992) proposed a linear transformation for which the nodal test has maximum power, that is, minimum of failures to identify existing gross errors, which are type II errors. The linear transformation consists of using the new transformed residuals as follows:

$$w = J^{-1}\, r \tag{5.19}$$

and therefore the maximum power test statistics is given by

$$|Z_i| = \frac{|w_i|}{\sqrt{(J^{-1})_{ii}}} \tag{5.20}$$

The application of the test is identical to that of the usual nodal test.

Measurement Test

The measurement test (MT) (Mah and Tamhane, 1982) is based on the vector of measurement adjustments (or corrections) denoted by a and defined by

$$a = F_R^{+} - \tilde{F}_R \tag{5.21}$$

The test is based on the assumption that the random errors for measurements are independently and normally distributed with zero mean. Under the null hypothesis H_0, the expected value of a is

$$E(a) = 0 \tag{5.22}$$

Now, the variance-covariance matrix of a is given by

$$\mathrm{Cov}(a) = \tilde{S}_R \tag{5.23}$$

Thus, the following variable is expected to follow a normal distribution:

$$|Z_i| = \frac{|a_i|}{\sqrt{(\tilde{S}_R)_{ii}}} \tag{5.24}$$

If no gross error is present, the above value should be lower than a critical value.

Therefore, the MT is summarized as follows: The *i*th measurement contains a gross error if

$$|Z_i| > Z_{\alpha'/2} \tag{5.25}$$

Maximum Power Measurement Test

Almasy and Sztano (1975) derived the maximum power (MP) test for the linear case, based on the normal distribution of the measurements, which has the greatest probability of correctly detecting a single gross error in a measurement, when only one is present. Tamhane (1982) extended the method for a nondiagonal covariance matrix S_R. Tamhane and Mah (1985) show that a linear transformation uses the new transformed adjustments as follows:

$$d = S_R^{-1}\, a \tag{5.26}$$

and therefore the maximum power test statistics is given by

$$|Z_i| = \frac{|d_i|}{\sqrt{W_{ii}}} \tag{5.27}$$

This test is then used the same way as the regular measurement test.

Generalized Likelihood Ratio

The alternative hypothesis in this test consists of a particular bias in a stream associated to a node or a leak (Narasimhan and Mah, 1987), that is, $H_1{:}\mu_r = b\, h_i$, where μ_r is the expected value of the node residual, g_i a vector in the direction of a bias ($h_i = E_R e_i$), or in the direction of a leak ($h_i = m_i$) and b is the size of this gross error. This alternative hypothesis contains an unknown number (b) and an unknown vector h_i is equivalent to $H_0{:}\mu_r = 0$ and $H_1{:}\mu_r \neq 0$. The test is based on finding the supremum of the likelihoods of each hypothesis, that is,

$$\lambda = \operatorname{Sup}\frac{\Pr\{r|H_1\}}{\Pr\{r|H_0\}} \tag{5.28}$$

where the supremum is computed over all possible bias and leaks. When the probability distributions used are normal, the test consists of computing

$$T = 2\ln\lambda = \operatorname*{Sup}_{\forall i}\frac{\left(h_i^T Q_R^{-1} r\right)^2}{h_i^T Q_R^{-1} h_i} \tag{5.29}$$

which is compared with the corresponding threshold. The next step is to determine the size of the bias or the leak. We omit the details here and cover size estimation in Chap. 7.

Principal Component Test

Tong and Crowe (1995, 1996, 1997) proposed the use of principal components, that is, the eigenvalue decomposition of the variance of the residuals $(E_R Q_R E_R^T)$, which is given by

$$\Lambda_r = U_r^T \left(E_R Q_R E_R^T\right) U_r \tag{5.30}$$

where U_r is the matrix of orthonormalized eigenvectors of $E_R Q_R E_R^T$. Accordingly, the principal components are

$$p_r = \left(\Lambda_r^{-1/2} U_r\right)^T r \tag{5.31}$$

which are tested against the normal distribution. Once a suspect has been identified, the error is traced back to the corresponding node by looking at the largest contributions to this component.

Remark: Narasimhan and Mah (1987) and also Crowe (1989) showed that the GLR is equivalent and have maximum power for the case of one gross error. Tong and Crowe (1995, 1996) proposed the use of principal components, that is, eigenvector decomposition of the residuals to detect gross errors. Industrial applications of PCMT were reported by Tong and Bluck (1998), who indicated that tests based on PCA are more sensitive to subtle gross errors than others, and have larger power to correctly identify the variables in error than the conventional nodal, measurement, and global tests. The books by Narasimhan and Jordache (2000) and Sánchez and Romagnoli (2000) have discussed the relationships between these models as well as the power. Finally, Johnston and Kramer (1995) proposed a Bayesian approach, which has unique features. The work of Chen et al. (2004) in this regard, although developed for dynamic data reconciliations makes use of Bayesian theory as well.

Remark: Harikumar and Narasimhan (1993) and Narasimhan and Harikumar (1993) applied the GLR methodology to the problem of gross error detection when bounds on errors exist.

Remark: Jordache et al. (1985) studied the power of various of the aforementioned methods. They include the performance as a function of the ratio of the size of the gross error over the standard deviation (δ/σ). Finally, Maquin and Ragot (1991) make some meaningful comparisons.

Multiple Gross Errors and Gross Error Elimination

In the elimination strategy a selected test is coupled with an elimination strategy. If the test flags the existence of errors, then a strategy is proposed to identify one or more variables, which are the "most suspected ones." The measurements of these variables are eliminated and the test is run again, even if that implies performing the data reconciliation again. A few of these strategies are described next.

Serial Elimination Strategy Based on the Global Test

Serial elimination was originally proposed by Ripps (1965). Measurements are deleted sequentially in groups of size t, $t = 1, 2, ... t_{max}$ based on which the biggest reduction of the global test (GT) statistics is provided. After each deletion the GT is again applied. This procedure continues until among all sets of measurements that are deleted, one can find one for which the GT indicates no gross error, or until t_{max} is reached. In this last case, the set that produces the largest reduction in test statistics is declared as the set of gross errors.

The above identification scheme stops when the GT shows no gross error, or when a set is identified that renders the biggest reduction in the GT statistics. Such a set may include measurements that do not contain gross errors, as indicated by small reductions in the test statistics obtained when they are deleted individually. On the other hand, such a set may exclude measurements that contain gross errors, as indicated by significant reductions obtained by deleting them. In practice, due to these problems, measurements are deleted one at a time, and those rendering the largest reductions are tested in groups. This serial elimination strategy has not been found effective in identifying gross errors.

Remark: Nogita (1972) later modified this approach by proposing to eliminate the measurement that reduces the objective function the most and stopping when the objective function increases or the maximum of deletions has been reached. Romagnoli and Stephanopoulos (1981) proposed a method to reevaluate the objective function without having to solve the reconciliation again and obtain an estimate of the gross errors. Additionally, Crowe et al. (1983) proposed a strategy where the global test is used in conjunction with other test statistics.

Rosenberg et al. (1987) presented an extension of the method where instead of only one measurement being deleted at a time, sets of measurements of different size are deleted. We now present Rosenberg et al.'s version of the algorithm.

The basis of the method is that measurements are deleted sequentially in groups of size t, $t = 1, 2, ... t_{max}$. After each deletion, the process constraints are projected and the global test is again applied. This procedure continues until among all sets of measurements that are deleted one can find one for which the global test indicates no gross error, or until t_{max} is reached. In this last case, the set that produces the largest reduction in test statistics is declared as the set of gross errors.

The above identification scheme stops when the global test shows no gross error, or when a set is identified that renders the biggest reduction in the GT statistics. Such a set may include measurements that do not contain gross errors, as indicated by small reductions in the test statistics obtained when they are deleted individually. On the other hand such a set may exclude measurements that contain gross errors, as indicated by significant reductions obtained by deleting them. In practice, due to these problems, measurements are deleted one at a time, and those rendering the largest reductions are tested in groups. This serial elimination strategy has not been found effective in identifying gross errors.

Rosenberg et al. (1987) also evaluated the performance of the GT and some other gross error detection schemes and concluded that its performance depends on the magnitude of the gross error, the network configuration and process constraints, and the position of the gross error. For the same average number of type I errors, the GT has superior selectivity in comparison with that of the MT, with a slight decrease in power.

Serial Elimination Based on the Measurement Test

The measurement test cannot be used alone, and needs to be coupled with elimination procedures because of the following problems:

- As it was shown when analyzing the smearing effect, a large-magnitude gross error can cause several related measurement tests to flag, and therefore falsely identifying gross error that may not exist.
- Likewise, the interaction of two or more gross errors can obscure the true sources of gross errors.
- The measurement test is not designed to detect leaks, although it is sensitive to them.

Different strategies have been proposed to mitigate the above problems with various degrees of success. Some of these methods are summarized next.

(a) Collective elimination: Eliminate all measurements that have z-statistics above the critical value (Mah and Tamhane, 1982).

(b) Serial elimination (Ripps, 1965): Eliminate one measurement that has failed the measurement test at a time and run the reconciliation again, repeating the procedure until there is no failure of the test for any variable. The criteria to pick which variable to eliminate gives rise to different variants of this strategy.
   - Eliminate the measurements that have the largest z-statistics above the critical value. This is implemented in some commercial software.
   - Eliminate the variable that reduces the objective function the most.

Remark: Crowe (1988) proposed an identification step where candidates for elimination are identified based on the change on the objective function that would result from the deletion of suspect measurements.

Remark: The serial elimination procedure is also known as iterative measurement test (IMT) and it has the problem that it may violate existing variable bounds. Serth and Heenan (1986) used the modified iterative measurement test (MIMT), where bounds are incorporated to the IMT. Rosenberg et al. (1987) proposed the extended measurement test (EMT), which is based on the serial elimination procedure, adding bound violation checking. To ameliorate some limitations due to the use of heuristics in the previous methods, Harikumar and Narasimhan (1993) added the bounds formally and solved the resulting quadratic data reconciliation. Narasimhan and Jordache (2000) presented a generalized version of this technique. Finally, Zhang et al. (2001) incorporated redundancy analysis to pick which gross error can be eliminated.

Remark: The use of combinations of tests is also worthwhile mentioning. Just to cite one of the latest, Mah et al (1976) presented a strategy of combining nodes and performed the nodal test on these to avoid the fact that a measurement bias usually affects more than one node. Wang et al. (2004) proposed an innovative combination of measurement and nodal tests in an iterative fashion. There are many others, which were reviewed by Narasimhan and Jordache (2000).

Remark: Rollins (1995) proposed a linear combination technique. Testing many hypotheses may increase the effect of "error cancellation" (Serth and Heenan, 1986). One way to reduce the tendency of "error cancellation" is to minimize the number of hypothesis tests involving material or energy balances around combinations of unit operation (i.e., nodes). That is, the goal of the LCT is to test the minimum number of hypotheses necessary to reach the correct conclusions for all measured variables in the network. More specifically, the LCT seeks to identify gross errors by testing linear combinations of nodal balances. Rollins et al. (1996) also used an intelligent selection of tests.

Remark: No matter which strategy one chooses for data collection, it is important to recognize that sampling should be limited. If the number of samples N_M is too large, the power (i.e., probability of detecting a nonzero bias δ or leak λ) will be too large (Chen et al., 1998). Since all measured streams are likely to have some degree of bias, very high power will declare that all streams have bias which will not be a helpful conclusion. Thus, one should attempt to control power by selecting the sample size a priori based on control of the type II error (i.e., false detection).

Remark: Many papers deal with the combined data reconciliation and gross error detection. The technique is based on modifying the least squares objective function to add a penalty for larger deviations (attributed to gross errors). This technique has been called *robust data reconciliation* by certain authors. For example, Tjoa and Biegler (1991), one of the first on this line of work, proposed a mixture distribution as the objective (likelihood) function

$$\text{Min}\frac{1}{\sigma\sqrt{2\pi}}\left[(1-p)e^{\frac{0.5\varepsilon^2}{\sigma^2}}+\frac{p}{b}e^{\frac{0.5\varepsilon^2}{\sigma^2 b^2}}\right] \tag{5.32}$$

where p is the probability associated with the absence of gross errors, and σ^2, the variance of such errors, and $b^2\sigma^2$ the variance of the biases. Finally ε represents the difference between the estimator and the measurement. At convergence, each measurement error can be tested against the combined distribution. If the probability associated with an error that is suspected to be an outlier is greater than that of random errors, then the measurement can be identified as an outlier. Thus the distribution function that is used for the objective function can also be used as a rational basis for a gross error detection test. For example, if the residual ε_μ satisfies the inequality

$$\frac{p}{b}e^{\frac{0.5\varepsilon^2}{\sigma^2 b^2}} > (1-p)e^{\frac{0.5\varepsilon^2}{\sigma^2}} \tag{5.33}$$

or,

$$|\varepsilon| > \sigma\sqrt{\frac{2b^2}{b^2-1}\ln\left[\frac{b(1-p)}{p}\right]} \tag{5.34}$$

then ε is identified as an outlier.

The shortcomings of this approach were addressed by Albuquerque and Biegler (1996) and Mah (1997), who proposed the use of the Fair function.

$$\rho(\varepsilon) = c^2\left[\frac{|\varepsilon|}{c} - \log\left(1+\frac{|\varepsilon|}{c}\right)\right] \tag{5.35}$$

where c is a parameter and has the advantage of being convex with continuous first and second derivatives.

In turn, Johnston and Kramer (1995) proposed to use the Lorentzian distribution

$$\rho(\varepsilon) = \frac{1}{1+\frac{\varepsilon^2}{2}} \tag{5.36}$$

which is a robust estimator because it has the ability to filter large gross errors.

All these methods are centered on outlier detection due to a large variance and not on bias detection. Later, Soderstrom et al. (2001), Arora and Biegler (2001), Wang and Romagnoli (2003), Ozyurt and Pike (2004), Zhao et al. (2004), Ragot et al. (2005), Morad et al. (2005), and Alhaj-Dibo et al. (2008) presented alternative robust schemes, some based on Bayesian theory, others using adaptive estimators and influence functions or "contaminated" distributions. Much earlier, Yamamura et al (1988), used the Aikake's information criteria (one based on the logarithmic likelihood and the number of parameters in the model). Arora and Biegler (2001) and Cong-li et al. (2007) followed up with a mathematical programming implementation that ameliorates the combinatorial problems reported by Yamamura et al. (1988).

Remark: Bagajewicz et al. (2000) proposed a technique based on mixed integer linear programming (MILP) to detect, identify, and estimate gross errors (measurement biases and leaks) in linear steady-state processes. The MILP-based gross error detection and identification model is constructed aiming at identifying the minimum number of gross errors and their sizes. One significant advantage of the method is that the detection, identification, and estimation of gross errors can be performed simultaneously.

Remark: Another type of methods are recursive and make use of determining the size of gross errors. Such technique is called a *compensation technique* and it consists of identifying a gross error, determining its size, add subtract from the measurement, and reconcile/detect gross errors again if needed in a serial/iterative manner. Narasimhan and Mah (1987) proposed such technique using the ability of GLR to estimate the sizes of leaks and biases. Later Keller et al. (1994) proposed a few modifications where only the type and location of the gross errors from previous iterations is assumed. In turn, Rollins and Davis (1992) proposed a method that is based on a collective compensation of biases and leaks. They followed up with a paper where the method is extended to the case where the variance-covariance matrix is not known (Rollins and Davis, 1993). Finally, Sanchez et al. (1999) proposed a method where gross errors are estimated and its size is used to identify new ones and Jiang and Bagajewicz (1999) proposed a serial elimination method based on compensation (error estimation) that has proven very effective. Gross error size estimation and some of these methods are covered in detail in Chap. 7.

Remark: Several other strategies with various success rates have been developed: for example, principal component tests have proven to be less efficient than the measurement test when used for multiple gross error identification (Bagajewicz et al., 1999, 2000), powerful and statistically correct combinations of nodal tests have been introduced by Rollins et al. (1996), Yang et al. (1995) proposed a strategy based on a combination of measurement test and nodal test and Devanathan et al. (2000) developed a technique called *imbalance correlation strategy* with which they can overcome known problems of the UBET technique (Rollins and Davis, 1992, 1993) and the GLR technique (Narasimhan and Mah, 1987) where individual measurements and the corresponding nodal imbalance are used. Finally, Tamhane et al. (1988a,b) introduced a Bayesian approach to gross error detection. The book by Narasimhan and Jordache (2000) discusses in detail the assumptions, the prior and the posterior probabilities, the instances in which this method outperforms others, and provides a detailed implementation strategy.

Remark: Another technique that is not based on hypothesis tests is one that uses neural networks (Gupta and Narasimhan, 1993; Terry and Himmelblau, 1993; Aldrich and van Deventer, 1994, 1995; Reddy and Mavrovouniotis, 1998). This is part of a class of methods that require training using known data and therefore prone to fail when existing data does not conform to previous patterns.

Remark: Ragot et al. (1991) discusses some parity space approach to data reconciliation and make comparisons to the standard least squares approach. Recently Maronna and Arcas (2009) showed that the data reconciliation problem can be represented as a standard regression problem. They also found the connections between the probability of detecting a gross error and the redundancy.

Failures of Tests

In many cases (actually very often) the measurement test gives the same value for many variables. We illustrate this next.

Example 5.7: We now consider the system introduced in Chap. 3 and consider that only S_1, S_8, and S_{11} are measured (Fig. 5.3). The measurements are given in Table 5.3.

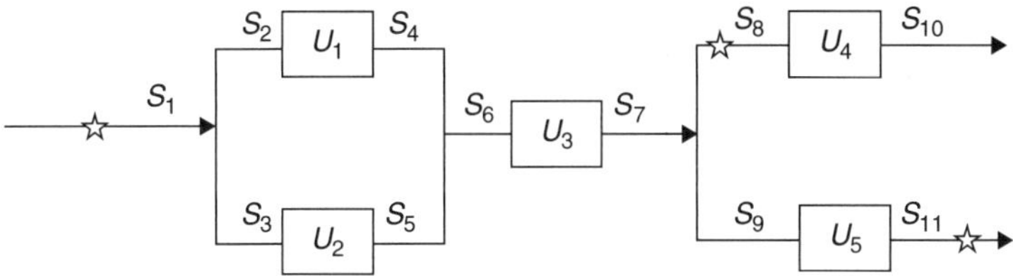

FIGURE 5.3 Flowsheet for Example 5.3.

Stream	Measured Flowrate	Variance
S_1	10.03	0.1
S_8	5.99	0.03
S_{11}	3.99	0.16

TABLE 5.3 Data for Example 5.3

If one adds a gross error $\delta = 5$ to stream S_1, making the measurement 15.03, reconciliation gives

$$\tilde{F}_R = \begin{bmatrix} 13.29 \\ 6.51 \\ 6.78 \end{bmatrix} \tag{5.37}$$

The detailed calculations are

$$S_R E_R^T = \begin{bmatrix} \sigma_1^2 & & \\ & \sigma_8^2 & \\ & & \sigma_{11}^2 \end{bmatrix} \begin{bmatrix} 1 \\ -1 \\ -1 \end{bmatrix} = \begin{bmatrix} \sigma_1^2 \\ -\sigma_8^2 \\ -\sigma_{11}^2 \end{bmatrix} = \begin{bmatrix} 0.1 \\ -0.03 \\ -0.16 \end{bmatrix} \tag{5.38}$$

Then,

$$E_R S_R E_R^T = \begin{bmatrix} 1 & -1 & -1 \end{bmatrix} \begin{bmatrix} \sigma_1^2 \\ -\sigma_8^2 \\ -\sigma_{11}^2 \end{bmatrix} = \begin{bmatrix} 1 & -1 & -1 \end{bmatrix} \begin{bmatrix} 0.1 \\ -0.03 \\ -0.16 \end{bmatrix} = 0.29 \tag{5.39}$$

and the inverse is $[E_R S_R E_R^T]^{-1} = \dfrac{1}{0.29}$. Thus,

$$S_R E_R^T \left(E_R S_R E_R^T\right)^{-1} E_R = \begin{bmatrix} 0.1 \\ -0.03 \\ -0.16 \end{bmatrix} \frac{1}{0.29} \begin{bmatrix} 1 & -1 & -1 \end{bmatrix} = \frac{1}{0.29} \begin{bmatrix} 0.1 & -0.1 & -0.1 \\ -0.03 & 0.03 & 0.03 \\ -0.16 & 0.16 & 0.16 \end{bmatrix} \tag{5.40}$$

Finally

$$\tilde{F}_R = \left[I - S_R E_R^T \left(E_R S_R E_R^T\right)^{-1} E_R\right] F_R^+ = \begin{bmatrix} 0.6552 & 0.3448 & 0.3448 \\ 0.1034 & 0.8966 & -0.1034 \\ 0.5517 & -0.5517 & 0.4483 \end{bmatrix} \begin{bmatrix} 15.03 \\ 5.99 \\ 3.99 \end{bmatrix} = \begin{bmatrix} 13.2888 \\ 6.5122 \\ 6.7761 \end{bmatrix} \tag{5.41}$$

and therefore,

$$\tilde{F}_R - F_R^+ = \begin{bmatrix} 13.2888 \\ 6.5122 \\ 6.7761 \end{bmatrix} - \begin{bmatrix} 15.03 \\ 5.99 \\ 3.99 \end{bmatrix} = \begin{bmatrix} -1.7412 \\ 0.5222 \\ 2.7861 \end{bmatrix} \tag{5.42}$$

The variance is given by

$$\hat{S}_R = S_R E_R^T \left(E_R S_R E_R^T\right)^{-1} E_R S_R = \frac{1}{0.29} \begin{bmatrix} 0.01 & -0.003 & -0.016 \\ -0.003 & 0.009 & 0.0048 \\ -0.016 & 0.0048 & 0.0256 \end{bmatrix} \tag{5.43}$$

We now calculate the z-statistics.

$$\begin{aligned} Z_1 &= \frac{|a_1|}{\sqrt{\left(\tilde{S}_R\right)_{11}}} = 9.3776 \\ Z_7 &= \frac{|a_7|}{\sqrt{\left(\tilde{S}_R\right)_{77}}} = 9.3776 \\ Z_8 &= \frac{|a_8|}{\sqrt{\left(\tilde{S}_R\right)_{88}}} = 9.3776 \end{aligned} \tag{5.44}$$

which shows that the three values are equal.

The above example illustrates the uncertainty as of which variable should be eliminated in the case where the strategy is based on the elimination of the variable with largest Z-statistics. When two variables have the same Z-statistics, they also produce the same value of objective function when they are eliminated. This poses a problem for the serial elimination strategy because there is no criterion to determine which one of these three measurements should be eliminated.

This phenomenon was first pointed out by Iordache et al. (1985) and is now well understood (Bagajewicz and Jiang, 1998). In the case of two gross errors, it happens because the corresponding columns of ER are proportional. Indeed, the last two columns (corresponding to S_8 and S_{11}) are equal, and the proportionality constant between the first and the two last columns is (–1). For cases involving more streams, this is explained in the next chapter. As it will be shown, if one understands the nature of these uncertainties, the serial elimination strategies are not flawed (in this sense) after all.

Remark: Most current commercial software use serial elimination. The strategy that has been implemented in commercial software more often is the one where the largest measurement test (MT) is used to determine the measurement to eliminate.

References

Albuquerque, J. S. and L. T. Biegler, "Data reconciliation and Gross-Error Detection for Dynamic Systems," *AIChE J.* **42**(10):2841 (1996).

Aldrich, C. and J. S. J Van Deventer, "Identification of Gross Errors in Material Balance Measurements By Means of Neural Nets," *Chem. Eng. Sci.* **49**:1357–1368 (1994).

Aldrich, C. and J. S. J Van Deventer, "Comparison of Different Neural Nets for the Detection and Location of Gross Errors in Process Systems," *Ind. Eng. Chem. Res.* **34**:216–224 (1995).

Alhaj-Dibo, M., D. Maquin, and J. Ragot, "Data Reconciliation: A Robust Approach Using a Contaminated Distribution," *Control Engineering Practice* **16**(2):159–170 (2008).

Almasy, G. A. and T. Sztano, "Checking and Correction of Measurements on the Basis of Linear System Model," *Problems of Control and Information Theory* **4**(1):57–69 (1975).

Arora, N. and L. T. Biegler, "Redescending Estimators for Data Reconciliation and Parameter Estimation," *Comput. Chem. Eng.* **25**:1585–1599 (2001).

Bagajewicz, M. and Q. Jiang, "Gross Error Modeling and Detection in Plant Linear Dynamic Reconciliation," *Comput. Chem. Eng.* **22**(12):1789–1810 (1998).

Bagajewicz, M. and Q. Jiang, "A Mixed Integer Linear Programming-Based Technique for the Estimation of Multiple Gross Errors in Process Measurements," *Chem. Eng. Commun.* **177**:139–155 (2000).

Bagajewicz, M., Q. Jiang, and M. Sánchez, "Performance Evaluation of PCA Tests for Multiple Gross Error Identification," *Comput. Chem. Eng.* **23** supp.:S589–S592 (1999).

Bagajewicz, M., Q. Jiang, and M. Sánchez, "Performance Evaluation of PCA Tests in Serial Elimination Strategies for Gross Error Identification," *Chem. Eng. Commun.* **183**:119–139 (2000).

Bagajewicz, M., Q. Jiang, and M. Sánchez, "Removing Singularities and Assessing Uncertainties in Two Efficient Gross Error Collective Compensation Methods," *Chem. Eng. Commun.* **178**:1–20 (2000).

Buzzi Ferraris G., "Statistical Tests and Confidence Intervals," *Comput. Chem. Eng.* **24**:2037–2039 (2000).

Chen, W., B. R. Bakshi, P. K. Goel, and S. Ungarala, "Bayesian Estimation via Sequential Monte Carlo Sampling: Unconstrained Nonlinear Dynamic Systems," *Ind. Eng. Chem. Res.* **43**:4012–4025 (2004).

Chen, V. C. P., M. Melendez, and D. K. Rollins, "The Problem of Too Much Power in Detecting Biases in Real Chemical Processes," *ISA Transactions* **37**:329–336 (1998).

Cong-li, M., S. Hong-ye, and C. Jian, "Detection of Gross Errors Using Mixed Integer Optimization Approach in Process Industry," *Journal of Zhejiang University—Science A* **8**:904–909 (2007).

Crowe, C. M., "Recursive Identification of Gross Errors in Linear Data Reconciliation," *AIChE J.* **34**(4):541–550 (1989).

Crowe, C. M., "Test of Maximum Power for Detection of Gross Errors in Process Measurements," *AIChE J.* **35**:869–872 (1989).

Crowe, C. M., "The Maximum-Power Test for Gross Errors in the Original Constraints in Data Reconciliation," *Canadian Journal of Chemical Engineering* **70**:1030–1036 (1992).

Devanathan, S., D. K. Rollins, and S. B. Vardeman, "A New Approach for Improved Identification of Measurement Bias," *Comput. Chem. Eng.* **24**(12) 1:2755–2764 (2000).

Gupta, G., and S. Narasimhan, "Application of Neural Networks for Gross Error Detection," *Ind. Eng. Chem. Res.* **32**:1651–1657 (1993).

Harikumar, P. and S. Narasimhan, "A Method to Incorporate Bounds in Data Reconciliation and Gross Error Detection-II. Gross Error Detection Strategies," *Comput. Chem. Eng.* **17**:1121–1128 (1993).

Jordache, C., R. Mah, and A. Tamhane, "Performance Studies of the Measurement Test for Detection of Gross Errors in Process Data," *AIChE J.* **31**:1187 (1985).

Jiang, Q. and M. Bagajewicz, "On a Strategy of Serial Identification with Collective Compensation for Multiple Gross Error Estimation in Linear Data Reconciliation," *Ind. Eng. Chem. Res.* **38**(5):2119–2128 (1999).

Jiang, Q., M. Sánchez, and M. Bagajewicz, "On The Performance of Principal Component Analysis in Multiple Gross Error Identification," *Ind. Eng. Chem. Res.* **38**(5):2005–2012 (1999).

Johnston, L. P. M. and M. A. Kramer, "Maximum Likelihood Data Rectification. Steady State Systems," *AIChE J.* **41**:11 (1995).

Keller, J. Y., M. Darouach, and G. Krzakala, "Fault Detection of Multiple Biases or Process Leaks in Linear Steady State Systems," *Comput. Chem. Eng.* **18**:1001–1004 (1994).

Madron, F., V. Veverka, V. Vanecek, "Statistical Analysis of Material Balance of a Chemical Reactor," *AIChE J.* **23**:482 (1977).

Madron, F., "A New Approach to the Identification of Gross Errors in Chemical Engineering Measurements," *Chem. Eng. Sci.* **40**:1855–1860 (1985).

Mah, R. S. H., "Chemical Process Structures and Information Flows," Butterworths, (1990).

Mah, R. S. H., "Letter to the Editor," *Comput. Chem. Eng.*, **21**(9):1069 (1997).

Mah, R. S. H. and A. C. Tamhane, "Detection of Gross Errors in Process Data," *AIChE J.* **28**:828–830 (1982).

Mah, R. S. H., G. Stanley, and D. Downing, "Reconciliation and Rectification of Process Flow and Inventory Data," *Ind. Eng. Chem Res. Proc. Des. Dev.* **15**:175–183 (1976).

Maronna, R., J. Arcas, "Data Reconciliation and Gross Error Diagnosis Based on Regression," *Comput. Chem. Eng.* **33**:65–71 (2009).

Maquin, D. and J. Ragot, "Comparison of Gross Error Detection Methods in Process Data," *Proc. 30th Conf on Decision and Control*, Brighton, England:2254–2261 (1991).

Morad, K., B. R. Young, and W. Y. Svrcek, "Rectification of Plant Measurements Using a Statistical Framework," *Comput. Chem. Eng.* **29**:919–940 (2005).

Narasimhan, S. and C. Jordache, "Data Reconciliation and Gross Error Detection. An Intelligent Use of Process Data," Gulf Publishing Company (2000).

Narasimhan, S. and R. S. H. Mah, "Generalized Likelihood Ratio Method for Gross Error Identification," *AIChE J.* **33**:1514–1521 (1987).

Narasimhan, S. and P. Harikumar, "A Method to Incorporate Bounds in Data Reconciliation and Gross Error Detection-II. Gross Error Detection Strategies," *Comput. Chem. Eng.* **17**:1121–1128 (1993).

Nogita, S., "Statistical Test and Adjustment of Process Data," *Ind. Eng. Chem. Process Des. Develop* **2**:197 (1972).

Ozyurt, D. B. and R. W. Pike, "Theory and Practice of Simultaneous Data Reconciliation and Gross Error Detection of Chemical Processes," *Comput. Chem. Eng.* **28**:381–402 (2004).

Ragot, J., A. Aitouche, F. Kratz, D. Maquin, "Detection and Location of Gross Errors in Instruments Using Parity Space Technique," *Int J. of Mineral Processing* **31**:281–299 (1991).

Ragot, J., M. Chadli, and D. Maquin, "Mass Balance Equilibration: A Robust Approach Using Contaminated Distribution," *AIChE J.* **51**(5) (2005).

Reddy, V. N. and M. L. Mavrovouniotis, "An Input-Training Neural Network Approach for Gross Error Detection and Sensor Replacement," *Trans. IChemE* **76** Part A:478–489 (1998).

Reilly, P. M. and R. E. Carpani, "Application of Statistical Theory of Adjustments to Material Balances," *13th Can. Chem. Eng. Conf.*, Montreal, Quebec (1963).

Ripps, D. L., "Adjustment of Experimental Data," *Chem. Eng. Progr. Symp. Ser. no. 55* **61**:8–13 (1965).

Rollins, D. K. and J. F. Davis, "Unbiased Estimations of Gross Errors in Process Measurements," *AIChE J.* **38**:563–572 (1992).

Rollins, D. K., J. F. Davis, "Gross Error Detection when Variance-Covariance Matrices are Unknown," *AIChE J.* **39**(8):1335–1341 (1993).

Rollins, D. K., Y. Cheng, and S. Devanathan, "Intelligent Selection of Hypothesis Tests to Enhance Gross Error Identification," *Comput. Chem. Eng.* **5**:517–530 (1996).

Romagnoli, J. A., and G. Stephanopoulos, "Rectification of Process Measurement Data in the Presence of Gross Errors," *Chem. Eng. Sci.* **36**:1849–1863 (1981).

Rosenberg, J., R. S. H. Mah, and C. Lordache, "Evaluation of Schemes for Detecting and Identifying Gross Errors in Process Data," *Ind. Eng. Chem. Res.* **26**:555 (1987).

Sánchez, M. and J. Romagnoli, "Data Processing and Reconciliation for Chemical Process Operations," Academic Press (2000).

Sánchez, M., J. Romagnoli, Q., Jiang, and M. Bagajewicz, "Simultaneous Estimation of Biases and Leaks in Process Plants," *Comput. Chem. Eng.* **23**(7):841–858 (1999).

Serth, R. W. and W. A. Heenan, "Gross Error Detection and Data Reconciliation in Steam-Metering Systems," *AIChE J.*. **32**:733–742 (1986).

Sidak, Z., "Rectangular Confidence Regions for the Means of Multivariate Normal Distribution," *Journal of American Statistics Association* **62**:626–633 (1967).

Soderstrom, T. A., D. M. Himmelblau, T. F. Edgar, "A Mixed Integer Optimization Approach for Simultaneous Data Reconciliation and Identification of Measurement Bias," *Control Engineering Practice* **9**(8):869–876 (2001).

Tamhane, A. C., "A Note on the Use of Residuals for Detecting an Outlier in Linear Regression," *Biometrika* **69**:488 (1982).

Tamhane, A. C. and R. S. H. Mah, "Data Reconciliation and Gross Error Detection in Chemical Process Networks," *Technometrics* **27**(4):409 (1985).

Tamhane, A. C., C. Lordache, and R. S. H. Mah, "A Bayesian Approach to Gross Error Detection in Chemical Process Data. Part I. Model Development," *Chemometrics and Intel. Lab. Sys.* **4**:33–45 (1988a).

Tamhane, A. C., C. Lordache, and R. S. H. Mah, "A Bayesian Approach to Gross Error Detection in Chemical Process Data. Part II. Simulation Results," *Chemometrics and Intel. Lab. Sys.* **4**:131–146 (1988b).

Terry, P. A. and D. M. Himmelblau, "Data Rectification and Gross Error Detection in Steady-State Process via Artificial Neural Networks," *Ind. Eng. Chem. Res.* **32**:3020–3028 (1993).

Tjoa, I. B. and L. T. Biegler, "Simultaneous Strategies for Data Reconciliation and Gross Error detection of Nonlinear Systems," *Comput. Chem. Eng.* **15**(10): 679–690 (1991).

Tong, H. and C. M. Crowe, "Detection of Gross Errors in Data Reconciliation Using Principal Component Analysis," *AIChE J.* **41**:1712–1722 (1995).

Tong, H. and C.M. Crowe, "Detecting Persistent Gross Errors by Sequential Analysis of Principal Components," *Comput. Chem. Eng.* **20** suppl.:S733–S739 (1996).

Tong, H. and C. M. Crowe, "Detecting Persistent Gross Errors by Sequential Analysis of Principal Components," *AIChE J.* **43**:1242–1249 (1997).

Tong, H. and D. Bluck, "An Industrial Application of Principal Component Test to Fault Detection and Identification," *IFAC Workshop on On-Line-Fault Detection and Supervision in the Chemical Process Industries*, Solaize (Lyon), France (1998).

Wang, D. and J. A. Romagnoli, "A Framework for Robust Data Reconciliation Based on a Generalized Objective Function," *Ind. Eng. Chem. Res.* **42**(13):3075–3084 (2003).

Wang, F., X. Jia, S. Zheng, and J. Yue, "An Improved MT–NT Method for Gross Error Detection and Data Reconciliation," *Comput. Chem. Eng.* **28**:2189–2192 (2004).

Yamamura, K., M. Nakajima, and H. Matsuyama, "Detection of Gross Errors in Process Data Using Mass and Energy Balances," *Int. Chem. Eng.* **28**:91–98 (1988).

Yang, Y., R. Ten, and L. Jao, "A Study of Gross Error Detection and Data Reconciliation in Process Industries," *Comput. Chem. Eng.* **19**:S217–S222 (1995).

Zhang, P., G. Rong, and Y. Wang, "A New Method of Redundancy Analysis in Data Reconciliation and its Application," *Comput. Chem. Eng.* **25**:941–949 (2001).

Zhao, W., D. Chen, S. Hu, "Detection of Outlier and a Robust BP Algorithm Against Outlier," *Comput. Chem. Eng.* **28**:1403–1408 (2004).

CHAPTER 6

Equivalency of Gross Errors

Definition

Two sets of gross errors are equivalent when they have the same effect in data reconciliation, that is, when eliminating either one, leads to the same value of the objective function. Therefore, the equivalent sets of gross errors are theoretically undistinguishable. In other words, when a set of gross errors is identified, there exists an equal possibility that the true locations of gross errors are in one of its equivalent sets. We illustrate this now through an example.

Example 6.1: Consider the case of Fig. 6.1 and the measurements given in Table 6.1. Random errors have been eliminated in this table so that the phenomenon is clearly visible.

We now analyze the set $\{S_2, S_4, S_5\}$. As it is shown in Table 6.1, three different sets of gross errors and reconciled values can explain these measurements (cases 1, 2, and 3). Therefore, without a priori additional knowledge as of where the gross errors might be, or some additional information about the reconciled values, the three cases are impossible to distinguish.

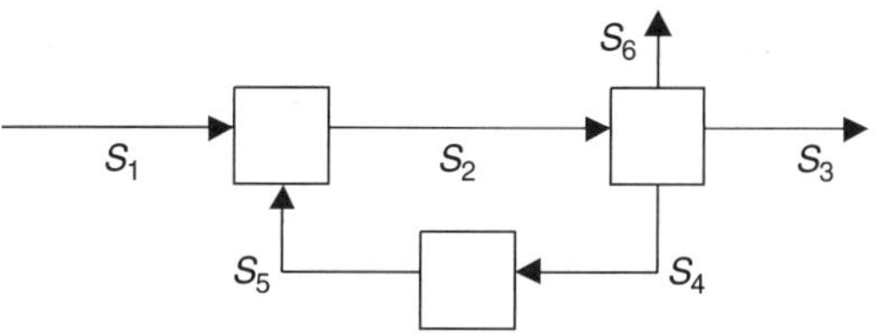

FIGURE 6.1 Flowsheet of Example 6.1.

Stream		S_1	S_2	S_3	S_4	S_5	S_6
Measurement		**12**	**18**	**10**	**4**	**7**	**2**
Case 1 (bias in S_4, S_5)	Reconciled data	12	18	10	6	6	2
	Estimated biases				−2	1	
Case 2 (bias in S_2,S_4)	Reconciled data	12	19	10	7	7	2
	Estimated biases		−1		−3		
Case 3 (bias in S_2,S_5)	Reconciled data	12	16	10	4	4	2
	Estimated biases		2			3	

Table 6.1 Illustration of Equivalent Sets in $\{S_2, S_4, S_5\}$ of Fig. 6.1

Practical Consequences

This has very important consequences in practice. For example, if the gross errors are in reality in $\{S_2,S_4\}$, any strategy for gross error detection can identify with equal probability any of the three cases. The impact of this is not only in the location and size of the gross errors, but it is also in the reconciled values. The importance of this in production accounting is paramount.

Cardinality of Equivalent Sets

In the case of the example of Table 6.1, we noticed that every case has two gross errors. This is no coincidence. These are called *equivalent sets.* Every equivalent set has a certain minimum number of gross errors that can represent all situations. Indeed, there are infinite numbers of sets of three gross errors that can be represented by two gross errors. The next example illustrates this fact.

Example 6.2: Table 6.2 illustrates how different sets of two gross errors can be formed. However, it is not possible to find one gross error that will be equivalent to any of these situations.

Stream		S_1	S_2	S_3	S_4	S_5	S_6
Measurement		**12**	**18**	**10**	**4**	**7**	**2**
Case 1 (bias in S_4, S_5)	Reconciled data	12	18	10	6	6	2
	Estimated biases				−2	1	
Case 4 (bias in S_2, S_4, S_5)	Reconciled data	12	17	10	5	5	2
	Estimated biases		1		−1	2	
Case 5 (bias in S_2, S_4, S_5)	Reconciled data	12	16.5	10	4.5	4.5	2
	Estimated biases		1.5		−0.5	2.5	

Table 6.2 Illustration of Equivalent Sets in $\{S_2, S_4, S_5\}$ of Fig. 6.1

Definition The gross error cardinality of a set of variables is the minimum number of gross errors that are required to represent all possible sets of gross errors in the variables.

The cardinality is easy to determine: It is equal to the number of units involved in the loop that the set forms minus one.

Basic Subset of an Equivalent Set

Definition A set of variables constitutes a basic subset of a system, when every set of gross errors is equivalent to a set of gross errors in the basic set.

The number of elements in a basic subset is equal to the equivalent set cardinality. In the example above any of the sets of case 1, 2, and 3 are basic sets.

Determination of Equivalent Sets

Equivalent sets can be identified using the method presented by Bagajewicz and Jiang (1998). The method is based on obtaining the largest set Λ that contains the set of n identified gross errors Ψ, such that the gross error cardinality remains equal to n, the number of element (set cardinality) of Ψ. The set Λ can be constructed by adding all streams that are in a loop with all subset elements of Ψ. Thus, the list of equivalent basic sets $\{\varphi_i\}$ is constructed by identifying all subsets of n variables in Λ, such that the streams in $\{\varphi_i\}$ do not form a loop.

Example 6.3: Figure 6.2 consists of a flowsheet where the environmental node has been added for the purpose of identifying loops.

Assume that a set of gross errors: $\Psi = \{S_1, S_2\}$ ($n = 2$) is identified. Since Leak1 is forming a loop with S_1 and Leak2 is in a loop with $\{S_1, S_2\}$, one can construct Λ by adding Leak1 and Leak2 to Ψ, that is, $\Lambda = \{S_1, S_2, \text{Leak1}, \text{Leak2}\}$. Thus all equivalent sets of Ψ can be obtained. They are: $\{S_1, \text{Leak2}\}$, $\{S_2, \text{Leak1}\}$, $\{S_2, \text{Leak2}\}$, and $\{\text{Leak1}, \text{Leak2}\}$. The set $\{S_1, \text{Leak1}\}$, for example, is not an equivalent set since S_1 and Leak1 form a loop.

In general, the following loops that do not include leaks can be identified:

- $\{S_3, S_6\}$
- $\{S_2, S_4, S_5\}$

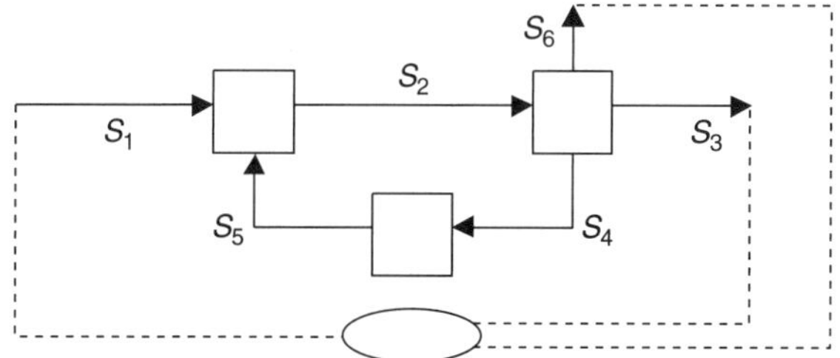

Figure 6.2 Illustration of gross error equivalency.

- $\{S_1, S_2, S_3\}$
- $\{S_1, S_2, S_6\}$

Therefore, if one disregards leaks, any identified set of gross errors should be tested if it belongs to any of these sets.

Practical Consequence

When a set of gross errors has been identified, then uncertainty exists regarding to what equivalent set it belongs. Such uncertainty can only be resolved with new information.

Degeneracy

There are cases where a certain number of gross errors can be represented by a number of gross errors lower than the gross error cardinality. Two such examples are shown next.

Example 6.4: Consider the flowsheet of Fig. 6.2. In Table 6.3 a set of two gross errors (case 1) is equivalent to one gross error (case 2). These cases are rare because they require that the gross errors have certain particular sizes. In this example, the two real gross errors have equal sizes.

		S_1	S_2	S_3	S_4	S_5	S_6
Measurement		**12**	**18**	**10**	**7**	**7**	**2**
Case 1 (bias in S_4, S_5)	Reconciled data	12	18	10	6	6	2
	Estimated biases				1	1	
Case 2 (bias in S_2)	Reconciled data	12	19	10	7	7	2
	Estimated biases		−1				

Table 6.3 Illustration of Degenerate Cases in $\{S_2, S_4, S_5\}$ of Fig. 6.1

Example 6.5: Consider the process depicted in Fig. 6.3. The true values for this system are $x = [1, 2, 3, 2, 1, 1, 1, 0.4, 0.6]$ and the standard deviations are 2% of each measurement. Once again, random errors have been eliminated and the measurements are equal to the true values plus the gross errors simulated.

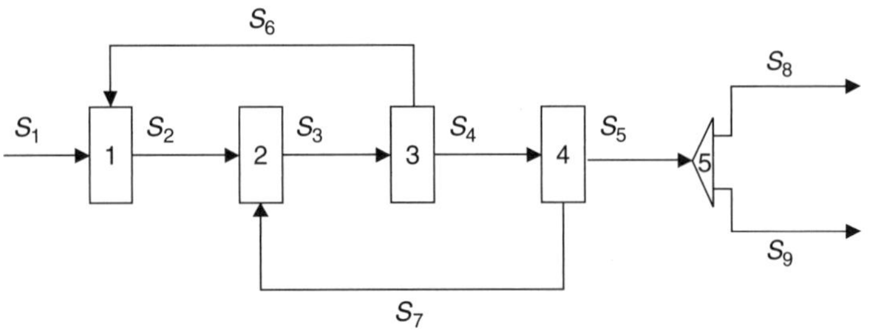

Figure 6.3 Illustration of gross error degeneracy.

Now consider the cases in Table 6.4. In case 1, the system consisting of S_1, S_4, S_5, S_6, S_9 has a cardinality $\Gamma = 4$. The system cardinality is 4 in case 2 and 5 in case 3. The number of gross errors identified can be larger (case 1), smaller (case 2), or equal (case 3) to the number of gross errors introduced.

Quasi-Degeneracy

In practice many numbers are considered equal if they are close enough within a certain tolerance. Thus situations similar to degeneracy may happen. The next example illustrates this.

Example 6.6: Table 6.5 provides some cases that illustrate quasi-degeneracy.

Consider case 1: first note that the set $\{S_2, S_3, S_6\}$ has gross error cardinality 2, that is, any set of gross errors in this set is equivalent to one with no more than two gross errors. Therefore, the real set of gross errors is equivalent to $\delta_2 = 0.030$ and $\delta_6 = -0.380$. Now, note that the value for δ_2 in this new equivalent set is within the size of the standard deviation of the measurement. Therefore, it is absorbed as a random error.

Quasi-Equivalency

Quasi-equivalency occur when only a subset of the identified gross errors is equivalent to the real gross errors, and in addition the non-equivalent gross errors are of small size. We illustrate this through the following example:

Example 6.7: Assume that in Fig. 6.3, a gross error exists is introduced in stream S_1 of size $\delta_1 = +1$. Assume also that the gross error identification finds two gross errors in S_1 and S_2, with sizes $\delta_1 = +0.98$, $\delta_2 = +0.05$. Technically, one should say that the gross error detection scheme that was used to obtain this result has failed. However, the "failure" consists of a correct identification accompanied with a small size estimate. In principle, even though the result is based on the usage of statistical tests, one is tempted to disregard δ_2 and declare the identification successful. One important observation in this case is that S_1 and S_2 are not a basic set of any subset of the graph. In other words, no degeneracy or equivalency can apply.

Detection of Leaks

If one considers a leak as another stream, a leak forms at least one loop with some streams or other leaks in the augmented graph. Therefore, it will be represented with at least one equivalent set of biases with the model. Let us illustrate this issue with two examples.

Example 6.8: Consider the process in Fig. 6.4. Assume that there is a leak with the size of +5. Corresponding to this leak the measurements for flowrates of S_1 and S_2 are 100 and 95, respectively. Any bias identification scheme will identify either a bias of −5 in S_2 or a bias of +5 in S_1.

Consider now the process of Fig. 6.5. The equivalencies are shown in Table 6.6.

Stream		S_1	S_2	S_3	S_4	S_5	S_6	S_7	S_8	S_9
True Value		**1**	**2**	**3**	**2**	**1**	**1**	**1**	**0.4**	**0.6**
Case 1	Gross error introduced					–0.235				–0.235
	Gross error estimated	0.235			0.235		–0.235			
	Reconciled data	0.765	2.000	3.000	1.765	0.765	1.235	1.000	0.400	0.365
Case 2	Gross error introduced	0.380	0.380			0.380				
	Gross error estimated							0.380	–0.380	
	Reconciled data	1.380	2.380	3.000	2.000	1.380	1.000	0.620	0.780	0.600
Case 3	Gross error introduced	0.389		0.389		0.389				
	Gross error estimated		–0.389		–0.389				–0.389	
	Reconciled data	1.389	2.389	3.389	2.389	1.389	1.000	1.000	0.789	0.600

TABLE 6.4 Illustration of Degenerate Cases of Fig. 6.3

Stream		S_1	S_2	S_3	S_4	S_5	S_6	S_7	S_8	S_9
True Value		**1**	**2**	**3**	**2**	**1**	**1**	**1**	**0.4**	**0.6**
Case 1	Gross error introduced		0.410	0.380						
	Gross error estimated						–0.401			
	Reconciled data	1.000	2.401	3.400	1.998	1.000	1.401	0.998	0.400	0.600
Case 2	Gross error introduced	0.350			0.350	0.400				
	Gross error estimated						0.371		–0.373	
	Reconciled data	1.373	2.001	2.997	2.368	1.373	0.629	0.996	0.773	0.600
Case 3	Gross error introduced					0.235				0.24
	Gross error estimated	–0.238			–0.238		0.238			
	Reconciled data	1.238	2.000	3.000	2.238	1.238	0.762	1.000	0.400	0.839

TABLE 6.5 Illustration of Quasi-Degenerate Cases of Fig. 6.3

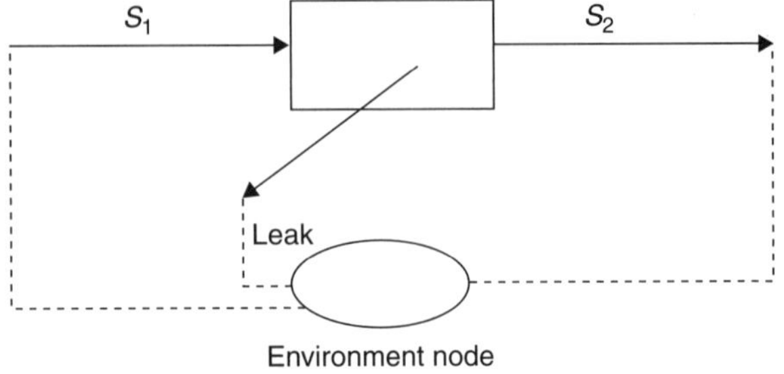

Figure 6.4 A simple process with a leak.

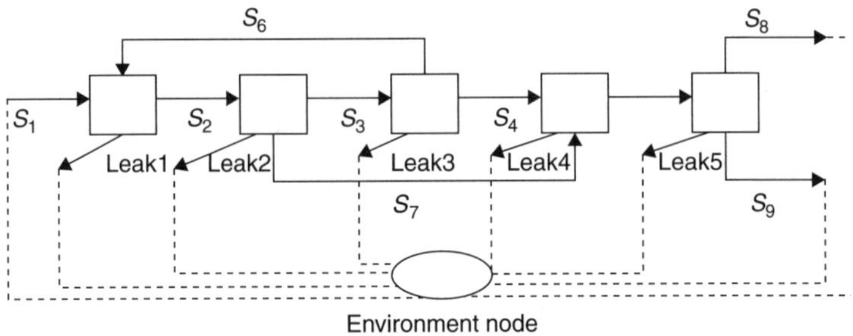

Figure 6.5 A process with several candidate leaks.

Case No.	Gross Error Introduced	Gross Error Identified	Equivalent Sets
1	Leak1	S_1	{Leak1}
2	Leak2	Same size biases in S_1, S_2	{Leak2}, {Leak1, S_2}
3	Leak5	S_8	{Leak5}, {S_9}
4	Leak1, Leak2	Different size biases in S_1, S_2	{Leak1, S_2}, {Leak1, Leak2} {S_1,Leak2}, {S_2,Leak2}
5	Leak1, S_3	S_1, S_3	{Leak1, S_3}

Table 6.6 Identification of Leaks

Thus, any leak is equivalent to a set of equal-size biases in a set of streams connecting to the unit where the leak occurs and the environment. This is an important result and leads to the following conclusion:

Since a leak is equivalent to a set of biased streams, any steady-state method that contains a test to detect biased instruments can be used in conjunction with the equivalency theory to assess the existence of leaks.

Practical Approach to Equivalency

While it is easy to understand the results when the real location of gross errors is known, when the real location is not known the notion that there are several equivalent sets is troublesome. This is especially bad when the number of existing gross errors is large.

Example 6.9: Table 6.7 depicts three situations that illustrate how serious the situation can become. The table depicts real gross errors introduced, the gross errors identified, and the corresponding number of equivalent sets.

Real Gross Errors	Gross Errors Detected	Equivalent Sets
$S_1(+1)$, $S_4(+2)$, $S_8(+1.5)$, $L_3(+1)$	$S_4(-0.409)$, $S_5(-1.982)$, $S_6(0.953)$	None
$S_1(-1.0)$, $S_2(+1.0)$, $S_4(+1.5)$, $S_8(+0.5)$	$S_2(2.032)$, $S_4(1.565)$, $S_8(0.478)$, $L_2(0.964)$	10
$S_1(-2)$, $S_2(+3)$, $S_4(+1.75)$, $S_5(-1)$, $S_8(-3.5)$	$S_2(3.978)$, $S_4(2.831)$, $S_6(-1.063)$, $S_8(-2.519)$, $L_2(0.924)$	429

Table 6.7 Number of Equivalent Sets

These sets of gross errors belong to different equivalent sets that have common streams. Through these common streams and the use of degeneracy, the results can be explained. We will omit this explanation, as our intention is to illustrate other issues.

- The same or similar amount of gross errors can have from none to a large number of equivalent sets.
- Even in the cases when there are no equivalent sets, the uncertainty persists.

Now that awareness exists about the uncertainty associated with any gross error identification, we present two strategies that can be implemented to reduce the uncertainty.

1. Pick the equivalent set that has a gross error closer to the battery limits. This is in many cases the place where a larger frequency of missing transactions exists.
2. Establish a database of failure frequency of instruments and pick the equivalent sets with larger frequencies.

Remark: Equivalency has an impact on the assessment of power of the gross error identification. Indeed, in assessing the power, the usual procedure is to introduce an error in a certain measurement with a random size and verifying how well the method identify the error. As it turns out, almost all assessment methods compute a failure when an equivalent set is identified. Bagajewicz et al. (2000) showed how these uncertainties and singularities can be treated. In fact, they showed that when an equivalent set has been identified, then it has to count as a success, all this regardless of the gross error detection method used. Thus, the power of several methods is actually larger than what was reported.

References

Bagajewicz, M. and Q. Jiang, "Gross Error Modeling and Detection in Plant Linear Dynamic Reconciliation," *Comput. Chem. Eng.* **22**(12):1789–1810 (1998).

Bagajewicz, M., Q. Jiang, and M. Sánchez, "Removing Singularities and Assessing Uncertainties in Two Efficient Gross Error Collective Compensation Methods," *Chem. Eng. Commun.* **178:**1–20 (2000).

CHAPTER 7

Gross Error Size Elimination and Estimation

The Compensation Model

A linear steady-state data reconciliation problem was presented in Chap. 4.

$$\left.\begin{aligned} &\operatorname{Min}[\tilde{F}_R - F_R^+]^T S_R^{-1}[\tilde{F}_R - F_R^+] \\ &\text{s.t.} \\ &E_R \tilde{F}_R = 0 \end{aligned}\right\} \tag{7.1}$$

Assume now a set of instrument biases represented by the vector **d**. Therefore,

$$F_R^+ = \tilde{F}_R + \delta \tag{7.2}$$

where δ is a vector that contains biases for bias candidates. Since we are interested in considering only a limited number of biased streams, the rest of the elements of δ that are not candidates are zero. This is accomplished by writing δ as follows:

$$\delta = L\hat{\delta} \tag{7.3}$$

where

$$\hat{\delta} = \begin{bmatrix} \delta_1 \\ \delta_2 \\ \vdots \\ \delta_{nb} \end{bmatrix} \qquad L = [e_1 \quad e_2 \quad \cdots \quad e_{nb}] \tag{7.4}$$

and $\mathbf{e}_i$ is a vector with unity in a position corresponding to a bias candidate and zero elsewhere.

The reconciliation and gross error estimation model in terms of redundant variables is

$$\left.\begin{aligned} &\text{Min}\left[\tilde{F}_R + L\tilde{\delta} - F_R^+\right]^T S_R^{-1}\left[\tilde{F}_R + L\tilde{\delta} - F_R^+\right] \\ &\text{s.t.} \\ &\qquad E_R\tilde{F}_R = 0 \end{aligned}\right\} \tag{7.5}$$

The solution to this problem is given by

$$\tilde{\delta} = \left[(H+L)^T S_R^{-1}(H+L)\right]^{-1} H(H+L)^T S_R^{-1}\left(F_R^+ - \hat{F}_R\right) \tag{7.6}$$

$$\tilde{F}_R = \hat{F}_R + H\tilde{\delta} \tag{7.7}$$

where $\hat{F}_R$ is the solution for the reconciliation problem assuming no biases, that is,

$$\hat{F}_R = \left[I - S_R E_R^T \left(E_R S_R E_R^T\right)^{-1} E_R\right] F_R^+ \tag{7.8}$$

and H is given by

$$H = [S_R C^T (C S_R C^T)^{-1} C - I]L \tag{7.9}$$

The above model was originally suggested by Romagnoli (1983), discussed in its dynamic form by Bagajewicz and Jiang (1998), and used by Jiang and Bagajewicz (1999). In several cases $(H+L)^T S_R^{-1}(H+L)$ is singular (cannot be inverted), which is a result of picking a set of candidate gross errors that is an equivalent set (Bagajewicz and Jiang, 1998). Equivalent sets were discussed in Chap. 6. Thus, a strategy is needed to determine which candidate set to pick. As in the case of the serial elimination, a serial determination of candidates is presented next. It makes use of equivalent theory to avoid picking sets that have cardinality lower than the number of streams.

Serial Identification with Collective Compensation Strategy (SICC)

This strategy is serial in the sense that uses the measurement test at each step to identify a candidate gross error, which is added to the list. However, in contrast with other serial procedures where the sizes of the gross errors identified are determined at each step, this procedure evaluates the size of all the gross errors in the candidate list at each step.

The steady-state version of this strategy (Jiang and Bagajewicz, 1999) can be summarized as follows:

1. Run the data reconciliation and calculate the measurement test (MT) for all measured variables. If no MT flags, declare no gross error and stop. Otherwise go to step 2.
2. Construct a list of candidates (LC) by including all variables that failed the MT. If any two members in LC form a loop, erase one of them. Create a list of confirmed gross errors (LCGE). This list is empty at this stage.
3. Run the data reconciliation with gross error estimation model simulating a gross error in all the members of the LCGE and one member of the LC at a time.
4. Determine which member of the LC leads to the smallest value of the objective function. Add that variable to the LCGE.
5. Calculate MT for the run chosen in step 4. Erase all elements of the LC and place the latest flagged variables in LC. If there are any two members in LC forming a loop with any member(s) in LCGE, erase one of them. If LC is empty, go to step 6. Otherwise, go to step 3.
6. Determine all equivalent sets and corresponding gross error sizes. Declare all members in LCGE in suspect and stop.

This strategy is based on the use of the measurement test to perform the identification of the gross error candidates. It does not contain any combinatorial search, and as such it can be applied to large systems.

The Unbiased Estimation Model (UBET)

The UBET method was developed by Rollins and Davis (1992). Consider the balance residuals

$$r = E_R F_R^+ \tag{7.10}$$

and its expected value

$$\mu_R = E_R \delta + M\gamma \tag{7.11}$$

where δ are the suspected biases, γ the suspected leaks, and M the matrix that accounts for all the leaks in each balance equation. Clearly, if each row of ER represents a balance around one unit only, then M is a permutation of the identity matrix.

By partitioning E_R, M, δ, and γ and assuming there are always q gross errors, one can get

$$\mu_r = \begin{bmatrix} E_{R,11} & 0 \\ E_{R,21} & M_{22} \end{bmatrix} \begin{bmatrix} \delta_1 \\ \gamma_2 \end{bmatrix} = C_1 \theta_1 \tag{7.12}$$

Finally, by introducing

$$l_i^T = e_i^T C_1^{-1} \tag{7.13}$$

one obtains

$$l_i^T \mu_r = e_i^T \theta_1 = \theta_i \tag{7.14}$$

Thus, $l_i^T r (i = 1,\ldots, q)$ are unbiased estimators of the components of δ and γ contained in θ_1.

The procedure for applying UBET can be then summarized as follows:

1. Use a gross error identification strategy, like the nodal strategies to isolate the suspect nodes and also construct the candidate bias/leak list from the suspect nodes.
2. Obtain θ_1 with elements no more than the number of constraint equations (q).
3. Construct C_1 with rank equal to q.
4. Obtain the size estimation for the elements in θ_1.
5. Use hypothesis testing to identify the gross errors. Rollins and Davis (1992) suggested adding Bonferroni confidence intervals (see Chap. 5, Nodal Test section).

This method has been identified, together with the SICC as one of the most powerful for multiple gross error detection.

Remark: For cases where the UBET and the GLR method fail to perform accurately, Devanathan et al. (2000) proposed a technique that makes use of information contained in the relationship between individual measurements and the corresponding nodal imbalance.

Conversion between Equivalent Sets

Once all the equivalent sets have been identified, one may want to determine what are the sizes of the gross errors in one equivalent set, when the sizes in another equivalent set are known. In addition, one wants to know what are the new values of the reconciled streams.

We recall that in Chap. 6 it was showed that

- Each leak is equivalent to a set of biases, and consequently,
- Leaks are part of equivalent sets.

Thus, since leaks and biases are interchangeable, they should also be part of the data reconciliation and gross error estimation model. Thus we extend Eq. (7.5) to include leaks as follows:

$$\left.\begin{aligned} &\text{Min } [\tilde{F}_R + L\tilde{\delta} - F_R^+]^T S_R^{-1} [\tilde{F}_R + L\tilde{\delta} - F_R^+] \\ &\text{s.t.} \\ &E_R \tilde{F}_R = K\tilde{\mu} \end{aligned}\right\} \tag{7.15}$$

where $K = [e_1 \; e_2 \cdots e_{nsl}]$ and $\tilde{\mu}$ is a vector of leaks.

We now introduce the following change of variables:

$$f = \tilde{F}_R + L\tilde{\delta} \tag{7.16}$$

with which one can rewrite Eq. (7.15) as follows:

$$\left.\begin{array}{l} \text{Min } [f - F_R^+]^T S_R^{-1} [f - F_R^+] \\ \text{s.t.} \\ \quad Cf - CL\tilde{\delta} - K\tilde{\mu} = 0 \end{array}\right\} \tag{7.17}$$

Assume that one already obtained a set of gross errors and its sizes $\tilde{\delta}_1$ and $\tilde{\mu}_1$ and wants to know the gross error sizes $\hat{\delta}_2$ and $\tilde{\mu}_2$ of an equivalent set. The following holds between two equivalent sets:

$$f_1 = f_2 \tag{7.18}$$

Thus one gets

$$CL_1\hat{\delta}_1 + K_1\tilde{\mu}_1 = CL_2\hat{\delta}_2 + K_2\tilde{\mu}_2 \tag{7.19}$$

or

$$[CL_1 \quad K_1]\begin{bmatrix} \hat{\delta}_1 \\ \tilde{\mu}_1 \end{bmatrix} = [CL_2 \quad K_2]\begin{bmatrix} \hat{\delta}_2 \\ \tilde{\mu}_2 \end{bmatrix} \tag{7.20}$$

Let P be a matrix, such that $P[CL_2 \quad K_2]$ is of canonical form, that is,

$$P[CL_2 \quad K_2] = \begin{bmatrix} I \\ 0 \end{bmatrix} \tag{7.21}$$

Thus, premultiplying both $[CL_1 \quad K_1]$ and $[CL_2 \quad K_2]$ by P, one can obtain the new gross error sizes $\hat{\delta}_2$ and $\tilde{\mu}_2$.

$$\begin{bmatrix} \hat{\delta}_2 \\ \tilde{\mu}_2 \\ 0 \end{bmatrix} = P[CL_1 \quad K_1]\begin{bmatrix} \hat{\delta}_1 \\ \tilde{\mu}_1 \end{bmatrix} \tag{7.22}$$

Example 7.1: To illustrate this procedure consider the system in Fig. 6.3. The true values for this system are $x = [1,2,3,2,1,1,1,0.4,0.6]$ and the standard deviations are 2% of each measurement.

Suppose that the set of gross error $\{S_2, S_3\}$ with sizes $\{1.5, 2.5\}$ is identified and one wants to know the sizes in one of its equivalent sets, for example $\{S_3, S_6\}$.

One has

$$C=\begin{bmatrix}1 & -1 & 0 & 0 & 0 & 1 & 0 & 0 & 0\\ 0 & 1 & -1 & 0 & 0 & 0 & 1 & 0 & 0\\ 0 & 0 & 1 & -1 & 0 & -1 & 0 & 0 & 0\\ 0 & 0 & 0 & 1 & -1 & 0 & -1 & 0 & 0\\ 0 & 0 & 0 & 0 & 1 & 0 & 0 & -1 & -1\end{bmatrix} \tag{7.23}$$

$$L_1=\begin{bmatrix}0 & 0\\ 1 & 0\\ 0 & 1\\ 0 & 0\\ 0 & 0\\ 0 & 0\\ 0 & 0\\ 0 & 0\\ 0 & 0\end{bmatrix} \quad K_1=0 \quad L_2=\begin{bmatrix}0 & 0\\ 0 & 0\\ 1 & 0\\ 0 & 0\\ 0 & 0\\ 0 & 1\\ 0 & 0\\ 0 & 0\\ 0 & 0\end{bmatrix} \quad K_2=0 \tag{7.24}$$

Thus, omitting leaks (the original and equivalent sets contain only biases) we obtain

$$CL_1=\begin{bmatrix}-1 & 0\\ 1 & -1\\ 0 & 1\\ 0 & 0\\ 0 & 0\end{bmatrix} \quad CL_2=\begin{bmatrix}0 & 1\\ -1 & 0\\ 1 & -1\\ 0 & 0\\ 0 & 0\end{bmatrix} \tag{7.25}$$

Premultiplying both sides by P

$$P=\begin{bmatrix}0 & -1 & 0 & 0 & 0\\ 0 & -1 & -1 & 0 & 0\\ 0 & 0 & 0 & 0 & 0\\ 0 & 0 & 0 & 0 & 0\\ 0 & 0 & 0 & 0 & 0\end{bmatrix} \tag{7.26}$$

We get

$$P\,CL_1=\begin{bmatrix}-1 & 1\\ -1 & 0\\ 0 & 0\\ 0 & 0\\ 0 & 0\end{bmatrix} \quad P\,CL_2=\begin{bmatrix}1 & 0\\ 0 & 1\\ 0 & 0\\ 0 & 0\\ 0 & 0\end{bmatrix} \tag{7.27}$$

Let $\hat{\delta}_2$ be given by

$$\hat{\delta}_2=\begin{bmatrix}\hat{\delta}_{2,3}\\ \hat{\delta}_{2,6}\end{bmatrix} \tag{7.28}$$

Then,

$$P\,CL_2\hat{\delta}_2=\begin{bmatrix}\hat{\delta}_{2,3}\\ \hat{\delta}_{2,6}\\ 0\\ 0\\ 0\end{bmatrix} \tag{7.29}$$

Thus, Eq. (7.22) becomes

$$\begin{bmatrix} \hat{\delta}_{2,3} \\ \hat{\delta}_{2,6} \\ 0 \\ 0 \\ 0 \end{bmatrix} = P\, CL_1 \hat{\delta}_1 = \begin{bmatrix} -1 & 1 \\ -1 & 0 \\ 0 & 0 \\ 0 & 0 \\ 0 & 0 \end{bmatrix} \begin{bmatrix} \hat{\delta}_{1,2} \\ \hat{\delta}_{1,3} \end{bmatrix} = \begin{bmatrix} -\hat{\delta}_{1,2} + \hat{\delta}_{1,3} \\ -\hat{\delta}_{1,2} \\ 0 \\ 0 \\ 0 \end{bmatrix} \hat{\delta}_1 = \begin{bmatrix} 1 \\ -1.5 \\ 0 \\ 0 \\ 0 \end{bmatrix} \tag{7.30}$$

The gross error sizes of $\{S_3, S_6\}$ are $\{1.0, -1.5\}$.

Remark: There are other collective compensation schemes also based on successive serial strategies. Sanchez et al. (1999) relies on using the global test for identification, and as in the SICC method, compensation is made at each step. Finally, Bagajewicz and Jiang (2000) provided a collective strategy based on mixed integer linear programming that does not need any iterative strategy.

References

Bagajewicz, M. and Q. Jiang, "Gross Error Modeling and Detection in Plant Linear Dynamic Reconciliation," *Comp. Chem. Eng.* **22**(12):1789–1810 (1998).

Bagajewicz, M. and Q. Jiang, "A Mixed Integer Linear Programming-Based Technique for the Estimation of Multiple Gross Errors in Process Measurements," *Chem. Eng. Comm.* **177**:139–155 (2000).

Devanathan, S., D. K. Rollins, and S. B. Vardeman, "A New Approach for Improved Identification of Measurement Bias," *Comp. Chem. Eng.* **24**:2755–2764 (2000).

Jiang, Q. and M. Bagajewicz, "On a Strategy of Serial Identification with Collective Compensation for Multiple Gross Error Estimation in Linear Data Reconciliation," *Ind. Eng. Chem. Research* **38**(5):2119–2128 (1999).

Rollins, D. K. and J. F. Davis, "Unbiased Estimations of Gross Errors in Process Measurements," *AIChE J.* **38**:563–572 (1992).

Romagnoli, J. A., "On Data Reconciliation-Constraints Processing and Treatment of Bias," *Chem. Eng. Sci.* **38**:1107–1117 (1983).

Sánchez, M., J. Romagnoli, Q. Jiang, and M. Bagajewicz, "Simultaneous Estimation of Biases and Leaks in Process Plants," *Comp. Chem. Eng.* **23**(7):841–858 (1999).

CHAPTER 8

Nonlinear Data Reconciliation

Component Balances

In addition to material balances, component balances can be performed. Assume there are P components. Then the system equations are

$$E_R F_R = 0 \tag{8.1}$$

$$E_R Z_j = 0 \qquad j = 1, \ldots, p \tag{8.2}$$

where the vector Z_j is a vector of products of flowrate and composition, that is, component flow.

Example 8.1: Consider the two streams S_1 and S_2 in Fig. 8.1 and two components

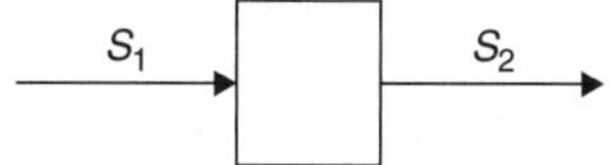

Figure 8.1 Flowsheet of Example 8.1.

The balance equations are

$$\left.\begin{aligned} F_1 &= F_2 \\ F_1 x_{1,1} &= F_2 x_{2,1} \\ F_1 x_{1,2} &= F_2 x_{2,2} \end{aligned}\right\} \tag{8.3}$$

If we write

$$Z_j = \begin{bmatrix} F_1 x_{1,j} \\ F_2 x_{1,j} \end{bmatrix} \qquad j = 1,2 \tag{8.4}$$

then the following represents a balance of components.

$$Z_1 = Z_2 \tag{8.5}$$

The data reconciliation problem for the above example can be written symbolically as follows:

$$\left.\begin{aligned} &\text{Min}\left\{\left[\tilde{F}_R - F_R^+\right]^T S_R^{-1}\left[\tilde{F}_R - F_R^+\right] + \left[\tilde{x}_R - x_R^+\right]^T P_R^{-1}\left[\tilde{x}_R - x_R^+\right]\right\} \\ &\text{s.t.} \\ &\qquad E_R\tilde{F}_R = 0 \\ &\qquad E_R\tilde{Z}_j = 0 \qquad j = 1,\ldots,p \\ &\qquad \tilde{Z}_j = \tilde{F}_R \otimes \tilde{x}_{R,j} \qquad j = 1,\ldots,p \end{aligned}\right\} \tag{8.6}$$

where ⊗ indicates the Hadamard product of the two vectors, that is, $a \otimes b = [a_1b_1 \quad a_2b_2 \ldots \quad a_nb_n]$.

Splitters

A simplification of Eq. (8.6) is possible. Note that component flow balances around splitters reduce to an equation where all the exit concentrations are the same and equal to the concentration of the inlet stream. Thus, one can simplify the constraint equations in Eq. (8.6) to include only those component flow balances that are not made around splitters and introduce equalities for the concentrations around splitters. This is shown next.

$$\left.\begin{aligned} E_R\tilde{F}_R &= 0 \\ K\left[\tilde{F}_R \otimes \tilde{x}_{R,j}\right] &= 0 \qquad j = 1,\ldots,p \\ K_S\tilde{x}_{R,j} &= 0 \qquad j = 1,\ldots,p \end{aligned}\right\} \tag{8.7}$$

where K represents the rows of E_R that do not contain balances around splitters and K_S is a matrix that reflects the equalities of concentration between the inlet stream and the outlet streams of all splitters.

Example 8.2: We use the flowsheets of Fig. 8.2 to illustrate these constraints. These systems represent a flash unit followed by a mixer for Fig. 8.2*a* and a splitter for Fig. 8.2*b*.

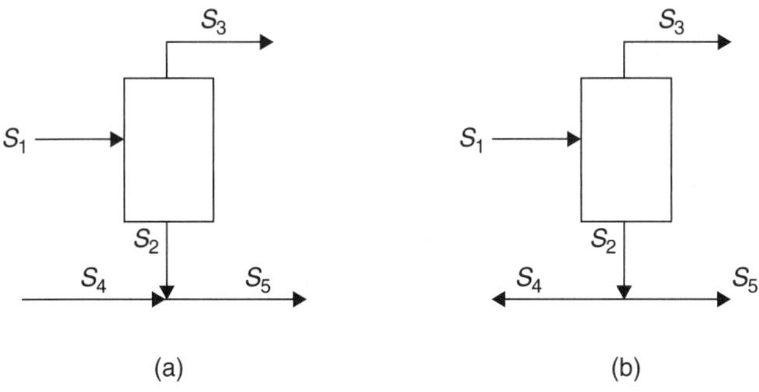

Figure 8.2 Flowsheets of Example 8.2.

The matrices E_R for both systems are

$$E_{R,a} = \begin{array}{c} \begin{array}{ccccc} S_1 & S_2 & S_3 & S_4 & S_5 \end{array} \\ \left[\begin{array}{ccccc} 1 & -1 & -1 & & \\ & 1 & & 1 & -1 \end{array}\right] \end{array} \qquad E_{R,b} = \begin{array}{c} \begin{array}{ccccc} S_1 & S_2 & S_3 & S_4 & S_5 \end{array} \\ \left[\begin{array}{ccccc} 1 & -1 & -1 & & \\ & 1 & & -1 & -1 \end{array}\right] \end{array} \tag{8.8}$$

For the system in Fig. 8.2*a*, we have $K = E_R$ and matrix K_S does not exist because there is no splitter. However, for system in Fig. 8.2*b*, matrices K and K_S are

$$K = \begin{array}{c} \begin{array}{ccccc} S_1 & S_2 & S_3 & S_4 & S_5 \end{array} \\ \left[\begin{array}{ccccc} 1 & -1 & -1 & & \end{array}\right] \end{array} \qquad K_S = \begin{array}{c} \begin{array}{ccccc} S_1 & S_2 & S_3 & S_4 & S_5 \end{array} \\ \left[\begin{array}{ccccc} & 1 & & -1 & \\ & 1 & & & -1 \end{array}\right] \end{array} \tag{8.9}$$

Number of Components

If all components are to be reconciled, then the solution of problem Eq. (8.6) does not guarantee that the compositions of all components will add to one. Thus, normalization equations of the following type should be added:

$$\sum_{j=1}^{p} \tilde{x}_{R,j} = 0 \qquad j = 1, \ldots, p \tag{8.10}$$

Measurement Pattern

One should also notice that the equations in Eq. (8.6) contain redundant flowrates and measured concentrations for the same redundant flowrates. This is not always the case. Indeed, concentrations may be measured in streams for which flows are not measured. The fact that problem Eq. (8.6) includes redundant measured flows only is because an observability analysis on linear systems (Chap. 3) that allows the elimination of observable and unobservable variables.

One can easily show that some measurements of concentration can also help obtain the value of flowrates that would otherwise be unobservable. The converse, however, is not true. This would not be a major problem if redundancy would not be affected, which is not true. The new observable flow can induce a new set of redundant equations, including component flows. Thus redundancy is affected.

Mah (1990) presents a graph approach to the observability analysis of these systems. A detailed analysis of observability for bilinear systems can be found in Bagajewicz (2000). Sánchez and Romagnoli (2000) review all the procedures based on matrix projection (Crowe, 1986; 1989) and Q-R decompositions (Sánchez and Romagnoli, 1996). Narasimhan and Jordache (2000) devote a whole chapter to bilinear data reconciliation.

The issue could be easily settled by rewriting Eq. (8.6) using all variables, including the unmeasured with a large variance (Chapter 4 outlines a method to avoid the classification step). However, this practice has two disadvantages:

a. The presence of unobservable variables could lead to numerical difficulties, including some singularities.

b. Too many variables make the optimization problem too large.

Successive linearizations and straight nonlinear programming techniques (sequential quadratic programming, reduced gradient) are reviewed by Sánchez and Romagnoli (2000) and the impact of a priori classification is discussed. Some commercial softwares perform observability analysis and use nonlinear techniques to solve the problem.

Energy Balances

In addition to material and component balances, energy balances can be added. The energy balance equations are based on enthalpy and are therefore similar to Eq. (8.7).

Typically, temperatures are measured. Thus, the relation between enthalpy flows and temperature is needed. This relation is sometimes expressed assuming constant and known specific heats as follows:

$$h = cp \otimes T \tag{8.11}$$

Thus, a system of equations similar to Eq. (8.7) can be written as follows:

$$\left.\begin{aligned} E_R\tilde{F}_R &= 0 & & \\ K\left[\tilde{F}_R \otimes cp_{R,j} \otimes \tilde{T}_{R,j}\right] &= 0 & \quad j &= 1,\ldots,p \\ K_S\tilde{T}_{R,j} &= 0 & \quad j &= 1,\ldots,p \end{aligned}\right\} \tag{8.12}$$

Heat Exchangers

Heat exchangers obey entirely different balances and cannot be represented by the same flow balance equations. In particular, one has to write two algebraic balance equations for each heat exchanger and one algebraic heat balance for it. Thus, assume

$$C = \begin{bmatrix} C_F \\ C_D \end{bmatrix} \tag{8.13}$$

where C_F corresponds to heat exchangers and any other equipment where heat is exchanged indirectly and C_D corresponds to the rest of the units in the system where heat is directly exchanged, except splitters. Notice that, unlike in flow systems, each heat exchanger requires two rows of C_D, one material balance for each exchanger side. We now construct a new matrix, M_H, by simply adding the two rows containing material balances on heat each exchanger. Then the material and energy equations corresponding to units, where indirect heat transfer takes places, are written as follows:

$$M_D F = 0 \tag{8.14}$$

$$M_H [F \otimes h] = 0 \tag{8.15}$$

Example 8.3: We illustrate the structure of this matrix through the following small example. Consider one heat exchanger.

In this case the corresponding matrices are

$$\begin{matrix} & S_1 & S_2 & S_3 & S_4 \end{matrix}$$

$$C_D = \begin{bmatrix} 1 & -1 & & \\ & & 1 & -1 \end{bmatrix} \tag{8.16}$$

$$\begin{matrix} & S_1 & S_2 & S_3 & S_4 \end{matrix}$$

$$C_H = \begin{bmatrix} 1 & -1 & -1 & 1 \end{bmatrix} \tag{8.17}$$

which will render one energy balance.

Example 8.4: We now illustrate how successive linearization works. Consider a heat exchanger, as in Fig. 8.3. The equations of the model are

$$F_3 = F_4 \tag{8.18}$$

$$F_1 = F_2 \tag{8.19}$$

$$Q = F_3 Cp_3 (T_3 - T_4) \tag{8.20}$$

$$Q = F_1 Cp_1 (T_2 - T_1) \tag{8.21}$$

$$Q = UAF_c \left[\frac{(T_3 - T_2) - (T_1 - T_4)}{\ln \dfrac{(T_3 - T_2)}{(T_1 - T_4)}} \right] \tag{8.22}$$

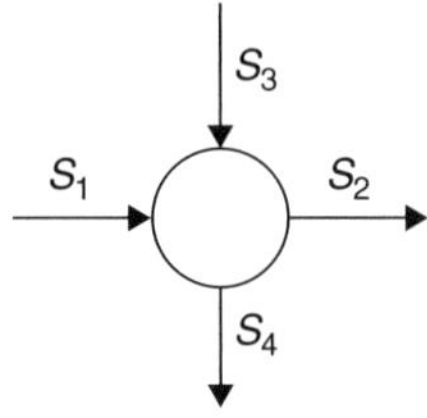

Figure 8.3 Heat exchanger.

Let us focus on Eq. (8.22) first. In this equation the heat transfer coefficient U is unknown and impossible to measure. Therefore, if all the four temperatures are known, its value can be estimated. Therefore, the heat transfer coefficient is at the most observable, but never a redundant variable. We therefore omit Eq. (8.22) from further consideration.

If all temperatures and all flowrates are measured, this system has all its remaining equations redundant. Indeed, we eliminate Q from Eqs. (8.20) and (8.21).

$$F_1 Cp_1 (T_2 - T_1) = F_3 Cp_3 (T_3 - T_4) \tag{8.23}$$

Linearization of the left-hand side gives

$$\begin{aligned} F_1 Cp_1 (T_2 - T_1) = {} & \tilde{F}_1^{(k)} Cp_1 \left(\tilde{T}_2^{(k)} - \tilde{T}_1^{(k)}\right) + \tilde{F}_1^{(k)} \left[\left(T_2 - \tilde{T}_2^{(k)}\right) - \left(T_1 - \tilde{T}_1^{(k)}\right)\right] \\ & + \left(F_1 - \tilde{F}_1^{(k)}\right) Cp_1 \left(\tilde{T}_2^{(k)} - \tilde{T}_1^{(k)}\right) \end{aligned} \tag{8.24}$$

where $\{\tilde{F}_1^{(k)}, \tilde{F}_3^{(k)}, \tilde{T}_1^{(k)}, \tilde{T}_2^{(k)}, \tilde{T}_3^{(k)}$, and $\tilde{T}_4^{(k)}\}$ are the reconciled values at iteration (k). A similar expression can be obtained for the right-hand side. If both expressions are substituted in Eq. (8.23), we obtain

$$\tilde{F}_1^{(k)} Cp_1 (T_2 - T_1) + F_1 Cp_1 \left(\tilde{T}_2^{(k)} - \tilde{T}_1^{(k)}\right) = \tilde{F}_3^{(k)} Cp_3 (T_3 - T_4) + F_3 Cp_3 \left(\tilde{T}_3^{(k)} - \tilde{T}_4^{(k)}\right) \tag{8.25}$$

The linearized reconciliation problem becomes

$$\left.\begin{aligned} & \text{Min}\left\{\left[\tilde{F}_R^{(k+1)} - F_R^+\right]^T S_R^{-1}\left[\tilde{F}_R^{(k+1)} - F_R^+\right] + \left[\tilde{T}_R^{(k+1)} - T_R^+\right]^T P_R^{-1}\left[\tilde{T}_R^{(k+1)} - T_R^+\right]\right\} \\ & \text{s.t.} \\ & \tilde{F}_3^{(k+1)} = \tilde{F}_4^{(k+1)} \\ & \tilde{F}_1^{(k+1)} = \tilde{F}_2^{(k+1)} \\ & \tilde{F}_1^{(k+1)} Cp_1 \left(\tilde{T}_2^{(k)} - \tilde{T}_1^{(k)}\right) + \tilde{F}_2^{(k)} Cp_1 \left(\tilde{T}_2^{(k+1)} - \tilde{T}_1^{(k+1)}\right) = \tilde{F}_3^{(k+1)} Cp_1 \left(\tilde{T}_3^{(k)} - \tilde{T}_4^{(k)}\right) \\ & \qquad + \tilde{F}_3^{(k)} Cp_3 \left(\tilde{T}_3^{(k+1)} - \tilde{T}_4^{(k+1)}\right) \end{aligned}\right\} \tag{8.26}$$

This system can be solved with the tools of Chap. 4. Successive applications of the same minimization problem should render the proper solution.

Full Nonlinear Systems

A general nonlinear system of equations can be classified using reduction to canonical forms, matrix projection or Q-R factorization and other methods. If one linearizes the system ($F(x) = 0$) around the design or operating point and ignores terms of higher order (Swartz, 1989; Romagnoli, 1983; Joris and Kalitventzeff, 1987; Maquin et al., 1991; Meyer et al., 1993; Crowe, 1996), one obtains

$$\left(\frac{\partial F(x)}{\partial x}\right)_{x=x_0} x = F(x_0) - \left(\frac{\partial F(x)}{\partial x}\right)_{x=x_0} x_0 \tag{8.27}$$

which can be rewritten symbolically as follows

$$D\,x \cong b \tag{8.28}$$

where matrix D represents the Jacobian of $F(x)$ around x_0 and b is the corresponding constant.

$$b = F(x_0) - \left(\frac{\partial F(x)}{\partial x}\right)_{x=x_0} x_0 \tag{8.29}$$

Remark: Defining $z = x - \alpha$, where α is such that

$$\left(\frac{\partial F(x)}{\partial x}\right)_{x=x_0} \alpha = F(x_0) - \left(\frac{\partial F(x)}{\partial x}\right)_{x=x_0} x_0 \tag{8.30}$$

The system of Eq. (8.27) becomes

$$\left(\frac{\partial F(x)}{\partial x}\right)_{x=x_0} z = 0 \tag{8.31}$$

which has the required form that allows using the equations of Chap. 4. To avoid numerical problems, the new variables, z, can be scaled so that $\underset{\forall j}{\text{Max}} \left|A_{ij}\right| = 1, \quad \forall i.$

Example 8.5: We consider the CSTR process which was introduced by Bhushan and Rengaswamy (2000) shown in Fig. 8.4.

Material balance equation in the reactor for reactant i is

$$\frac{F_i}{V}(c_{Ai} - c_A) - c_d c_A k_0 e^{-\frac{E}{RT}} = 0 \tag{8.32}$$

Energy balance equation in the reactor is

$$\frac{F_i}{V}(T_i - T) + \frac{c_d c_A k_0 e^{-\frac{E}{RT}}(-\Delta H)}{\rho C_p} - \frac{UA(T - T_c)}{V\rho C_p} = 0 \tag{8.33}$$

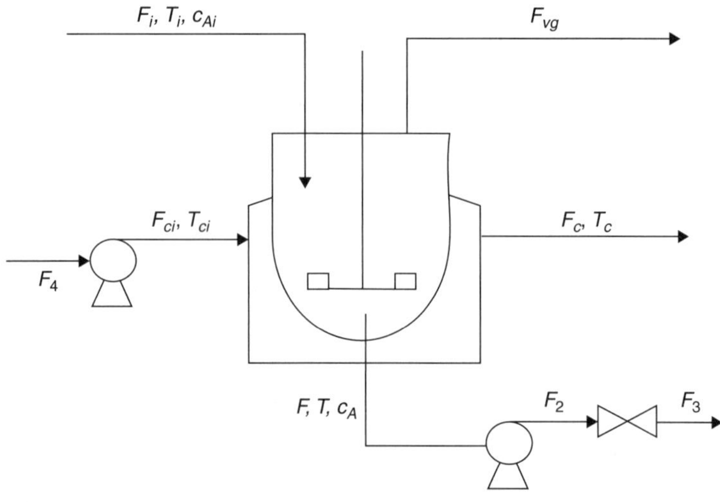

FIGURE 8.4 The CSTR problem.

Energy balance equation in the jacket is

$$\frac{F_c}{V_j}(T_{ci} - T_c) + \frac{UA(T - T_c)}{V_j \rho_j C_{pj}} = 0 \tag{8.34}$$

Reaction rate is

$$c_d c_A k_0 e^{-\frac{E}{RT}} V - F_{vg} = 0 \tag{8.35}$$

Overall material balance equation in the reactor is

$$F_i - F = 0 \tag{8.36}$$

Material balance equation in the pump is

$$F_2 - F = 0 \tag{8.37}$$

Material balance equation in the valve is

$$F_3 - F_2 = 0 \tag{8.38}$$

Material balance equation in the jacket is

$$F_4 - F_c = 0 \tag{8.39}$$

The nominal operation conditions are given in Table 8.1 (value of flowrate is given in ft³/h, temperature: °R, concentration: lb · mole/ft³).

Variable	F_i	F_c	F_{vg}	F	F_2	F_3	F_4
Value	40	56.626	10.614	40	40	40	56.626
Variable	c_{Ai}	c_A	T	T_i	T_c	T_{ci}	
Value	0.5	0.2345	600	530	590.51	530	

TABLE 8.1 Nominal Operating Condition for the CSTR Example

The linearized model matrix is

$$
D = \begin{pmatrix}
F_i & c_{Ai} & c_A & T & T_i & T_c & F_c & T_{ci} & F_{vg} & F & F_2 & F_3 & F_4 \\
-0.00531 & -0.8333 & 1.7763 & 0.00923 & 0 & 0 & 0 & 0 & 0 & 0 & 0 & 0 & 0 \\
1.4583 & 0 & -754.4 & 5.9503 & -0.8333 & -125 & 0 & 0 & 0 & 0 & 0 & 0 & 0 \\
0 & 0 & 0 & -93.8067 & 0 & 108.5 & 15.7169 & -14.708 & 0 & 0 & 0 & 0 & 0 \\
0 & 0 & -45.2612 & -0.443 & 0 & 0 & 0 & 0 & 1 & 0 & 0 & 0 & 0 \\
1 & 0 & 0 & 0 & 0 & 0 & 0 & 0 & 0 & -1 & 0 & 0 & 0 \\
0 & 0 & 0 & 0 & 0 & 0 & 0 & 0 & 0 & 0 & 1 & -1 & 0 \\
0 & 0 & 0 & 0 & 0 & 0 & 0 & 0 & 0 & 1 & -1 & 0 & 0 \\
0 & 0 & 0 & 0 & 0 & 0 & -1 & 0 & 0 & 0 & 0 & 0 & 1
\end{pmatrix}
$$

Matrix D may be partitioned in submatrices D_M and D_U, which are related to the vector of measured variables (x_M) and unmeasured parameters and state variables (x_U), respectively.

$$
[D_M \quad D_U]\begin{bmatrix} x_M \\ x_U \end{bmatrix} = b \tag{8.40}
$$

Thus, if one uses Eq. (8.28), the system can be solved using the methods of Chap. 4. Successive linearizations can be used to adjust the value of b.

The reconciliation problem is written as follows:

$$
\left.\begin{aligned}
&\text{Min}\,\frac{1}{2}(\tilde{x}_M - x_M^+)^T Q^{-1}(\tilde{x}_M - x_M^+) \\
&\text{s.t.} \\
&\quad D_M\tilde{x}_M + D_U\tilde{x}_U = b
\end{aligned}\right\} \tag{8.41}
$$

where x_M^+ and Q are the measured flowrates and their variance, respectively. The assumption here is that unmeasured and unobservable quantities have been removed from x_U, that is, variable classification was performed. The solution to this problem is

$$
\tilde{x}_M = x_M^+ - QA^T(I - G^{-1}BH^{-1}B^T)G^{-1}\left(-b + D_M x_M^+\right) \tag{8.42}
$$

$$
\tilde{x}_U = H^{-1}B^T G^{-1}\left(-b + D_M x_M^+\right) \tag{8.43}
$$

where $G = D_M Q D_M^T$ and $H = D_U^T G^{-1} D_U$. Since matrix D_U has been assumed to have a rank equal to the number of unmeasured quantities, H is nonsingular. If not, there are unobservable variables in x_U, and a model reduction has to be performed to obtain a canonical form (Madron, 1992; Veverka and Madron, 1997). The variance of the reconciled results is

$$
\tilde{Q}_U = H^{-1} \tag{8.44}
$$

$$
\tilde{Q}_M = Q - QD_M^T(I - G^{-1}BH^{-1}B^T)G^{-1}D_M Q \tag{8.45}
$$

and the precision of each variable is

$$
\sigma_i = \begin{cases} \sqrt{\left[\tilde{Q}_M\right]_{ii}} & \text{if } x_i \text{ is measured} \\ \sqrt{\left[\tilde{Q}_U\right]_{ii}} & \text{otherwise} \end{cases} \tag{8.46}
$$

Remark: Very early, Stanley and Mah (1977) proposed a method to obtain flows and temperatures in systems featuring material and energy balances using a quasi-steady-state estimator based on the Kalman filter.

Historical Note: According to Narasimhan and Jordache (2000) the first to propose the nonlinear data reconciliation problem were Knepper and Gorman (1980).

Remark: Romagnoli (1983) also discussed the simultaneous treatment of bias. Many of the aforementioned methods use an iterative procedure with updates on the Jacobian. However, Kyriakopoulou and Kalitvenzeff (1996) proposed the use of a sequential quadratic algorithm. Kelly (1998) uses a successive linearization approach to solve the nonlinear problem and discusses several implementation issues. As in Chap. 5, the method does not require prior classification and can distinguish observables from unobservables. Finally, Ragot and Maquin (2004) analyze the case of nonlinear data reconciliation when there are detection limits.

Remark: Observability and redundancy analysis can be made using several of the methods cited for linear systems in Chap. 3. Methods dealing directly with nonlinear systems involve the work of Kretsovalis and Mah (1987, 1988a, b) and Ragot et al. (1996) who use graph theory, Ragot et al. (1990) and Albuquerque and Biegler (1996) who use a matrix algebra-based methods after linearization. In addition, as we mentioned in Chap. 3, the approaches of Ponzoni et al. (1999) based on structural rearrangement of the occurrence matrix, Ferraro et al. (2002) with their symbolic derivation to perform redundancy analysis, followed by the work of Ponzoni et al. (2004) and Carballidoa et al. (2009) provide meaningful alternatives to linearization. Finally, we mentioned in Chap. 3 that Kelly (1999) proposed an alternative method that allows easy updating when measurements are removed. This method is useful for the nonlinear case also.

Remark: While observability analysis in nonlinear systems is better made using the direct methods proposed by Ponzoni et al. (1999), Ferraro et al. (2002), Ponzoni et al. (2004), Carballidoa et al. (2009), Domancich et al. (2009), and Kelly (1998), one can always attempt to use the method of assigning a large variance to unmeasured variables (explained in Chap. 4), which avoids classification.

Remark: Aside from all the "robust" estimators cited in Chap. 5 with applications to linear systems, which can be used for the nonlinear case, there is specific work for nonlinear systems (Hu and Shao, 2006).

Remark: Some additional efforts to perform data reconciliation in nonlinear systems efficiently are worth mentioning: Gau and Stadtherr (2002) focused on obtaining the global optimum. They used the interval-Newton approach, and although typically regarded as being applicable only to very small problems, they successfully applied it to problems with over 200 variables in examples such as a reactor, for vapor-liquid equilibrium, and in a heat exchanger network. Kim et al. (1997) proposed to use the modified iterative measurement test (MIMT), developed by Serth and Heenan (1986) in the context of nonlinear data reconciliation. The MIMT simply makes sure that there are not unreasonable estimators produced and made corrections. The algorithm was tested on a CSTR example and shows improved robustness compared to existing gross error detection algorithms. Therefore this enhanced algorithm appears to be quite promising for data reconciliation and gross error detection of highly nonlinear processes in chemical engineering.

Remark: Bilinear problems deserve a special mention: Simpson et al. (1988) proposed a method that eliminates all constraints and an equal number of variables. The compositions in the objective function (least squares) are replaced by the corresponding stream flowrate and component flowrate by using a first-order Taylor series approximation. All flowrate variables are substituted in terms of the corresponding stream component flows by using the normalization equation. Finally, a set of independent component flow variables is chosen, and all component flows in the objective function are written in terms of these by using the component balances. Flament and Hodouin (1985) presented a material balance technique that was later used recursively for steady state data reconciliation in mineral processing (Makni and Hodouin, 1994). Rao and Narasimhan (1996) tested this method against the matrix projection method for nonlinear cases (Crowe, 1986) and showed it is superior in accuracy and efficiency. Dovi and Del Borghi (1999) looked into the bilinear case where detection limits exist. Problems with bilinear terms also offer the opportunity to obtain analytical expressions and other functions for Jacobians. This was attempted by Schraa and Crowe (1998) and later by Kellly (2004a).

Remark: Kelly (2004b) presented some useful techniques to help practitioners solve industrial and nonlinear reconciliation problems in the process industries. Because many of these problems are large and contain bilinear and nonconvex terms, four different solving methods are proposed to overcome these difficulties. Two different ways to generate starting values for the unmeasured values were introduced as well as two simple techniques to overcome ill-conditioning of the kernel matrix and poor convergence problems due to inconsistency were highlighted.

Remark: There are several techniques based on the use of principal component analysis and so-called "residuals," which we will briefly mention in Chap. 9 because they employ dynamic data.

Gross Error Detection in Nonlinear Systems

Gross error detection in nonlinear systems is usually performed assuming that the model has been linearized, and therefore all techniques presented in Chap. 5 can be implemented. There are exceptions: Rollins and Roelfs (1992) treated the case of gross error detection when bilinear terms are present. Departing from making the linearization assumption, Renganathan and Narasimhan (1999) presented a technique similar to the GLR technique that, although lacks statistical basis, seems to work well. For many errors they employ a serial compensation strategy with better success than a serial elimination one. Amand et al. (2001) used data reconciliation and principal component analysis for simultaneous fault detection.

There is also a whole class of alternative methods introduced in Chap. 5, like neural network-based methods (Gupta and Narasimhan,

1993; Terry and Himmelblau, 1993; Aldrich and van Deventer, 1994, 1995; Du et al., 1997; Reddy and Mavrovouniotis, 1998). Finally, Sánchez et al. (2008) introduced a new data driven technique based on a reformulation of the D-Statistic (Alvarez et al., 2007). The statistics is given by $D = (z - \bar{z})^T S^{-1}(z - \bar{z})$, where z is the vector of measurements, $\bar{z}$ the average of a certain window and S the corresponding variance-covariance. The method makes use of a critical value of D for detection, which is also data-driven.

Robust estimators have also become popular (Kim et al., 1990; Tjoa and Biegler, 1991; Soderstrom et al., 2001; Wang and Romagnoli, 2003; Ozyurt and Pike, 2004; Zhao et al., 2004; Ragot et al., 2005; Morad et al., 2005; Alhaj-Dibo et al., 2008). In particular, Chen et al. (1998) showed that isolated outliers can be identified by directly examining the measurement distribution. They use a nonlinear limiting transformation on the data set to eliminate or reduce the influence of these outliers. Arora and Biegler (2001) consider maximum likelihood estimators (M estimators), namely, Huber estimators and Hampel estimators. The first one significantly reduces the effect of large outliers and the second nullifies their effect. They use an influence function, to quantify their effect. They compare the use of the three part redescending estimator of Hampel with a Huber estimator. They use the Akaike information criterion, (AIC) to tune the redescending estimators to the particular observed system data through an iterative procedure, a procedure that is not easily afforded by the Huber estimators. Later, a generalized objective function was introduced by Wang and Romagnoli (2003) and used on a heat exchanger network. Morad et al. (2005) followed up with a Bayesian approach that renders the product of a prior probability density function (pdf) of process states and the probability distribution of measurements given the true process states. The first pdf, given the measured data condition, is estimated by the expectation–maximization (EM) algorithm or adaptive mixtures. The second pdf is modeled by the product of bimodal Gaussian distributions one for each sensor. Wongrat et al. (2005) used a genetic algorithm on a redescending estimator to solve the steady-state nonlinear data reconciliation (DR) problem using

Remark: Except from methods that are based on bivariate distributions, there is some doubt expressed about the use of some of these estimators. Mah (1997) expressed some of these doubts in regard to the approach of Tjoa and Biegler (1991). Biegler and Tjoa (1993) responded addressing the statistical basis of his method, and agreeing on the need for such statistical basis in these methods.

Parameter Estimation

Models that are used for operations optimization require the value of parameters. Examples of such parameters in process plants are catalysts' activities/selectivities, compressor efficiencies, heat exchanger fouling, plate efficiencies in columns, etc. Because all these are not measured, they must be inferred from other measurements, or better, from estimators such as those obtained through data reconciliation, when the parameters are observable. One could attempt to solve the data reconciliation problem using a linearized model or a full blown nonlinear model. McDonald and Howat (1988) proposed a nonlinear data reconciliation of a single stage flash unit. There are, however, earlier attempts focused on parameter estimation and error-in-variables methods (Deming, 1943; Britt and Luecke, 1973) as well as methods after 1980 in the same area (Reilly and Patino-Leal, 1981; Valko and Vadja, 1987). Ricker (1984) evaluated the efficiency of three maximum likelihood estimators. Pai and Fisher (1988) proposed a successive linearization method based on Broyden's updates that works better than the Gaus-Newton-based method proposed by Knepper and Gorman (1980). This was followed by the work of Stephenson and Shewchuck (1986) and Serth et al. (1987), who used a newton-raphson iterative method. Later Tjoa and Biegler (1991) used successive quadratic programming. Stewart et al. (1992) review several Bayesian and likelihood methods for parameter estimation, which can be used for planta parameter estimation (they illustrated their model using a chemical kinetic network). Improving over their earlier work, Biegler and Tjoa (1993) used a special implementation of an SQP based on a decomposition of the problem. Pages et al. (1994) used the aid of simulation to infer the value of parameters. Departing from the use of data reconciliation and classical gross error detection methods, such as thodse presented in Chap. 5, Krishnan et al. (1992, 1993) used a series of so-called "sieves" based on structural analysis (observability) and singular value analysis to get the confidence region of the parameters.

References

Albuquerque, J. S. and L. T. Biegler, "Data Reconciliation and Gross-Error Detection for Dynamic Systems," *AIChE J.* **42**(10):2841 (1996).

Aldrich, C. and J. S. J van Deventer, "Identification of Gross Errors in Material Balance Measurements by Means of Neural Nets," *Chem. Eng. Sci.* **49:**1357–1368 (1994).

Aldrich, C. and J. S. J van Deventer, "Comparison of Different Neural Nets for the Detection and Location of Gross Errors in Process Systems," *Ind. Eng. Chem. Res.* **34:**216–224 (1995).

Alhaj-Dibo, M., D. Maquin, and J. Ragot, "Data reconciliation: A robust approach using a contaminated distribution," *Control Eng. Practice* **16**(2):159–170 (2008).

Alvarez, R., A. Brandolin, and M. Sánchez, "On the Variable Contributions to the D-Statistic," *Chemometr. Intell. Lab. Syst.* **88:**89–196 (2007).

Amand, Th., G. Heyen, and B. Kalitventzeff, "Plant Monitoring and Fault Detection: Synergy between Data Reconciliation and Principal Component Analysis," *Comp. Chem. Eng.* **25:**501–507 (2001).

Arora, N. and L. T. Biegler, "Redescending Estimators for Data Reconciliation and Parameter Estimation," *Comp. Chem. Eng.* **25:**1585–1599 (2001).

Bagajewicz, M., *Design and Upgrade of Process Plant Instrumentation*, ISBN:1-56676-998-1, Technomic Publishing Company (http://www.techpub.com) (2000).

Bhushan, M. and R. Rengaswamy, "Design of Sensor Location Based on Various Fault Diagnosis Observability and Reliability Criteria," *Comp. Chem. Eng.* **24:**735 (2000).

Biegler, L. T. and I. Tjoa, "A Parallel Implementation for Parameter Estimation with Implicit Models," *Ann. Oper. Res.* **42:**1–23 (1993).

Britt, H. I. and R. H. Luecke, "The Estimation of Parameters in Non-Linear Implicit Models," *Technometrics* **15:**223–247 (1973).

Carballidoa, J. A., I. Ponzoni, and N. B. Brignole, "SID-GA: An Evolutionary Approach for Improving Observability and Redundancy Analysis in Structural Instrumentation Design," *Comp. Ind. Eng.* **56:**1419–1428 (2009).

Chen, J., A. Bandoni, and J. A. Romagnoli, "Outlier Detection in Process Plant Data," *Comp. Chem. Eng.* **22:**641–646 (1998).

Crowe, C. M., "Reconciliation of Process Flowrates by Matrix Projection. Part II: The Nonlinear Case," *AIChE J.* **32:**616–623 (1986).

Crowe, C. M., "Observability and Redundancy of Process Data for Steady State Reconciliation," *Chem. Eng. Sci.* **44:**2908–2917 (1989).

Crowe, C. M., "Data Reconciliation—Progress and Challenges," *J. Proc. Control* **6:**89–98 (1996).

Deming, W. E., *Statistical Adjustment of Data*, Wiley, New York (1943).

Domancich, A. O., M. Durante, S. Ferraro, P. Hoch, N. B. Brignole, and I. Ponzoni, "How to Improve the Model Partitioning in a DSS for Instrumentation Design," *Ind. Eng. Chem. Res.* **48**:3513-3525 (2009).

Dovi, V. G. and A. Del Borghi, "Reconciliation of Process Flow Rates when Measurements Are Subject to Detection Limits: The Bilinear Case," *Ind. Eng. Chem. Res.* **38**:2861–2866 (1999).

Du, Y., D. Hodouin, and J. Thibault, "Use of a Novel Autoassociative Neural Network for Nonlinear Steady State Data Reconciliation," *AIChE J.* **43:**1785–1796 (1997).

Ferraro, S. J., I. Ponzoni, M. C. Sánchez, and N. B. Brignole, "A Symbolic Derivation Approach for Redundancy Analysis," *Ind. Eng. Chem. Res.* **41**(23):5692–5701 (2002).

Flament, F. and D. Hodouin, "BILMAT Computer Program for Material Balance of Mineral Processing Circuits," *SPOC Manuals, CANMET, Special Report* SP85-31/E, Energy, Mines, and Resources Canada, Ottawa, Ontario **140** (1985).

Gau, C. and M. A. Stadtherr, "Deterministic Global Optimization for Error-in-Variables Parameter Estimation," *AIChE J.* **48:**1192–1197 (2002).

Gupta, G. and S. Narasimhan, "Application of Neural Networks for Gross Error Detection," *Ind. Eng. Chem. Res.* **32:**1651–1657 (1993).

Hu, M. and H. Shao, "Theory Analysis of Nonlinear Data Reconciliation and Application to a Coking Plant," *Ind. Eng. Chem. Res.* **45:**8973–8984 (2006).

Joris, P. and B. Kalitventzeff, "Process Measurement Analysis and Validation," Proceedings of CEF'87: *Computers and Chemical Engineering*, pp. 41–46 (1987).

Kelly, J. D., "A Regularization Approach to the Reconciliation of Constrained Data Sets," *Comp. Chem. Eng.* **22:**1771–1788 (1998).

Kelly, J. D., "Reconciliation of Process Data Using Other Projection Matrices," *Comp. Chem. Eng.* **23:**785–789 (1999).

Kelly, J. D., "Formulating Large-Scale Quantity–Quality Bilinear Data Reconciliation Problems," *Comp. Chem. Eng.* **28:**357–362 (2004a).

Kelly, J. D., "Techniques for Solving Industrial Nonlinear Data Reconciliation Problems," *Comp. Chem. Eng.* **28:**2837–2843 (2004b).

Kim, I., W. Liebmann, and T. F. Edgar, "Robust Error-in-Variable Estimation Using Nonlinear Programming Techniques," *AIChE J.* **36:**985–993 (1990).

Kim, I., M. S. Kang, S. Park, and T. F. Edgar, "Robust Data Reconciliation and Gross Error Detection: The Modified MIMT Using NLP," *Comp. Chem. Eng.* **21**(7):775–782 (1997).

Knepper, J. C. and J. W. Gorman, "Statistical Analysis of Constrained data," *AIChE J.* **26:**260–264 (1980).

Kretsovalis, A. and R. S. H. Mah, "Observability and Redundancy Classification in Multicomponent Process Networks," *AIChE J.* **33:**70–82 (1987).
Kretsovalis, A. and R. S. H. Mah, "Observability and Redundancy Classification in generalized Process Networks II. Theorems," *Comp. Chem. Eng.* **12:**671–687 (1988a).
Kretsovalis, A. and R. S. H. Mah, "Observability and Redundancy Classification in Generalized Process Networks II. Algorithms," *Comp. Chem. Eng.* **12:**689–703 (1988b).
Krishnan, S., G. W. Barton, and J. D. Perkins, "Robust Parameter Estimation in Online Optimization—Part I. Methodology and Simulated Case Study," *Comp. Chem. Eng.* **16:**545–562 (1992).
Krishnan, S., G. W. Barton, and J. D. Perkins, "Robust Parameter Estimation in Online Optimization—Part I. Application to an Industrial Process," *Comp. Chem. Eng.* **17:**663–669 (1993).
Kyriakopoulou, D. J. and B. Kalitvenzeff, "Validation of Measurement Data Using an Interior Point SQP," *Comp. Chem. Eng.* **20**(suppl.):S563–S568 (1996).
Madron, F., *Process Plant Performance,* Ellis Horwood, Chichester, England (1992).
Mah, R. S. H., *Chemical Process Structures and Information Flows,* Butterworths, Boston (1990).
Mah, R. S. H., "Letter to the Editor," *Comp. Chem. Eng.* **21:**1069 (1997).
Makni, S. and D. Hodouin, "Recursive Bilmat Algorithm: An On-Line Extension of Data Reconciliation Techniques for Steady-State Bilinear Material Balance" *Minerals Engineering* **7**:1179-1191 (1994).
Maquin, D., G. Bloch, and J. Ragot, "Data Reconciliation of Measurements," *Diagnostic et Sûreté de Fonctionnement* **1:**145–181 (1991).
McDonald, R. J. and C. S. Howat, "Data Reconciliation and Parameter Estimation in Plant Performance Analysis," *AIChE J.* **34:**1–8 (1988).
Meyer, M., B. Koehert, and M. Enjalbert, "Data Reconciliation on Multicomponent Network Processes," *Comp. Chem. Eng.* **17:**807–817 (1993).
Morad, K., B. R. Young, and W. Y. Svrcek, "Rectification of Plant Measurements Using a Statistical Framework," *Comp. Chem. Eng.* **29:**919–940 (2005).
Narasimhan, S. and C. Jordache, Data Reconciliation and Gross Error Detection: An Intelligent Use of Process Data, Gulf Publishing Company, Houston (2000).
Ozyurt D. B., R. W. Pike "Theory and Practice of Simultaneous Data Reconciliation and Gross Error Detection of Chemical Processes," *Comp. Chem. Eng.* **28:**381–402 (2004).
Pages, A., H. Pingaud, M. Meyer, and X. Joulia, "A Strategy for Simultaneous Data Reconciliation and Parameter Estimation on Process Flowsheets," *Comp. Chem. Eng.* **18**(suppl.):S223–S227 (1994).
Pai, C. C. D. and G. D. Fisher, "Application of Broyden's Method to Reconciliation of Nonlinearly Constrained Data," *AIChE J.* **34:**873–876 (1988).
Ponzoni, I., M. C. Sánchez, and N. B. Brignole, "A New Structural Algorithm for Observability Classification," *Ind. Eng. Chem. Res.* **38**(8):3027–3035 (1999).
Ponzoni, I., M. C. Sánchez, and N. B. Brignole, "Direct Method for Structural Observability Analysis," *Ind. Eng. Chem. Res.* **43**(2):577–588 (2004).
Ragot, J., M. Luong, and D. Maquin, "Observability of Systems Involving Flow Circulation," *Int. J. Mineral Processing* **4:**125–140 (1996).
Ragot, J., D. Maquin, G. Bloch, and W. Gomolka, "Observability and Variables Classification in Bilinear Processes," *Benelux Quarterly J. Automatic Control, Journal A* **31**:17–23 (1990).
Ragot, J. and D. Maquin, "Reformulation of Data Reconciliation Problem with Unknown but-Bounded Errors," *Ind. Eng. Chem. Res.* **43:**1530–1536 (2004).
Ragot, J., M. Chadli, and D. Maquin, "Mass Balance Equilibration: A Robust Approach Using Contaminated Distribution," *AIChE J.* **51:**1569–1575 (2005).
Rao, R. and S. Narasimhan, "Comparison of Techniques for Data Reconciliation of Multicomponent Processes," *Ind. Eng. Chem. Res.* **35:**1362–1368 (1996).
Reddy, V. N. and M. L. Mavrovouniotis, "An Input-Training Neural Network Approach for Gross Error Detection and Sensor Replacement," *Trans. IChemE* **76**(part A):478–489 (1998).

Reilly, P. M. and H. Patino Leal, "A Bayesian Study on the Error-in-Variable Models," *Technometrics* **23:**221–231 (1981).

Renganathan, T. and S. Narasimhan, "A Strategy for Detection of Gross Errors in Nonlinear Processes," *Ind. Eng. Chem. Res.* **38:**2391–2399 (1999).

Ricker, N., "Comparison of Methods for Nonlinear Parameter Estimation," *Ind. Eng. Chem. Process Des. Dev.* **23:**283–286 (1984).

Rollins, D. K. and S. D. Roelfs, "Gross Error Detection when Constraints Are Bilinear," *AIChE J.* **38**(8):1295–1298 (1992).

Romagnoli, J. A., "On Data Reconciliation—Constraints Processing and Treatment of Bias," *Chem. Eng. Sci.* **38:**1107–1117 (1983).

Sánchez, M. and J. Romagnoli, "Use of Orthogonal Transformations in Data Classification-Reconciliation," *Comp. Chem. Eng.* **20:**483–493 (1996).

Sánchez, M. and J. Romagnoli, *Data Processing and Reconciliation for Chemical Process Operations*, Academic Press, San Diego (2000).

Sánchez, M., C. R. Alvarez, and A. Brandolin, "A Multivariate Statistical Process Control Procedure for Bias Identification in Steady-State Processes," *AIChE J.* **54:**2082–2088 (2008).

Schraa, O. J. and C. M. Crowe, "The Numerical Solution of Bilinear Data Reconciliation Problems Using Constrained Optimization Methods," *Comp. Chem. Eng.* **22:** 1215–1228 (1998).

Serth, R. W. and W. A. Heenan, "Gross Error Detection and Data Reconciliation in Steam-Metering Systems," *AIChE J.* **32:**733–742 (1986).

Serth, R. W., C. M. Valero, and W. A. Heenan, "Detection of Gross Errors in Nonlinearly Constrained Data: A Case Study," *Chem. Eng. Comm.* **51:**89–104 (1987).

Simpson, D. E., M. G. Everett, and V. R. Voller, "Reducing the Number of Unknowns in a Constrained Minimisation Problems an Application to Material Balances," *Appl. Math. Modeling* **12:**204–212 (1988).

Soderstrom T. A., D. M. Himmelblau, T. F. Edgar, "A Mixed Integer Optimization Approach for Simultaneous Data Reconciliation and Identification of Measurement Bias," *Control Engineering Practice* **9**(8)**:**869–876 (2001).

Stanley, G. M. and R. H. S. Mah, "Estimation of Flows and Temperatures in Process Networks," *AIChE J.* **23:**642–650 (1977).

Stephenson, G. R. and C. F. Shewchuck, "Reconciliation of Process Data with Process Simulation," *AIChE J.* **32:**247–254 (1986).

Stewart, W. E., M. Caracotsios, and J. P. Sorensen, "Parameter Estimation from Multiresponse Data," *AIChE J.* **38:**641–650 (1992).

Swartz, C. L. E., "Data Reconciliation for Generalized Flowsheet Applications," *In Proceedings of the American Chemical Society National Meeting*, Dallas, TX (1989).

Terry, P. A. and D. M. Himmelblau, "Data Rectification and Gross Error Detection in Steady-State Process via Artificial Neural Networks," *Ind. Eng. Chem. Res.* **32:**3020–3028 (1993).

Tjoa, I. and L. T. Biegler, "Simultaneous Strategies for Data Reconciliation and Gross Error Detection of Nonlinear Systems," *Comp. Chem. Eng.* **15:**679–690 (1991).

Valko, P. and S. Vadja, "An Extended Marquardt-Type Procedure for Fitting Error-in- Variable Models," *Comp. Chem. Eng.* **11:**37–43 (1987).

Veverka, V. and F. Madron, *Material and Energy Balancing in Process Industries: From Microscopic Balances to Large Plants*, Elsevier, Amsterdam (1997).

Wang, D. and J. A. Romagnoli, "A Framework for Robust Data Reconciliation Based on a Generalized Objective Function," *Ind. Eng. Chem. Res.* **42:**3075–3084 (2003).

Wongrat, W., T. Srinophakun, and P. Srinophakun, "Modified Genetic Algorithm for Nonlinear Data Reconciliation," *Comp. Chem. Eng.* **29:**1059–1067 (2005).

Zhao W., D. Chen, and S. Hu, "Detection of Outlier and a Robust BP Algorithm against Outlier," *Comp. Chem. Eng.* **28:**1403–1408 (2004).

CHAPTER 9

Dynamic Data Reconciliation

Filtering

Early work in dynamic data reconciliation is rooted in the problem of process state estimation using the concept of filtering. Lately, the problem has been solved using the concept of model-based data smoothing.

Consider the three types of state estimation problems that are illustrated in Fig. 9.1 (Gelb, 1974). Assume an estimation of the state of the system is desired at time t.

When only measurement values prior to the time of prediction t are used, including the measurement at time t, the estimation is called *filtering*. When time t is not included, the estimation is called *prediction*, and finally, when data for times larger than t is used, the estimation process is called *smoothing*. Finally, when discrete measurements are used, the estimators are called *discrete estimators*.

Recursive Filters

Recursive filters are filters that do not need to store all the information of past measurements to compute estimators. They make use of recursive formulas to make the estimation at time t as a function of the estimate at time t-Δt and the measurement at t.

For example, consider the estimation of a scalar value x, based on k measurements z_i. The estimate is

$$\hat{x}_k = \frac{1}{k}\sum_{i=1}^{k} z_i \tag{9.1}$$

If one measurement is added, then

$$\hat{x}_{k+1} = \frac{1}{(k+1)}\sum_{i=1}^{k+1} z_i \tag{9.2}$$

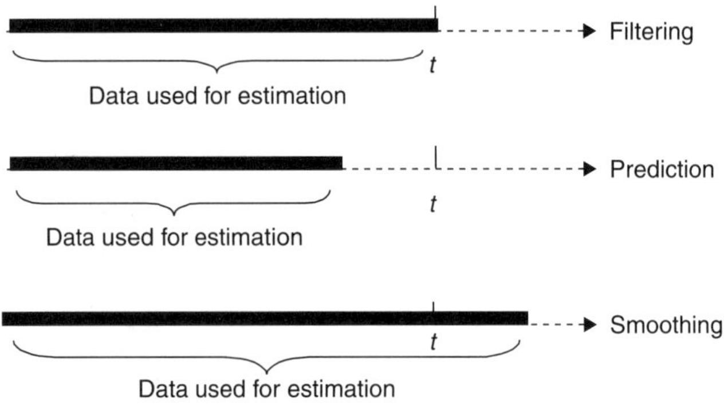

FIGURE 9.1 Different state estimation cases.

The following manipulation gives

$$\hat{x}_{k+1} = \frac{k}{(k+1)}\left\{\frac{1}{k}\sum_{i=1}^{k} z_i\right\} + \frac{1}{(k+1)} z_{k+1} = \frac{k}{(k+1)}\hat{x}_k + \frac{1}{(k+1)} z_{k+1} \tag{9.3}$$

which is the formula needed. This formula can be further rewritten in the following recursive form:

$$\hat{x}_{k+1} = \hat{x}_k + \frac{1}{(k+1)}(z_{k+1} - \hat{x}_k) \tag{9.4}$$

Linear Estimators

Consider now a discrete system whose behavior is given by the following model:

$$x_k = \Phi_{k-1} x_{k-1} + w_k \tag{9.5}$$

where x_k is a vector of system states and w_k is called *process noise* and has zero mean and variance R_k. For example, the discrete model for a constant is $x_k = x_{k-1}$. Consider also the following measurement model:

$$z_k = H_k x_k + v_k \tag{9.6}$$

where v_k is a zero mean random error with variance S_k and z_k is the vector of measurements at time t_k, namely, $z_k = [z_{k,1}\, z_{k,2} \ldots z_{k,m}]^T$.

A linear estimator is constructed as follows:

- The a priori estimate at time t_k is obtained using the system equation

$$\hat{x}_k^{(-)} = \Phi_{k-1}\hat{x}_{k-1}^{(+)} \tag{9.7}$$

where we have ignored the process noise.

- Given the estimate at time t_k denoted by $\hat{x}_k^{(-)}$, we seek an update estimate $\hat{x}_k^{(+)}$ based on z_k using the following linear, recursive form:

$$\hat{x}_k^{(+)} = K_k^{(-)}\hat{x}_k^{(-)} + K_k^{(z)}z_k \tag{9.8}$$

Discrete Kalman Filter

We now present the well-known Kalman filter (Kalman, 1960). Let

$$\hat{x}_k^{(+)} = x_k + \tilde{x}_k^{(+)} \tag{9.9}$$

$$\hat{x}_k^{(-)} = x_k + \tilde{x}_k^{(-)} \tag{9.10}$$

Substitute Eqs. (9.6), (9.9), and (9.10) in Eq. (9.8) to obtain

$$\tilde{x}_k^{(+)} = (K_k^{(-)} + K_k^{(z)}H_k - I)x_k + K_k^{(-)}\tilde{x}_k^{(-)} + K_k^{(z)}v_k \tag{9.11}$$

Note first that the expected value $E(v_k) = 0$ (by definition). In addition, we seek an unbiased estimator, so we impose the condition that $E(\tilde{x}_k^{(+)}) = 0$ under the assumption that $E(\tilde{x}_k^{(-)}) = 0$. This gives

$$\tilde{x}_k^{(+)} = (I - K_k^{(z)}H_k)\tilde{x}_k^{(-)} + K_k^{(z)}v_k \tag{9.12}$$

The next step is to establish conditions on the filter that will allow the determination of $K_k^{(z)}$. This derivation relies on minimizing the diagonal elements of the variance of the estimate. Assume the variance of the estimate $\tilde{x}_k^{(-)}$ is $P_k^{(-)}$, which is obtained as follows:

$$P_k^{(-)} = P_{k-1}^{(+)} + R_k \tag{9.13}$$

Then the derivation of $K_k^{(z)}$ (which is omitted) renders

$$K_k^{(z)} = P_k^{(-)}H_k^T(H_kP_k^{(-)}H_k^T + S_k)^{-1} \tag{9.14}$$

which is called the Kalman gain matrix. The following update of the variance is used

$$P_k^{(+)} = (I - K_k^{(z)}H_k)P_k^{(-)} \tag{9.15}$$

Quasi-Steady State Estimator

If one assumes that the system is at steady state, then the model is

$$x_k = x_{k-1} \tag{9.16}$$

where the process noise is again assumed zero. The measurement model is given by $H_k = I$. The Kalman filter applied to this situation results in the following procedure:

1. Obtain $\hat{x}_k^{(-)}$ as follows:

$$\hat{x}_k^{(-)} = \hat{x}_{k-1}^{(+)} \tag{9.17}$$

2. Obtain the variance of the new estimate

$$P_k^{(-)} = P_{k-1}^{(+)} \tag{9.18}$$

3. Obtain the Kalman gain matrix

$$K_k^{(z)} = P_k^{(-)}(P_k^{(-)} + S)^{-1} \tag{9.19}$$

4. Obtain the new estimate $\hat{x}_k^{(+)}$

$$\hat{x}_k^{(+)} = (I - K_k^{(z)})\hat{x}_k^{(-)} + K_k^{(z)} z_k \tag{9.20}$$

5. Obtain the variance of the new estimate $P_k^{(+)}$

$$P_k^{(+)} = (I - K_k^{(z)})P_k^{(-)} \tag{9.21}$$

The above steps can be summarized in two single formulas:

$$\hat{x}_k = (I - P_{k-1}(P_{k-1} + S)^{-1})\hat{x}_{k-1} + P_{k-1}(P_{k-1} + S)^{-1} z_k \tag{9.22}$$

$$P_k = (I - P_{k-1}(P_{k-1} + S)^{-1})P_{k-1} \tag{9.23}$$

A Balance-Based Quasi-Steady State Estimator

If in addition one assumes that the system is at steady state, one also assumes that the model satisfies the balance equations, then the model is

$$x_k = x_{k-1} \tag{9.24}$$

$$Cx_k = 0 \tag{9.25}$$

Now assume that we partition C as follows:

$$C = [C_1\ C_2] \tag{9.26}$$

where C_1 is nonsingular. This is equivalent to expressing a set of variables in terms of the others. Thus we partition $x = [x_1\ x_2]^T$ accordingly. Then the model becomes a model based on independent variables x_2.

$$x_{2,k} = x_{2,k-1} \tag{9.27}$$

$$x_{1,k} = -C_1^{-1} C_2 x_{2,k-1} \tag{9.28}$$

Therefore, the QSS Kalman filter (Stanley and Mah, 1977) becomes

1. Obtain $\hat{x}_k^{(-)}$ as follows:

$$\hat{x}_{2,k}^{(-)} = \hat{x}_{2,k-1}^{(+)} \tag{9.29}$$

$$\hat{x}_{1,k}^{(-)} = -C_1^{-1}C_2\hat{x}_{2,k-1}^{(+)} \tag{9.30}$$

2. Obtain the variance of the new estimate

$$P_k^{(-)} = P_{k-1}^{(+)} \tag{9.31}$$

3. Obtain the Kalman gain matrix

$$K_k^{(z)} = P_k^{(-)}(P_k^{(-)} + S)^{-1} \tag{9.32}$$

4. Obtain the new estimate $\hat{x}_k^{(+)}$

$$\hat{x}_k^{(+)} = (I - K_k^{(z)})\hat{x}_k^{(-)} + K_k^{(z)}z_k \tag{9.33}$$

5. Obtain the variance of the new estimate $P_k^{(+)}$

$$P_k^{(+)} = (I - K_k^{(z)})P_k^{(-)} \tag{9.34}$$

The above steps can be summarized in two single formulas:

$$\hat{x}_k = (I - P_{k-1}(P_{k-1} + S)^{-1})\begin{bmatrix} -C_1^{-1}C_2 \\ I \end{bmatrix}\hat{x}_{2,k-1} + P_{k-1}(P_{k-1} + S)^{-1}z_k \tag{9.35}$$

$$P_k = (I - P_{k-1}(P_{k-1} + S)^{-1})P_{k-1} \tag{9.36}$$

Remark: Almasy (1990) also proposed a Kalman filter approach based on a state space representation of the process. Later, Singhal and Seborg (2000) proposed a new methodology based on the Kalman filter that rectifies noise and removes gross errors. It uses the expectation-maximization algorithm to find the maximum likelihood estimatesof the variables based on a state-space model, past data, and measurements.

Difference Estimators

We now briefly present a discrete estimator called singular or generalized dynamic estimator (Darouach and Zasadzinski, 1991). Consider the following balance equations:

$$B_R \frac{dw_R}{dt} = A_R f_R \tag{9.37}$$

$$C_R f_R = 0 \tag{9.38}$$

where f_R represents flows and w_R tank's hold-ups.

The discrete version of Eqs. (9.37) and (9.38) is

$$B_R(w_{R,i+1} - w_{R,i}) = A_R f_{R,i+1} \tag{9.39}$$

$$C_R f_{R,i+1} = 0 \tag{9.40}$$

which can be expressed as follows:

$$E_R z_{R,i+1} + G_R z_{R,i} = 0 \tag{9.41}$$

where

$$G_R = \begin{bmatrix} 0 & B_R \\ 0 & 0 \end{bmatrix} \tag{9.42}$$

Unfortunately, E_R is usually nonsquare and therefore, singular. Therefore, the discrete Kalman filter cannot be applied. Instead, all estimates for all times up to t_N are reconciled at the same time using Eq. (9.41) as constraints, that is,

$$\left.\begin{array}{l} \text{Min} \sum_{k=0}^{N} [z_{R,k} - z_{R,k}^{+}]^T Q_R^{-1} [z_{R,k} - z_{R,k}^{+}] \\ \text{s.t.} \\ E_R z_{R,k+1} + G_R z_{R,k} = 0 \qquad \forall k = 0, N-1 \end{array}\right\} \tag{9.43}$$

Remark: Rollins and Devanathan (1993) improved this technique offering an unbiased estimator of dynamic states, using a recursive dynamic data reconciliation technique that obtains very accurate estimators. When compared with the work of Darouach and Zasadzinski (1991) in terms of computational speed and accuracy of estimators, their technique showed to be computationally faster, but not as precise when variances of process measurements are large. However, the precision of the proposed estimators is shown to approach that of the recursive technique by iteratively recalculating estimates and when measurement variances decrease.

Remark: Narasimhan and Jordache (2000) showed that a particular version of Eq. (9.43) where the variance covariance matrices are the same as those obtained by the Kalman filter renders the same answer as a particular case of Kalman filtering. In other words, if disturbances in state variables are ignored, both methods are equivalent.

Historical Note: Hlavacek (1977) was among the first to make a review of the state of the art in data reconciliation and gross error detection for transient behavior in process plants. Gertler (1998) followed with one of the first-time invariant recursive filters to obtain process input and output estimators, optimal in the least squares sense.

Integral Approach

We now present a smoothing approach based on polynomial representations and integration of the differential equations (Jiang and Bagajewicz, 1997). Consider the following s-order polynomial representation of f_R and w_R:

$$f_R \approx \sum_{k=0}^{s} \alpha_k^R t^k \tag{9.44}$$

$$w_R \approx w_{R0} + \sum_{k=0}^{s} \omega_{k+1}^R t^{k+1} \tag{9.45}$$

Therefore,

$$\int_0^t f_R(\xi)d\xi \approx \sum_{k=0}^{s} \frac{\alpha_k^R}{k+1} t^{k+1} \tag{9.46}$$

Thus, Eq. (9.37) is equivalent to

$$B_R[w_R - w_{R0}] = B_R \sum_{k=0}^{s} \omega_{k+1}^R t^{k+1} = A_R \sum_{k=0}^{s} \frac{\alpha_k^R}{k+1} t^{k+1} \tag{9.47}$$

Since this equation is valid for all t, then

$$B_R \omega_{k+1}^R = \frac{A_R \alpha_k^R}{k+1} \qquad k = 0, \ldots, s \tag{9.48}$$

In turn, Eq. (9.38) is equivalent to the following set of equations:

$$C_R \alpha_k^R = 0 \qquad k = 0, \ldots, s \tag{9.49}$$

For k+1 measurements, the dynamic data reconciliation problem assumes the following form:

$$\left.\begin{aligned} &\text{Min} \sum_{k=1}^{N} [z_{R,k} - z_{R,k}^{+}]^T Q_R^{-1} [z_{R,k} - z_{R,k}^{+}] \\ &\text{s.t.} \\ &B_R(w_{R,k} - w_{R,0}) = A_R \int_0^t f_R(\xi)d\xi \qquad (k = 0, \cdots, N) \\ &\qquad C_R f_{R,k} = 0 \qquad (k = 0, \cdots, N) \end{aligned}\right\} \tag{9.50}$$

Example 9.1: Typical solutions of these types of models are shown in Fig. 9.2.

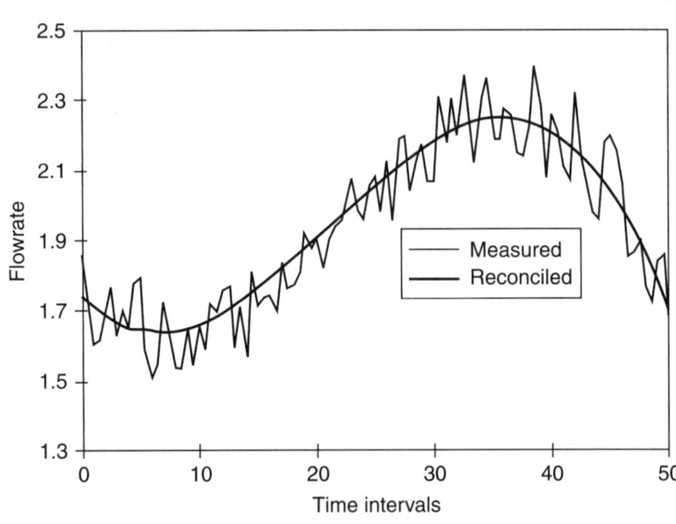

FIGURE 9.2 Typical results obtained with integral dynamic data reconciliation.

Remark: Benqlilou et al. (2001) used the above technique on a moving window with some small variations and applied it to a polimerization reactor.

Nonlinear Case

The nonlinear case, which we do not cover in detail in this book, has several different approaches: (1) The extended Kalman filter (EKF) approach (Jazwinski, 1970), (2) the geometric observer (GO) approach (Bestle and Zeitz, 1983; Krener and Isidori, 1983; Krener and Respondek, 1985; Xia and Gao, 1989), (3) the high gain (HG) approach Tsinias, 1989; Gauthier et al., 1992; Deza et al., 1992, 1993; Ciccarella et al., 1993, (4) the sliding-mode (SM) approach (Slotine et al., 1987; Walcott et al., 1987), and (5) the data reconciliation approach presented above and extended to nonlinear cases (Albuquerque and Biegler, 1995, 1996; Barbosa et al., 2000). The first four are restricted to open-loop plants that are observable; the EKF being the most popular one in chemical process engineering. Albuquerque et al. (1997) deal with parameter estimation using error-in-variable inference.

Following the work of Muske and Edgar (1997), Narasimhan and Jordache (2000) also reviewed the use of the extended Kalman filter to obtain recursive estimators of variables. Alvarez and Lopez (1999), in turn, extended their P-estimator to the nonlinear case and provided convergence as well as a systematic tuning procedure. Narasimham and Jordache (2000) concluded that, in general, EK-based methods are more efficient than nonlinear optimization in moving horizons methods. Nonlinear data reconciliation is appropriate, however, when inequalities are included.

The moving horizon data reconciliation problem to solve is the following:

$$\text{Min} \sum_{k=0}^{N} \left[z_{R,k} - z_{R,k}^{+}\right]^{T} Q_R^{-1} \left[z_{R,k} - z_{R,k}^{+}\right] \tag{9.51}$$

s.t.

$$\frac{dx}{dt} = F(x,u,p,t) \tag{9.52}$$

$$z = G(x,u,p,t) \tag{9.53}$$

$$x_0 = x(t=0) \tag{9.54}$$

where z are the reconciled measurements, x the states, u the inputs, p the parameters. In turn Eq. (9.52) is the system and Eq. (9.53) the measurement model. In many cases Eq. (9.52) is presented in a much more general way as an implicit equation on the derivative, that is,

$$F\left(\frac{dx}{dt}, x, u, p, t\right) = 0 \tag{9.55}$$

Bellman et al. (1967) used quasi-linearization in the context of direct integration of Eq. (9.52). They did it in a sequential manner, that is, used the parameter values to determine the states, which in turn allows determining the objective and its gradients. Hwang and Seinfeld (1972) improved this algorithm by adding sensitivities to the parameters and Birardello et al (1993) used a similar approach but employing Gauss-Newton steps. At the same time Ganguly et al. (1993) tested several variants of these techniques concluding that the dynamic data reconciliation is superior.

Linearization and discretization of Eqs. (9.52) and (9.53) renders:

$$x_k = A_{k-1}x_{k-1} + B_{k-1}u_{k-1} \tag{9.56}$$

$$z_k = C_k x_k + D_k u_k \tag{9.57}$$

One can easily verify that this is of the same form as Eq. (9.43). Jang et al. (1986) were among the first to compare the EKF to the moving-horizon data reconciliation approach. Notwithstanding the differences with other more statistically based least squares, they show that, although much slower, the data reconciliation approach is superior in the presence of errors in modeling and noise. Kim et al. (1991) also attempted such an approach. Later, Ramamurthi et al. (1993) uses a successively linearized horizon-based estimation approach to estimate states and parameters and tested the technique using open-loop and closed-loop simulations. They also decoupled the estimation of the inputs, from the estimation of the states and the parameters and reducing computing time. They also compared their technique favorably against the EKF. Robertson et al. (1996) also compare the EKF to a moving horizon least squares algorithm.

Semino et al. (1996) showed that updating the parameters by considering them as "fictitious states" improves the EKF performance. This does not address, however, the failures of the EKF to the presence of unmodeled disturbances or modeling errors. Sirohi and Choi (1996) also compared the EKF to the data reconciliation approach in a polymerization reactor. Finally, Rao and Rawlings (1998) explain that if each horizon is treated independently, the method may introduce errors and miss converging to the true values and suggest that somewhat information from previous horizons is beneficial. Following this idea, Soderstrom et al (2000) add the difference between current and previous estimates of certain variables in the objective function.

Departing from a linearization approach, Liebman et al. (1992) proposed a collocation method. In essence, the method proposes to represent $z(t)$ as follows:

$$x(t) = \sum_{i=1}^{N} l_i(t) x_i(t_i) \tag{9.58}$$

where $l_i(t)$ are the Lagrange interpolationg polynomials.

$$l_i(t) = \prod_{j=1, j\neq i}^{N} \frac{t - t_j}{t_i - t_j} \tag{9.59}$$

When measurements are made at regular intervals, then at the interpolating points, the following is true:

$$\frac{dx}{dt} = \frac{1}{\Delta t} \sum_{i=1}^{N} \frac{dl_i(t)}{dt} x(t_i) \tag{9.60}$$

Substituting Eq. (9.60) in Eq. (9.52) one obtains:

$$D^r x^r - \Delta t \; s_i(x^r) = 0 \tag{9.61}$$

where x^r is the vector of state values at the collocation points, D^r the matrix of first-derivative collocation point weights obtained evaluating $dl_i(t)/dt$ and $s_i(x^r)$ the vector of values of the right hand side of Eq. (9.52) at the collocation points. The problem is now reduced to minimize Eq. (9.51) subject to Eq. (9.61). Liebman et al. (1992) discuss implementation and numerical challenges.

Mingfang et al. (2000) also presented an integral approach that integrates a finite element collocation method, a filtering technique based on a moving average and a robust method based on influence functions to deal with gross errors.

Esposito and Floudas (2000) compared two methods: (1) converting the algebraic-differential equation system into a pure algebraic one through the use of a collocation method and (2) Following Pontryagin (1962) they convert the algebraic-ODE system of equations into just a set of ODEs. Next, the set of ODEs is converted into a set of twice differentiable functions of time and the parameters. They solve both problems to global optimality and applied it to reaction systems that are relatively small. It is then unclear how this would work computationally in much larger systems.

Remark: Chen et al. (2004) point out that most existing methods rely on simplifying assumptions to obtain a tractable but approximate solution. For example, they say, "extended Kalman filtering linearizes the process model and assumes Gaussian prior and noise. Moving horizon-based least-squares estimation also assumes Gaussian or other fixed-shape prior and noise to obtain a least-squares optimization problem." To ameliorate these problems they introduce a sequential Monte Carlo sampling for Bayesian estimation. They show that their methodology is more computationally efficient than and at least as accurate as moving-horizon-based least-squares. Kong et al. (2004) deal with simultaneous data reconciliation and gross error detection for nonlinear methods. They use the full differential equation of the process as constraints and derive a gross error identifiable condition. Morad et al. (2005) also used a Bayesian approach to assess adjustment to measurement, but also adjustment to the process state through two different probability distribution functions (pdfs). It is a recursive nonparametric method that fits a mixture of Gaussian pdfs to the data. Tona et al. (2005) proposed to use trend analysis using wavelets as a preliminary step for data reconciliation in linear dynamic systems. Vachhani et al. (2005) presented recursive nonlinear dynamic data reconciliation (RNDDR) and a combined predictor-corrector optimization (CPCO) method for efficient state and parameter estimation in nonlinear systems, combining the efficiency of EKF and the ability of NDDR to handle algebraic inequality and equality constraints. In addition, the CPCO technique includes parameter variation, which is a restriction of the EKF. They show that the RNDDR performs as well as the two traditional approaches, and that the CPCO formulation is more accurate, being only marginally more computationally expensive. Finally, Chen et al. (2008) pointed out that the EKF is less applicable in cases where the state and measurement functions are highly nonlinear or where the posterior distribution of the states is non-Gaussian. They proposed an alternative approach in which particle filters are utilized for dynamic data rectification. Particle filters are a class of statistical tools that sequentially estimate the system states from a state-space model. They approximated the distribution of the states, given the measurement sequence using a set of random samples (also called particles). Thus, the particle filters generate Monte Carlo samples from the posterior distribution of the system states, and thus provide the basis for rectifying the process measurements.

Remark: Departing from all of the above line of work, Karjala and Himmelblau (1994) and Himmelblau and Karjala (1996) used neural networks. Also, Karjala and Himmelblau (1994, 1996) proposed the use of neural networks in conjunction with the extended Kalman Filter. In turn, Martinez Prata et al. (2009) used particle swarm optimization to perform simultaneous data reconciliation and parameter estimation on a moving window. Finally, Bai et al. (2007) pointed out that the use of the Kalman filter "suffers from two restrictive conditions: (1) it requires state-space models and (2) it has to be tuned online to achieve its best performance." They combine autoassociative neural networks (AANNs) and dynamic data reconciliation claiming that this overcomes the need for online tuning of the Kalman filter, and the online optimization required by conventional dynamic data reconciliation methods. They applied the method to a distillation column showing it is robust to changes of the noise level in plant measurements and the loss of measurements.

Remark: Although there is a lot of work that has been performed in the field of dynamic data reconciliation its implementation in industrial sites is incipient. Software vendors are still considering dynamic data reconciliation as too computationally intensive and logistically an effort in programming that is not worthwhile. Ultimately, it will be the pressure from the practitioners, who will be not satisfied with the steady state based approach that will force the commercial implementation of these techniques. Applications in practice are possible. For example, Soderstrom et al. (2000) uses approximately 50 differential and algebraic equations with 100 measured variables, resulting in 2700 variables subject to 1300 constraint equations over a horizon of 60 min with data sampled at 2 min intervals. Almost 10 years later, with all the advances in computing, these techniques are ready for larger plant size challenges.

Remark: Alici and Edgar (2002) presented two efficient approaches for dynamic data reconciliation using model identification tools and commercial dynamic simulation software. The first one is based on an analogy to the nonlinear dynamic data reconciliation method developed by Liebman et al. (1992). The second approach uses time series analysis to generate a simplified model of the plant. A simplified process model is generated by a model identification method to replace the simulation software.

Gross Error Detection

Gross error detection in the context of dynamic data reconciliation is not abundant. Narasimhan and Jordache (2000) reviewed efforts to incorporate bias in the measurement model of the Kalman filter and showed that a Global test as well as the generalized likelihood ratio (GLR) test can be constructed. One of the earliest attempts is the work by Kao et al. (1990) in which the measurement test (MT) is used assuming that the errors are serially correlated. The test is applied only after correcting the variance, or obtaining uncorrelated residuals using prewhitening. Narasimhan and Mah (1988) followed with an application of their GLR method to dynamic systems. In their follow-up paper, Kao et al. (1992) proposed a composite test procedure for detecting and identifying gross errors making use of the CUSUM test first to detect gross errors and follow with the GLR method to identify as well as to estimate the magnitudes of the gross errors. They also resort to prewhitening before applying their test. Fillon et al. (1995, 1996) also presented a GLR based gross error detection method. Albuquerque and Biegler (1996) used a Bayesian approach and robust estimators. McBrayer and Edgar (1995) incorporated the bias as a parameter in the data reconciliation procedure. Sanchez and Romagnoli (2000) presented these and a few other methodologies (Friedland, 1969; Wilsky and Jones, 1976; Caglayan, 1980; Romagnoli and Gani, 1983) in detail. Tjoa and Biegler (1996) proposed to incorporate the bias to the objective function of the data reconciliation by using a contaminated objective function. Finally, Rollins et al. (2002) combined the unbiased estimation method (Rollins and Devanathan, 1993) with a bias detection technique and Kongsjahju et al. (2000) extended the UBET method (see Chap. 7), to serially correlate the data.

A large class of methods exists that are based on structured residuals. These methods are based on the following premise: When there is a fault, the measurement $x(t)$ is represented by

$$x(t) = x^*(t) + \Xi_i f_i(t) \tag{9.62}$$

where $x^*(t)$ is the normal value, $f_i(t)$ the vector of fault magnitude, and Ξ_i the matrix of fault directions. The residual $e(t)$ is then obtained through

$$e(t) = Be(t) = Bx^*(t) + B\Xi_i f_i(t) = e^*(t) + B\Xi_i f_i(t) \tag{9.63}$$

Thus, the objective of sensor validation is to detect the onset of the fault, estimate the fault direction matrix, Ξ_i and the fault magnitude $f_i(t)$. Qin and Li (1999, 2001) proposed a fault detection index:

$$d(t) = e^T(t) R_e^{-1} e(t) \tag{9.64}$$

where $R_e = \mathrm{E}\{e^*(t)e^{*T}(t)\}$ is the covariance matrix of $e^*(t)$. Thus, without faults, $d(t)$ follows a Chi-squared distribution and therefore, its value can be used to detect the presence of gross errors when it is larger than a threshold (as in the Global test). Lin and Qin (2005) offered a way of identifying the fault directions and the fault magnitudes. This class of methods is rooted in the knowledge of the normal value of the measurement $x^*(t)$, which needs to be determined beforehand (see Gertler, 1998). Another class of methods that is also based on comparing with "normal" conditions is one that uses principal component analysis (Dunia et al., 1996). We pointed out that these residual-based and PCA-based methods and similar ones do not rely on models and therefore, they only apply to the particular set of operating conditions chosen, making them inapplicable to others. We do not hide our preference for model-based methods and therefore we do not pursue covering these classes of methods in any more detail.

Multiscale methods have also been proposed by Luo et al. (1998). They consist of using wavelet decomposition to denoise the signals followed by statistical tests to determine gross errors, trends, etc. Bakshi et al. (1997) decomposed the signals first using wavelets and then applying PCA methods. Nounou and Bakshi (1999) also used wavelet filtering on a moving window to detect errors and remove them. Ungarala and Bakshi (2000) applied a Bayesian error-in-variable approach after wavelet decomposition to linear systems and Bakshi et al. (2001) extended the method to nonlinear systems.

References

Albuquerque, J. S. and L. T. Biegler, "Decomposition Algorithms for On-line Estimation with Nonlinear DAE Models," *Comp. Chem. Eng.* **19:**1031 (1995).

Albuquerque, J. S. and L. T. Biegler, "Data Reconciliation and Gross-Error Detection for Dynamic Systems," *AIChE J* **42:**2841 (1996).

Albuquerque, J. S., L. T. Biegler, and R. E. Kaas, "Inference in Error-in-Variable Measurement Problems," *AIChE J.* **43:**986–996 (1997).

Alici, S. and T. F. Edgar, "Nonlinear Dynamic Data Reconciliation via Process Simulation Software and Model Identification Tools," *Ind. Eng. Chem. Res.* **41:**3984–3992 (2002).

Almasy, G., "Principles of Dynamic Balancing," *AIChE J.* **36**(9):1321–1330 (1990).

Alvarez, J. and T. Lopez, "Robust Dynamic State Estimation of Nonlinear Plants," *AIChE J.* **45**(1):107–123 (1999).

Bakshi, B. R., P. Bansal, and M. N. Nounou, "Multiscale Rectification of Random Errors without Fundamental Process Models," *Comp. Chem. Eng.* **21** (Suppl.): S1167–S1172 (1997).

Bakshi, B. R., M. N. Nounou., P. Goel, and X. Shen, "Multiscale Bayesian Rectification of Data Form Linear Steady State and Dynamic Systems without Accurate Models," *Ind. Eng. Chem. Res. Comp. Chem. Eng.* **40:**261–274 (2001).

Bai, S., D. D. McLean, and J. Thibault, "Autoassociative Neural Networks for Robust Dynamic Data Reconciliation," *AIChE J.* **53**(2):438–448 (2007).

Barbosa, V. P., M. R. M. Wolf, and R. Maciel Fo, "Development of Data Reconciliation for Dynamic Nonlinear System: Application the Polymerization Reactor," *Comp. Chem. Eng.* **24:**501–506 (2000).

Bellman, R., J. Jacquez, R. Kalaba, and S. Schwimmer, "Quasilinearization and the Estimation of Chemical Rate Constraints from Raw Kinetic Data," *Math. Biosci.* **1:**71 (1967).

Benqlilou, C. R., V. Tona, A. Espuña, and L. L. Puigjaner, "On-Line Application of Dynamic Data Reconciliation," *Proc. Pres'01, 4th Conference: Process Integration, Modeling and Optimisation for Energy Savings and Pollution preventions*. Florence, Italy 403–406 (2001).

Bestle, D. and M. Zeitz, "Canonical Form Observer Design for Non-Linear Time-Variable Systems," *Int. J. Cont.* **38**(2):419 (1983).

Birardello, P., X. Joulia, J. LeLann, H. Delams, and B. Koehret, "A General Startegy for Parameter Estimation in Differential-Algebraic Systems," *Comp. Chem. Eng.* **17:**517 (1993).

Chen, W., B. R. Bakshi, P. K. Goel, and S. Ungarala, "Bayesian Estimation via Sequential Monte Carlo Sampling: Unconstrained Nonlinear Dynamic Systems," *Ind. Eng. Chem. Res.* **43:**4012–4025 (2004).

Chen, T., J. Morris, E. Martin, "Dynamic Data Rectification Using Particle Filters," *Comp. Chem. Eng.* **32:**451–462 (2008).

Ciccarella, G., M. Dalla Mora, and A. Germani, "A Luenberguer-Like Observer for Nonlinear Systems," *Int. J. Cont.* **57**(3):537 (1993).

Darouach, M. and M. Zasadzinski, "Data Reconciliation in Generalized Linear Dynamic Systems," *AIChE J.* **37**(2):193 (1991).

Deza, F., E. Busvelle, J. P. Gauthier, and D. Rakotopara, "High Gain Estimation for Nonlinear Systems," *Syst. Control Lett.* **18:**295 (1992).

Deza, F., D. Bossanne, E. Busvelle, J. P. Gauthier, and D. Rakotopara, "Exponential Observers for Nonlinear Systems," *IEEE Trans. Automat. Contr.* **AC-38**(3):482 (1993).

Dunia, R., S. J. Qin, T. F. Edgar, and McAvoy, T. J, "Identication of Faulty Sensors Using Principal Component Analysis," *AIChE J.* **42:**2797–2812 (1996).

Esposito, W. R. and C. A. Floudas, "Global Optimization for the Parameter Estimation of Differential-Algebraic Systems," *Ind. Eng. Chem. Res.* **39:**1291–1310 (2000).

Fillon, M., M. Meyer, H. Pingaud, and X. Joulia, "Data Reconciliation Based on Elemental Balances applied to Batch Experiments," *Comp. Chem. Eng.* (Suppl.) **19:**S293–S298 (1995).

Fillon, M., M. Meyer, H. Pingaud, and M. Enjalbert, "Efficient Formulation for Batch Reactor Data Reconciliation," *Ind. Eng. Chem. Res.* **35:**2288–2298 (1996).

Friedland, B., "Treatment of Bias in Recursive Filtering," *IEEE Trans. Autom. Cont.* **AC-14:**359–367 (1969).

Ganguly, S., V. Sairam, and D. N. Saraf, "Nonlinear Parameter Estimation for Real-Time Analytical Distillation Models," *Ind. Eng. Chem. Res.* **32:**99–107 (1993).

Gauthier, J. P., H. Hammouri, and S. Othman, "A Simple Observer for Nonlinear Systems. Applications to Bioreactors," *IEEE Trans. Automat. Contr.* **AC-37**(6):875 (1992).

Gelb, A., *Applied Optimal Estimation*, MIT Academic Press, Cambridge, Mass. (1974).

Gertler, J., "A Constrained Minimum Variance Input-Output Estimator for Linear Dynamic Systems," *Automatica* **15:**353–358 (1979).

Gertler, J., *Fault Detection and Diagnosis in Engineering Systems*, Marcel Dekker: New York (1998).

Himmelblau, D. M. and T. W. Karjala, "Rectification of Data in a Dynamic Process Using Artificial Neural Networks," *Comp. Chem. Eng.* **20:**805–812 (1996).

Hlavacek, V., "Analysis of a Complex Plant Steady State and Transient Behavior," *Comp Chem. Eng.* **1:**75–100 (1977).

Hwang, M. and J. H. Seinfeld, "A New Algorithm for the Estimation of Parameters in Ordinary Differential Equations," *AIChE J.* **18:**90 (1972).

Jang, S., B. Joseph, and H. Mukai, "Comparison of Two approaches to On-line Parameter and State Estimation of Nonlinear Systems," *Ind. Eng. Chem. Process Des. Dev.* **25:**809–814 (1986).

Jiang, Q. and M. Bagajewicz, "An Integral Approach to Dynamic Data Reconciliation," *AIChE J.* **43**(10) (Oct.):2546 (1997).

Jazwinski, A. H., *Stochastic Processes and Filtering Theory*, Academic Press, New York (1970).

Kalman, R. E., "New Approach to Linear Filtering and Prediction Problems," *J. Basic Eng. ASME* **82D:**35 (1960).

Kao, C. S., A. C. Tamhane, and R. S. H. Mah, "Gross Error Detection in Serially Correlated Process Data," *Ind. Eng. Chem. Res.* **29:**1004–1012 (1990).

Kao, C. S., A. C. Tamhane, and R. S. H. Mah, "Gross Error Detection in Serially Correlated Process Data 2 Dynamic Systems," *Ind. Eng. Chem. Res.* **31:**254 (1992).

Karjala, T. W. and D. M. Himmelblau, "Dynamic Data Rectification by Recurrent Neural Networks vs. Traditional Methods," *AIChE J.* **40:**1865–1875 (1994).

Karjala, T. W. and D. M. Himmelblau, "Dynamic Rectification of Data via Recurrent Neural Nets and the Extended Kalman Filter," *AIChE J.* **42:**2225 (1996).

Kim, I. W., M. J. Liebman, and T. F. Edgar, "A Sequential Error-in-Variables Method for Nonlinear Dynamic Systems," *Comp. Chem. Eng.* **15:**663–670 (1991).

Kong M., B. Chen, X. He, S. Hu., "Gross Error Identification for Dynamic System," *Comp Chem. Eng.* **29:**191–197 (2004).

Kongsjahju, R., D. K. Rollins, and M. B. Bascuñana, "Accurate Identification of Biased Measurements under Serial Correlation," *Chem. Eng. Res. Des.* **78:**1010–1018 (2000).

Krener, A. and A. Isidori, "Linearization by Output Injection and Nonlinear Observers," *Syst. Control Lett.* **3**(47) (1983).

Krener, A. and W. Respondek, "Nonlinear Observers with Linearizable Error Dynamics," *SIAM J. Control Optim.* **23**(2):197 (1985).

Liebman, M. J., T. F. Edgar, and L. S. Lasdon, "Efficient Data Reconciliation and Estimation for Dynamic Process Using Nonlinear Programming Techniques," *CES* **16**(10/11):963 (1992).

Lin W. and S. J. Qin, "Optimal Structured Residual Approach for Sensor Fault Diagnosis," *Ind. Eng. Chem. Res.* **44:**2117–2124 (2005).

Luo, R., M. Misra, S. J. Qin, R. Barton, and D. M. Himmelblau, "Sensor Fault Detection via Multiscale Analysis and Nonparametric Statistical Inference," *Ind. Eng. Chem. Res.* **37:**1024–1032 (1998).

Martinez Prata, D., M. Schwaab, E. L. Lima[a], and J. C. Pinto, "Nonlinear Dynamic Data Reconciliation and Parameter Estimation through Particle Swarm Optimization: Application for an Industrial Polypropylene Reactor," *Chem. Eng. Sci.* **64:**3953–3967 (2009).

McBrayer, K. F. and T. F. Edgar, "Bias Detection and Estimation Ion Dynamic Data Reconciliation," *J. Proc. Contr.* **15:**285–289 (1998).

Mingfang, K., C. Bingzhen, and L. Bo, "An Integral Approach to Dynamic Data Rectification," *Comp. Chem. Eng.* **24:**749–753 (2000).

Morad K., B. R. Young, and W. Y. Svrcek, "Rectification of Plant Measurements Using a Statistical Framework," *Comp. Chem. Eng.* **29:**919–940 (2005).

Muske, K. R. and T. F. Edgar, "Nonlinear State Estimation," in M. A. Henson and D. E. Seborg (eds.), Prentice Hall, New Jersey:311–370 (1997).

Narasimhan, S. and R. S. H. Mah, "Generalized Likelihood Ratios for Gross Error Identification in Dynamic Processes," *AIChE J.* **34:**1321 (1988).

Narasimhan, S. and C. Jordache, *Data Reconciliation and Gross Error Detection. An Intelligent Use of Process Data*, Gulf Publishing Company (2000).

Nounou, M. N. and B. R. Bakshi, "Online Multiscale Filtering of Random & Gross Error Without Proc Models," *AIChE J.* **45:**1041–1058 (1999).

Pontryagin, L. S, *Ordinary Differential Equations*, Addison-Wesley, Reading, Mass., (1962).

Qin, S. J. and W. Li, "Detection, Identification, and Reconstruction of Faulty Sensors with Maximized Sensitivity," *AIChE J.* **45:**1963–1976 (1999).

Qin, S. J. and W. Li, "Detection and Identification of Faulty Sensors in Dynamic Processes," *AIChE J.* **47:**1581–1593 (2001).

Ramamurthi, Y., P. B. Sistu, and B. W. Bequette, "Control Relevant Dynamic Data Reconciliation and Parameter Estimation," *Comp. Chem. Eng.* **17:**41–59 (1993).

Rao, C. V. and J. B. Rawlings, "Moving Horizon State Estimation," in *Nonlinear Predictive Control*; Allgower, F. and A. Zheng, (eds.), Progress in Systems and Control Theory Series; Birkhauser Verlag:Basel, **26**. (2000).

Robertson, D. G., J. H. Lee, and J. B. Rawlings, "A Moving Horizon-Based Approach for Least-Squares Estimation," *AIChE J.* **42**(8) (Oct.):2209–2224 (1996).

Rollins, D. K. and S. Devanathan, "Unbiased Estimation in Dynamic Data Reconciliation," *AIChE J.* **39**(8) (1993).

Rollins, D. K., S. Devanathan, and M. V. B. Bascuñana, "Measurement Bias Detection in Linear Dynamic Systems," *Comp. Chem. Eng.* **26:**1201–1211 (2002).

Sánchez, M. and J. Romagnoli, *Data Processing and Reconciliation for Chemical Process Operations*, Academic Press (2000).

Semino, D., M. Morreta, and C. Scali, "Parameter Estimation in Extended Kalman Filters for Quality Control in Polymerization Reactors," *Comp. Chem. Eng.* **20**(Suppl.):S913–S918 (1996).

Singhal, A. and D. E. Seborg, "Dynamic Data Rectification using the Expectation Maximization Algorithm," *AIChE J.* **46:**1556–1565 (2000).

Sirohi, A. and K. Y. Choi, "Online Parameter Estimation in a Continuous Polymerization Process," *Ind. Eng. Chem. Res.* **35:**1332–1343 (1996).

Slotine, J. J. E., J. K. Hedrick, and E. A. Misawa, "On Sliding Observers for Nonlinear Systems," *J. Dyn. Syst., Meas. Cont.* **109:**245 (1987).

Soderstrom, T. A., T. F. Edgar, L. P. Russo, and R. E. Young, "Industrial Application of a Large-Scale Dynamic Data Reconciliation Strategy," *Ind. Eng. Chem. Res.* **39:**1683–1693 (2000).

Stanley, G. M. and R. H. S. Mah, "Estimation of Flows and Temperatures in Process Networks," *AIChE J.* **23**(5):642 (1977).

Tjoa, I. B. and L. T. Biegler, "Simultaneous Strategies for Data Reconciliation and Gross Error Detection of Nonlinear Systems," *Comp. Chem. Eng.* **15**(10):679–690 (1991).

Tona, R. V., C. Benqlilou, A. Espuña, and L. Puigjaner, "Dynamic Data Reconciliation Based on Wavelet Trend Analysis," *Ind. Eng. Chem. Res.* **44:**4323–4335 (2005).

Tsinias, J., "Observer Design for Nonlinear Systems,"*Syst. Control Lett.* **13:**13 (1989).

Ungarala, S. and B. R. Bakshi, "A Multiscale, Bayesian and Error in Variables Approach for Linear Dynamic Data Rectification," *Comp. Chem. Eng.*, **24:**445–451 (2000).

Vachhani, P., R. Rengaswamy, V. Gangwal, and S. Narasimhan, "Recursive Estimation in Constrained Nonlinear Dynamical Systems," *AIChE J.* **51:**946–959 (2005).

Walcott, B. L., M. L. Corless, and S. H. Zak, "Comparative Study of Nonlinear State-Observation Techniques," *Int. J. Cont.* **45**(6):2109 (1987).

Wilsky, A. S. and L. Jones, "A Generalized Likelihood Ratio Approach to the Detection and Estimation of Jumps in Linear Systems," *IEEE Trans. Autom. Cont.* **AC-21:**108–112 (1976).

Xia, X. and W. Gao, "Nonlinear Observer Design by Observer Error Linearization," *SIAM J. Control Optim.* **27:**199 (1989).

CHAPTER 10

Accuracy of Estimators

Having reviewed the basic concepts of data reconciliation, we are now aware of the ability of this procedure to filter biases that are sufficiently large to be detected and to increase the precision of the estimators.

As it has been discussed in Chap. 2, there are still biases stemming from conformity curves that can exist as well as biases that show up through time at the instrument by loss of calibration, or through the electronic processing of the signal afterward.

In view of the existence of biases that data reconciliation cannot detect and identify efficiently, one is left to wonder what is the real accuracy of the information that one is receiving. This is the objective of this chapter.

Accuracy of Measurements

Accuracy of an instrument is defined as the sum of the systematic error plus the precision of the instrument (Miller, 1996).

$$a_i = \delta_i + \sigma_i \tag{10.1}$$

where a_i, δ_i, and σ_i are the accuracy, systematic error, and precision (square root of variance) of the mean of a certain number of repeated measurements made by a meter on variable *i*.

The problem with this definition is that it is useless in practice. Indeed, on one hand, if one knows the systematic error, either the instrument is immediately recalibrated, and therefore the systematic error is eliminated, or the measurement is adjusted. On the other hand, if one does not know the value of the systematic error (the usual case), such systematic error cannot be inferred directly from the measurements of this instrument, only the precision can. Thus, although the definition has conceptual value, it has no practical implications. We know that in the absence of data reconciliation, these biases cannot be detected unless another instrument is installed to measure the same variable. Such option (called "hardware redundancy") would be too costly.

Because hardware redundancy is very costly, software redundancy is the most popular way of identifying biases. In particular, data reconciliation renders estimators of process variables that satisfy balance constraints and is capable of filtering biases to some extent.

Remark: Literature on data reconciliation often uses the term accuracy when referring to the variance $\tilde{Q}$. However, literature on sensors (Liptak, 2003) refers to the accuracy of an instrument as the degree of conformity with a standard or true value, that is, the variance obtained through calibration. Moreover, the distinction between accuracy (as per the above definition) and precision is made specifically (Liptak, 2003). Accuracy and precision are only equivalent in the absence of systematic errors or biases. In data reconciliation, one should also add model inaccuracy as another source, since leaks can affect the accuracy of reconciled data. We also discussed issues related to the *Guide to the Expression of Uncertainty in Measurement* (GUM) in Chap. 2.

We now concentrate on defining accuracy using a new concept: the induced bias.

Induced Bias

Let the measurement vector be described by

$$y = \mu + \delta + \varepsilon \tag{10.2}$$

where y is the vector of measurements, μ the vector of true values, δ the vector of biases, and ε the vector of random errors. We now use the common assumption that $\varepsilon \sim N_p(0, \Sigma)$ and because $E_R\mu = 0$, the maximum likelihood estimate of μ is given by

$$\hat{\mu} = y - S_R E_R^T \left(E_R S_R E_R^T\right)^{-1} Ey \tag{10.3}$$

with expected value given by

$$E[\hat{\mu}] = \mu + \delta - S_R E_R^T \left(E_R S_R E_R^T\right)^{-1} E\delta \tag{10.4}$$

We note that the estimator is unbiased only in the absence of systematic errors, that is, $E[\hat{\mu}] = \mu$ only when $\delta = 0$. Bagajewicz (2005) defined *induced bias* as the bias that is observed in all reconciled values due to undetected biases. From Eq. (10.4) we conclude that the vector of induced biases is given by

$$\hat{\delta} = \left[I - S_R E_R^T \left(E_R S_R E_R^T\right)^{-1} E_R\right]\delta \tag{10.5}$$

We note that even when only one measurement is bias, the induced bias is manifested in all the estimators smearing them. This was illustrated in Chap. 5.

Any bias detection technique is only effective to detect gross errors of a size that is above a certain threshold. Below such threshold, biases are not detected and they remain as induced biases in the reported estimators. Thus, small induced biases will always persist. This leads to a new definition of accuracy which is presented next.

Accuracy of Estimators

For consistency, it is proposed to extend Eq. (10.1) to define accuracy of an estimate in the same form. Thus, the accuracy of an estimator (or software accuracy) is defined as the sum of the maximum undetected induced bias plus the precision of the estimator, that is,

$$\hat{a}_i = \hat{\sigma}_i + \delta_i^* \tag{10.6}$$

where $\hat{a}_i$, δ_i^*, and $\hat{\sigma}_i$ are the accuracy, the maximum undetected induced bias, and the precision (square root of variance $\hat{S}_{ii}$) of the estimator, respectively. The accuracy of the system can be defined in various ways, for example, making an average of all accuracies or taking the maximum among them. Since this involves comparing the accuracy of measurements of different magnitudes, relative values like the following are used:

$$\tilde{\alpha}_S = \underset{\forall i}{\mathrm{Max}} \left\{ \frac{\tilde{a}_i}{F_i} \right\} \tag{10.7}$$

where $\tilde{\alpha}_S$ is the accuracy of the system.

Maximum Undetected Induced Bias

Assume now that the maximum power measurement test is used. The maximum power test statistics for stream k under the presence of gross error in stream s ($Z_{k,s}^{\mathrm{MP}}$) is given by the following expression:

$$Z_{k,s}^{\mathrm{MP}} = \frac{\left| S_R^{-1}(\hat{\mu} - y)_k \right|}{\sqrt{W_{kk}}} = \frac{\left| e_k^T W \,(\mu + \delta_s \, e_s + \varepsilon) \right|}{\sqrt{W_{kk}}} \tag{10.8}$$

where $W = E_R^T (E_R S_R E_R^T)^{-1} E_R$. The hypothesis test consists of comparing the statistic with a critical value associated with the chosen degree of confidence. Thus, the usual assumption of white noise and confidence $p = 95\%$ renders a critical value. One can, of course, add the Bonferroni correction. Thus, the expected value of the maximum power test is

$$E[Z_{k,s}^{\mathrm{MP}}] = \frac{\left| W_{ks} \, \delta_s \right|}{\sqrt{W_{kk}}} \tag{10.9}$$

This test is consistent, that is, under the assumption of one gross error, it points to the right location. In other words,

$$E[Z_{s,s}^{\mathrm{MP}}] > E[Z_{k,s}^{\mathrm{MP}}] \qquad \forall k \neq s \tag{10.10}$$

Therefore, a gross error in variable s (δ_s) will be detected to be in that variable (and not elsewhere) if its value is larger than a threshold value $\delta_{\mathrm{crit},s}^{(p)}$ given by

$$\delta_{\mathrm{crit},s}^{(p)} = \frac{Z_{\mathrm{crit}}^{(p)}}{\sqrt{W_{ss}}} \tag{10.11}$$

In turn, the corresponding maximum undetectable induced bias in variable i ($\hat{\delta}^{(p)}_{\text{crit},i,s}$) (always under the assumption of one gross error) due to the undetected gross error in variable s is

$$\hat{\delta}^{(p)}_{\text{crit},i,s} = \left[[I - S_R W] e_s \delta^{(p)}_{\text{crit},s} \right]_i = [(I - S_R W)_{is}] \delta^{(p)}_{\text{crit},s}$$

$$= \frac{[(I - S_R W)_{is}]}{\sqrt{W_{ss}}} Z^{(p)}_{\text{crit}} \tag{10.12}$$

Thus, the maximum induced bias that will be undetected is given by (Bagajewicz, 2005)

$$\hat{\delta}^{(p,1)}_i = \underset{\forall s}{\text{Max}}\, \hat{\delta}^{(p)}_{\text{crit},i,s} = Z^{(p)}_{\text{crit}} \underset{\forall s}{\text{Max}} \frac{[(I - S_R W)_{is}]}{\sqrt{W_{ss}}} \tag{10.13}$$

We note here that although the maximum power test is consistent, that is, it will point to the right location of the bias, this does not mean that the maximum undetectable induced value from a location of a gross error in some other variable cannot be larger.

Maximum Power Measurement Test-Based Software Accuracy

One can now present the definition of software accuracy of order one as follows:

The accuracy of the estimator of a variable i, under the assumption of the use of the maximum power (MP) *test with confidence p and the presence of only one gross error is given by*

$$\tilde{a}^{\text{MP}(p,1)}_i = \sqrt{[\hat{S}_R]_{ii}} + Z^{(p)}_{\text{crit}} \underset{\forall s}{\text{Max}} \frac{[I - (S_R W)_{is}]}{\sqrt{W_{ss}}} \tag{10.14}$$

or in relative terms

$$\tilde{\alpha}^{\text{MP}(p,1)}_i = \frac{\sqrt{[\hat{S}_R]_{ii}}}{F_i} + \frac{Z^{(p)}_{\text{crit}}}{F_i} \underset{\forall s}{\text{Max}} \frac{[I - (S_R W)_{is}]}{\sqrt{W_{ss}}} \tag{10.15}$$

We call it software because we are focusing on the maximum value that will go undetected, that is, the worst-case scenario.

Higher-order accuracies can be defined the same way: Consider now the presence of n_T gross errors. The maximum power statistics will render

$$E[Z^{\text{MP}}_{k,T}] = \frac{\left| \sum_{\forall s \in T} W_{ks}\, \delta_s \right|}{\sqrt{W_{kk}}} \tag{10.16}$$

where T is the set of n_T gross errors located in specific variables.

We assume that the gross error detection and identification technique will point to the right location of the gross errors. The same can be said by the uncertainty of the gross error location arising from equivalency theory (Bagajewicz and Jiang, 1998) covered in Chap. 7.

The MP test will flag for any combination of gross errors such that

$$Z_{\text{crit}}^{(p)} \leq \frac{\left| \sum_{\forall s \in T} W_{ks}\, \delta_s \right|}{\sqrt{W_{kk}}} \tag{10.17}$$

Therefore, the sets of critical values of a particular set of gross errors T is not unique and cannot be uniquely determined as in the case of a single gross error. Consider one such set of critical values. We then write

$$\left| \sum_{\forall s \in T} W_{ks}\, \delta_{\text{crit},s}^{(p)} \right| = Z_{\text{crit}}^{(p)} \sqrt{W_{kk}} \qquad \forall k \tag{10.18}$$

The corresponding induced bias in variable i is

$$\hat{\delta}_i^{(p)} = \left[\left[I - S_R W \right] \delta_{\text{crit}}^{(p)} \right]_i = \delta_{\text{crit},i}^{(p)} - \sum_{s \in T} (S_R W)_{is}\, \delta_{\text{crit},s}^{(p)} \tag{10.19}$$

where $\delta_{\text{crit}}^{(p)}$ is the vector containing a critical value of the gross error size in the selected positions corresponding to the set T at the confidence level p. In order to find the maximum possible undetected induced bias, one has to explore all possible values of gross errors in the set. Thus, for each set T we obtain the maximum induced and undetected bias by solving the following problem:

$$\left.\begin{array}{l} \hat{\delta}_i^{(p)}(T) = \underset{\forall s \in T}{\text{Max}} \left| \delta_{\text{crit},i} - \sum_{s \in T} (S_R W)_{is}\, \delta_{\text{crit},s} \right| \\ s.t \\ \left| \sum_{\forall s \in T} W_{ks}\, \delta_{\text{crit},s} \right| \leq Z_{\text{crit}}^{(p)} \sqrt{W_{kk}} \qquad \forall k \end{array}\right\} \tag{10.20}$$

This problem should be solved for each set of n_T gross errors located in specific variables given by the set T and the unknowns are the components of the vector δ_{crit}, where the inequality is written for convenience (the constraint is binding at the optimum).

We formally define the location of the gross errors through the use of a vector of binary variables q_T, that is, $q_{T,s} = 1$ if a gross error of set T is located in variable s and zero otherwise. Clearly $\sum_{\forall s} q_{T,s} = n_T$. In addition, we can drop the absolute value from the objective because one can invert the signs of all gross errors and obtain the same answer. Thus, we can write

$$\left.\begin{aligned}
&\hat{\delta}_i^{(p)}(T) = \text{Max}\left\{\delta_{\text{crit},i} q_{T,i} - \sum_{\forall s} (SW)_{is}\,\delta_{\text{crit},s} q_{T,s}\right\} \\
&s.t. \\
&\qquad \left|\sum_{\forall s} \frac{W_{ks}}{\sqrt{W_{kk}}}\delta_{\text{crit},s} q_{T,s}\right| \le Z_{\text{crit}}^{(p)} \qquad \forall k \\
&\qquad -\sum_{\forall s} \frac{W_{ks}}{\sqrt{W_{kk}}}\delta_{\text{crit},s} q_{T,s} \le Z_{\text{crit}}^{(p)} \qquad \forall k \\
&\qquad q_{T,k}\left|\delta_{\text{crit},k}\right| \ge 0 \qquad \forall k
\end{aligned}\right\} \tag{10.21}$$

After the absolute value is replaced by the following two standard inequalities:

$$\sum_{\forall s} \frac{W_{ks}}{\sqrt{W_{kk}}}\delta_{\text{crit},s} q_{T,s} \le Z_{\text{crit}}^{(p)} \qquad \forall k \tag{10.22}$$

$$-\sum_{\forall s} \frac{W_{ks}}{\sqrt{W_{kk}}}\delta_{\text{crit},s} q_{T,s} \le Z_{\text{crit}}^{(p)} \qquad \forall k \tag{10.23}$$

the problem is linear for a given instrumentation network (the vector q_T is a constant here).

Therefore, since one has to consider all possible combinations of bias locations, we write

$$\hat{\delta}_i^{(p,n_T)} = \underset{\forall T}{\text{Max}}\,\delta_i^{(p)}(T) \tag{10.24}$$

which implies that one has to solve problem in Eq. (10.21) for all combinations of n_T gross errors in the system. One can convert this into a simple MILP problem by recognizing that it contains products of integer and continuous variables for which standard transformations are available. Indeed, any product of a continuous (positive) variable x and a binary variable y can be linearized as follows:

$$z = x \cdot y \quad \Leftrightarrow \quad \begin{cases} z - y\,\Gamma \le 0 \\ z \ge 0 \\ (x - z) - (1 - y)\,\Gamma \le 0 \\ (x - z) \ge 0 \end{cases} \tag{10.25}$$

where Γ is an upper bound on x. Indeed, when $y = 0$, then the first two equations render $z = 0$ and the last two are trivially satisfied (the third is trivial because it renders $x \leq \Gamma$ and the fourth renders another trivial equation). In turn, when $y = 1$, then the last two equations render $x = z$, the correct value and the first two trivially satisfied ($z = x \leq \Gamma$ and $z = x \geq 0$).

Therefore, one can now complete the definition of accuracy as follows:

Definition *The accuracy of the estimator of a variable i, under the assumption of the use of the maximum power test and the presence of t gross errors is given by*

$$\tilde{a}_i^{\mathrm{MP}(p,n_T)} = \sqrt{[\hat{S}_R]_{ii}} + \hat{\delta}_i^{(p,n_T)} \tag{10.26}$$

Example 10.1: Consider the simple example of Fig. 10.1, which we use to clarify the intricacies of software accuracy.

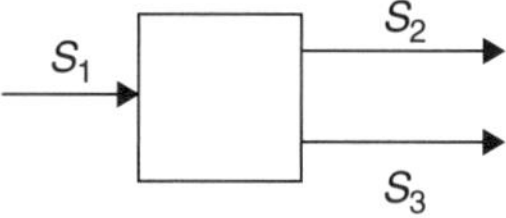

FIGURE **10.1** System for Example 10.1.

The accuracy of the different streams in the presence of one gross error and for all measurements with equal precision: $\sigma_i^2 = 1$ is shown in Table 10.1. The maximum undetected gross error is 3.395 and it renders the same value of the maximum power statistics, regardless of whether it is present in stream S_1, S_2, or S_3. This is a result of gross error equivalency (see Chap. 6). The precisions of the estimators are all the same and, because of the fact that all variances are the same, the software accuracy is the same for all cases. Interestingly, while the maximum undetected bias is 3.395 regardless of location, it gives rise to the corresponding value of accuracy for each stream when such bias is on that stream. In other words, the maximum undetected bias for stream S_1 takes place when the real bias is actually in stream S_1. When this takes place, streams S_2 and S_3 have lower induced biases (1.948 for both). Note that, unlike other examples, the values of precision and accuracy for Example 10.1 are reported in absolute values not percentage ones.

Stream	F_i	σ_i	$\hat{\sigma}_i$	$\tilde{a}_i^{MP(0.95,1)}$	$\delta_{crit,s}^{(p)}(s)$
S_1	100	1	0.816	3.079	3.395 (1)
S_2	50	1	0.816	3.079	3.395 (2)
S_3	50	1	0.816	3.079	3.395 (3)

TABLE **10.1** Accuracies of Order One for Example 10.1 (All Measurements with the Same Variance)

Things are different when measurement variances are not equal. In Table 10.2, we summarize the results when the variance of the three streams is different, namely $\sigma_i^2 = 1, 2$, and 3, respectively. In this case, because the sensors in streams

Stream	F_i	σ_i	$\hat{\sigma}_i$	$\tilde{a}_i^{MP(0.95,1)}$	$\delta_{crit,s}^{(p)}(s)$
S_1	100	1	0.913	4.914	4.801 (1)
S_2	50	1.414	1.155	4.355	4.801 (2)
S_3	50	1.732	1.225	3.625	4.801 (3)

Table 10.2 Accuracies of Order One for Example 10.1 (Measurements with Different Variance)

S_2 and S_3 are now of worse precision, the ability of the MP test to detect a bias also deteriorates and the maximum undetected bias (equal for all locations because of equivalency) is 4.801. However, because of the different variances, accuracy is now different for each stream, because the induced bias changes. For example, the maximum undetected bias occurring in stream S_1 gives rise to an induced bias of 4.00 in the same stream, whereas it renders smaller induced biases in streams S_2 and S_3 (1.6 and 2.4, respectively). The same can be observed when the measurement bias occurs in stream S_2, that is, the maximum induced bias will show in stream S_2 and smaller induced biases will show in the other two streams.

We conclude that, despite equivalencies, that is, inability to determine the bias location, accuracy can still be calculated. This is not true when there are two or more gross errors in the system. In fact, any two or three gross errors participate in a loop (see Chap. 7) and therefore render an unbounded solution to the accuracy problem.

Example 10.2: We illustrate the above concepts using the following process of Fig. 10.2. We recognize this to be a subset of the process shown in Fig. 7.3, where its equivalent sets were identified. We consider the case where all measurement variances are $\sigma_i^2 = 1$, except $\sigma_2^2 = 0.2$. The values of precision and accuracy of the different variables for order one are given in Table 10.3.

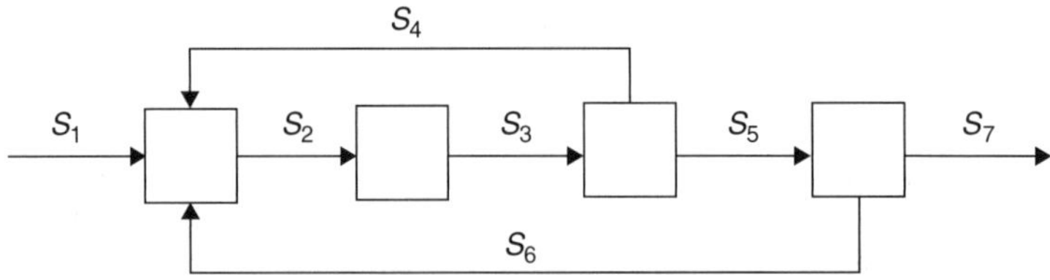

Figure 10.2 Flowsheet for Example 10.2.

The table illustrates that flows F_4 and F_6 have a reasonable precision (3.3 and 3.6%, respectively) but a very bad accuracy (of the order of 10%), whereas others do not. If for example, one is only interested in F_1 or F_7, then the accuracy is an acceptable value of 1.56%. That is, only induced gross errors of size smaller than 0.95 will be undetected.

Table 10.4 shows the results for the accuracy of order two. The last column shows the values and locations of gross errors that produce the largest undetected bias.

We first note that F_1 and F_7 have the same accuracy because they are part of an equivalent set (see Chap. 7). Second, we note that all but one stream have the

Stream	F_i	σ_i	σ_i/F_i (%)	$\hat{\sigma}_i$	$\hat{\sigma}_i/F_i$ (%)	$\tilde{a}_i^{MP(0.95,1)}$	$\tilde{\alpha}_i^{MP(0.95,1)}$ (%)	$\delta_{crit,s}^{(p)}(s)$
S_1	100	1	1	0.614	0.614	1.56	1.56	2.509 (1)
S_2	140	0.447	0.319	0.389	0.278	1.74	1.24	1.788 (2)
S_3	140	1	0.714	0.389	0.278	1.74	1.24	1.788 (2)
S_4	20	1	5	0.659	3.294	1.80	9.00	2.632 (4)
S_5	120	1	0.833	0.629	0.525	1.64	1.37	2.548 (5)
S_6	20	1	5	0.713	3.569	2.15	10.77	2.827 (6)
S_7	100	1	1	0.614	0.614	1.56	1.56	2.509 (7)

Table 10.3 Accuracies of Order One for Example 10.2

Stream	$\tilde{a}_i^{MP(0.95,2)}(\%)$	$\tilde{\alpha}_i^{MP(0.95,2)}(\%)$	$\delta_{crit,s_1}^{(p)}, \delta_{crit,s_2}^{(p)}$ (s_1, s_2)
S_1	2.748	2.748	3.696, 1.958 (1,7)
S_2	2.771	1.980	2.317, 4.208 (2,3)
S_3	2.771	1.980	2.317, 4.208 (2,3)
S_4	3.374	16.869	1.891, 4.208 (2,4)
S_5	2.709	2.258	–1.903, 3.619 (4,5)
S_6	3.145	15.724	1.841, 3.818 (5,6)
S_7	2.748	2.748	1.958, 3.696 (1,7)

Table 10.4 Accuracies of Order Two for Example 10.2

largest undetected induced bias contributed by biases in their own measurements. Finally, we confirm that the smearing of data reconciliation renders induced errors smaller than the actual errors.

We now turn to analyze the case of three gross errors. In this case the definition fails to produce a bounded value of induced bias. Indeed, all streams in this flowsheet participate in a loop of three streams, which raises the possibility of a set of three gross errors in these streams to go completely undetected. Indeed, take for example, streams F_1, F_6 and F_7. In this set, consider a set of three gross errors of the same size but different sign as follows: δ_1, $\delta_6 = -\delta_1$ and $\delta_7 = \delta_1$. Such a set renders the maximum power test unable to find gross errors. We illustrate this case later.

Graphical Representation of Undetected Biases

Although we rely on problem in Eq. (10.21) to calculate the software accuracy, which is an extreme (worst case) value, we still need to understand how multiple gross errors that are undetected can combine. When one bias or gross error is present, then it will go undetected if Eqs. (10.22) and (10.23) hold. If, however, one has more than one gross errors, different (infinite actually) combinations can go undetected. Sometimes, one bias will go undetected and the other will be detected, etc.

We start with two gross errors: To determine what are the combinations that will go undetected, we assume we use MP measurement test statistics for two measurements i_1, i_2. These are given by

$$Z_1^{MP} = \frac{\left|W_{i_1 i_1}\delta_1 + W_{i_1 i_2}\delta_2\right|}{\sqrt{W_{i_1 i_1}}} \tag{10.27}$$

$$Z_2^{MP} = \frac{\left|W_{i_2 i_1}\delta_1 + W_{i_2 i_2}\delta_2\right|}{\sqrt{W_{i_2 i_2}}} \tag{10.28}$$

In the serial elimination strategy, if one gross error has been detected, the corresponding measurement is eliminated and a new reconciliation is made rendering a new statistics. Which gross error is detected (or detected first) depends on which test statistic is larger (Z_1 or Z_2). Therefore, it is necessary to identify the regions where we have $Z_1 \geq Z_2$ and $Z_2 \geq Z_1$.

Now, $Z_1 = Z_2$ when

$$\frac{W_{i_1i_1}\delta_1 + W_{i_1i_2}\delta_2}{\sqrt{W_{i_1i_1}}} = \frac{W_{i_2i_1}\delta_1 + W_{i_2i_2}\delta_2}{\sqrt{W_{i_2i_2}}} \tag{10.29}$$

in quadrants I and III, or

$$\frac{W_{i_1i_1}\delta_1 + W_{i_1i_2}\delta_2}{\sqrt{W_{i_1i_1}}} = -\frac{W_{i_2i_1}\delta_1 + W_{i_2i_2}\delta_2}{\sqrt{W_{i_2i_2}}} \tag{10.30}$$

in quadrants II and IV. Then, we have two lines

$$\left(\sqrt{W_{i_1i_1}} - \frac{W_{i_2i_1}}{\sqrt{W_{i_2i_2}}}\right)\delta_1 = \left(\sqrt{W_{i_2i_2}} - \frac{W_{i_1i_2}}{\sqrt{W_{i_1i_1}}}\right)\delta_2 \text{ (line 1)} \tag{10.31}$$

$$\left(\sqrt{W_{i_1i_1}} + \frac{W_{i_2i_1}}{\sqrt{W_{i_2i_2}}}\right)\delta_1 = -\left(\sqrt{W_{i_2i_2}} + \frac{W_{i_1i_2}}{\sqrt{W_{i_1i_1}}}\right)\delta_2 \text{ (line 2)} \tag{10.32}$$

These two lines divide the plane into two regions: region A, where we have $Z_1 \geq Z_2$ and region B, where we have $Z_2 \geq Z_1$ (Fig. 10.3).

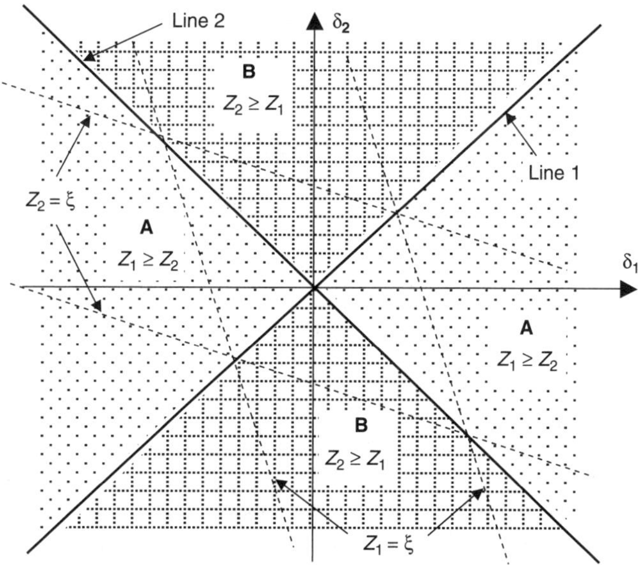

FIGURE 10.3 Identifying regions where $Z_1 \geq Z_2$ or vice versa $Z_2 \geq Z_1$.

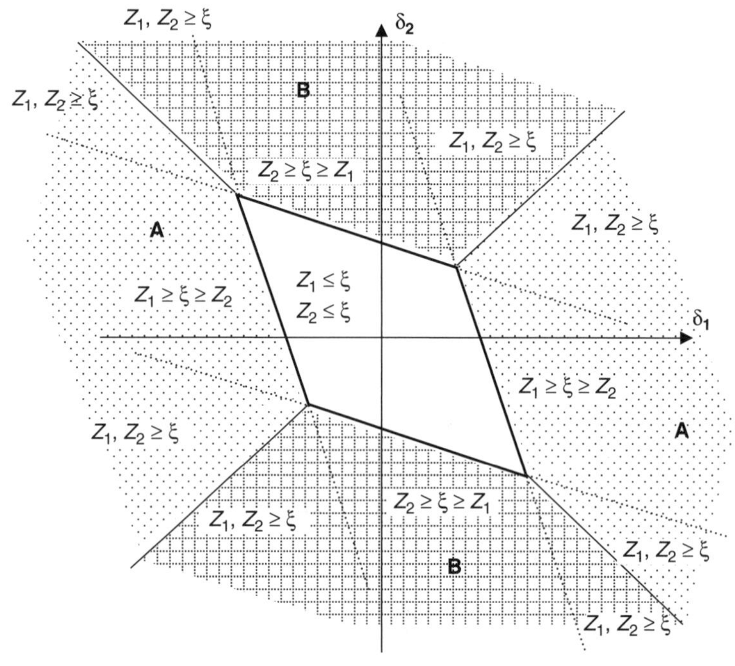

FIGURE 10.4 Regions corresponding to different values of Z_1 and Z_2.

The lines $Z_1 = \xi$ and $Z_2 = \xi$ (where ξ is the threshold value for the MT test statistics and is equal to 1.96 at a level of confidence of 95%) define the rhombus shown in Fig. 10.4. In this figure we assume that $W_{ij} > 0$. Inside the rhombus, we have $Z_1 \leq \xi$ and $Z_2 \leq \xi$; that is, none of the two gross errors are detected because they are too small to make the MT flag positive. Outside the rhombus, at least one of the two test statistic Z_1, Z_2 is greater than ξ, then at least one gross error is detected.

Now consider the case where gross error δ_1 is detected first, that is, the combination of two gross errors is outside the rhombus and in region A where we have $Z_1 \geq Z_2 \geq \xi$ or $Z_1 \geq \xi \geq Z_2$. Following the serial elimination strategy, the corresponding measurement i_1 is eliminated. Then, at the next stage, the new test statistic for the measurement i_2 is $Z_2' = |W''_{i_2 i_2}\delta_2| / \sqrt{W''_{i_2 i_2}}$ (here, W'' is the updated matrix W after measurement i_1 has been eliminated). Therefore, gross error δ_2 is also detected when $Z_2' = |W''_{i_2 i_2}\delta_2| / \sqrt{W''_{i_2 i_2}} \geq \xi$ or $|\delta_2| \geq \xi / \sqrt{W''_{i_2 i_2}}$ (that is, both gross errors are detected and gross error δ_1 is detected first). The limit $\xi / \sqrt{W''_{i_2 i_2}}$ is denoted by δ_2^*. As a consequence, only gross error δ_1 is detected when $|\delta_2| < \delta_2^*$. Therefore, the lines $|\delta_2| = \delta_2^*$ divide the regions under consideration (in region A and outside the rhombus, i.e., gross error δ_1 is detected first) into two smaller regions: region A_1 where we have $|\delta_2| < \delta_2^*$, that is, both gross errors are detected; and region A_2

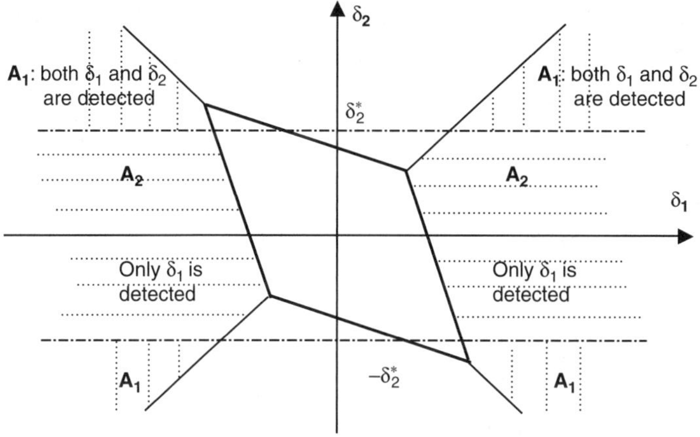

FIGURE 10.5 Regions corresponding to the detection of δ_2 given δ_1 is detected.

where we have $|\delta_2| < \delta_2^*$, that is, only gross error δ_1 is detected. These regions are depicted in Fig. 10.5.

Similarly, if gross error δ_2 is detected first (outside the rhombus and in region B) and the corresponding measurement i_2 is eliminated, gross error δ_1 is also detected when $Z_1' = |W''_{i_1i_1}\delta_1|/\sqrt{W''_{i_1i_1}} \geq \xi \Rightarrow |\delta_1| \geq \xi/\sqrt{W''_{i_1i_1}} = \delta_1^*$ (W' is the updated matrix W after measurement i_2 has been eliminated). Finally, all the regions corresponding to the possibilities of the presence of undetected gross errors are given in Fig. 10.6 for the case $W_{ij} > 0$.

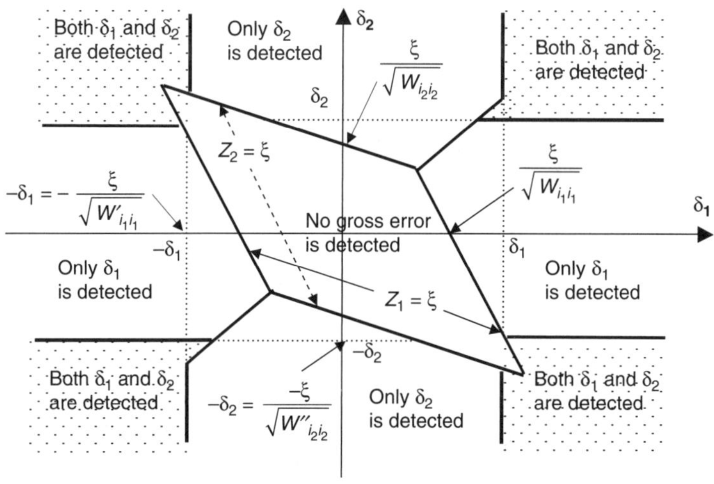

FIGURE 10.6 Different regions when two gross errors are present in the system.

Note that we have $\delta_2^* = \xi/\sqrt{W''_{i_2i_2}} \geq \xi/\sqrt{W_{i_2i_2}}$ and $\delta_1^* = \xi/\sqrt{W''_{i_1i_1}} \geq \xi/\sqrt{W_{i_1i_1}}$. These expressions stem from the fact that $W_{i_2i_2} \geq W''_{i_2i_2}$ and $W_{i_1i_1} \geq W''_{i_1i_1}$. These inequalities are somewhat intuitive. Indeed, they indicate that in the presence of a larger redundancy, the threshold for detection of a gross error is smaller. The formal proof can be found in the appendix of the article by Nguyen et al. (2006).

One can proceed with three gross errors and depict rhombus-type regions in three dimensions and so on. We omit depicting them as they exhibit the same features, that is, a core region where all errors are not detected, regions where only one gross error is detected, regions where two out of three errors are detected, and finally, the outside region where all errors will be detected. This geometry will become important when analyzing the economic value of accuracy in the next chapter.

Effect of Equivalency of Errors

Worth noticing that in some cases Eqs. (10.22) and (10.23) can be linearly dependent. This happens when the set of gross errors proposed is an equivalent set (see Chap. 6). We illustrate this for the system of Fig. 10.1.

Assume that one wants to compute the induced bias in stream S_1 by considering two biases in S_2 and S_3. This leads to two parallel lines, as illustrated in Fig. 10.7.

Thus, if the bias in stream S_2 and a bias in stream S_3 are equal but in opposite sign (such that the balance $S_1 = S_2 + S_3$ is still valid), then those biases cannot be detected no matter how big they are. This phenomenon is not unlikely. In fact, it can take place often for two gross errors.

Gross errors are not, however, unbounded, and although they can be larger, they become obvious by simple observation after a certain

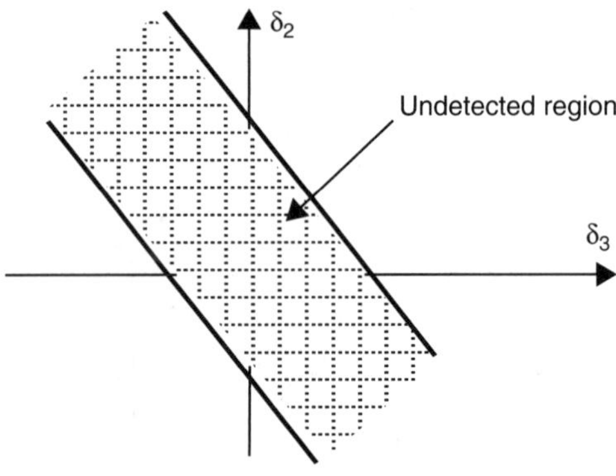

FIGURE 10.7 Graphical consequence of biases equivalency.

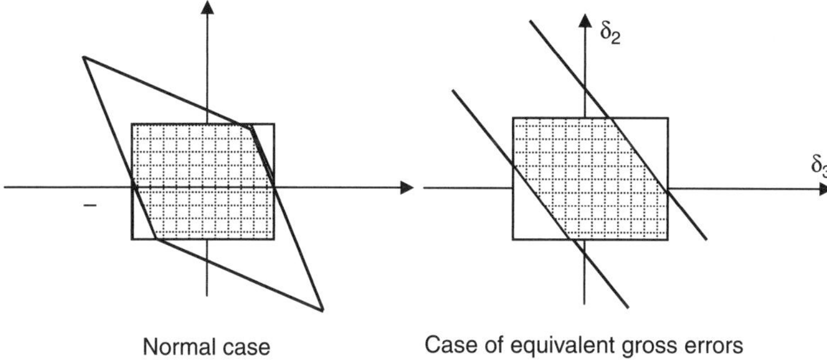

Figure 10.8 Box and rhombus constraints defining undetected biases.

size. Indeed, if a bias in a measurement passes a certain threshold, which is usually a certain percentage of the normal value of the variable, by common sense the operators can tell that there is bias in the measurement. The threshold above which bias is detected by the operators' judgment is used as (upper) limit for bias. Figure 10.8 shows such limits for both cases.

Thus, the accuracy calculations supported by the optimization problem given by Eq. (10.21) need to be made using additional box constraints on each variable.

$$-\delta_{\max,s} \le \delta^{(p)}_{\text{crit},s} \le \delta_{\max,s} \quad \forall s \tag{10.33}$$

Example 10.3: If accuracy of order two (two gross errors are considered $n_T = 2$) is to be calculated for the example of Fig. 10.1, then, without considering the aforementioned natural bounds, the maximum undetected biases $\delta^{(p)}_{\text{crit},s}$ for any set of two biases are infinite. Clearly, it can be seen that all the test statistics $E[Z^{\text{MP}}_{k,T}]$ takes zero value when: (i) biases in streams (S_1, S_2) or (S_1, S_3) are equal, (ii) biases in streams (S_2, S_3) are equal but in opposite sign. This fact can be deducted by investigating the material balance for the system in Fig. 10.1: a bias in stream S_1 can be detected (and can take any value) if there is another bias with the same magnitude in stream S_2 or S_3 because the material balance is fully satisfied in such situation. We show this in more detail through the equations. Matrices W and I-SW for the case where the variance of the three streams is different (σ_i^2 = 1, 2, and 3, respectively) are

$$W = \begin{pmatrix} 0.1667 & -0.1667 & -0.1667 \\ -0.1667 & 0.1667 & 0.1667 \\ -0.1667 & 0.1667 & 0.1667 \end{pmatrix} \tag{10.34}$$

$$I - S_R W = \begin{pmatrix} 0.8333 & 0.1667 & 0.1667 \\ 0.3333 & 0.6667 & -0.3333 \\ 0.5 & -0.5 & 0.5 \end{pmatrix} \tag{10.35}$$

Consider the two biases locations to be (S_1, S_2). Then Eqs. (10.22) and (10.23) for S_1 become

$$0.40825\,\delta_{crit,1} - 0.40825\,\delta_{crit,2} \le Z_{crit}^{(p)} \tag{10.36}$$

$$-0.40825\,\delta_{crit,1} + 0.40825\,\delta_{crit,2} \le Z_{crit}^{(p)} \tag{10.37}$$

For S_2 they become

$$-0.40825\,\delta_{crit,1} + 0.40825\,\delta_{crit,2} \le Z_{crit}^{(p)} \tag{10.38}$$

$$0.40825\,\delta_{crit,1} - 0.40825\,\delta_{crit,2} \le Z_{crit}^{(p)} \tag{10.39}$$

In other words, they are the same and they give rise to the situation depicted in Fig. 10.7. Quite clearly, there are combination of values $\delta_{crit,1}$ and $\delta_{crit,2}$ that can be unbounded but still satisfy the inequalities. Suppose they lay on a line that goes through the origin and is parallel to these lines. Then the relation between $\delta_{crit,1}$ and $\delta_{crit,2}$ will be

$$\delta_{crit,1} = \delta_{crit,2} \tag{10.40}$$

Under this condition, the accuracies can also be unbounded. Indeed, $\tilde{a}_1^{(MP,2)}$, the accuracy of stream S_1 is given by

$$\begin{aligned}\tilde{a}_1^{(MP,2)} &= \hat{S}_{11} + [I-SW]_{11}\delta_{crit,1}^{(p)} + [I-SW]_{12}\delta_{crit,2}^{(p)} \\ &= 1 + 0.8333 \times \delta_{crit,1}^{(p)} + 0.16667 \times \delta_{crit,2}^{(p)}\end{aligned} \tag{10.41}$$

If accuracy of order three ($n_T = 3$) is to be evaluated, the unboundedness is also observed if there are three biases, and the sum of biases in streams S_2 and S_3 is equal to the bias in stream S_1. The result is that the accuracy is also unbounded.

We now consider adding constraints [Eq. (10.33)]. If the bound on undetected biases is assumed to be 5 times the standard deviation of measurements (namely, 5.0, 7.07, and 8.66 for streams S_1, S_2, and S_3, respectively), then the accuracy of order two and three assume finite values. The accuracy of order two for stream S_1 is calculated for the case where the gross errors in streams S_1 and S_3 are

$$\begin{aligned}\tilde{a}_1^{(MP,2)} &= \hat{S}_{11} + \sum_{s=1,3} [I-SW]_{1s}\delta_{max,s} \\ &= 1 + 0.8333 \times 5.0 + 0.16667 \times 8.66 = 6.61\end{aligned} \tag{10.42}$$

Maximum values in other combinations of two streams render a smaller induced bias. The case of three gross errors is more interesting. We first recognize that the largest induced bias for stream S_1 will be achieved by the largest possible value for all three positive biases (or all three negative). This comes from the fact that the first row of *I-SW* has all positive values. If all three maximum values $\delta_{max,s}$ are used, however, the MP measurement test renders a value that is larger than the threshold (1.96). Thus certain combination of large values, but not all three at their maximum, renders a maximum undetected bias. These combinations are: $\delta_1 = 5$, $\delta_2 = 7.07$, $\delta_3 = 2.73$, and $\delta_1 = 5$, $\delta_2 = 1.14$, $\delta_3 = 8.66$. Both render an induced bias of 5.8 and an accuracy $\tilde{a}_1^{(MP,3)} = 6.8$. These two values of $\tilde{a}_1^{(MP,2)}$ and $\tilde{a}_1^{(MP,3)}$ are not too much larger than the one for order one $\tilde{a}_1^{(MP,1)}$.

Example 10.4: Downs and Vogel (1993) introduced the Tennessee Eastman (TE) process as an industrial challenge problem. Its flow diagram is shown in Fig. 10.9. A simplified TE model described by Ricker and Lee (1995a, 1995b).

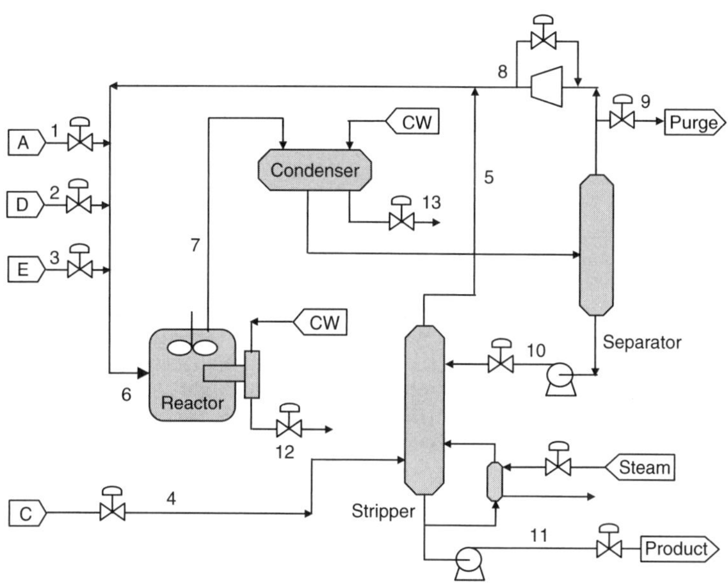

Figure 10.9 Tennessee Eastman process (*following Downs and Vogel, 1993*).

The steady-state equations used (Ricker and Lee, 1995a, 1995b) are given next. We start with the component balances

$$y_{i,6}F_6 - y_{i,7}F_7 + \sum_{j=1}^{3} \nu_{ij}R_j = 0 \qquad i = A, B, \ldots, H \tag{10.43}$$

$$y_{i,7}F_7 - y_{i,8}(F_8 + F_9) - x_{i,10}F_{10} = 0 \qquad i = A, B, \ldots, H \tag{10.44}$$

$$z_{i,1}F_1 + z_{i,2}F_2 + z_{i,3}F_3 + F_{i,5} + y_{i,8}F_8 + F_i^* - y_{i,6}F_6 = 0 \qquad i = A, B, \ldots, H \tag{10.45}$$

$$(1-\phi_i)x_{i,10}F_{10} - x_{i,11}F_{11} \qquad i = G, H \tag{10.46}$$

where ϕ_i ($i = G,H$) = separation factor of component i in the stripper, $z_{i',j'}$ $y_{i',j}$ and $x_{i',j}$ = molar fraction of chemical i in stream j, which can be feed stream ($z_{i,j}$), liquid stream ($x_{i,j}$), or gas stream ($y_{i,j}$); ν_{ij} = stoichiometry factor of chemical i in reaction j. The reaction rates R_j are given by the following expressions:

$$R_1 = \beta_1 V_{Vr} \exp\left[44.06 - \frac{42{,}600}{RT_r}\right] p_{A,r}^{1.08} p_{C,r}^{0.311} p_{D,r}^{0.874} \tag{10.47}$$

$$R_2 = \beta_2 V_{Vr} \exp\left[10.27 - \frac{19,500}{RT_r}\right] P_{A,r}^{1.15} P_{C,r}^{0.370} P_{D,r}^{1.00} \tag{10.48}$$

$$R_3 = \beta_3 V_{Vr} \exp\left[59.50 - \frac{59,500}{RT_r}\right] P_{A,r} (0.77 P_{D,r} + P_{E,r}) \tag{10.49}$$

where β_j is "tuning" factor of reaction j; $V_{v,r}$ is liquid volume in the reactor; T_r = temperature in the reactor, and $P_{i,r}$ is partial pressure of chemical i in the reactor.

The process has 47 streams and 28 equations. The precision of all the sensors is 2%. The bound on undetected biases is 5 times the precision of measurements. To determine the incidence matrix the nonlinear system of equations was linearized, as explained in Chap. 8.

The variables, their nominal operating conditions, and costs of associated sensors are given in Table 10.5. Values of flowrates F_i are given in kmol/h, P_r, P_s: pressure in reactor and separator, respectively (KPa); T_r, T_s: temperature in reactor and separator, respectively (K); subscripts A, B, C, D, E, F, G, H denote components; subscripts 6, 7, 8, 9, 10, 11 denote stream number. The variables listed in Table 10.5 are considered as candidates for measurements; other variables in the TE process (e.g., input flowrates F_1, F_2, F_3) are assumed to be either known by measurements (forced measurements) or of little importance for consideration.

Table 10.5 also shows the software accuracies of order one, two, and three corresponding to the case where all the variables are measured. These values deteriorate when fewer variables are measured and improve if more precise instruments are used. In this table, we indicate with an adjacent asterisk the variables that reach the limit (5 times the precision of measurements). Even though a bound is imposed on the undetected bias, the software accuracy for some estimators is still too high like accuracy for estimator F_6; this problem can be addressed by calculating accuracy as an average of all possible values instead of the maximum one. This is what we know now as stochastic accuracy. In this table, the case that the software accuracy is obtained when at least one of the undetected biases is hitting its limit (5 times the precision of measurements) is indicated by an asterisk, which is shown to have a high chance when accuracy of high order is considered. Moreover, even though a bound is imposed on the undetected bias, a small deviation (undetected bias) in measurements of variables with large magnitude can induce a large error (i.e., induced bias) in some other estimators with small magnitude through smearing effect of data reconciliation. This explains the large accuracy of estimator of flow F_6, which has a large induced bias due to a bias in estimator (that has a large magnitude) of reactor temperature T_r. These problems reveal the shortcomings of software accuracy.

Stochastic Software Accuracy

One of the criticisms of the definition of software accuracy is that it considers the worst-case scenario, that is, the case in which the undetected bias or gross error is very close to the detectability limit. Sensor failures, however, are such that smaller biases can exist. We do not have, however, the probability distribution of biases, so in the lack of it several different assumptions can be made; for example, it is gaussian, or uniform, within a certain interval, etc. Such data simply does not exist and needs to be gathered.

Variable	Value	Software Accuracy of Order One $\tilde{a}_i^{MP(0.95,1)}$ (%)	Software Accuracy of Order Two $\tilde{a}_i^{MP(0.95,2)}$ (%)	Software Accuracy of Order Three $\tilde{a}_i^{MP(0.95,3)}$ (%)
F_6	1.890	1.43	15.28	529.94
F_7	0.322	2.36	15.31	21.29*
F_{10}	0.089	3.28	5.10	11.22*
F_{11}	0.264	2.21	4.39	7.86*
$Y_{A,6}$	0.069	2.10	3.01*	4.14*
$Y_{B,6}$	0.187	2.57	4.38	8.74*
$Y_{C,6}$	0.016	8.23	10.46	14.54*
$Y_{D,6}$	0.035	5.46	11.73*	12.78*
$Y_{E,6}$	0.017	8.61	14.90*	15.60*
$Y_{F,6}$	1.475	1.49	6.78	12.92*
$Y_{G,6}$	0.272	3.82	7.48	12.10*
$Y_{H,6}$	0.114	3.30	5.11	318.73
$Y_{A,7}$	0.198	3.94	8.33	12.21*
$Y_{B,7}$	0.011	12.06	16.38*	16.38*
$Y_{C,7}$	0.177	3.37	6.77	11.23*
$Y_{D,7}$	0.022	6.33	8.18	13.82*
$Y_{E,7}$	0.123	1.72	3.62*	5.10*
$Y_{F,7}$	0.084	1.94	3.18*	4.34*
$Y_{G,7}$	0.330	3.84	7.41	12.26*
$Y_{H,7}$	0.138	3.15	4.86	11.33*
$Y_{A,8}$	0.240	4.00	8.34	43.31*
$Y_{B,8}$	0.013	11.01	15.09*	15.09*
$Y_{C,8}$	0.186	4.36	7.64	12.95*
$Y_{D,8}$	0.023	6.63	8.54	13.24*
$Y_{E,8}$	0.048	5.61	12.19*	13.27*
$Y_{F,8}$	0.023	8.08	14.75*	15.39*
$Y_{G,8}$	0.330	0.83*	0.97*	0.99*
$Y_{H,8}$	0.138	0.72	0.72	0.72
$Y_{A,9}$	0.240	1.14*	1.33*	3.69*
$Y_{B,9}$	0.013	17.53*	17.92*	18.20*
$Y_{C,9}$	0.186	6.16	48.13*	50.48*
$Y_{D,9}$	0.023	8.60	49.96*	52.23*
$Y_{E,9}$	0.048	12.03*	12.26*	12.36*
$Y_{F,9}$	0.023	14.33*	14.46*	14.52*
$Y_{G,9}$	0.258	2.04	4.48	5.64*
$Y_{H,9}$	0.002	59.99*	60.08*	60.13*
$Y_{D,10}$	0.136	3.97	10.33*	15.39*
$Y_{E,10}$	0.016	8.99	15.44*	16.12*
$Y_{F,10}$	0.472	2.07	4.12*	5.68*
$Y_{G,10}$	0.373	2.05	5.92*	6.33*
$Y_{H,10}$	0.2113	2.76	4.36	11.16*
$Y_{G,11}$	0.537	2.77	4.37*	11.16*
$Y_{H,11}$	0.438	2.76	4.36*	11.16*
P_r	2.806	0.1	0.1	0.1
T_r	393.6	0.1	0.1	0.1*
P_s	2.7347	4.35	11.60*	11.60*
T_s	353.3	4.21	11.22*	11.22*

*The variable has reached its preestablished limit (5 times the standard deviation).

TABLE 10.5 Software Accuracies of the TE Process

The failure of the sensor is a random event. After the initial burn-out period a sensor enters a constant failure rate period that lasts long enough before wear and tear as well as lack of proper preventive maintenance makes the failure rate increase. Most of the time, maintenance maintains the sensor within this constant failure rate period.

Thus, each sensor has its own failure frequency which is independent of what happens to other sensors. Thus, at a specific point t in time, the induced bias and the accuracy is the function of the number, location, and the sizes of undetected biases and also the number of eliminated measurements through the data reconciliation procedure. This is what is called the "state of the system." Therefore, the state of the system varies with time.

A Monte Carlo simulation can be used to simulate failure events for each sensor (within a specified time horizon), sampling the failure probabilities, which are obtained using sensor reliability data. When the conditions (e.g., failed or functioning) of all sensors in the system are available, the condition of the system is then obtained as a combination of the conditions of the individual sensors at each time. Information on the condition of the system within a specified time horizon, obtained from the Monte Carlo simulation, is then used to calculate the accuracy; hence the name stochastic-based accuracy.

After the sampling is performed, sensor failure times and also corrective actions are identified. Thus, in the interval of time in between each of the failure times and/or corrective actions of all sensors, there is no change of the system, and therefore an accuracy value can be obtained using the real random values of the undetected biases in that interval. More precisely, the accuracy value of estimator i within a time interval (t_s, t_{s+1}), $a_i, (t_s, t_{s+1})$, is calculated by

$$a_i,(t_s,t_{s+1}) = [\hat{\sigma}_i + \tilde{\delta}_i^{(p,n_T)}] = \hat{\sigma}_i + \left|\left([I - S_R W]\delta_{n_T}\right)_i\right| \tag{10.50}$$

where $\tilde{\delta}_i^{(p,n_T)}$ is the induced bias due to a specific set T of n_T undetected biases existing within the time interval (t_s, t_{s+1}), δ_{n_T} is the vector of bias sizes for the set T of n_T undetected biases, and $\hat{\sigma}_i$ is the precision of estimator i. In case one or more measurements are eliminated, $\hat{\sigma}_i$ is replaced by $\hat{\sigma}_i^R$, which is the residual precision after elimination of measurements.

The average value of accuracy within the time horizon T_h, $\bar{a}_i$, is calculated as the average value of all accuracy values in time intervals using the duration of the time intervals as weights.

$$\bar{a}_i = \sum_t \frac{a_i,(t_s,t_{s+1}).(t_{s+1} - t_s)}{T_h} \tag{10.51}$$

This average value $\bar{a}_i$ is also a random number. One Monte Carlo simulation attempt gives one value for $\bar{a}_i$. Finally, the expected value of accuracy of estimator i, $E[\hat{a}_i]$, is calculated as the mean value of $\bar{a}_i$ after N_{sim} simulations

$$E[\hat{a}_i] = \frac{1}{N_{sim}} \sum_{n=1}^{N_{sim}} \bar{a}_{i,n} \tag{10.52}$$

where $\bar{a}_{i,n}$ is average accuracy value obtained at simulation n. Next, we present the Monte Carlo simulation procedure.

Monte Carlo Sampling

We recall that in Chap. 2 three types of biases were introduced:

- Sudden fixed value bias
- Randomly emerging drift
- Deterministic drift

The first two are random events. Moreover, with the sudden fixed value bias, the bias magnitude (A) is also a random number and is here assumed without loss of generality to follow a normal distribution $h_k(\theta_k, \bar{\delta}_k, \rho_k)$, but it can have any distribution without affecting the sampling procedure steps. For randomly emerging drifts, the bias size increases continuously with time. The same is stated for the bias magnitude of deterministic drifts. However, deterministic drifts appear right at the beginning, when the sensor is put in use.

We now describe the sampling procedure for sudden fixed value biases (Bagajewicz and Nguyen, 2008): We assume that, at time $t = 0$, all sensors are as good as brand new. As time elapses, sensors may degrade and fail. Within the constant failure rate period of the bathtub curve, the failure of a sensor (sensor k) is a random event and its cumulative failure probability is given by (Bagajewicz, 2000)

$$f_k(t) = 1 - e^{-r_k t} \tag{10.53}$$

where r_k is failure rate of the sensor k.

To sample the time of failure, first we sample randomly the probability of failure $f_{k,1}(t)$ and calculate the failure time as follows:

$$t = -\frac{1}{r_k}\ln(1 - f_{k,1}(t)) \tag{10.54}$$

Obviously, a sensor with higher failure rate will have shorter time to failure.

The next step is to sample the bias size according to its distribution (recall that, without loss of generality, we use a normal distribution). If the bias is large enough to be detected then the corresponding measurement is eliminated and the failed sensor is repaired (repair time is R_k). This sensor may fail again, so we sample the next failure of this

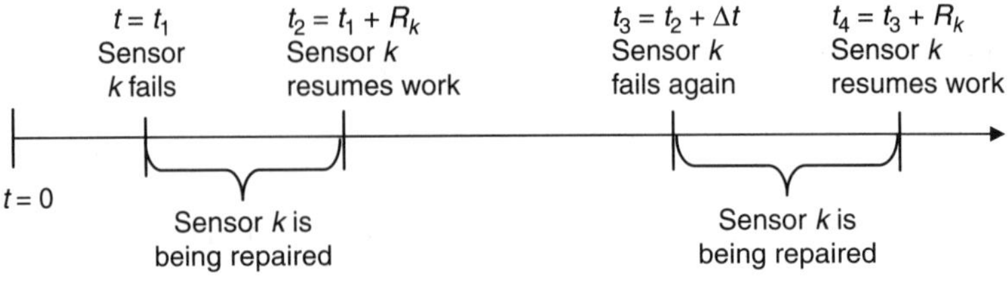

FIGURE 10.10 Result of sampling two consecutive detected failures.

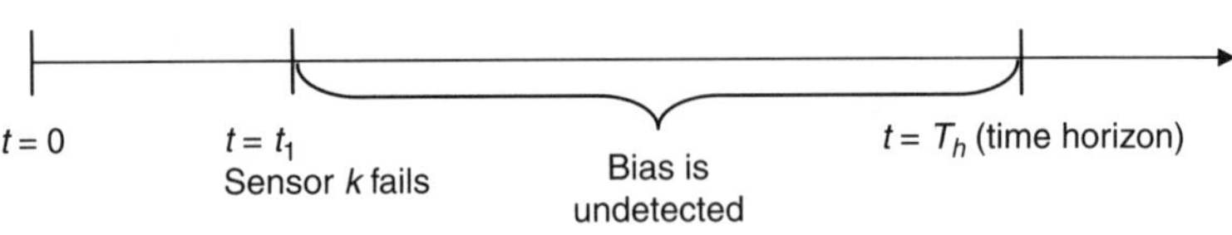

FIGURE 10.11 Sampling of undetected failure.

sensor and the time from the repair to the next failure is obtained. The magnitude of the bias is then sampled and if this bias is large enough to be detected, we have the situation depicted in Fig. 10.10.

If t_4 (Fig. 10.10) is still within the time horizon under consideration, we may have another failure, so we sample another failure time and bias size, and so on. If a bias is undetected and if no preventive maintenance, which will remove hidden biases, is available, it continues undetected for the rest of the time horizon. This possibility is illustrated in Fig. 10.11.

Consider now a randomly emerging drift. The time at which a bias appears is a random event. Because the size of the bias increases with time, the bias is eventually detected when the bias size reaches a threshold value. Then the sensor is repaired and resumes work. After that, we sample another failure of that sensor and the same cycle repeats. The described cycle is illustrated in Fig. 10.12.

With deterministic drifts, the situation is almost the same as with randomly emerging drifts. The difference is that bias appears right at the beginning when the sensor is put into use and no sampling is needed.

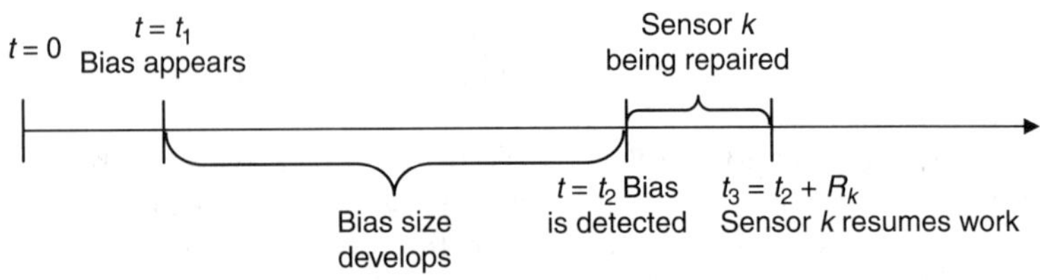

FIGURE 10.12 Sampling of failure for randomly emerging drifts.

Time (days) 0 12.2 54.8 174.5 177.5 318.9 365						
Sensor 1	No bias	Contains undetected bias S_1				
Sensor 2	No bias		Contains undetected bias S_2			
Sensor 3	No bias			Being repaired and measurement is eliminated	No bias	Contains undetected bias S_3
Sensor 4	no bias		Contains undetected bias S_4 (bias size increases with time)			
Sensor 5	Contains undetected bias S_5 (bias size increases with time)					
System condition	Contains bias in S_5	Contains biases in S_1 and S_5	Contains biases in S_1, S_2, S_4, and S_5	Contains biases in S_1, S_2, S_4, S_5, and one eliminated measurement	Contains biases in S_1, S_2, S_4, and S_5	Contains biases in S_1, S_2, S_3, S_4, and S_5

Table 10.6 Sampling Procedure Illustration

After the sampling of failure events for all sensors has been performed, the condition of the sensors system is then obtained as the combination of the conditions of all the sensors. One example of such scenario is illustrated in Table 10.6 for a system with five sensors in which three sensors (sensors 1, 2, and 3) are subjected to sudden fixed value bias, one sensor (sensor 4) is subjected to a randomly emerging drift, and the last one (sensor 5) is subjected to a deterministic drift.

The failure times and bias sizes for each kind of biases for each sensor are sampled separately and the overall bias is then calculated as combination of the three types of biases. When the bias is detected and sensor is repaired, we assume that the sensor is as good as brand new and samplings of the three types of biases are restarted in a new cycle. This situation is illustrated in Fig. 10.13.

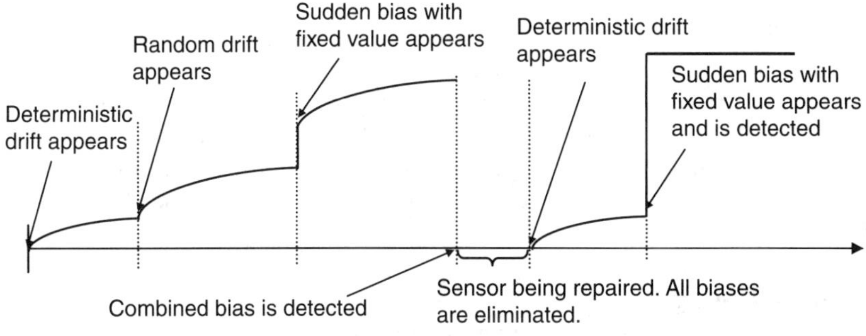

Figure 10.13 Occurrence of three types of biases in a sensor.

Constructing the sample shown in Table 10.6 is complicated by the fact that the detection of a sensor failure is a function of the presence of biases in other sensors. Therefore, one cannot know if a sensor bias will be detected until all sensors are sampled. Take, for example, sensors 1 and 2 in Table 10.6. At times 12.2 and 54.8, respectively, these sensors develop a bias, which the table says are undetected. This is not known until all the other sensors are sampled. In turn, sensor 3 develops a bias at time 174.5, which is detected. When the original sampling was performed, it was not known that this bias would be detected. If it was not, then the bias would stay undetected until the end of the horizon. In such case, we do not perform a new sampling for the same type of failure. This is part of our simplifying assumptions: We do not consider a second failure of a sensor that has already failed. To do this, one would need to resort to a different reliability function, corresponding to an already failed sensor. Thus, to consider the interactions between biases in measurements, the sampling was conducted in the following way:

- Failure times and bias sizes for every sensor in the system are sampled and recorded until the end of time horizon is reached.
- The time intervals between failures in the system are obtained by combining the failure times of all sensors, as illustrated in Table 10.7.
- At each failure time in the system, the maximum power measurement test (MPMT) is performed and the sensors that are detected being biased are singled out.
- If the MPMT cannot detect any bias, no action is needed. We then move on to investigate the next time interval until the end of time horizon is reached.
- If the MPMT flags the presence of biases, then, for each sensor with a detected bias a repair time is added and a new sampling for sudden fixed value biases and emerging drifts are added. Deterministic drifts are added if present.

Instantaneous Testing

In the above procedure the instantaneous testing is assumed. This means that the reconciliation is performed in real time and the gross errors are detected as soon as they show if they have enough size to be detected. Thus, the sampling procedure requires marching through the time horizon and continuously resampling the sensors whose failure has been detected.

Periodic Testing

This consists of applying the MPMT at fixed intervals ($t = 20, 40, 60, 80,$ etc.). This procedure reflects the fact that in practice, the gross errors detection (e.g., the measurement test) may be applied at periodic time intervals. This reduces the number of times resampling is needed.

Example 10.5: We now return to our three streams example of Example 10.1 (Fig. 10.1) so that we can compare with software accuracy results. We use instruments with variances $\sigma_i^2 = 1, 2$, and 3, respectively, as above. We also assume that the biases have zero mean and standard deviation $\rho_k = 2, 4$, and 6, respectively, failure rates of 0.025, 0.015, 0.005 (1/day) and repair time of 0.5, 2, and 1 day, respectively. The time horizon used was 5 years. As discussed, the system is barely redundant: Only one gross error can be determined, and when it is flagged by the measurement test, hardware inspection is needed to obtain its exact location. We focus on the accuracy of stream S_3 and for simplicity we only consider the situation that only one type of bias can occur in a sensor. We consider two cases. Undetected biases can take any value (no bound) and only one type of bias, sudden fixed value biases, is considered. The other case we consider is to impose a limit on undetected biases to be 5 times the precision of measurements. For comparison, the software accuracy under the assumption that undetected biases are limited to be lower than 5 times the precision of measurements is also calculated.

We now illustrate the sampling procedure for these three sensors. In the MP measurement test, due to gross error equivalency, the test statistics for three measurements are the same, so if the test statistics are greater than threshold value, we assume that the measurement with biggest bias size is the detected one. Figure 10.14 illustrates how the sampling procedure of the sudden fixed

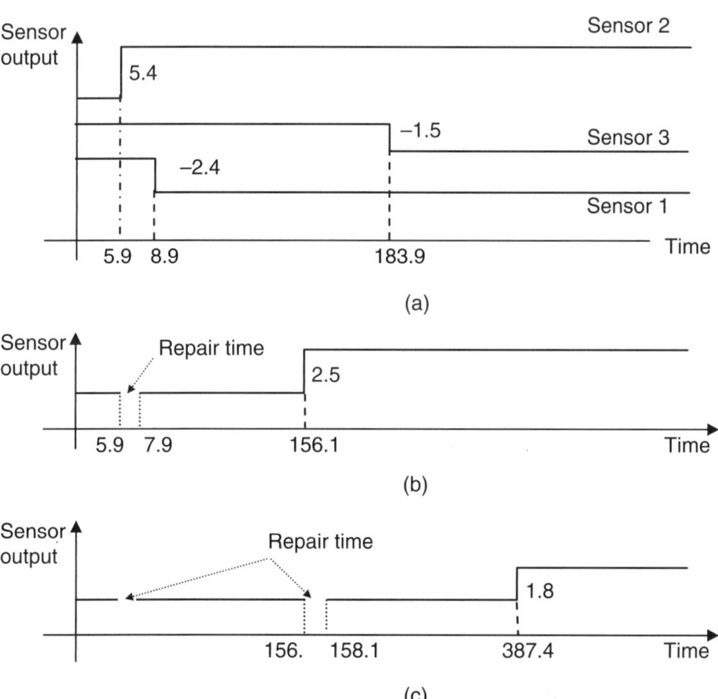

Figure 10.14 Illustration of sampling procedure with instantaneous testing (a) original sampling; (b) sensor two, first resampling; (c) sensor two, second resampling.

value bias with instantaneous testing is conducted. In this figure, the change (rather than absolute value) of sensor output (i.e., measurement) with time is depicted. The italic number right next to the step change is bias size. Figure 10.14*a* shows the original sampling for the three sensors, indicating that at t = 5.9 the failure in sensor 2 (bias size = 5.4) was detected as being higher than the threshold (4.8).

Biases in sensor 1 (taking place at t = 8.9) and sensor 3 (taking place at t = 183.9) are below their corresponding threshold values and are undetected. Figure 10.14*b* shows the repair time and a new sampling for sensor 2 with a bias of 2.5 at t = 156.1. The error is again detected and therefore a new sampling is performed. Figure 10.14*c* shows the corresponding repair time for sensor 2 and a third sampling for it.

Note that the threshold value to detect a bias in sensor 2 without the presence of other biases is 4.8; it is due to interaction of biases that bias size of 2.5 (<4.8) in measurement two can still be detected. However, this observation cannot be generalized because it is possible that the threshold value for detecting biases is larger (not just smaller) due to interactions. Clearly, we see that if the biases in measurements 1, 2 (or 1, 3) are equal to each other, the MT test cannot detect biases whatever sizes they have. Now, after the second resampling of sensor 2, the MP measurement test cannot detect any bias in the rest of the time horizon.

We now describe the sampling as applied for periodic testing in Fig. 10.15. In this case, the original sampling is the same as in the case of instantaneous testing (Fig. 10.15*a*). Because the MPMT is conducted at a scheduled time in periodic

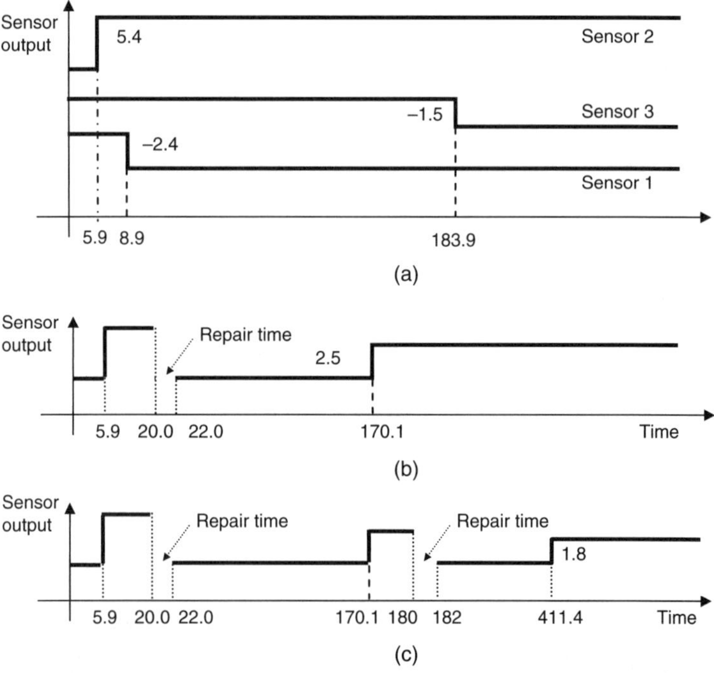

FIGURE 10.15 Illustration of sampling procedure with periodic testing (a) original sampling; (b) sensor two, first resampling; (c) sensor two, second resampling.

testing, the bias in sensor 2 will be detected at the next scheduled time of MPMT after it appears, that is, at $t = 20$ (an interval time of 20 is used), as shown in Fig. 10.15*b*. This means that the undetected biases contribute to a worse accuracy than in the instantaneous testing, as one would expect. Figure 10.15*b* shows the repair time of sensor 2 and a new sampling. Assuming the sampling had rendered the same numbers, one must observe that the new time for failure of sensor 2 is at $t = 170.2$ and the bias will be detected at $t = 180.0$, because of the regularity in testing. Although the same failure time samples as in instantaneous testing were used, the change is due to the fact that the sensor was now repaired at $t = 20$ and not at $t = 5.9$.

The results of the average accuracy values are shown in Fig. 10.16 for 1000 sampling attempts and for both types of testing types. Table 10.7 shows that additional sampling does not alter significantly the result. Clearly we see that when number of sampling $N \geq 10^4$, the expected value of accuracy converges.

It should be noted that because the system cannot identify where is the bias because of equivalency issues, the maximum undetected bias is larger than the corresponding threshold of the z-statistics (4.08), a situation that has already been pointed out earlier when discussing software accuracy with bounds. Indeed, equivalency issues and how to respond to them pose a significant problem. If one does not eliminate any measurement, then one gets the result reported in Fig. 10.16. However, if one eliminates one measurement at random, one could be eliminating the wrong measurement and therefore incur an even bigger induced bias.

In all cases of stochastic accuracy, the lowest accuracy value obtained was around 1.23 (very near to the precision value of 1.225). The lowest value corresponds to the case where all biases are detected, or undetected biases have very small sizes. When no bounds are used, the largest values observed are 12.5 for instantaneous testing and 16.5 for periodic testing (these values are obtained after 10^6 simulations; hence they are not seen in Fig. 10.16). When bounds are used, the corresponding values are 10.6 and 10.7, respectively. These values correspond to the cases where almost all biases are undetected and the sizes of the biases are large.

From Table 10.7 it is also obvious that, due to the delay in detecting the measurement biases of the periodic testing case, the expected value of accuracy obtained for periodic testing case is larger than the value obtained for instantaneous testing case. The stochastic accuracy with bounds on undetected biases is smaller than the stochastic accuracy without bounds although the difference is small, which suggests that, for this example, the chance that the undetected biases go beyond limit is small. Moreover, the largest stochastic values with bounds (10.6 and 10.7) are smaller than the software accuracy of order three with bounds (11.59), which highlights the fact that using maximum undetected bias is too conservative.

Consider now calculating the accuracy value at specific times, instead of calculating the average value of accuracy for the whole horizon. The results are shown in Fig. 10.17 (the number of samplings is 10^5). We see that, for small t, the sensors in the system are in good condition and few sensors fail. When time elapses, more and more sensors fail and that makes the accuracy worse (the accuracy value increases). This tells us that the stochastic accuracy is a function of time and therefore the real valuable tool is to know what is the expected value for different values of time, rather than one expected value over the whole (arbitrary) horizon. This information can be utilized in planning/scheduling the preventive maintenance activities to keep accuracy of estimators below certain values. This is possible because of the ability of preventive maintenance operations to detect all hidden biases so that the accuracy value returns back to normal value

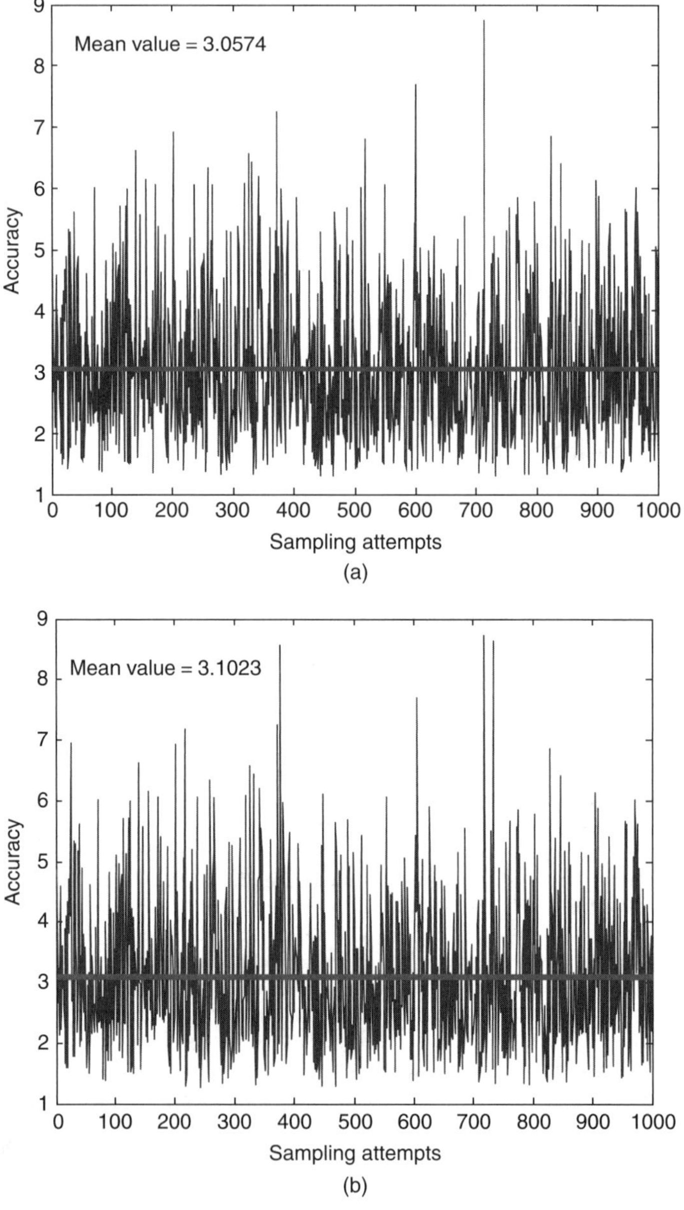

FIGURE 10.16 Average accuracy for Example 10.5 (a) instantaneous testing; (b) periodic testing.

(i.e., precision value). For example, for the case of instantaneous testing, if we want to keep expected value of accuracy of estimator lower than 2.7, it is necessary to perform preventive maintenance every 210 days (7 months).

Both the expected value over the whole time horizon and the value at a specific time of the instantaneous testing are smaller than the corresponding values of

Number of Sampling		10^3	10^4	10^5	10^6
Stochastic accuracy, without bounds	Instantaneous testing	3.0574	3.0774	3.0752	3.0761
	Periodic testing	3.1023	3.1308	3.1255	3.1303
Stochastic accuracy, with bounds	Instantaneous testing	3.0567	3.0640	3.0609	3.0600
	Periodic testing	3.0908	3.1048	3.0961	3.0992
Software accuracy, order one, with bounds		3.625	3.625	3.625	3.625
Software accuracy, order two, with bounds		9.09	9.09	9.09	9.09
Software accuracy, order three, with bounds		11.59	11.59	11.59	11.59

TABLE 10.7 Expected Value of Accuracy for Stream S_3 in Example 10.5

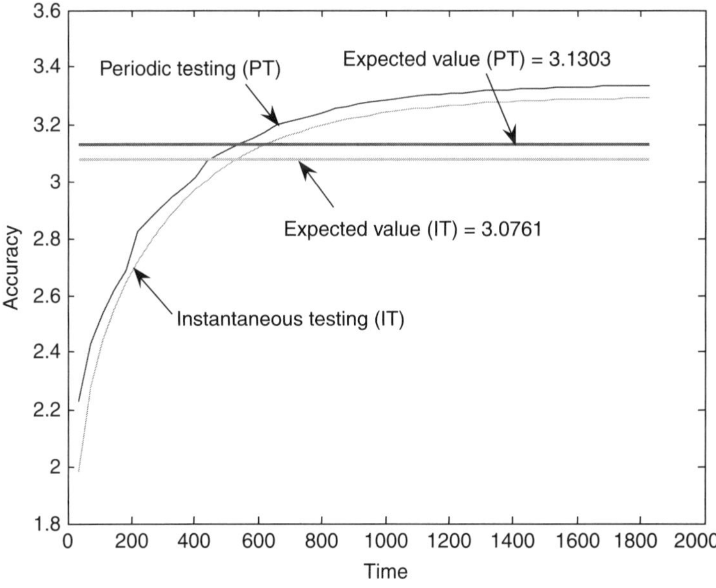

FIGURE 10.17 Expected accuracy at specific points in time (corrective maintenance only).

the periodic testing, as clearly depicted in Table 10.8 and Fig. 10.17. This is due to the delay in detecting biases of the periodic testing as explained above.

Another interesting feature to observe is the distribution of accuracy in any time interval. Figures 10.18 and 10.19 show this distribution for accuracy at specific times $t = 400$ and $t = 800$ for both testing types.

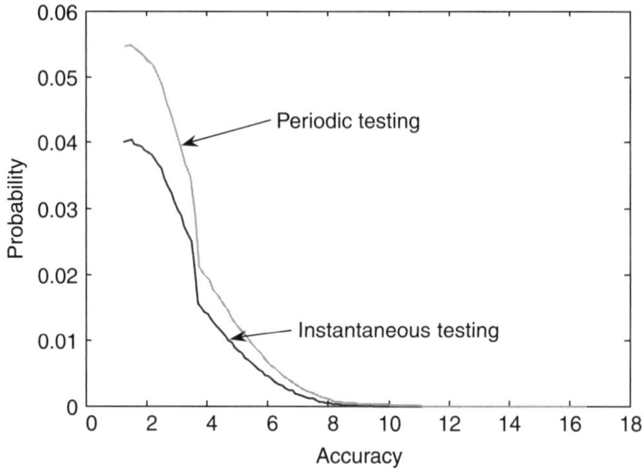

FIGURE 10.18 Distribution of accuracy at time $t = 400$.

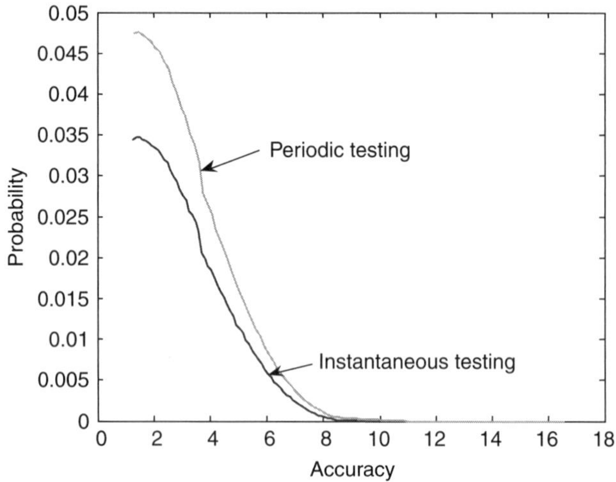

FIGURE 10.19 Distribution of accuracy at time $t = 800$.

The distribution of accuracy value at a specific point in time is a monotonic decreasing function, that is, a small accuracy value has a higher probability than a large accuracy value. The probability that the system contains few biases with small bias sizes (i.e., small accuracy value) is larger than the probability that the system contains many biases with large bias sizes (i.e., large accuracy value) because the latter is the extreme case. As time increases, the distribution function shifts slightly to the right and the slope decreases, that is, the probability for high accuracy value increases at the expense of lower probability for small accuracy value. As a consequence, the expected value of accuracy at a specific point in time increases when time increases. Figure 10.17 confirms this result. Figure 10.20 shows the distribution of the accuracy obtained for the whole time horizon (the average value $\overline{a}_i$).

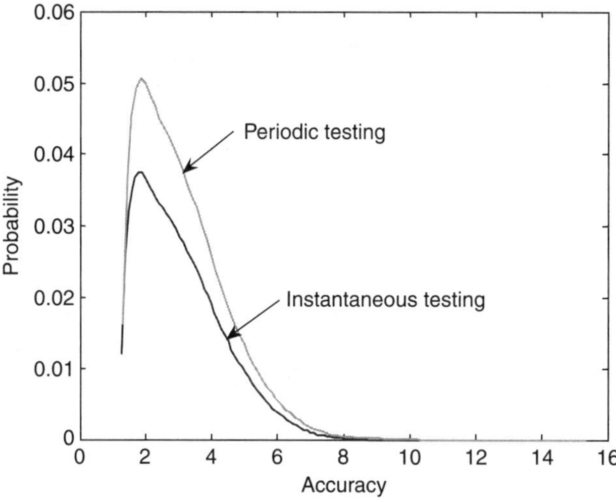

Figure 10.20 Distribution of the average accuracy over the whole time horizon.

The distribution of the average accuracy value is a combination of distributions of the accuracy values in all time intervals within the time horizon (that shift to the right as time increases). Figure 10.20 shows that, unlike the monotonic decreasing distribution function of accuracy at a time or in a time interval, the distribution of the average accuracy exhibits a peak. This can be explained by the fact that low accuracy values in the left-hand side of the peak have low probability at the end of time horizon while high accuracy values in the right-hand side of the peak have low probability in the whole time horizon. The expected value of accuracy is slightly to the right of this peak because the distribution function has a rather long tail on the right side.

Example 10.6: We now present the results of stochastic accuracy for the system of Example 10.2. In this case, we consider failure rate as follows. Sensors 1, 3, 5, 7: 0.01 (1/day) and sensors 2, 4, 6: 0.02 (1/day). The repair time is 1 day for sensors 1, 3, 5, 7 and 2 days for sensors 2, 4, 6. Finally, the biases are considered normally distributed with zero mean and variance equal to 4 times the standard deviation of measurements (for all sensors). Undetected biases are limited to be lower than 5 times the standard deviation of measurements. Table 10.8 shows the results. We include accuracies of order one and two for comparison. As in the previous small example, the stochastic accuracy is smaller and more realistic than the software accuracy (as expected). What one discovers is that software accuracy is not necessarily an unreasonable upper bound.

Example 10.7: Consider one large-scale example process, as illustrated in Fig. 10.21. Assume that all streams are measured with the flowrates given in Table 10.9 and with 2.5% precision. The failure rate is $r_i = 0.01$ (1/day), $i = 1, 3, 5,\dots, 23$ and $r_i = 0.02$ (1/day), $i = 2, 4, 6,\dots, 24$. Finally, the repair time is $R_i = 1$ day, $i = 1, 3, 5,\dots, 23$ and $R_i = 2$ days, $i = 2, 4, 6,\dots, 24$. Biases with sudden fixed values are assumed to be normally distributed with zero mean and variance equal to 4 times the standard deviation of measurements.

Stream	σ_i	Stochastic Accuracy $\hat{a}_i$ (%)	Software Accuracies	
			$\tilde{a}_i^{MP(0.95,1)}$ (%)	$\tilde{a}_i^{MP(0.95,2)}$ (%)
S_1	1	1.15	1.56	2.748
S_2	0.447	0.62	1.24	1.98
S_3	1	0.62	1.24	1.98
S_4	1	6.93	9	16.869
S_5	1	1.06	1.37	2.258
S_6	1	7.27	10.77	15.724
S_7	1	1.15	1.56	2.748

TABLE 10.8 Stochastic Accuracies for Example 10.2

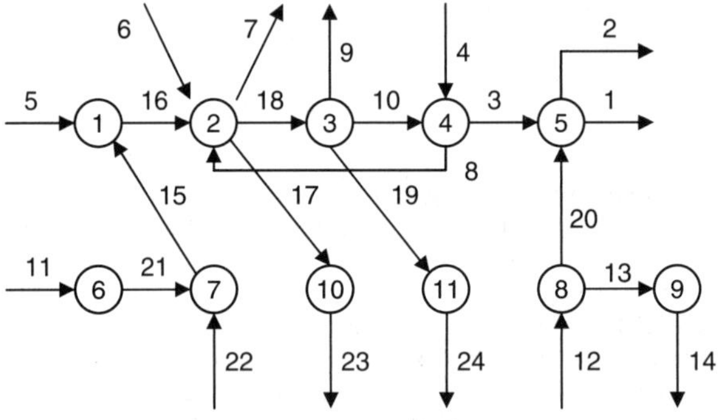

FIGURE 10.21 Flow sheet for Example 10.7.

Stream	Flow	Stream	Flow	Stream	Flow
S_1	140	S_9	10	S_{17}	5
S_2	20	S_{10}	100	S_{18}	135
S_3	130	S_{11}	80	S_{19}	45
S_4	40	S_{12}	40	S_{20}	30
S_5	10	S_{13}	10	S_{21}	80
S_6	45	S_{14}	10	S_{22}	10
S_7	15	S_{15}	90	S_{23}	5
S_8	10	S_{16}	100	S_{24}	45

TABLE 10.9 Flowrates for Example 10.2

No information regarding the slope of deterministic drifts and the shape and slope of randomly emerging drifts is available, so the following reasonable assumptions are made. The deterministic drifts with asymptotic shape is represented by the following expression: bias size $= A^*(1 - \exp(-t/T))$, with $A = 10$, $T = 500$. The randomly emerging drifts with assumed concave shape are represented by the following expression: bias size $= A^*\ln(1 + (t/T))$ with $A = 10$, $T = 200$. With these assumed parameters and when no other biases interfere with the detection of the drifts, usually the drifts will be detected around 1 year (for deterministic drifts) and 3 months (for randomly emerging drifts) after they start, which are reasonable numbers. The number of sampling attempts is 1000, time horizon is 365 (days). Finally, an upper bound of 5 times the standard deviation of measurements is imposed on all three types of biases.

In this example, we want to show the effect of the different types of biases on software accuracy. This is shown in Tables 10.10, 10.11, and 10.12 for sudden fixed value biases, emerging drifts, and deterministic drifts, respectively. Table 10.13 shows the case where all three types of biases are considered.

Streams	$\hat{\sigma}(\%)$	$\hat{a}(\%)$ Instantaneous Testing	$\hat{a}(\%)$ Periodic Testing	Streams	$\hat{\sigma}(\%)$	$\hat{a}(\%)$ Instantaneous Testing	$\hat{a}(\%)$ Periodic Testing
S_1	1.10	2.02	2.27	S_{13}	1.75	3.23	3.47
S_2	2.48	5.71	5.71	S_{14}	1.75	3.23	3.47
S_3	1.08	2.02	2.24	S_{15}	1.08	1.91	2.07
S_4	2.35	5.21	5.37	S_{16}	0.98	1.75	1.91
S_5	2.48	5.01	5.14	S_{17}	1.77	3.00	3.23
S_6	2.20	4.68	5.04	S_{18}	0.88	1.74	1.92
S_7	2.47	5.05	5.16	S_{19}	1.67	3.09	3.35
S_8	2.49	5.65	5.80	S_{20}	1.99	3.92	4.38
S_9	2.49	5.09	5.09	S_{21}	1.22	2.13	2.32
S_{10}	1.24	2.37	2.66	S_{22}	2.48	5.78	5.80
S_{11}	1.22	2.13	2.32	S_{23}	1.77	3.00	3.23
S_{12}	1.51	2.97	3.31	S_{24}	1.67	3.09	3.36

TABLE 10.10 Precision and Accuracy of Estimators for Example 10.7 (Sudden Fixed Value Biases)

The results shown in Table 10.10 (accuracy for sudden fixed value biases) and the tables that follow (accuracy for other types of biases) show that some estimators are sensitive to undetected biases in the system and the accuracy for such estimators is higher than the others. Usually this problem applies to the estimators of low flowrates such as the estimators of streams S_5, S_8, S_9, S_{17}, S_{22}, and S_{23}. On the other hand, some estimators are not very sensitive to the presence of undetected biases such as estimators of streams S_1, S_3. The estimators of these streams are said to be more robust to the presence of undetected biases. Usually the estimators of high flowrates exhibit this characteristic. The reason is that, with high flowrates, the induced bias is relatively small when

Streams	$\hat{\sigma}(\%)$	$\hat{a}(\%)$ Instantaneous Testing	$\hat{a}(\%)$ Periodic Testing	Streams	$\hat{\sigma}(\%)$	$\hat{a}(\%)$ Instantaneous Testing	$\hat{a}(\%)$ Periodic Testing
S_1	1.10	1.77	1.97	S_{13}	1.75	1.88	2.72
S_2	2.48	4.37	4.93	S_{14}	1.75	1.88	2.72
S_3	1.08	1.67	1.82	S_{15}	1.08	1.80	2.00
S_4	2.35	3.35	3.60	S_{16}	0.98	1.74	1.97
S_5	2.48	4.38	5.25	S_{17}	1.77	1.81	2.99
S_6	2.20	4.24	4.69	S_{18}	0.88	1.57	1.74
S_7	2.47	3.97	4.52	S_{19}	1.67	2.31	2.65
S_8	2.49	5.75	6.63	S_{20}	1.99	3.77	4.55
S_9	2.49	4.42	5.36	S_{21}	1.22	2.27	2.60
S_{10}	1.24	2.13	2.36	S_{22}	2.48	5.80	6.94
S_{11}	1.22	2.27	2.60	S_{23}	1.77	1.81	2.99
S_{12}	1.51	2.85	3.51	S_{24}	1.67	2.31	2.65

TABLE 10.11 Precision and Accuracy of Estimators for Example 10.7 (Emerging Drifts)

compared with the flowrate value, which is not the case for small flowrates where induced bias can be many times larger than the flowrate value. It can also be seen that the accuracy for the case regular testing is higher than the accuracy for the case instantaneous testing, the difference ranges from about 7 (sudden fixed value biases) to 50% (simultaneous occurrence of three types of biases). In turn,

Streams	$\hat{\sigma}(\%)$	$\hat{a}(\%)$ Instantaneous Testing	$\hat{a}(\%)$ Periodic Testing	Streams	$\hat{\sigma}(\%)$	$\hat{a}(\%)$ Instantaneous Testing	$\hat{a}(\%)$ Periodic Testing
S_1	1.10	1.82	1.88	S_{13}	1.75	1.81	1.89
S_2	2.48	8.84	8.98	S_{14}	1.75	1.81	1.89
S_3	1.08	1.53	1.59	S_{15}	1.08	3.37	3.37
S_4	2.35	4.17	4.17	S_{16}	0.98	3.18	3.26
S_5	2.48	4.08	5.60	S_{17}	1.77	4.85	8.52
S_6	2.20	7.48	7.48	S_{18}	0.88	1.78	1.93
S_7	2.47	7.07	7.33	S_{19}	1.67	2.78	2.97
S_8	2.49	4.22	5.94	S_{20}	1.99	6.27	6.40
S_9	2.49	3.86	5.66	S_{21}	1.22	4.00	4.11
S_{10}	1.24	2.24	2.44	S_{22}	2.48	4.51	6.00
S_{11}	1.22	4.00	4.11	S_{23}	1.77	4.85	8.52
S_{12}	1.51	4.72	4.82	S_{24}	1.67	2.78	2.97

TABLE 10.12 Precision and Accuracy of Estimators for Example 10.7 (Deterministic Drifts)

Table 10.11 shows the results when only emerging drifts are considered. The results in Table 10.11 show that, generally, accuracy values of estimators for emerging drifts are comparable to those values for sudden fixed value biases. Next, Table 10.12 shows the results when only deterministic drifts are considered.

With deterministic drifts, the biases appear right after sensors are put into use. The biases keep undetected until their sizes reach threshold values. The biases are then identified and sensors are repaired. When sensors resume work, biases appear again and the same cycle repeats. The presence of this type of bias is possible if the fluid is highly corrosive and dirty such that it affects the sensors right after sensors are put into use. This means that the measurements always contain undetected biases so it is expected that the accuracy values for type **c** biases (shown in Table 10.12) are worse (larger) than the accuracy values for sudden fixed value biases and random emerging drifts (shown in Table 10.10 and 10.11), which is shown to be true by calculation results.

Finally, consider the case that three types of biases can occur in a sensor (all sensors contain simultaneously three types of biases), the same parameters for three kinds of biases as stated above are used and the calculated results are given in Table 10.13.

The calculations show that the accuracy value obtained when three types of biases can occur in a sensor (Table 10.13) is larger than the accuracy value when only one type of bias can take place (Tables 10.10 and 10.11). This is due to the fact that more types of biases that can occur in a sensor lead to a larger chance for the sensor to be biased; thus, the accuracy value increases. Considering two cases, this case (all three biases, Table 10.13) and the case that only type **c** bias is present (Table 10.12), in both cases the measurement sets always contain some undetected biases. Type **c** bias is always present in measurement; other types of biases (sudden fixed value biases and randomly emerging drifts) may add up

Streams	$\hat{\sigma}$(%)	$\hat{a}$(%) Instantaneous Testing	$\hat{a}$(%) Periodic Testing	Streams	$\hat{\sigma}$(%)	$\hat{a}$(%) Instantaneous Testing	$\hat{a}$(%) Periodic Testing
S_1	1.10	2.28	2.89	S_{13}	1.75	3.37	5.97
S_2	2.48	6.65	7.96	S_{14}	1.75	3.37	5.97
S_3	1.08	2.24	2.80	S_{15}	1.08	2.40	3.04
S_4	2.35	4.88	5.94	S_{16}	0.98	2.41	3.18
S_5	2.48	6.95	10.07	S_{17}	1.77	5.39	15.21
S_6	2.20	4.95	6.07	S_{18}	0.88	1.92	2.46
S_7	2.47	7.13	8.49	S_{19}	1.67	3.19	4.25
S_8	2.49	6.45	9.89	S_{20}	1.99	4.39	6.16
S_9	2.49	6.78	9.86	S_{21}	1.22	2.95	3.93
S_{10}	1.24	2.68	3.49	S_{22}	2.48	6.86	9.98
S_{11}	1.22	2.95	3.93	S_{23}	1.77	5.39	15.21
S_{12}	1.51	3.31	4.73	S_{24}	1.67	3.19	4.25

Table 10.13 Calculated Precision and Accuracy of Estimators for Example 10.7, Simultaneous Occurrence of Three Types of Biases

to increase the deterministic drift bias size (total bias size > deterministic drift bias size), or contrarily it reduces the deterministic drift bias size (cancellation effect). So there is no way to say a priori which one renders better accuracy, as seen from Table 10.12 and 10.13. Finally note that the failure rate is determined experimentally and only one value is reported, so there is no such specific failure rate data available associated with a specific failure mechanism (i.e., specific type of bias). However, one can assume that the failure rates for all types of biases that occur suddenly (sudden fixed value biases and randomly emerging drifts) are the same, which is the assumption used in our calculation.

Table 10.14 shows the computation time for the two cases: instantaneous testing and periodic testing. It is clear that the computation time of the former is much longer than that of the latter because the former requires more computation as explained above.

Computation of Accuracy Value in the Case of	Instantaneous Testing	Periodic Testing
Sudden fixed value bias only	2 min 49 s	1 min 8 s
Randomly emerging drifts only	9 min 25 s	1 min 46 s
Deterministic drifts only	4 min 11 s	1 min 19 s
All three types of biases	26 min 28 s	3 min 11 s

Table 10.14 Comparison of Computation Time for the Two Cases

Example 10.8: We return to the Tennessee Eastman process to compare software accuracy with stochastic accuracy. All variables are measured and undetected biases are limited to be lower than 5 times the precision of measurements. The following data is used in obtaining the stochastic accuracy: Sensor failure is assumed to follow exponential distribution with failure rate of 0.01/day, repair time = 1 day; biases are assumed to follow normal distribution with zero mean and variance equal to 4 times the sensor precision, number of samplings = 1000. Software accuracy and stochastic accuracy for all variables is shown in Table 10.15. We notice that stochastic accuracy is much smaller than software accuracy of order two, as expected, but only slightly lower than software accuracy of order one in many cases, but sometimes significantly lower. This is a reflection of the fact that, even when there is a bias present, it is not always the one that will induce the larger error in the variable in question.

Historical Note: Connections to Gross Error Detectability and Resilience

Accuracy embeds the notion that gross errors might be present together with white noise. In turn, gross error detectability is defined as the ability of a network to detect a gross error of a certain minimum size or larger. This concept was introduced by Bagajewicz (1997, 2000) and he proposed the minimum size to be given by

$$\hat{\delta}_k(m,\alpha,\beta) = \rho(m,\alpha,\beta)\frac{\sigma_k^2}{(\sigma_k^2 - \tilde{\sigma}_k^2)^{1/2}} \tag{10.55}$$

which is the smallest size of gross error that can be detected with probability β. Typical values of β are 50 and 90% and tables for $\rho(m, \alpha, \beta)$ as well as an empirical expression for large number of degrees of freedom (m) are given by Madron (1992). This expression is obtained by realizing that the formula above is the

Variables	Value	Software Accuracy of Order One $\tilde{a}_i^{MP(0.95,1)}$ (%)	Software Accuracy of Order Two $\tilde{a}_i^{MP(0.95,2)}$ (%)	Stochastic Accuracy (%)
F_6	1.890	1.43	15.28	1.25
F_7	0.322	2.36	15.31	1.64
F_{10}	0.089	3.28	5.10	1.94
F_{11}	0.264	2.21	4.39	1.56
$Y_{A,6}$	0.069	2.10	3.01	1.25
$Y_{B,6}$	0.187	2.57	4.38	1.65
$Y_{C,6}$	0.016	8.23	10.46	1.97
$Y_{D,6}$	0.035	5.46	11.73	2.17
$Y_{E,6}$	0.017	8.61	14.90	2.19
$Y_{F,6}$	1.475	1.49	6.78	1.27
$Y_{G,6}$	0.272	3.82	7.48	2.12
$Y_{H,6}$	0.114	3.30	5.11	1.99
$Y_{A,7}$	0.198	3.94	8.33	2.17
$Y_{B,7}$	0.011	12.06	16.38	2.31
$Y_{C,7}$	0.177	3.37	6.77	1.99
$Y_{D,7}$	0.022	6.33	8.18	1.89
$Y_{E,7}$	0.123	1.72	3.62	1.38
$Y_{F,7}$	0.084	1.94	3.18	1.32
$Y_{G,7}$	0.330	3.84	7.41	2.13
$Y_{H,7}$	0.138	3.15	4.86	1.91
$Y_{A,8}$	0.240	4.00	8.34	2.18
$Y_{B,8}$	0.013	11.01	15.09	2.41
$Y_{C,8}$	0.186	4.36	7.64	2.26
$Y_{D,8}$	0.023	6.63	8.54	2.09
$Y_{E,8}$	0.048	5.61	12.19	2.46
$Y_{F,8}$	0.023	8.08	14.75	2.54
$Y_{G,8}$	0.330	0.83	0.97	0.21
$Y_{H,8}$	0.138	0.72	0.72	0.1
$Y_{A,9}$	0.240	1.14	1.33	0.28
$Y_{B,9}$	0.013	17.53	17.92	3.61
$Y_{C,9}$	0.186	6.16	48.13	2.62
$Y_{D,9}$	0.023	8.60	49.96	2.53
$Y_{E,9}$	0.048	12.03	12.26	3.70
$Y_{F,9}$	0.023	14.33	14.46	3.69
$Y_{G,9}$	0.258	2.04	4.48	1.51
$Y_{H,9}$	0.002	59.99	60.08	3.69
$Y_{D,10}$	0.136	3.97	10.33	2.25
$Y_{E,10}$	0.016	8.99	15.44	2.24
$Y_{F,10}$	0.472	2.07	4.12	1.53
$Y_{G,10}$	0.373	2.05	5.92	1.52
$Y_{H,10}$	0.2113	2.76	4.36	1.77
$Y_{G,11}$	0.537	2.77	4.37	1.78
$Y_{H,11}$	0.438	2.76	4.36	1.78
P_r	2.806	0.1	0.1	0.1
T_r	393.6	0.1	0.1	0.1
P_s	2.7347	4.35	11.60	2.24
T_s	353.3	4.21	11.22	2.16

TABLE 10.15 Comparison of Software Accuracy and Stochastic Accuracy

relation between the minimum value $\delta_i^*(m,\alpha,\beta)$, and the noncentrality parameter $\rho(m,\alpha,\beta)$ of a chi-squared distribution that is followed by $r(C_R Q_R C_R^T)^{-1} r$ in the presence one gross error. Expressions for larger number of gross errors have not been developed.

The expression used for gross error detectability is based on the idea that one is using the global test, while the definition of accuracy above assumes that the maximum power measurement test is used. It is therefore claimed that the accuracy definition proposed here is much more appropriate to be used as a robustness measure than gross error detectability. Moreover, the former was extended to the presence of multiple gross errors, while the latter is limited in this regard.

Finally, resilience can be expressed as the difference between accuracy and precision. Indeed, resilence is the ability of a network to limit the smearing effect of gross errors. In other words, it is a number that sets an upper bound on the effect of undetected gross errors in estimators obtained using data reconciliation. Therefore, while the concept in itself can continue to be used, we see little purpose in using it, as accuracy contains everything one might need in robustness terms.

References

Bagajewicz, M., "Design and Retrofit of Sensor Networks in Process Plants," *AIChE J.* **43**(9):2300 Sept. (1997).

Bagajewicz, M., "Process Plant Instrumentation. Design and Upgrade," (ISBN: 1-56676-998-1), (Technomic Publishing Company) (http://www.techpub.com). Now CRC Press (http://www.crcpress.com) (2000).

Bagajewicz, M., DuyQuang Nguyen, "Stochastic-Based Accuracy of Data Reconciliation Estimators for Linear Systems," *Comput. Chem. Eng.* **32**(6): 1257–1269 (2008).

Bagajewicz, M., "On the Definition of Software Accuracy in Redundant Measurement Systems," *AIChE J.* **51**(4):1201–1206 (2005).

Bagajewicz, M. and Q. Jiang, "Gross Error Modeling and Detection in Plant Linear Dynamic Reconciliation," *Comput. Chem. Eng.* **22**(12):1789–1810 (1998).

Downs, J. J. and E. F. Vogel, "A Plant-Wide Industrial Process Control Problem," *Comput. Chem. Eng.* **17**(3):245–255 (1993).

Lipták, B. G., editor, Instrument Engineer's Handbook, 4th ed., **1**, CRC Press, Boca Raton, FL (2003).

Miller, R. W., "Flow Measurement Engineering Handbook," McGraw-Hill, USA (1996).

Madron, F., "Process Plant Performance," Ellis Horwood, Chichester, England (1992).

Nguyen Thanh, D. Q., K. Siemanond, and M. Bagajewicz, "On the Determination of Downside Financial Loss of Instrumentation Networks in the Presence of Gross Errors," *AIChE J.* **52**(11):3825, 3841 (2006).

Ricker, N. L. and J. H. Lee, "Nonlinear Model Predictive Control of the Tennessee Eastman Challenge Process," *Comput. Chem. Eng.* **19**(9):961–981 (1995a).

Ricker, N. L. and J. H. Lee, "Nonlinear Modeling and State Estimation for the Tennessee Eastman Challenge Process," *Comput. Chem. Eng.* **19**(9):983–1005 (1995b).

CHAPTER 11

Economic Value of Accuracy

Any engineer in a plant, when asked about the precision or accuracy he/she would like to have in the measurements, or in the variable estimators if software redundancy is available (through data reconciliation, Kalman filtering, or some other method), then the engineer will prefer accuracy to be as good as possible, that is, the smaller, the better. It is very likely, however, that appropriate decision making regarding improvements can be done to make the accuracy pass a certain threshold. This threshold is linked to the impact the changes in accuracy have on economics.

Thus, a link between accuracy and profit must be established. Once this link is established one can justify increasing accuracy, through the combined use of data reconciliation and instrumentation upgrade. We start with precision.

Value of Precision

A typical refinery consists of several tank units that receive the crude, several processing units, and several tanks where products are stored. All this is summarized in Fig. 11.1 where H represents holdups and m flowrates. To help introducing concepts, consider for the time being the following scenario:

1. Raw materials are purchased and stored at the beginning of operations for a short period of time and measured with relatively high precision. Thus, H_p is likely to be known with very good accuracy.
2. Sales take place at the end of the processing period and are also very accurately measured. In other words, $H_S(T)$ is known, but only at time T, not before.
3. Steady state.

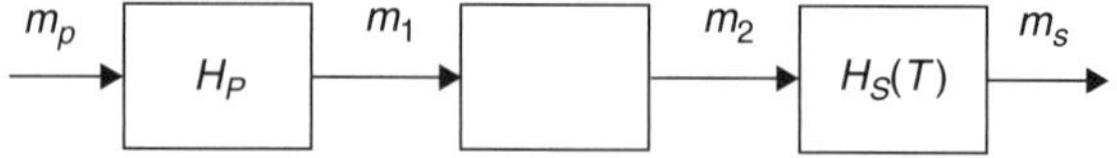

FIGURE 11.1 Material balance in a refinery.

In the absence of leaks, the expected value of $H_S(T)$ is only related to the true value of m_2 and m_s throughout time. However, since m_2 is only known through its measurements or through its estimators throughout time, this true value is not known exactly. We assume m_s is the planned delivery and is well-known.

Consider now that a target production H_s^* at the end of the time period T is pursued. Therefore, only if m_2 on average is equal to $m_2^* = H_s^*/T$, there is certainty that the target will be met. Otherwise, the probability of not meeting the targeted production is $P\{H_s(T) \le H_s^*\}$, which in turn can be rewritten as $P\{m_2(t) \le m_2^*\}$, that is, it is equal to the probability of the true value of m_2 being smaller than the measurement. Suppose only measurements of m_2 are performed with variance $\hat{\sigma}_{2,m}^2$. Since only a finite number of measurements are performed, then, assuming these measurement follows a normal distribution, an estimate of m_2, say $\hat{m}_2$, has a normal distribution with variance $\hat{\sigma}_{2,m}^2 = \sigma_{2,m}^2/(N+1)$, where N is the number of such measurements. Thus $\hat{m}_2$ is the estimate one has of the true value of m_2. Consider then that production is adjusted to meet the targeted value, based on the estimate. Thus, if $\hat{m}_2 < m_2^*$, production is increased and vice versa; if $\hat{m}_2 > m_2^*$, production is decreased.

We now look at only a period of time T, say a day or a week, before the decision to increase or decrease production is made. Consider the case $\hat{m}_2 \ge m_2^*$, that is, the measurement indicates that the target has been met. In such a situation, the operator would not do any correction to the set points. The probability of being wrong is given by the following conditional probability $P\{m_2 \le m_2^* \mid \hat{m}_2 \ge m_2^*\}$, that is, the probability of having missed the target, given that the estimator is larger than the target. Because these are independent, the above probability is equal to $P\{m_2 \le m_2^*\}P\{\hat{m}_2 \ge m_2^*\}$. Now,

$$P\{m_2 \le m_2^*\} = \int_{-\infty}^{m_2^*} g_P(\xi; m_2^*, \sigma_2)\, d\xi \tag{11.1}$$

where $g_P(\xi; m_2^*, \sigma_2)$ is the probability distribution of the process values around the mean m_2^*. We assume here that the control system is such that the mean is the target value. Clearly, for any symmetric distribution, including the usual common choice of normal distributions, we have $P\{m_2 \le m_2^*\} = 0.5$. However, due to the fact that this distribution is intimately tied to the type of control system used, some other than the normal distribution, possibly nonsymmetric, is possible. In turn, the probability of the measurement being larger than the target, $P\{\hat{m}_2 \geq m_2^*\}$, depends on both the value of m_2 and the quality of the measurement, which depends on the precision of the estimator ($\hat{\sigma}_{2,m}$) and is given by its probability distribution $g_M(\xi;m_2,\hat{\sigma}_{2,m})$ around the real value m_2. Thus, for a fixed value of m_2

$$P\{\hat{m}_2 \geq m_2^* \mid m_2 fixed\} = \int_{m_2^*}^{\infty} g_M(\xi;m_2,\hat{\sigma}_{2,m})d\xi \tag{11.2}$$

Now, since m_2 follows a distribution, we have

$$P\{\hat{m}_2 \geq m_2^* \mid m_2\} = P\{\hat{m}_2 \geq m_2^* \mid m_2 fixed\}\, g_P(m_2;m_2^*,\sigma_2) \tag{11.3}$$

or, integrating over all possible values of the true rate

$$P\{\hat{m}_2 \geq m_2^*\} = \int_{-\infty}^{m_2^*}\left\{\int_{m_2^*}^{\infty} g_M(\xi;m_2,\hat{\sigma}_{2,m})d\xi\right\} g_P(m_2;m_2^*,\sigma_2)dm_2 \tag{11.4}$$

Notice that the integral is taken over all possible values of m_2 below the target because of the underlying assumption for m_2 that it is lower than the target. When both distributions are normal, we obtain

$$P\{\hat{m}_2 \geq m_2^*\} = \frac{1}{4} + \frac{1}{2\sqrt{\pi}}\int_0^{\infty} erfc(z\sigma_2/\hat{\sigma}_{2,m})\, e^{-z^2} dz \tag{11.5}$$

This value depends on the standard deviations of the true value and the measurement. When $\sigma_2/\hat{\sigma}_{2,m} \to 0$, the above probability has a limit of 0.5. When $\sigma_2/\hat{\sigma}_{2,m} \to \infty$ $P\{\hat{m}_2 \geq m_2^*\} \to 1/4$. However, for reasonable large values of $\sigma_2/\hat{\sigma}_{2,m}$, the values drop between 0.35 and 0.5. For example, for $\sigma_2/\hat{\sigma}_{2,m} = 2$, we have $P\{\hat{m}_2 \geq m_2^*\} = 0.387$.

We now turn into determining what is the expected financial loss associated to this probability. The distributions of the true value and the measurement are depicted in Fig. 11.2 for one instance of the true value m_2.

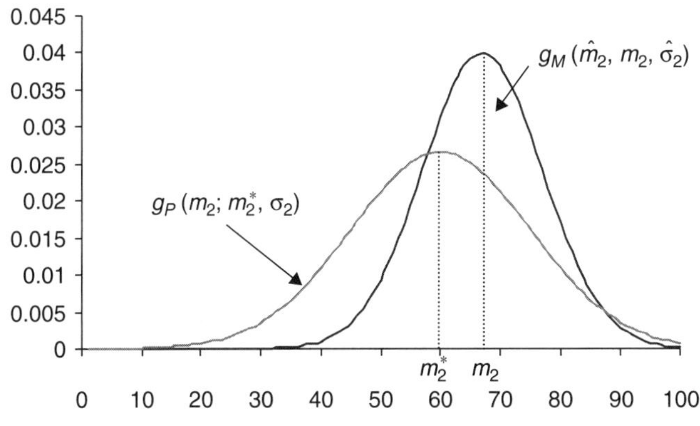

FIGURE 11.2 Example of process and measurement distributions.

The downside financial loss incurred for a fixed value of m_2 is given by $K_S T(m_2^* - m_2)$, where K_S is the value of the products sold and T is the period of time under consideration. However, inventory is usually used to absorb variations. In such a case, the value of K_S needs to be reassessed to reflect the cost of keeping inventory.

One could easily integrate this expression to obtain an expected value as follows: $K_S T \int_{-\infty}^{m_2^*} (m_2^* - m_2)\, g_P(m_2; m_2^*, \sigma_2) dm_2$. However, this would be wrong because it would just calculate an expected value for negative deviations from the target regardless of the information available, that is, not using the estimator $\hat{m}_2$ suggested by the measurement, directly or in general, by data reconciliation estimators or similar.

To incorporate the notion that a measurement is being made and therefore some information is added that allows the operator to make corrections, we write the integral over all possible values of m_2 below the target as follows:

$$DEFL(\hat{\sigma}_{2,m} | \hat{m}_2) = K_S T \int_{-\infty}^{m_2^*} (m_2^* - m_2)\, \hat{g}_L(m_2, \hat{\sigma}_{2,m} | \hat{m}_2) dm_2 \qquad (11.6)$$

where $\hat{g}_L(m_2, \hat{\sigma}_{2,m} | \hat{m}_2)$ is the likelihood function, that is, the probability distribution of the state m_2, given the estimator $\hat{m}_2$. But the likelihood function is equal to the distribution of the measurement, that is,

$$\hat{g}_L(m_2, \hat{\sigma}_{2,m} | \hat{m}_2) = g_M(\hat{m}_2; m_2, \hat{\sigma}_{2,m}) \qquad (11.7)$$

Now, since m_2 is not fixed, we need to integrate over all possible values of m_2 below the target m_2^* multiplied by the probability density of each state $g_P(m_2^*, \sigma_2, m_2)$. We integrate below the target because those values are the ones that will lead to financial loss. Thus,

$$DEFL(\hat{\sigma}_{2,m}, \sigma_2) = \int_{-\infty}^{m_2^*} g_P(m_2^*, \sigma_2, m_2) \left\{ K_S T \int_{-\infty}^{m_2^*} (m_2^* - \hat{m}_2)\, g_M(\hat{m}_2; m_2, \hat{\sigma}_{2,m}) d\hat{m}_2 \right\} dm_2 \qquad (11.8)$$

When both distributions are normal, the expected financial loss is given by

$$DEFL(\hat{\sigma}_{2,m}, \sigma_2) = \gamma K_S T \left\{ \frac{\hat{\sigma}_{2,m}}{\sqrt{\left(\frac{\sigma_2}{\hat{\sigma}_{2,m}}\right)^2 + 1}} + \sigma_2 \sqrt{\frac{1}{\left(\frac{\hat{\sigma}_{2,m}}{\sigma_2} + 1\right)}} \right\} \qquad (11.9)$$

where $\gamma = 1/2\sqrt{2\pi} \int_0^{\infty} \xi e^{-\xi^2} d\xi = 0.19947$ (see Bagajewicz et al., 2005). When the process variations are negligible, the probability density for the process $g_p(m_2^*, \sigma_2, m_2)$ reduces to one-half of the delta function, that is, $g_p(m_2^*, \sigma_2, m_2) = 0.5\delta(m_2 - m_2^*)$. The one-half factor comes from the fact that we are integrating in the interval $(-\infty, m_2^*)$. Conversely, when the process variations are large, that is, $\sigma_2 >> \hat{\sigma}_{2,m}$, the first term becomes irrelevant (i.e., measurements have no influence) and

$$DEFL(\hat{\sigma}_{2,m}, \sigma_2) \rightarrow \gamma K_S T \sigma_2 \tag{11.10}$$

Intermediate values were studied in more detail by Bagajewicz et al. (2005).

We recall that the assumption was made that the operator will not introduce corrective actions when the measurement is above the target. Therefore, the expected financial loss is associated with a given level of confidence given by

$$P\{m_2 \leq m_2^* \mid \hat{m}_2 \geq m_2^*\} = 0.5P\{\hat{m}_2 \geq m_2^*\} \tag{11.11}$$

which is a function of the measurements and the process variability. Thus, a given set of instruments provides both this level of confidence and its corresponding expected financial loss, which is only function of the estimator's precision. Therefore, under almost no process variations (there is always some), there is a 25% chance that a downside financial loss of $\gamma K_S T \hat{\sigma}_{2,m}$ on average is achieved. If corrections to set points are made when readings indicate that targeted productions are exceeded, the result would be even worse.

We notice that a good relation between the probability and the financial loss has not been established. Thus while knowing the probability is a good thing, there is not yet a theory developed that can quantify the associated financial loss. In other words, the same downside financial loss may correspond to different probabilities or vice versa. This ambiguity calls for further development of the theory. In the meantime, we identify the value of precision with the downside financial loss $DEFL(\hat{\sigma}_{2,m}, \sigma_2)$.

Value of Accuracy

When there is a bias, induced or not, it could go undetected, which means it has an absolute value size smaller than $\hat{\delta}_{p,\max}^{i1,\ldots,in_b}$, which is the maximum induced bias that goes undetected by the maximum power measurement test when there are n_b gross errors. This value is a function of the existing instrumentation precision and the method being used to detect gross errors. When there is no redundancy, this value is, theoretically, infinite, but in practical terms, when the bias reaches a certain value $\delta_p^{\#}$, it becomes truly apparent to the operator that there is a bias and hopefully, the instrument is calibrated and/or the measurement is ignored. When there is redundancy, the value is finite and depends on the method used. We therefore concentrate in redefining $g_M(\xi; m_p, \hat{\sigma}_p)$ to include the possibility of biases.

Assuming one gross error in variable i and none in the others (i.e., $n_b = 1$), we have

$$g_M = g_M^{R,i}\left(\xi; m_p, \hat{\sigma}_{p,i}^R\right) \qquad \left|\hat{\delta}_p^i\right| > \hat{\delta}_{p,\max}^i \tag{11.12}$$

where $\hat{\sigma}_{p,i}^R$ is the residual precision left after the measurement of variable i has been eliminated; $\hat{\delta}_p^i$ is the induced bias in the estimator of m_p, which is a function of the original bias, δ_i, that is, $\hat{\delta}_p^i = \hat{\delta}_p^i(\delta_i)$. We note that under the condition of one and only one bias present in the system, certain tests like the maximum power and GLR tests are consistent, that is, if one bias is present, then when these tests flag positive, they point to the correct location of the bias. This assumption is important because in the absence of consistency, the assumption implicit in Eq. (11.12) (that the right measurement will be eliminated) is no longer valid. From now on, we will assume consistency in gross error detection. In turn, when the induced gross error is smaller than the threshold of detection, then it is undetected, and therefore,

$$g_M = g_M\left(\xi; m_p + \hat{\delta}_p^i, \hat{\sigma}_p\right) \qquad \left|\hat{\delta}_p^i\right| \le \hat{\delta}_{p,\max}^i \tag{11.13}$$

Probabilities

Thus, the probability of the estimate to be higher than the true value, given a bias in measurement i, is now given by

$$P\{\hat{m}_p \ge m_p^* \mid \delta_i\} = \begin{cases} \displaystyle\int_{-\infty}^{m_p^*} \left\{ \int_{m_p^*}^{\infty} g_M(\xi; m_p, \hat{\sigma}_{p,i}^R)\, d\xi \right\} g_P(m_p; m_p^*, \sigma_p)\, dm_p & \left|\hat{\delta}_p^i\right| > \hat{\delta}_{p,\max}^i \\ \displaystyle\int_{-\infty}^{m_p^*} \left\{ \int_{m_p^*}^{\infty} g_M(\xi; m_p + \hat{\delta}_p^i, \hat{\sigma}_p)\, d\xi \right\} g_P(m_p; m_p^*, \sigma_p)\, dm_p & \left|\hat{\delta}_p^i\right| \le \hat{\delta}_{p,\max}^i \end{cases} \tag{11.14}$$

which is a direct extension of Eq. (11.4) to the presence of one gross error. Indeed, the integral is conditional to the fact that the bias has been detected ($|\hat{\delta}_p^i| > \hat{\delta}_{p,\max}^i$) or not ($|\hat{\delta}_p^i| \le \hat{\delta}_{p,\max}^i$). We notice that the first integral uses the same distribution of the estimator as in the case of no bias, but with a standard deviation of the estimator given by the residual precision ($\hat{\sigma}_{p,i}^R$) because it is assumed that the bias is sufficiently large to have been detected and the measurement was eliminated. In turn, the second integral (the case of an undetected induced bias), the measurement/estimator distribution g_M, is centered around a shifted mean value.

The above formula has been studied in detail for various cases, and extended to more than one gross error by Bagajewicz (2006). The main results are that the occurrence of one gross error, two or more simultaneous gross errors are independent events. Thus, let n_b be the maximum number of gross errors that occur simultaneously, then,

$$P\left\{\hat{m}_p \ge m_p^*\right\} = \sum_k P\left\{\hat{m}_p \ge m_p^* \mid n_b = k\right\} \tag{11.15}$$

We make the sum over all different instances of k gross errors, because they are independent events. Thus, for one bias we have

$$P\left\{\hat{m}_p \geq m_p^* \mid k=1\right\} = \sum_i P\left\{\hat{m}_p \geq m_p^* \mid \delta_i\right\} \tag{11.16}$$

Similarly, for two biases, we have

$$P\left\{\hat{m}_p \geq m_p^* \mid k=2\right\} = \sum_{i1,i2} P\left\{\hat{m}_p \geq m_p^* \mid \delta_{i1}, \delta_{i2}\right\} \tag{11.17}$$

In general, for many biases

$$P\left\{\hat{m}_p \geq m_p^* \mid k=n_b\right\} = \sum_{i1,i2,\ldots,i_{n_b}} P\left\{\hat{m}_p \geq m_p^* \mid \delta_{i1}, \delta_{i2}, \ldots, \delta_{in_b}\right\} \tag{11.18}$$

We now consider that when an instrument fails, which happens according to a certain probability $f_i(t)$ (a function of time), the size of the bias follows a certain distribution $h_i(\theta;\bar{\delta}_i,\rho_i)$ with mean $\bar{\delta}_i$ and variance ρ_i^2. Note that depending on the value of the measurement in the range of the instrument, the mean is very likely to be nonzero (Liptak, 2003; Bagajewicz, 2000). We are also assuming here that the gross error size distribution is independent of time. Thus, we now need to integrate over all possible values of the gross error and multiply by the probability of such bias to develop. Therefore, if we assume that one instrument fails at a time, then the probability of instrument i failing and the others not is given by $\Phi_i^1 = f_i(t)\prod_{s\neq i}[1-f_s(t)]$. This result stems from the fact that the failure events are independent so the probability of one sensor failing and the other not is given by the product of all these probabilities. Thus, the probability of the estimator to be larger than the target value is given by

$$P\left\{\hat{m}_p \geq m_p^* \mid i\right\} = \Phi_i^1 \int_{-\infty}^{\infty} P\left\{\hat{m}_p \geq m_p^* \mid \theta\right\} h_i\left(\theta;\bar{\delta}_i,\rho_i\right) d\theta \tag{11.19}$$

In this expression, we multiply the probability of one sensor failing by the integral over all possible values of gross errors (in principle from $-\infty$ to ∞) of the probability of the error having size θ, given by $h_i(\theta;\bar{\delta}_i,\rho_i)$ of rendering an estimator larger by the target. This last probability is the product of the probability of the error having size and the probability that such error renders an estimator larger than the target.

Generalizing, we have $\Phi_{i1,i2,\ldots,in_b}^{n_T} = f_{i1}(t)\ldots\ldots f_{in_b}(t) \prod_{s\neq i1,\ldots\ldots s\neq in_b} [1-f_s(t)]$, and we write

$$P\left\{\hat{m}_p \geq m_p^* \mid i1,\ldots,in_b\right\} = \Phi_{i1,\ldots in_b}^{n_T} \int_{-\infty}^{\infty}\ldots\int_{-\infty}^{\infty} P\left\{\hat{m}_p \geq m_p^* \mid \theta_1,\ldots,\theta_{in_b}\right\}$$
$$h_{i1}\left(\theta_1;\bar{\delta}_{i1},\rho_{i1}\right)\ldots h_{in_b}\left(\theta_{in_b};\bar{\delta}_{in_b},\rho_{in_b}\right) d\theta_1 \ldots d\theta_{in_b} \tag{11.20}$$

These integrals are in general very complex and we omit further analysis here (Bagajewicz, 2006 discusses those details). We only summarize some simple results. For example, for a normal distribution of biases with zero mean, negligible process variation and the case of zero mean ($\bar{\delta}_i = 0$), we have

$$P\left\{\hat{m}_p \geq m_p^* \mid i\right\} = \frac{\Phi_i^1}{4} \tag{11.21}$$

In turn, for a normal distribution of biases with zero mean and negligible process variation, we have

$$\lim_{\sigma_p/\hat{\sigma}_p \to 0} P\left\{\hat{m}_2 \geq m_2^*\right\} = \frac{\Phi^0}{4} + \frac{\Phi_i^1}{4} + \sum_{r=2}^{n} P\left\{\hat{m}_2 \geq m_p^* \mid r\right\} \tag{11.22}$$

Although the rest of the terms in the summation can be calculated, we omit them here (see Bagajewicz, 2006, for more details). We focus on the financial loss instead and we notice that, as in the case of precision, a good relation between the probability and the financial loss has not been yet established. We therefore identify the value of accuracy with the downside expected financial loss for the time being.

Downside Expected Financial Loss

To obtain a complete expression for the financial loss, we add the financial loss corresponding to the different mutually exclusive events multiplied by their frequency as follows:

$$DEFL = \Psi^0 DEFL^0 + \sum_i \Psi_i^1 \, DEFL^1\Big|i + \sum_{i1,i2} \Psi_{i1,i2}^2 \cdot DEFL^2\Big|i1,i2 + \cdots \tag{11.23}$$

We can do this just because these are mutually exclusive states, that is, a state with no biases does not intersect with the state of one bias, and so on. In the above expression, Ψ^0 is the average fraction of time the system is in the state without biases, Ψ_i^1 the average fraction of time the system has only one undetected bias only in stream *i*, etc. These values are in fact equal to the probabilities of each state Φ_i^j.

When there is one gross error present, the expected financial loss is given by the following equation:

$$\left.\begin{aligned} DEFL^1\Big|i = {} & \left[\int_{-\infty}^{m_p^*} K_S T \left\{\int_{-\infty}^{m_p^*} \left(m_p^* - \xi\right) g_M\left(\xi; m_p, \hat{\sigma}_{p,i}^R\right) d\xi\right\} g_P\left(m_p; m_p^*, \sigma_p\right) dm_p\right] \\ & \qquad \left[\begin{aligned} & \int_{-\infty}^{-\bar{\delta}_i^p} h_i\left(\theta; \bar{\delta}_i, \rho_i\right) d\theta + \\ & \int_{\bar{\delta}_i^p}^{\infty} h_i\left(\theta; \bar{\delta}_i, \rho_i\right) d\theta \end{aligned}\right] \\ & + \int_{-\bar{\delta}_i^p}^{\bar{\delta}_i^p} \left[\int_{-\infty}^{m_p^*} K_S T \left\{\int_{-\infty}^{m_p^*} \left(m_p^* - \xi\right) g_M\left(\xi; m_p + \hat{\delta}_p^i(\theta), \hat{\sigma}_p\right) d\xi\right\} g_P\left(m_p; m_p^*, \sigma_p\right) dm_p\right] \\ & \qquad h_i\left(\theta; \bar{\delta}_i, \rho_i\right) d\theta \end{aligned}\right\} \tag{11.24}$$

This equation is similar to Eq. (11.8), except that it is broken into three integrals, two of them in the region where the gross error/bias is detected and therefore the residual precision is used (as in the case of the probabilities) and the integral of the error size distribution in the range where the error is not detected and the distribution of the estimator is shifted by the induced bias.

For normal distributions of gross errors and zero mean in the biases ($\bar{\delta}_i = 0$), this expression reduces to

$$DEFL^1|i = DEFL^0 \left\{ \left[\frac{\hat{\sigma}_p^{R,i}}{\hat{\sigma}_p}\right] erfc\left\{\frac{\tilde{\delta}_i^p}{\sqrt{2}\rho_i}\right\} + \frac{1}{4\gamma\rho_i\sqrt{\pi r}}\left[1 - erfc\left\{\sqrt{r}\tilde{\delta}_i^p\right\}\right] + \frac{|\alpha_p^i|\rho_i}{2\gamma\hat{\sigma}_p\sqrt{2\pi}}\left[1 - e^{-\frac{(\tilde{\delta}_i^p)^2}{2\rho_i^2}}\right]\right\} \quad (11.25)$$

where $\tilde{\delta}_i^p$ is the maximum undetected induced bias in stream p (absolute value). Now, if the variance of the bias ρ_i is small compared to $\tilde{\delta}_i^p$ one can write

$$DEFL^1|i = DEFL^0 \left[\frac{\hat{\sigma}_p^{R,i}}{\hat{\sigma}_p}\right] - O(e^{-\frac{\tilde{\delta}_i^p}{\sqrt{2}\rho_i}} \tilde{\delta}_i^p/\rho_i) \qquad \rho_i << \tilde{\delta}_i^p \quad (11.26)$$

reflecting the fact that almost all errors are detected. The formula also indicates that when good accuracy is dependent on only a few instruments, then the residual precision can be a large number and therefore the financial loss increases considerably. For example, if one has a system with orifice meters (about 2% precision) and decides to install a coriolis meter in stream p, then even if this meter has small biases, its frequency of failure needs to be also very small. Otherwise the financial loss can make a big jump because the residual precision increases considerably. It is therefore better to install more instruments so that the residual precision does not deteriorate under one instrument failure so much.

We also recognize that in the case where no data reconciliation is performed, $\tilde{\delta}_i^p = 0 \ \forall i \neq p$ and $\tilde{\delta}_p^p = \delta_p^\#$ (usually very large). In addition, $\hat{\sigma}_p^{R,i} \to \infty$, (the estimator is lost), so we use $\sigma_p^\#$, the variance of the best estimate one can make.

$$DEFL^1 \mid p = DEFL^0 \left[\frac{\sigma_p^\#}{\sigma_p}\right] \quad (11.27)$$

which indicates that the downside financial loss can be sizable.

Returning to Eq. (11.25), when variable p has very high accuracy, then $\tilde{\delta}_i^p$ is small compared to ρ_i. Thus, in the limit we get a different expression from Eq. (11.26)

$$DEFL^1|i = DEFL^0 \left\{ \begin{array}{l} \left[\left[\dfrac{\hat{\sigma}_p^{R,i}}{\hat{\sigma}_p}\right] + \sqrt{\dfrac{2}{\pi}}\left\{\sqrt{2} - \left[\dfrac{\hat{\sigma}_p^{R,i}}{\hat{\sigma}_p}\right]\right\} x + \right. \\ \left. -\dfrac{\alpha_p^i \rho_i}{2\hat{\sigma}_p} x^2 + \dfrac{2}{3}\sqrt{\dfrac{2}{\pi}}\left\{\dfrac{1}{4}\left[\dfrac{\hat{\sigma}_p^{R,i}}{\hat{\sigma}_p}\right] - \rho_i^2 r\right\} x^3 \right] \end{array} \right\} + O(x^4) \quad \tilde{\delta}_i^p << \rho_i \tag{11.28}$$

where $x = \tilde{\delta}_i^p / \rho_i$

The first term reflects the loss in precision due to the fact that large gross errors have been detected and their measurements eliminated. A large value of variance of the biases (ρ_i^2) appears to make matters better. This is because a smaller portion of those biases will have value smaller than $\hat{\delta}_i^p$, and, the rest will be detected. However, one needs to remember that $\tilde{\delta}_i^p << \rho_i$ and, therefore, this case would correspond to precise instruments that are prone to have large biases, something not very common.

In the case of no data reconciliation, we get

$$DEFL^1|p = DEFL^0 \left\{ \left[\left[\frac{\sigma_p^\#}{\sigma_p}\right]\right]\left[1 - \sqrt{\frac{2}{\pi}}\frac{\delta_p^\#}{\rho_p}\right] + \frac{2}{\sqrt{\pi}}\frac{\delta_p^\#}{\rho_p} + O(\left[\delta_p^\# / \rho_p\right]^2) \right\} \quad \delta_p^\# << \rho_p \tag{11.29}$$

Because $\sigma_p^\# > \sigma_p$, this number can be significant.

Similar expressions can be written for more errors. We omit them here (see Bagajewicz, 2006) for details. In fact, these expressions are integrated numerically over the regions presented in Chap. 10 (see Figs. 10.4 to 10.8) using two different methods (Nguyen et al., 2006): a Monte Carlo integration approach and also a method that is based on linearization of some portions of the integral so that analytical expressions are possible (the average of bounds method).

Effect of the Failure Frequency

Now, if all instruments have the same failure and repair rate, we have

$$\Phi^j = \sum_i \Phi_i^j = \binom{n}{j}\left(\frac{1}{1+\lambda}\right)^j \left(\frac{1}{1+\lambda}\right)^{n-j} \tag{11.30}$$

where λ is the ratio of repair rate to failure rate. Because the probabilities of failure depend on the time elapsed since the last repair, $P\{\hat{m}_p \geq m_p^*\}$ is a function of time, which makes matters more complicated. Indeed, as time goes by, the above probability should increase until a failure occurs, because the probability $f_i(t)$ is not constant. We will assume a bathtub curve and ignore the burn-in period. This is discussed for example by Dhillon (1983), and specifically as applied to instruments by Bagajewicz (2000). Thus, we will assume that such rate is constant.

If the sensors are never repaired (or seldom repaired/calibrated), then $f(t)$ is related to the service reliability $R_i^s(t)$ as follows (Bagajewicz, 2000):

$$f_i(t) = 1 - R_i^s(t) = 1 - e^{-r_i t} \tag{11.31}$$

where r_i is the failure rate. However, when corrective maintenance is performed, $f(t)$ is related to the service availability function $A_i^s(t)$, as follows (Bagajewicz, 2000):

$$f(t) = 1 - A_i^s(t) = \frac{r_i}{r_i + \mu_i}(1 - e^{-(r_i+\mu_i)t}) \tag{11.32}$$

where μ_i is the repair rate. In both cases, we start with a zero probability of failure and build up to a level off at one if no repair is made, or at $r_i/(r_i+\mu_i)$ in case of repairs. When corrective maintenance is performed, the usual case, the ratio of time to reach 99% of the asymptotic value $(t_i^{0.99})$ to the repair time $(1/\mu_i)$, is given by

$$\chi_i^{0.99} = \frac{t_i^{0.99}}{(1/\mu_i)} = -\lambda_i \frac{\ln 0.01}{(1+\lambda_i)} \cong 4.61 \frac{\lambda_i}{(1+\lambda_i)} \tag{11.33}$$

where $\lambda_i = \mu_i / r_i$, the ratio of repair rate to failure rate. Thus, for small values of λ_i, the probability of failure levels off very quickly, whereas for $\lambda_i >> 1$, the probability of failure levels off at about 4.6 times the inverse of the repair rate to level off. Notice that the failure rate for flowmeters (r_i) ranges from 0.1 to 10 failures/10^6 h (Bloch and Geltner, 1999). However, these are referred to as functional failure. When a noticeable loss of calibration is also referred to as failure, which is the case we are interested in, the numbers are higher. For simplicity, we will use here the asymptotic constant value for the probability of failure. This puts us in the worse condition and makes the analysis conservative.

To assess in what range the values are, consider two values of λ_i, 50 and 25. To get an idea of what these numbers represent, assume that the failure rate of an instrument is one failure per year, and that the repair rate is close to 20 repairs a week, then $\lambda_i = 50$. For the same repair rate, $\lambda_i = 12.5$ corresponds to a failure rate of 3.5 failures per year. Table 11.1 shows the results of the sum of failure probabilities Φ^j when all instruments have the same failure and repair rate.

For 10 instruments or less, we have that $\Phi^0 + \Phi^1 > 0.99$, which indicates that the rest can be ignored. Larger plants, like the refinery included in the article by Bagajewicz et al. (2005), which has 49 instruments, the ratio of repair to failure rate is critical. For example, when $\lambda_i = 50$, we have that $\Phi^0 + \Phi^1 + \Phi^2 > 0.99$ for $n = 20$ and larger that 0.92 for $n = 50$. The situation is not so good for the case of $\lambda_i = 12.5$. In this case, the probability of eight sensors or less failing is larger

	Φ^0		Φ^1		Φ^2	
n	$\lambda_i = 50$	$\lambda_i = 12.5$	$\lambda_i = 50$	$\lambda_i = 12.5$	$\lambda_i = 50$	$\lambda_i = 12.5$
1	0.9804	0.9259	0.0196	0.0741		
2	0.9612	0.8573	0.0384	0.1372	0.0004	0.0055
3	0.9423	0.7938	0.0565	0.1905	0.0011	0.0152
5	0.9057	0.6806	0.0906	0.2722	0.0036	0.0436
10	0.8203	0.4632	0.1641	0.3706	0.0148	0.1334
20	0.6730	0.2145	0.2692	0.3433	0.0511	0.2609
50	0.3715	0.0213	0.3715	0.0853	0.1820	0.1672

Table 11.1 Failure Probabilities as a Function of the Number of Sensors (*)

than 0.99 (0.97 for seven sensors or less). In addition, notice that for 50 instruments, one has to go to very large repair rates to have a sizable probability of no failure. This is costly and therefore suggests that better instrumentation should be bought to reduce the failure rate and also reduce the financial loss through larger accuracy, as we shall discuss below.

Trade-Off between Value and Cost

In the case of buying a data reconciliation package, one would write

$$\text{NPV} = d_n\{\text{Change in } DEFL\} - \text{change in cost} \tag{11.34}$$

where d_n is the sum of discount factors for n years and NPV the net present value. The change in cost includes now the cost of the license and/or the cost of new instrumentation plus the increased maintenance cost.

We therefore write

$$\text{NPV} = d_n\left\{DEFL\left(\bar{\sigma}_{2,m}, \sigma_2\right) - DEFL\left(\hat{\sigma}_{2,m}, \sigma_2\right)\right\} - \text{cost of license} \tag{11.35}$$

where $\bar{\sigma}_{2,m}$ is the value of precision after data reconciliation and $\hat{\sigma}_{2,m}$ the actual value of precision obtained directly from measurements.

The cost of maintenance is a function of the expected number of repairs. For a given instrument, this is given by (Bagajewicz, 2000)

$$\Lambda_i(t) = \mu_i\left[\frac{1}{(1+\lambda_i)}t + \frac{1}{r_i[1+\lambda_i]^2}\left(e^{-(1+\lambda_i)r_i t} - 1\right)\right] \tag{11.36}$$

Using this function, one can construct the net present cost of maintenance for new instrumentation. In fact, larger maintenance will reduce $DEFL$ by reducing the frequency of failure.

In the case of instrumentation upgrade

$$\text{NPV} = d_n\left\{DEFL\left(\bar{\sigma}_{2,m},\sigma_2\right) - DEFL\left(\hat{\sigma}_{2,m},\sigma_2\right)\right\}$$
$$-\text{Cost of new instrumentation} \qquad (11.37)$$

where $\bar{\sigma}_{2,m}$ is the value of precision after new instrumentation is added and $\hat{\sigma}_{2,m}$ the actual value. In more complete formulations one would also discount the maintenance costs. Thus, in any instrumentation upgrade program, two elements play an important role: the increase in value of information and the cost of it (in this case, the cost of instrumentation).

It is quite clear that plants with large number of instruments, especially if the rate of repair is low, require the evaluation of a fairly large number of terms of Eq. (11.23). We will cover instrumentation upgrade in Chap. 15.

Example 11.1: Consider the simplified ammonia process shown by Ali and Narasimhan (1993), given in Fig. 11.3. The total flowrates are the variables of interest. The total flowrates for streams $\{S_1, S_2, S_3, S_4, S_5, S_6, S_7, S_8\}$ are {152.6, 152.6, 152.6, 86.4, 66.2, 96.6, 30.4, 56}, respectively. We assume that the stream S_7 is the main product stream and is the stream for which financial loss is calculated and that the existing sensor network comprises of four sensors placed on four streams S_5, S_6, S_7 and S_8. We want to upgrade the existing sensor network by adding one more sensor on one of the streams S_1, S_2, S_3 or S_4. Our objective is to find the best sensor location by calculating the NPV for each candidate sensor location.

For this example we use $f_i(t) = 0.1$ for all sensors. Assume next that all sensor costs are equal; then the best location for the added sensor would be the one with the largest reduction of financial loss. The calculated financial losses are given in Table 11.2.

Clearly, by adding a sensor to stream S_4, one more *independent* material balance equation ($F_4 + F_7 = F_8$) that involves the flowrate of the product stream (F_7) is obtained to be used for data reconciliation and gross error detection. This is not the case for other candidates (i.e., with other candidates, the new equation obtained does not involve F_7).

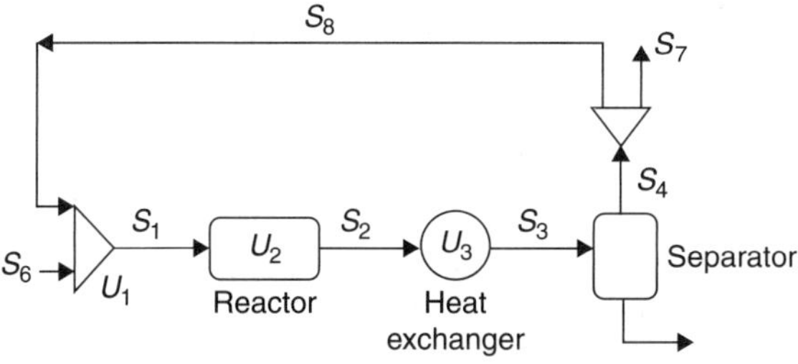

FIGURE 11.3 Simplified ammonia process.

System of Sensors	Sensor Added to the Original Sensor Network	Financial Losses (DEFL/K_s*T)	Improvement in Financial Loss due to Added Sensor
$\{S_5, S_6, S_7, S_8\}$		0.15813	—
$\{S_1, S_5, S_6, S_7, S_8\}$	S_1	0.15419	0.00394
$\{S_2, S_5, S_6, S_7, S_8\}$	S_2	0.15406	0.00407
$\{S_3, S_5, S_6, S_7, S_8\}$	S_3	0.15414	0.00399
$\{S_4, S_5, S_6, S_7, S_8\}$	S_4	0.14025	0.01788

TABLE 11.2 Financial Loss as a Function of Sensor Network

Example 11.2: We will now illustrate the results for a crude distillation unit (CDU) (Fig. 11.4), discussed by (Bagajewicz et al., 2005). Tables 11.3 and 11.4 indicate the streams and the measurement standard deviation. It is assumed that the instruments are fairly well maintained.

Value of performing data reconciliation: We first consider the case of precision only. Using the data listed in Table 11.3 and the costs of products of Table 11.4, the downside financial loss for the existing instrumentation without the aid of data reconciliation is around \$7.36 million, while after applying data reconciliation it reduces to \$7.12 million. This renders a net present value (over only 5 years) of \$236,817. This might justify the purchase of a data reconciliation package.

We now consider the value of accuracy: Assume $\lambda_i = 200$, which is fairly high. In such case, we can assume that the likelihood of two failures at the same time is smaller than 2.5%, rendering $\Psi^0 = 0.78$ and $\Psi_i^1 = 0.195$. To estimate the financial loss when there is no data reconciliation performed, we use Eq. (11.29) assuming that $\sigma_p^{\#} = 3\%$ (twice that of the measurement). For $\delta_p^{\#}/\rho_p = 0.25$, we get $\sum_i \text{DEFL}^1|_i = 2.2363\text{DEFL}^0$, which needs to be added for all the instruments (a total of 15). Thus, a lower bound for DEFL when no data reconciliation is made is about \$23.82 million (using a horizon of 5 years only).

If reconciliation is used we use Eq. (11.26) to evaluate the financial loss due to one gross error (we assume that the variance of the bias is very small, that is, we have good instruments and/or good maintenance). In this case, residual precision for single faults is about $\hat{\sigma}_p^{R,i} = 1.002\hat{\sigma}_p$ (there are several redundant measurements). This renders DEFL = 7.38 million. Thus the NPV of data reconciliation is 16.44 million. This proves that in plants of this size the financial loss due to biases is far larger than the one due to precision.

Net present value of new instrumentation to improve precision: We do not explore here any strategy to optimize the addition of new instrumentation. Rather, we illustrate the effect of adding an instrument of 1.5% standard deviation to some streams. The results are shown in Table 11.5. The first column of this table indicates which instruments have been added. The second column indicates the net present value of the project assuming no data reconciliation package has been installed, whereas the third column indicates the net present value for the case where the plant already has data reconciliation installed and in use.

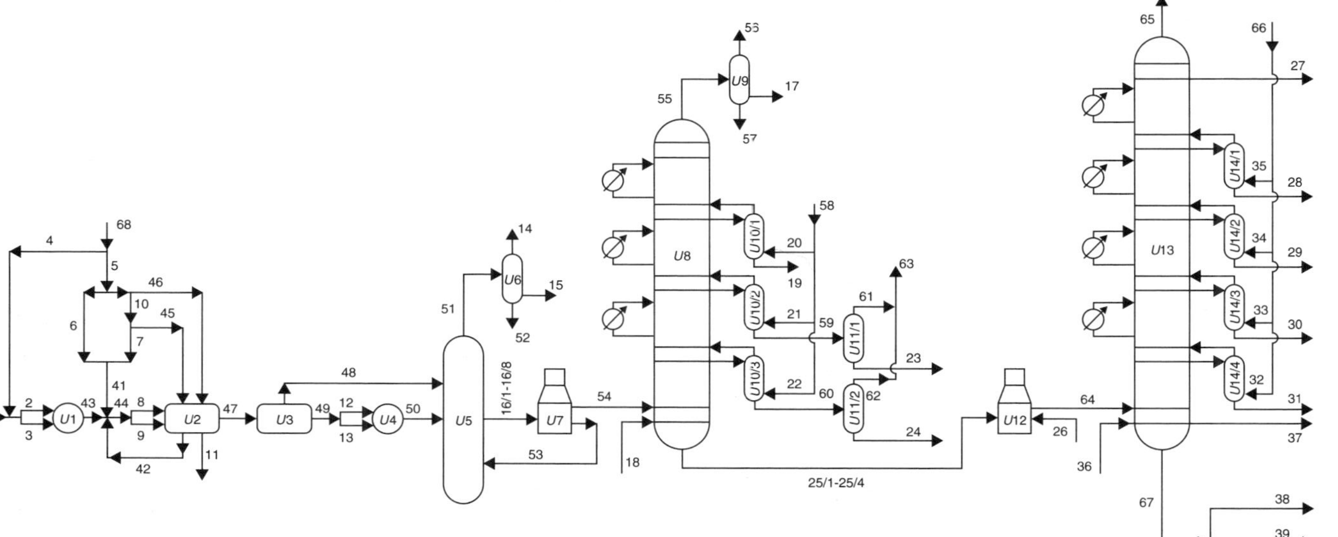

Figure 11.4 Flow sheet for crude distillation unit.

Units: *U*1—HEN; *U*2—desalting unit; *U*3—crude vessel; *U*4—HEN; *U*5—prefractionation tower; *U*6—condenser; *U*7—furnace; *U*8—atmospheric tower; *U*9—condenser; *U*10/1-*U*10/3—preflash column; *U*11/1, *U*11/2—atmospheric product dryer; *U*12—furnace; *U*13—vacuum tower; *U*14/1-*U*14/4—preflash column

Streams: crude oil: 1,2,3,8,9,43,44—desalted crude oil: 12,13,47,49,50—water 4,5,6,7,10,41,45,46,68—hydrocarbon vapors: 48,56,61,62,63,65—sour water: 11,42,48—oily water: 52,57—prefractionation products: 14,15—atmospheric products: 17,19,23,24—vacuum products: 27,28,29,30,31,37,67—steam: 18,20,21,22,26,32,33,34,35,36,58.

No. Stream	Measured Flowrate (kg/h)	Mass Flowrate from Balance (kg/h)	Standard Deviation (kg/h)	Flowrate after Reconciliation (kg/h)	Standard Deviation after Reconciliation (kg/h)	No. Stream	Measured Flowrate (kg/h)	Mass Flowrate from Balance (kg/h)	Standard Deviation (kg/h)	Flowrate after Reconciliation (kg/h)	Standard Deviation after Reconciliation (kg/h)
1	418,839	—	6,283	413,336	1,259	31	18,187	—	273	18,196	191
2	212,050	—	3,181	209,310	2,339	32	312	—	5	314	4.87
3	213,020	—	3,195	210,256	2,341	33	338	—	5	340	4.87
4	6,231	—	93	6,231	91	34	325	—	5	327	4.87
5	20,352	—	305	20,327	244	35	311	—	5	313	4.87
6	7,174	—	108	7,174	108	36	3,226	—	48	3,225	48
7	7,256	—	109	7,256	109	37	18,097	—	271	18,106	190
8	230,650	—	3,460	230,650	3,460	38	15,141	—	227	15,154	224
9	229,870	—	3,448	229,870	3,448	39	20,245	—	304	20,268	297
10	10,188	—	153	10,188	153	40	12,650	—	190	12,659	188
11	26,180	—	393	26,243	391	41	—	14,430	—	—	192
12	209,170	—	3,138	206,932	2,296	42	—	26,523	—	—	5,048
13	208,950	—	3,134	206,718	2,295	43	—	419,566	—	—	1,256
14	5,122	—	77	5,124	54	44	—	460,520	—	—	4,886
15	21,434	—	322	21,467	227	45	—	2,932	—	—	192
16/1	62,562	—	938	61,562	938	46	—	2,966	—	—	288
16/2	60,985	—	915	60,985	915	47	—	413,651	—	—	1,186
16/3	61,253	—	919	61,253	919	48	—	0	—	—	NA
16/4	61,490	—	922	61,490	922	49	—	413,651	—	—	1,191
16/5	61,009	—	915	61,109	915	50	—	413,651	—	—	1,200

16/6	60,796	—	912	60,796	912	51	—	27,068	—	—	192
16/7	62,012	—	930	62,012	930	52	478	—	7	478	7
16/8	60,413	—	906	60,413	906	53	—	103,938	—	—	2,849
17	45,680	—	685	45,829	478	54	—	386,582	—	—	1,171
18	4,275	—	64	4,272	64	55	—	57,169	—	—	494
19	26,084	—	391	26,133	275	56	7,130	—	107	7,137	107
20	155	—	2	156	1.98	57	4,200	—	63	4,202	63
21	256	—	4	260	3.83	58	795	—	12	759	5.85
22	337	—	5	343	4.65	59	—	73,900	—	—	752
23	73,319	—	1,100	73,704	753	60	—	50,852	—	—	538
24	50,533	—	758	50,716	528	61	196	—	3	196	3
25/1	45,721	—	686	45,902	615	62	136	—	2	136	2
25/2	45,698	—	685	45,878	615	63	—	332	—	—	3.6
25/3	45,747	—	686	45,928	615	64	—	185,593	—	—	721
25/4	45,671	—	685	45,851	615	65	4,512	—	68	4,513	68
26	2,035	—	31	2,035	31	66	1,322	—	20	1,293	9
27	38,515	—	578	38,557	392	67	—	48,081	—	—	245
28	18,921	—	284	18,931	198	68	26,583		399	26,559	249
29	19,835	—	298	19,846	208						
30	23,864	—	358	23,880	249						

Table 11.3 Stream Data for Example 11.2

Stream	K_s (\$/Kg)	Stream	K_s (\$/Kg)
14-15	0.13	28	0.13
17	0.14	29	0.12
19	0.25	30	0.11
23	0.24	31	0.1
24	0.23	37	0.08
27	0.15	38-39-40	0.06

TABLE 11.4 Stream Values or Inventory Costs

Location of New Sensors	NPV without Originally Installed Data Reconciliation	NPV with Originally Installed Data Reconciliation
14, 15, 17, 19, 23, 24, 27, 28, 29, 30, 31, 37, 38, 39, 40	$2,088,108	$1,851,290
14, 15, 17, 19, 23, 24, 27, 28, 29, 30, 31, 37, 38, 39	$2,098,762	$1,862,945
14, 15, 17, 19, 23, 24, 27, 28, 29, 30, 31, 37, 38	$2,071,202	$1,834,384
14, 15, 17, 19, 23, 24, 27, 28, 29, 30, 31, 37	$2,052,908	$1,816,091

TABLE 11.5 Effect of New Flowmeters on Savings

References

Ali, Y. and S. Narasimhan, "Sensor Network Design for Maximizing Reliability of Linear Processes," *AIChE J.* **39**(5):2237–2249 (1993).

Bagajewicz, M., "Process Plant Instrumentation. Design and Upgrade." (ISBN: 1-56676-998-1), (Technomic Publishing Company). Now CRC Press (2000).

Bagajewicz, M., "Value of Accuracy in Linear Systems," *AIChE J.* **52**(2):638–650 (2006).

Bagajewicz, M., M. Markowski, and A. Budek, "Economic Value of Precision in the Monitoring of Linear Systems," *AIChE J.* **51**(4):1304–1309 (2005).

Bloch, H. P. and Geltner F. K., "Use Equipment Failure Statistics," *Hydrocarbon Processing* January (1999).

Dhillon, B. S., "Reliability Engineering in Systems Design and Operation," Van Nostrand Reinholdt Company, New York (1983).

Lipták, B. G., editor, Instrument Engineer's Handbook, 4th ed., **1**, CRC Press, Boca Raton, FL (2003).

Nguyen Thanh D. Q., K. Siemanond, and M. Bagajewicz, "On the Determination of Downside Financial Loss of Instrumentation Networks in the Presence of Gross Errors," *AIChE J.* **52**(11):3825 –3841 (2006).

CHAPTER 12

Data Reconciliation Practical Issues

There are several data reconciliation commercial software packages in the market. Almost all

- Are steady-state data reconciliation packages.
- Assume that the measurements are samples of a normal distribution and do not test for this condition.
- Have no data conditioning (they assume that raw data is a single number for each measurement).
- Have simple bias/leak and gross error detection strategies, at times not very well implemented. The most common are simple nodal and measurement tests. Some feature simple serial elimination.
- Are mostly devoted to linear cases (material balances) and in many occasions have limited capabilities to deal with nonlinear models. A few packages include energy balances and pressure measurements, thus reconciling temperatures and pressures.
- Feature nice graphical user interface and databases capable of storing and handling historical data.

Despite their good attributes there are several additional implementation issues that need attention. We have already mentioned that commercial software do not have powerful enough bias detection procedures, especially when multiple gross errors (biases or leaks) are present. We consider this to be at the core of all problems in data reconciliation.

Data Preprocessing

The objective of data preprocessing for data reconciliation is to provide a measurement value. In addition for steady-state data reconciliation and some versions of dynamic data reconciliation, estimators of variance are needed. None of this data preprocessing is included in available commercial reconcilers.

Use of Filters

At the core of data reconciliation is the unaltered measurement, with the least manipulation possible. In many cases, however, a Kalman filter used on each measurement sent from the DCS to the historian, but only a few values, not all. There is disagreement, even among specialists as of whether this practice is proper.

- On one hand there are those that claim that the Kalman filter provides an estimator, with a variance, which in turn can be used to feed the data reconciliation.
- Although there is nothing in principle wrong with considering an estimator as a "measurement," it is unclear if this will maintain the Gaussian nature of the original signal. Another unanswered question is how the signal can be used to calculate the variance-covariance matrix S. As we shall see later, this is another issue we will discuss later.
- Another issue is when the corrections for temperature and pressure are made.

Remark: Removal of outliers in association with or prior to any signal treatment is of importance, especially when the number of outliers is large. We discussed methods for this in Chap. 2.

Steady-State Recognition and Variance

It has been a tradition to only reconcile values that correspond to steady state. Thus once the data has been recognized as corresponding to a steady state, they are averaged and eventually used to determine the variance. The former is current industrial practice in many process plants, while the latter is less frequent. In the former case, the assumption is that all measurements are independent. In the latter, it is assumed that there are some disturbances, mainly in the electronics and other perturbations that affect all measurement unequally. This is discussed later.

We now describe steady-state recognition and variance estimation. Then we discuss a procedure that can be used for the case in which a steady-state data reconciler is fed by dynamic data. This is particularly important if tank inventory is to be tracked daily.

Variance Estimation

In Chap. 2 we presented a method to determine the precision. It was based on direct observation of fluctuations around a steady-state value for each value.

Once a steady state is identified, one can estimate the variance as follows: We first calculate the mean value

$$\bar{f}_i = \frac{1}{n}\sum_{k=1}^{n} f^{+}_{i|k} \tag{12.1}$$

$$s_{i,j} = \mathrm{Cov}(f_i, f_j) = \frac{1}{n-1}\sum_{k=1}^{n}\left(\tilde{f}_{R,i|k} - \bar{f}_{R,i}\right)\left(\tilde{f}_{R,j|k} - \bar{f}_{R,j}\right) \tag{12.2}$$

This method is called the direct method and was proposed by Almasy and Mah (1984). Its advantage is that it is simple to implement.

Consider now the conventional exponentially weighted moving average of a signal

$$\hat{X}_i = \lambda_1 X_i + (1-\lambda_1)\hat{X}_{i-1} \tag{12.3}$$

where $0 < \lambda_1 \le 1$. Cao and Rhinehart (1995) proved that the following filter

$$\hat{v}_i^2 = \lambda_2 (X_i - \hat{X}_i)^2 + (1-\lambda_2)\hat{v}_{i-1}^2 \tag{12.4}$$

provides an unbiased estimate of the mean square deviation of the data, defined by

$$v^2 = E\{(X_i - \vec{X})^2\} \tag{12.5}$$

They also prove that the variance of the data, defined by

$$\hat{\sigma}^2 = \frac{1}{(N-1)}\sum_{i=1}^{N} (X_i - \vec{X}_N)^2 \tag{12.6}$$

is related to v^2 through the following relation:

$$\hat{\sigma}^2 = \frac{2-\lambda_1}{2}\hat{v}^2 \tag{12.7}$$

Then, one unbiased estimate of the noise variance can be obtained as follows:

$$\hat{s}_{1,i}^2 = \frac{2-\lambda_1}{2}\hat{v}_i^2 \tag{12.8}$$

Another estimate of the variance is obtained using the mean squared difference of successive data. Ideally, this is defined as follows:

$$\delta^2 = E\{(X_i - X_{i-1})^2\} \tag{12.9}$$

Cao and Rhinehart (1995) propose the following exponential weighted filter

$$\hat{\delta}_i^2 = \lambda_3 (X_i - X_{i-1})^2 + (1-\lambda_3)\hat{\delta}_{i-1}^2 \tag{12.10}$$

as an unbiased estimate of δ^2. It is easily shown that the second estimate of the variance is given by

$$\hat{s}_{2,i}^2 = \frac{\hat{\delta}_i^2}{2} \tag{12.11}$$

Most data reconciliation packages and reconciliation methods assume that the matrix is diagonal and allow only the input of one variance per measured variable. This is unlikely to change, even though it is not very difficult to implement in the reconciliation code.

Cao and Rhinehart (1995) offer a compromise set of values of $\lambda_1 = 0.2$, $\lambda_2 = 0.1$ and $\lambda_3 = 0.1$.

Example 12.1: We now test the above methods on two different signals. Consider the signal of Fig. 12.1. This signal was constructed using a normal distribution with means 100 and 85, and a variance of 3 for both cases. The step function occurs at $t = 125$.

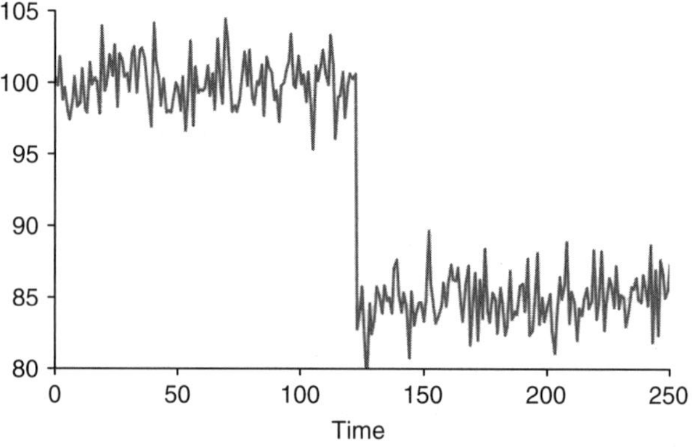

Figure 12.1 Signal with step and normal noise.

Using Almasy and Mah's method just for this signal renders mean values of 99.992 and 84.798, for the first and second portion with corresponding variances of 3.05 and 3.27 [Eqs. (12.1) and (12.2)]. Figure 12.2 shows the smoothed values obtained by the exponentially weighted filter [Eq. (12.3)].

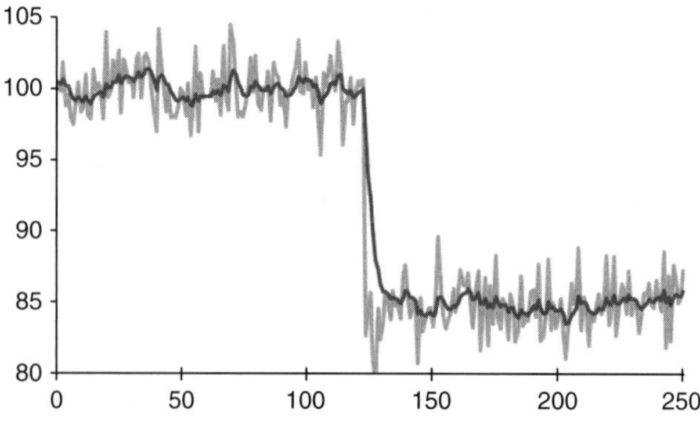

Figure 12.2 Estimator of average.

Figure 12.3*a* shows mean squared deviation of data [Eq. (12.8)] and Fig. 12.2*b* the mean squared difference of successive data [Eq. (12.10)]. These were constructed using a value of 3 for the first variance using the assumption that this is information provided by manufacturer.

In the first case (Fig. 12.3*a*), we see the effect of the filter going over the step. Because the average uses information over a larger portion of time, then the variance estimator takes a long time to stabilize. The averages of all these values were computed, rendering a value of 2.11 for the data before the step and 2.25 for the data after [in this case the average was taken in the interval (163, 250) to avoid the cumulative effect of the step].

In the second case (Fig. 12.3*b*), the effect of the filter is milder and the values are higher. The averages of all these values were computed, rendering a value of 2.97 for the data before the jump and 3.50 for the data after [in this case the average was also taken in the interval (163, 250)].

It seems that the mean squared deviation of data predicts values lower than the true value. Straight calculation of the variance in the intervals identified as being in steady state.

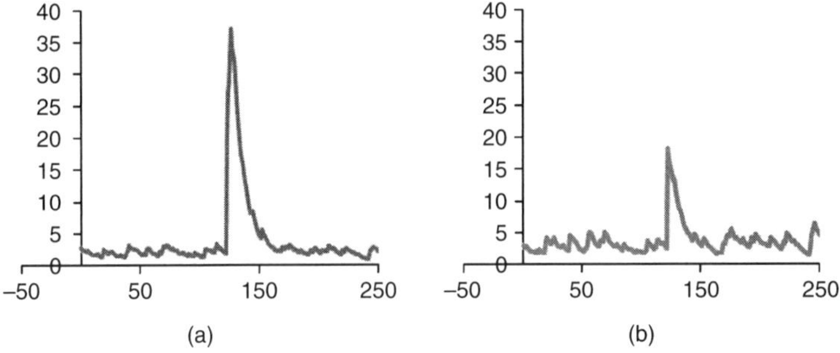

FIGURE 12.3 Variances for signal of Fig. 12.1: (*a*) Mean squared deviation of data, (*b*) mean squared difference of successive data.

Example 12.2: Consider now the more interesting signal of Fig. 12.4. This signal was constructed using a normal distribution with various means and a variance of 3, representing 4 steady states, with some smooth transitions. Figure 12.4 also includes the smoothed values obtained by the exponentially weighted filter [Eq. (12.3)].

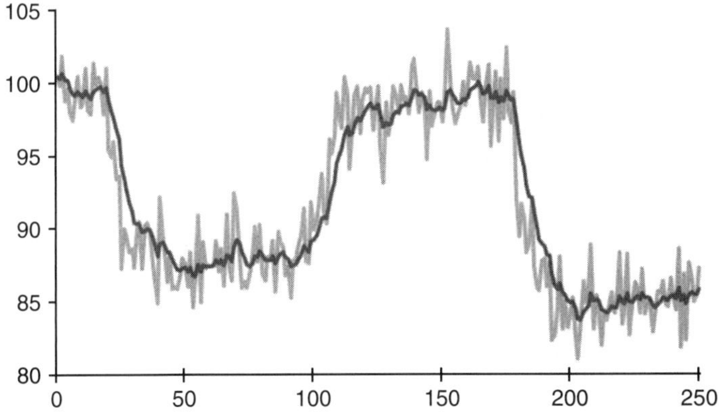

FIGURE 12.4 Signal with intermittent steady states and the average estimator.

Figure 12.5*a* shows mean squared deviation of data [Eq. (12.8)] and Fig. 12.5*b* the mean squared difference of successive data [Eq. (12.10)]. Quite clearly the former has larger peaks, giving an average value of 3.54. The average of the mean squared difference of successive data is 3.2. Quite clearly the overestimation includes the changes of steady state.

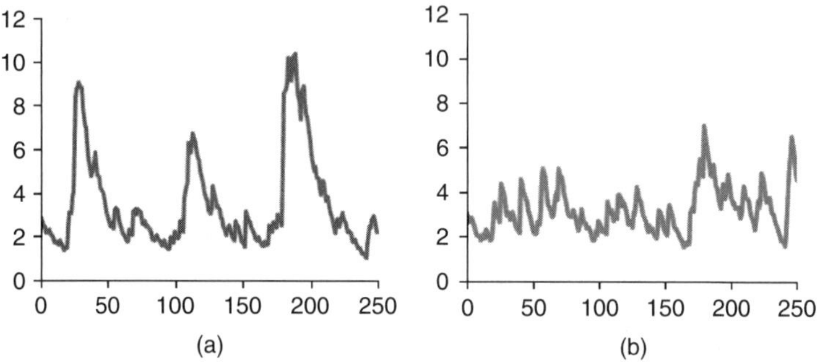

FIGURE 12.5 Variances for signal of Fig. 12.4: (*a*) Mean squared deviation of data, (*b*) mean squared difference of successive data.

Remark: There are other methods, called indirect methods to calculate the variance–covariance matrix (Darouach et al., 1989; Keller et al., 1992; Chen et al., 1997; Morad et al., 1999). In all these methods the main objective is to remove the influence of outliers. The reason for this is that variances are used as weights in the data reconciliation and gross error detection process. Outliers, when used to calculate the average of many measurements, a procedure used to feed steady-state data reconcilers, introduce a special type of bias, a bit different from a true continuous bias. Eventually, the gross error detection process is supposed to identify them. The outliers, if left in the determination of the variance, will tend to make it larger. Because gross error detection is based on comparing with the variance, we conclude that the outliers will be less likely to be detected. This is, to say it loosely, "ill-posed." The mechanism used by these indirect methods to ameliorate this problem is to construct M-estimators (Huber, 1981; 2004), which prevent large deviations from the mean (supposedly the outliers) influencing the estimation. They do that by putting larger weight on values close to the mean value. Details of these methods are discussed by Sánchez and Romagnoli (2000).

Remark: Zhou J. and R. H. Luecke (1995) proposed a method to estimate the covariances of process noise and measurement noise in dynamic systems in the context of the use of the Kalman filter. Heyen et al. (1996) for steady state and Ullrich et al. (2009) for dynamic state present methods to obtain the variance of the estimates. They follow the method shown in Chap. 4 (Eq. 4.28).

Remark: When processes are exempt from sharp step changes, polynomial interpolation can be used to obtain the polynomial coefficients (Jiang and Bagajewicz, 1997). The difference between measurements and polynomial predictions is used to obtain the variance. Spline polynomials are also an option.

Steady-State Detection

The examples of the previous section indicate that steady state needs are to be identified if a good estimate of the variance is wanted.

We now concentrate on discussing how to detect a steady state using the improved ratio test (Cao and Rhinehart, 1995) and the regression method.

Ratio Test

We first construct the ratio of the two variances introduced in Eqs. (12.8) and (12.11).

$$R_i = \frac{\hat{s}_{1,i}^2}{\hat{s}_{2,i}^2} = \frac{(2-\lambda_1)\hat{v}_i^2}{\hat{\delta}_i^2} \tag{12.12}$$

This test is comparing a long-term variance over a short-term one. The next step is to compare R_i with a critical value and determine whether there is a steady state or not. All this requires almost no storage and very few operations, which makes it very appealing to implement online.

The selection of the values of λ_1, λ_2, and λ_3 requires some trade-off analysis. Small values can reduce type II errors (false alarm). However, large values mean less filtering and can track the process more closely. Cao and Rhinehart (1997) offer a compromised set of values of $\lambda_1 = 0.2$, $\lambda_2 = 0.1$ and $\lambda_3 = 0.1$, and $R_{crit} = 2$. The recommendation, however, is to play around with this scheme. As per the number of data, they suggest to use a lag P such that the autocorrelation coefficient $\rho_P < 0.05$. Smaller lags may show a strongly autocorrelated data, and larger lags may incorporate the upsets from the past.

Remark: The autocorrelation describes the correlation between the signal values at different points in time. Let x_t be the value at time t and μ_t its mean value. Let also σ_t^2 be the corresponding variance. Then the autocorrelation coefficient is

$$\rho_{t,s} = \frac{E[(x_t - \mu_t)(x_s - \mu_s)]}{\sigma_t \sigma_s} \tag{12.13}$$

where E is the expected operator. We rule out a constant process where the variance would be zero. Usually, the autocorrelation ranges in the interval [−1, 1]. When the mean and variance are time independent, a true stationary process (called second-order stationary), one can make the autocorrelation a function of the time lag and independent of time. Thus, one would write

$$\rho(\tau) = \frac{E[(x_t - \mu)(x_{t+\tau} - \mu)]}{\sigma^2} \tag{12.14}$$

For a discrete stationary process, one would write

$$\hat{\rho}(k) = \frac{1}{(n-k)\sigma^2} \sum_{t=1}^{n-k} (x_t - \mu)(x_{t+k} - \mu) \qquad (k < n) \tag{12.15}$$

This concept will also be of importance in disturbance analysis and modeling (Chap. 13).

The method can be amended by disregarding the data prior to the declaration of steady-state change, which will allow the recognition of the new steady state very quickly. An additional problem is obtained when the signal is oscillating. In such case, the slope can give zero even though there is not a steady state.

Example 12.3: Figure 12.6 shows the original signal and the points where R_i exceeds the threshold. The method seems to identify these changes of steady state pretty well.

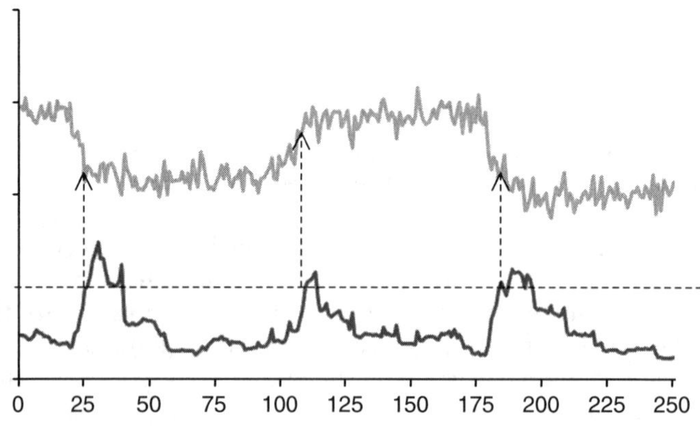

Figure 12.6 Signal of Example 12.2 and steady-state changes detection ratio.

Remark: Brown and Rhinehart (2000) presented a multivariable extension providing a single statistic and a test to determine when all variables are at steady state. The only requirements are that there is no autocorrelation between data points (during steady state), and no cross-correlation between variables (at steady state). They successfully applied the method to a distillation column.

Example 12.4: Figure 12.7 shows the slope of a regression line as a function of time for $N = 15$ (Fig. 12.7*a*) and $N = 30$ (Fig. 12.7*b*). Right when the slope becomes very negative, one can identify that there has been a loss of steady state. Prior to the step change and after it, the values of slope are fairly small. Figure 12.8 shows the same values for the signal of Example 12.2. When the number of points is small, it is difficult to determine the steady-state changes. However, when the number of points is $N = 30$, the method seems to identify well the three changes of steady state.

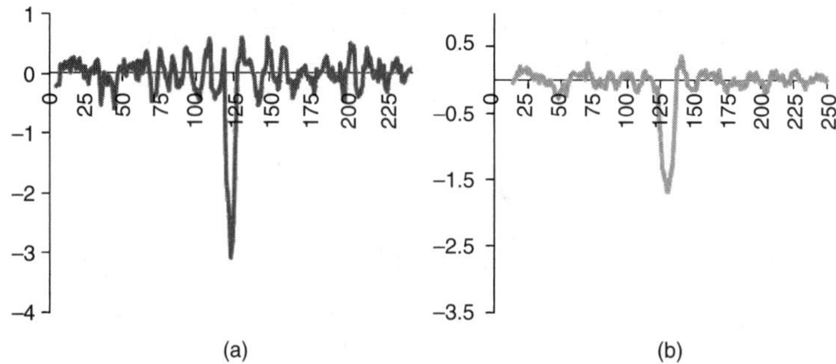

Figure 12.7 Variances for signal of Fig. 12.1. Using the regression method: (*a*) $N = 15$, (*b*) $N = 30$.

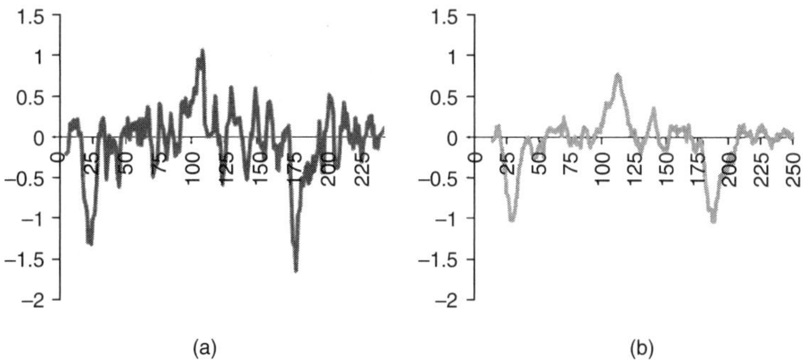

FIGURE 12.8 Variances for signal of Fig. 12.4. Using the regression method (*a*) $N = 15$, (*b*) $N = 30$.

Remark: Simpler versions of the moving average chart have been proposed. A threshold of 3σ is established above and below, and when the signal violates it an alarm is set off. This method has similar problems as the regression and ratio of variances, especially in relation to oscillating systems. For example, Loar (1992) uses a statistical process control (SPC) moving average chart. The problems encountered are that small oscillations cannot be caught, even using the 100-data window, which, if increased, may actually extend the period of analysis and lessen the ability to determine when the changes in steady state take place after all.

Remark: Narasimhan et al. (1986) proposed a two-step procedure called *composite statistical tests*, a small variation forms a method that can be found in Crow et al. (1955). They first propose the use of test of sample covariance matrices constructed using the data of a certain period. Next, depending on whether these covariances are statistically equal or not, they propose to use two different tests to determine if the moving averages are equal or not. If the process is at steady state, their ratio should be unity. However, due to limited sampling, the actual ratio of the variances is not exactly unity but a statistic that averages near unity. Thus, the drawbacks to this method are the user-required choices of data window length and a threshold for decision.

Remark: The determination of trends have been of particular importance to monitoring, even though they seem irrelevant to data reconciliation techniques. Mah et al. (1995) proposed the use of "piecewise linear online trending," a statistical technique that predicts the intervals in which new points will fall based on assumed trends. If they do not, they could be a new trend or an outlier, which is resolved using hypothetical testing. This method seem to perform much better than the earlier ones, the "Box Car with Backward Slope" (Hale and Sellars, 1981) and the "Swinging Door" (Bristol, 1990). In recent years the use of wavelet decomposition was proposed to determine process trends, including smoothing (Bakhtazad et al., 1998; Flehmig et al., 1998; Akbaryan and Bishnoi, 2000; Jiang et al., 2003). In particular, Jiang et al., (2003) state that "... process trends are extracted from the measured raw data via wavelet-based multi-scale processing. The process status is then measured using an index with value ranging from 0 to 1 according to the wavelet transform modulus of the extracted process signal. Finally, a steady state is identified if the computed index is small (close to zero)." Although elaborate examples presented seem to indicate that the method competes well with all other methods. An alternative approach for determining trends, which compares real trends to "semi-quantitative trends" stored in a dictionary of trends using an expert system was proposed by Sundarraman and Srinivasan (2003).

Remark: Bhat and Saraf (2004) modified the method of Cao and Rhinehart (1995) by optimizing the filter constants, minimizing type I and II errors, and simultaneously reducing the delay in state detection. A simple algorithm uses past measurements and the Kalman filter to detect the presence of gross error and estimate its magnitude. In addition they reconcile data. An industrial crude distillation unit was used to illustrate the results. A controversial feature of this technique is that because of a strong autocorrelation in the data, random noise was added to aid the steady-state identification.

Tanks and Steady-State Data Reconciliation

To add redundancy, hold-up changes should be modeled as pseudo-streams, a feature that many data reconciliation packages already include. This is illustrated in Fig. 12.9.

Quite clearly, the pseudo-stream variance will come from computing two level measurements. This is why it is critical that tank level measurements are precise enough.

Let the system of equations for material balances be represented by

$$\frac{dv}{dt} = Af \tag{12.16}$$

$$Cf = 0 \tag{12.17}$$

where v and f are the tank hold-ups and flows, respectively. Let f_i^+, $i = 1, \ldots, n$ be the vector of measurements of all stream flows for n instances of time and v_i^+ be the vector of tank hold-up measurements for the same times.

Steady-state data reconciliation is based on the average of the flow measurements $\overline{f}^+$ and the average $(\overline{f}^{v+})$ of a new artificial flowrate f_i^{v+} defined as follows:

$$f_i^{v+} = \frac{v_i^+ - v_0^+}{(t_i - t_0)} \tag{12.18}$$

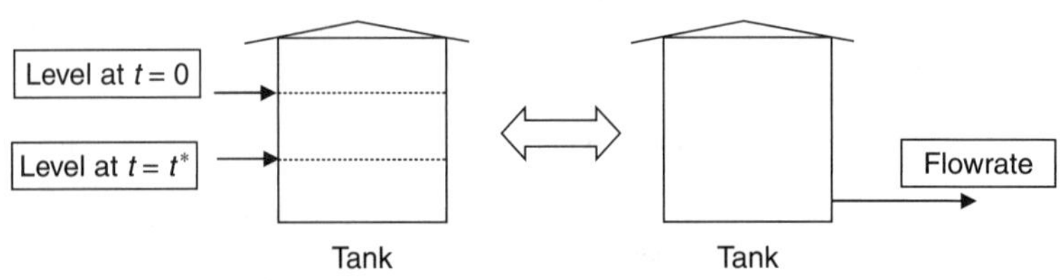

Figure 12.9 Tanks and pseudo-streams.

Thus, the system equations are now represented by

$$Dy = 0 \tag{12.19}$$

where

$$D = \begin{bmatrix} A & -I \\ C & 0 \end{bmatrix} \tag{12.20}$$

and

$$y = \begin{pmatrix} f \\ f^v \end{pmatrix} \tag{12.21}$$

Let R be the variance of all measurements. If all the flows are measured, then the solution to the problem is

$$\tilde{y} = \left\lfloor I - R\ D^T (DR\ D^T)^{-1} D \right\rfloor \bar{y}^+ \tag{12.22}$$

where $\bar{y}^+ = \begin{pmatrix} \bar{f}^+ \\ \bar{f}^{v+} \end{pmatrix}$.

Use of Dynamic Data in Steady-State Reconcilers

The theory of steady-state data reconciliation and its commercial implementation makes the assumption that one measurement is to be expected. However, in industrial practice, to get this one value, the average of many values is made. The question arises then as of what is the variance to be used. The correct answer is given by statistics (the variance of the average), but in practice the variance of the instrument is used. Assume that the measurement is a sample taken from a distribution with variance σ_i^2. Then the variance of the mean on n such measurements is σ_i^2/n. Let F_i^+ be the measured flowrate and $\bar{F}_i^+$ its average, and let $\bar{S}$ be the variance-covariance matrix of the measurement means. Thus, $\bar{S} = S/\sqrt{n}$ and therefore the data reconciliation model is

$$\left.\begin{aligned} &\text{Min}(\tilde{F} - F^+)^T \bar{S} (\tilde{F} - F^+) \\ &\text{s.t.} \\ &\qquad A\tilde{F} = 0 \end{aligned}\right\} \tag{12.23}$$

However,

$$(\tilde{F} - F^+)^T \bar{S} (\tilde{F} - F^+) = (\tilde{F} - F^+)^T \frac{S}{\sqrt{n}} (\tilde{F} - F^+) \tag{12.24}$$

This indicates that the real objective and the one used in practice only differ by a multiplicative constant, and therefore the result of the reconciliation does not change. This is indeed a fortunate result.

Remark: The assumption made above is that the variance-covariance of the measurements is given, not measured. If the variance is actually estimated from signal data, the multiplicative constant is not $1/n$, neither $1/(n-1)$ as Eq. (12.6) would suggest. The elements of S would already be the appropriate variance of the signal's mean.

Remark: Although arguing about the right value of variance is a moot point, as discovered above, it is not when variance is used for bias detection and elimination.

One connection between dynamic and steady-state data reconciliation was found by Bagajewicz and Jiang (2000). Specifically, it was proved that when the variance-covariance matrix is diagonal, and there are no tanks, that is, only flowrates are involved, the averages of reconciled values from dynamic data reconciliation are the same as the reconciled values from steady-state reconciliation based on dynamic data averages.

This claim was proved only for the integral dynamic data reconciliation model. Bagajewicz and Jiang (2000) also explore the effect of hold-up and conclude that the steady state still constitutes a reasonable approximation.

Example 12.5: Consider the system of Fig. 12.10. The calculated variances are given in Table 12.1.

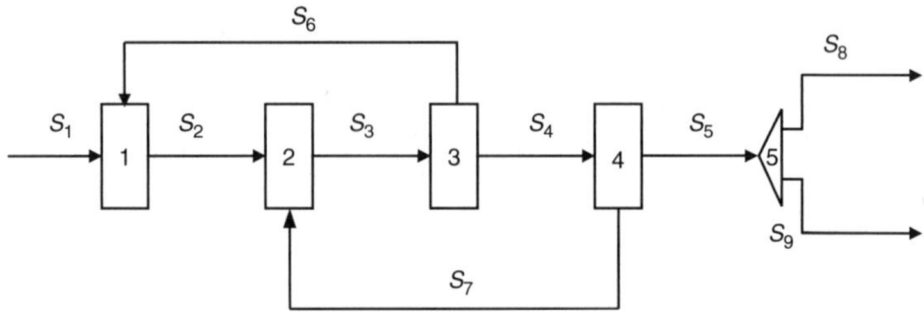

Figure 12.10 System of tanks.

Table 12.2 compares the reconciled flowrates (real and artificial) obtained performing steady-state data reconciliation with the average of the results obtained using the integral dynamic data reconciliation. Deviations from true values are shown.

By simply comparing the values of the estimates of the instrument variances (R_F and R_V) to the values of the variances used in steady-state data reconciliation,

	Variances of Flowrates									Variances of Hold-ups			
	1	2	3	4	5	6	7	8	9	1	2	3	4
Original	0.04	0.08	0.2	0.16	0.12	0.2	0.16	0.04	0.08	3.2	15.0	6.0	3.0
Estimated	0.045	0.082	0.21	0.18	0.15	0.22	0.16	0.042	0.078	3.85	19.99	5.93	3.04

TABLE 12.1 Variance of Flowrates and Hold-ups

	Averaged Flowrates (Including real and Artificial)												
	1	2	3	4	5	6	7	8	9	10*	11*	12*	13*
Reconciled at steady state	1.469	2.467	3.696	4.469	3.369	5.136	4.373	1.431	1.938	4.139	3.144	–5.910	–3.273
Average of reconciled at dynamic state	1.470	2.464	3.701	4.451	3.373	5.150	4.388	1.432	1.941	4.164	3.158	–5.909	–3.316
True value	1.477	2.477	3.688	4.477	3.377	5.177	4.377	1.431	1.965	4.183	3.173	–5.973	–3.283

*Flowrates 10 through 13 correspond to the artificial flowrates corresponding to tanks 1 through 3.

Table 12.2 Comparison of Steady-State and Dynamic-State Data Reconciliation

one finds that the relative values from one to another vary. Thus, estimates of variances of flowrates that exhibit larger flow variations are also relatively larger. The importance of picking the proper variance when using data reconciliation is thus highlighted.

Remark: Bagajewicz and Gonzales (2001) performed a similar experiment in the presence of accumulation (tanks hold-up), comparing results of steady-state data reconciliation using process averages, with the results of the dynamic data reconciliation method of Rollins and Devanathan (1993). They used the same example as both Darouach and Zasadinski (1991) and Rollins and Devanathan (a few interconnected tanks) and compared three situations: (a) high standard deviation of noise in hold-ups (16%) and small standard deviation of noise in flowrates measurements (3%), (b) high standard deviation of noise in flowrates (50%) and small standard deviation of noise in hold-ups measurements (0.01%), and (c) moderately high standard deviation of noise (5.4%) and in tank hold-ups (4.6%). In case (a) the differences are less than 0.02% for flows and less than 0.6% for tank volumes. In case (b) differences for flows are smaller than 1% whereas the differences for hold-ups are smaller than 0.05. Finally in case (c) the differences for flows are of the order of 2% whereas those of tank volumes are as large as 0.01%.

Clearly, one can conclude that even under conditions of very large variance in measurement (the variances used are far larger than those encountered in practice), the errors are fairly small. Although there has not been theoretical work that would back up this.

The above findings provide a reasonable argument for continuing using the technology available in commercial software.

Random Error Distributions

In Chap. 2, we demonstrated the use of *q-q* plots to determine if a set of numbers, a signal in our case, is best represented by a certain distribution. The question now is how the shape of the distribution affects the results of the data reconciliation.

We recall from Chap. 4 [Eq. (4.5)] that, when the variance-covariance matrix is diagonal the data reconciliation is based on minimizing the weighted sum of squared differences between the estimator and the measurements $[\tilde{F}_R - F_R^+]^T S_R^{-1}[\tilde{F}_R - F_R^+]$. It was also mentioned in a Remark that this minimization stands from the assumption that the data is distributed.

Bagajewicz (1996) studied the effect of dealing with data that is not distributed normally and showed there could be significant differences.

Multiple Measurements of the Same Variable

Most data reconciliation commercial software methods have steady-state data reconciliation calculation engines and consider only one flow measurement per stream. When multiple measurements are made of the same stream, one can consider including a pseudo-unit so that the reconciler sees one measurement per stream. This is frequent in plants that have different units, each one measuring its output and input, resulting in duplicate measurements when two units are connected. This is illustrated in Fig. 12.11.

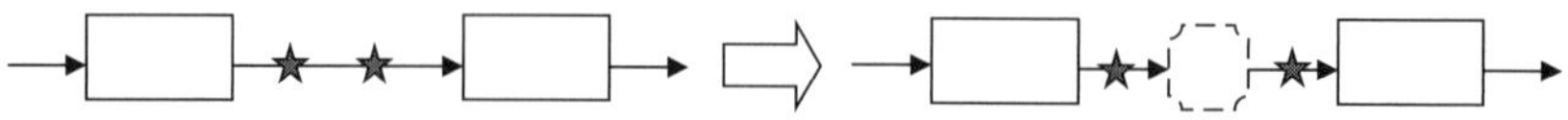

FIGURE 12.11 Multiple measurements and pseudo-units.

Excessive Number of Gross Errors

A plant with a large number of detected biases and leaks is a plant that will also have a large amount of undetected gross biases. At the same time the errors found have a great chance not to be correctly identified, according to the theory of equivalence (Chap. 6).

We conclude that when a large number of gross errors are present, then it is better not to do data reconciliation because these gross errors introduce large number of smearings. In other words, the induced errors are much larger than the expected biases in each measurement.

Quite clearly, one needs to determine if one wants to run the risk of smearing the estimators with induced errors or give up reconciling altogether.

Thus, the practical approach is to

- Identify one minimum redundancy system.
- Make sure these instruments are well maintained.
- Proceed to perform data reconciliation with this low redundancy set.
- Keep track of the accuracy of key variables.
- Start adding one or a few measurements at a time to the data reconciliation verifying if they flag gross errors.
- Adjust instrument maintenance practices to keep an eye on key measurements.
- If more than one gross error shows up, switch back to the minimum redundancy set, which is known to be well maintained.

Steps to Increase Accuracy

When one wants to implement data reconciliation in a plant, the measures to take are:

1. Verify the stochastic accuracy that is available in key variables.
2. Determine the value of the accuracy that might be obtained through data reconciliation (Chap. 11).

3. If a data reconciliation package is economically advantageous to be installed, then steps to increase accuracy need to be taken. These include a combination of the following:
   a. Incorporate in the data reconciliation the sets of measurements that produce better accuracy. In other words, do not use measurements that have high smearing effect on the key variables.
   b. Improve instrument maintenance (we cover this in Chap. 16).
   c. Place new instrumentation (covered in Chap. 15).

Hierarchy of Data Reconciliation

Figure 12.12 presents a schematic of the enterprise. It contains different plants, raw material reception, and distribution units. We assume all of them are interconnected through pipelines. Inside the plants one can distinguish units (○) and tanks (□).

Let us assume that data reconciliation is proposed to achieve different objectives, namely, more accurate production accounting, better parameter estimation in the units, etc. Let us focus first on production accounting and let us assume that this need is at the enterprise level.

In order to increase redundancy, one could create a big data reconciliation problem using *all* measurements available. Such a solution only works if in implementing it, there are no gross errors

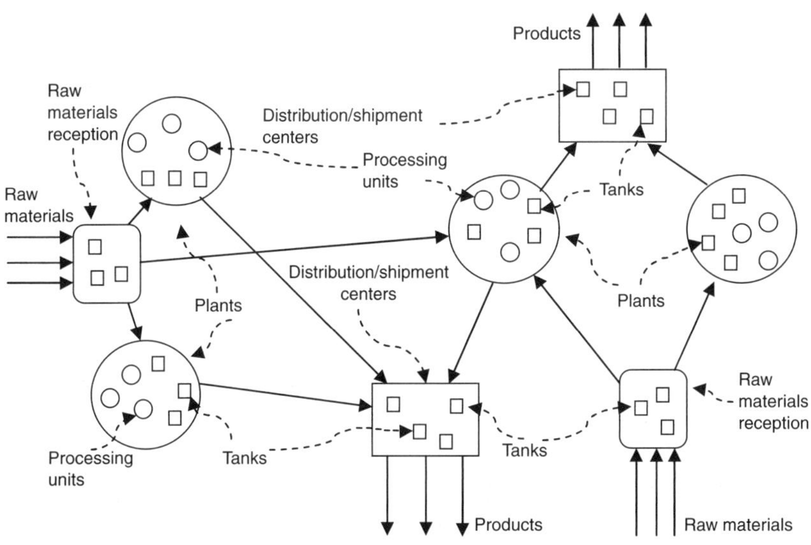

Figure 12.12 Connections in the enterprise.

to identify, which means that *all* instruments ought to be well maintained, something that does not match the reality of today's plants. In other words, one has to find ways of prioritizing maintenance and break down the data reconciliation task to meaningful components. We recall the results of the previous section on the amount of gross errors, so we conclude that some strategy needs to be implemented to make sure that the right maintenance program is implemented.

We therefore start with battery limits, and we suggest that the corresponding measurements are maintained bias free at all costs. In other words, the maintenance program needs to focus on these battery limit measurements as their highest priority. Figure 12.13 shows these sensors (★).

This system is not enough as it has several accumulation terms, namely, the tanks in each plant. Thus the next step is to start a good maintenance program on level measurements in the tanks. This will give a first set of data that can be reconciled at the enterprise level. We add therefore a pseudo-stream to indicate changes in the tanks. This is depicted in Fig. 12.14. Tank measurements are denoted by the symbol (✦).

In addition, the above set of measurements contains two measurements of streams that connect plants. This means that the model needs to be augmented with the pseudo-units that were discussed above. These pseudo-units are added and shown in Fig. 12.14.

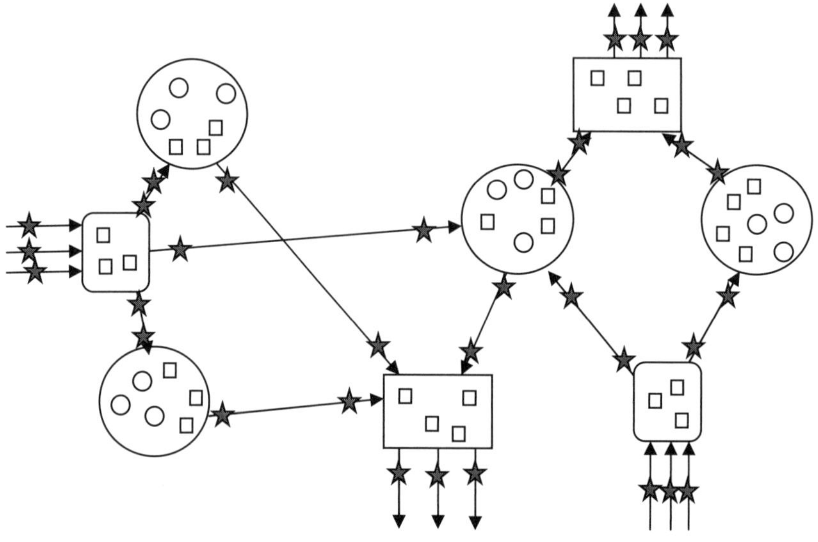

FIGURE 12.13 Sensors on battery limits.

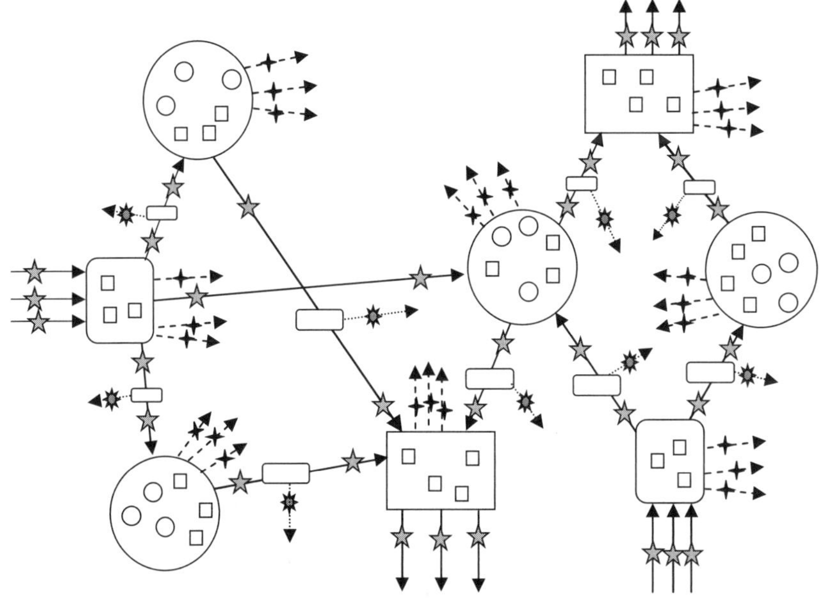

Figure 12.14 Enterprise-level data reconciliation with pseudo-units.

Remark: One of the alternatives to this scheme that builds redundancy at the battery limits first is one in which redundancy is built around equipments first, which seems more intuitive. However, one of the first problems one encounters is that the estimators of the flowrates of streams that connect different units do not coincide. Quiet clearly, the only way to remove this mismatch is to perform the data reconciliations with both units at the same time in one single run. But the number of gross errors could be too large, making the maintenance task even more costly. In addition, the objective to have material balances reconciled at the enterprise level would be delayed if not compromised.

After the frequency of systematic errors is reduced significantly, the next step is to start incorporating the flows inside the plants (interunits flows). This is shown in Fig. 12.15. This has to be done adding one plant at a time, and only after an independent data reconciliation and instrument maintenance program can guarantee a small number of gross errors. Pseudo-units need to be added to deal with the repeated measurements.

After this level has been accomplished, one can start incorporating the units within the plant.

Data Reconciliation Set-up Steps

When a package to perform data reconciliation is first purchased and installed, a few issues require attention.

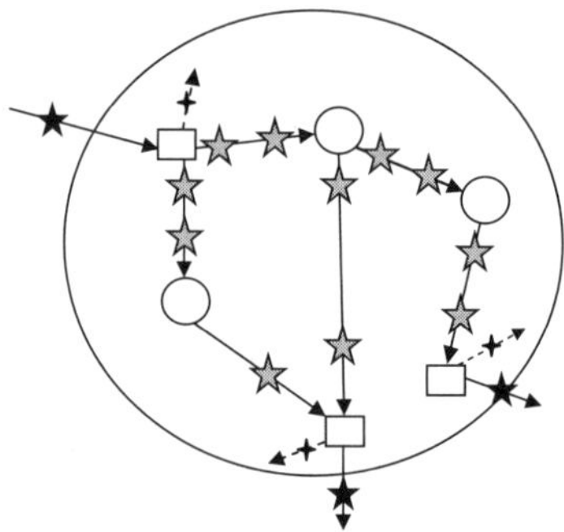

Figure 12.15 Plant-level data reconciliation.

- Manual correction of measurements by unit supervisors needs to be eliminated and discrepancies in reported values need to be encouraged. Such manual correction was encouraged in the past, and is still encouraged in certain plants because of pressures from production accounting, which in the absence of data reconciliation methods relies on process operators to perform this data adjustment, so the balance is closed.
- The hierarchy of implementation described above notwithstanding, at the plant level, there are issues regarding redundancy and biases.
  - If redundancy is low, something that can be tested before implementation, an instrument network retrofit might be necessary. Simply put, added redundancy increases the economic value of information, but it has an increased cost. The trade-off between value and cost is covered in Chap. 16.
  - If the number of biases is large, then a revision of instrument maintenance practices needs to be made. In Chap. 19 the value of information versus the cost is discussed, providing a tool to determine the optimal instrument preventive maintenance frequency.
- When instrument maintenance is not appropriate and too many biases are present, it is probably better to ignore the results of data reconciliation because of the high level of smearing (see Chap. 10) and concentrate on using only information on well-maintained instruments on battery limits and move on incorporating one well-calibrated instrument at a time. If there are no more elaborate methods used to make decisions, the rule of thumb recommended is to limit the biases detected to one, or two at the most.

Accounting Reports of Tank Volumes

Tank volumes are not reconciled, because they are obtained as the difference of two level measurements. Thus, after the reconciliation is done what one obtains is a reconciled change in tank inventory.

The question now is what should be reported as the final volume in the tank. Two procedures are possible:

- Add the volume change to the previous reported volume.
- Take the mean value of the tank volume measured and tank volume reconciled and add these to initial and final volumes. Report both volumes.

The advantage of the first option is that only one number is reported. The initial value is taken as the one reported in the previous period as final. The disadvantage is that errors can accumulate in such a way that they will be increasingly different from the measured ones.

In the second option, the calculated volume values are always close to the measured values. The disadvantage is that initial values of each period are different from the final values of the previous period, a report that may be unacceptable to production accounting.

Thus, the second option is the preferred one because if the first option is adopted severe discrepancies between measured and reported values may arise. To overcome any potential problems with production accounting, only the final volume can be reported.

There is a further alternative for tank forms, if reporting to production accounting is performed after several cycles of data reconciliation. One can include the time dimension, as depicted in Fig. 12.16. In this figure tank movements are performed in each period (among tanks and also to and from units), and volumes are transferred from period to period in the form of pseudo-streams.

In many companies these periods are of 1-day duration, and reporting to accounting is done once a month. The advantage of this cumulative data reconciliation is that the initial volume of each period is exactly equal to the final volume of the previous period.

One of the big advantages of this model is that the level of redundancy increases substantially.

Remark: Kelly (1999) discussed the issue of temporary flowrates (i.e., usually the so-called tank movements) in large plant-wide environments, like refineries. Together with the discussion of identifying bad measurements, he discussed the identification of potential missing or mis-specified material movements. Thus, he presents a two-phase binary subset deletion search strategy to detect and diagnose both material movement and measurement anomalies occurring daily in the plant-wide flowsheet. Later, he touches on the same issues (Kelly, 2000), introducing a mixed integer data reconciliation model to deal with missing flows.

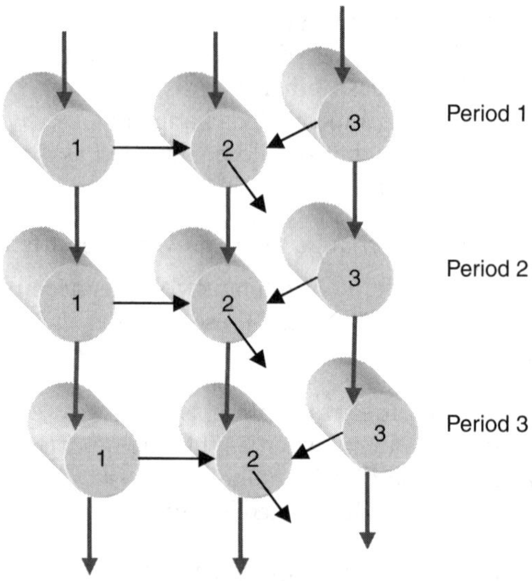

FIGURE 12.16 Multiperiod tank movements.

Remark: Tank volume measurement is of paramount importance to the success of all these techniques. While there is a large literature on level measurement and the associated devices, we point out a recent work by Binner et al. (2008) where they discuss issues of level/density measurement, calibration, and its reverification. They present a revised regression model to obtain a calibration from biased calibration data from one or more calibration runs. They show that cubic splines render a more accurate representation.

Remark: Leak detection techniques, as it has been seen in previous chapters are covered by several gross error detection techniques. With additional measurements, one could in principle add some model equations for emissions that can increase the redundancy and thus improve the leak detection capabilities. Rota et al. (2001) proposed a model that can be used for this purpose. Time series has also been attempted for this specific purpose (see Chmielewski et al., 2000). Later Chmielewski and Agrawal (2004), looked into the issue of scheduling personnel.

Remark: Another system where leak detection is of paramount importance is pipelines and networks of pipelines (gathering and distribution systems). We omit covering these because the book concentrates on process plants.

Gross Error Equivalency Problems

As it was pointed out in Chap. 6, equivalency of gross errors leads to several unwanted situations. One of these situations is that the gross error detection methods will identify the wrong set of errors forcing to remove errors in measurement that do not exist and leaving biased measurements that smear all other estimators. It is important to remember that this equivalency is independent of the method of gross error detection used.

Thus, one of the preventive measures that one should implement is to have in place a software that can identify the number of equivalent sets and determine the size of the biases and possible leaks of each set. This was covered in Chaps. 6 and 7.

Fortunately, if one calculates the accuracy of key important streams, one has either the worst case through software accuracy, and/or the expected value if stochastic accuracy is computed. Methods to improve accuracy include instrumentation upgrade and will be covered in Chap. 15.

Another way of improving accuracy and removing equivalencies is to add nonlinear equations. For example, leak detection in pipelines can be improved if the pressure drop relations are added to the reconciliation (Bagajewicz and Cabrera, 2003).

Most commercial softwares do not include the identification of equivalent sets, but the encouraging aspect is that such softwares can be implemented offline and be fed by the biases identified by commercial softwares. An additional alternative is to implement the gross error detection method separately altogether, especially because most methods in commercial software use gross error detection methods that are good to identify one gross error, but have poor performance in the presence of multiple errors.

Now, when such list of equivalent sets has been made, instrument maintenance needs to be informed immediately. This is a necessary alteration of the routine of instrument maintenance.

Simulation and Optimization for Data Reconciliation

Modern steady-state simulators contain optimization packages and can be used to perform data reconciliation. This option is not recommended for material balances as an analytical solution is available. More complex and nonlinear systems can be tried. A word of caution needs to be added. Sometimes these optimizers are not that reliable and automatic and need help from the user, which is not that advantageous if the reconciliation is to be done routinely. Now, assuming the optimization is successful, there is an additional issue: Because these are nonlinear problems, they could exhibit multiple local optima and the optimizers are not usually equipped with global optimality identification.

Reconciliation of Standard Volumetric Flows

There is a tradition in several industries (oil, gas, etc.) to use standard volumetric flows instead of mass flows. Most data reconciliation packages, however, use mass flows, so the conversion needs to be made. A few reconciliation packages allow the use of standard flows. We show the effect of doing this in the following examples:

Example 12.6: Consider the small three streams–one unit example of Fig. 12.17.

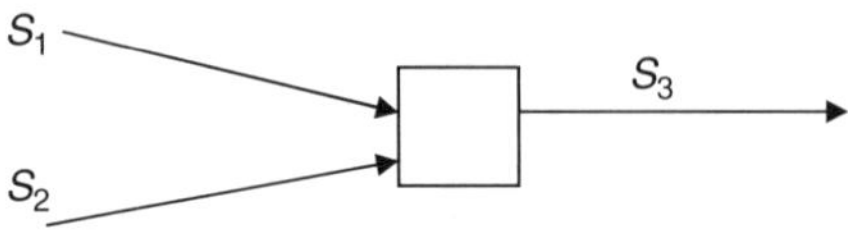

FIGURE 12.17 Three streams example.

We know that mass is conserved. Thus,

$$F_1 + F_2 = F_3 \tag{12.25}$$

However, $F_i = \rho_i F_i^V$, where ρ_i is the density and F_i^V the volumetric flow. But using a temperature and pressure corrections, one can write $F_i = \rho_i F_i^V = \rho_i^{(\text{std})} F_i^{V(\text{std})}$, where std refers to standard conditions. Thus, Eq. (12.20) can be rewritten as follows:

$$\rho_1^{(\text{std})} F_1^{V(\text{std})} + \rho_2^{(\text{std})} F_2^{V(\text{std})} = \rho_3^{(\text{std})} F_3^{V(\text{std})} \tag{12.26}$$

Dividing by r $\rho_1^{(\text{std})}$ we get

$$F_1^{V(\text{std})} + \frac{\rho_2^{(\text{std})}}{\rho_1^{(\text{std})}} F_2^{V(\text{std})} = \frac{\rho_3^{(\text{std})}}{\rho_1^{(\text{std})}} F_3^{V(\text{std})} \tag{12.27}$$

Now, densities in standard conditions are a function of composition. Thus, if they are similar, that is, $\rho_1^{(\text{std})} \approx \rho_2^{(\text{std})} \approx \rho_3^{(\text{std})}$, then the data reconciliation using standard volumes can be performed. In other words, if the compositions are not too different, it might work.

Example 12.7: Consider the case of a crude unit. The feed is oil and the products are various streams with different varied densities. One could write

$$\left.\begin{aligned} F_{\text{crude}}^{V(\text{std})} = {} & \frac{\rho_{\text{residue}}^{(\text{std})}}{\rho_{\text{crude}}^{(\text{std})}} F_{\text{residue}}^{V(\text{std})} + \frac{\rho_{\text{gas-oil}}^{(\text{std})}}{\rho_{\text{crude}}^{(\text{std})}} F_{\text{gas-oil}}^{V(\text{std})} + \frac{\rho_{\text{diesel}}^{(\text{std})}}{\rho_{\text{crude}}^{(\text{std})}} F_{\text{diesel}}^{V(\text{std})} \\ & + \frac{\rho_{\text{kerosene}}^{(\text{std})}}{\rho_{\text{crude}}^{(\text{std})}} F_{\text{kerosene}}^{V(\text{std})} + \frac{\rho_{\text{gasoline}}^{(\text{std})}}{\rho_{\text{crude}}^{(\text{std})}} F_{\text{gasoline}}^{V(\text{std})} + \frac{\rho_{\text{LPG}}^{(\text{std})}}{\rho_{\text{crude}}^{(\text{std})}} F_{\text{LPG}}^{V(\text{std})} \end{aligned}\right\} \tag{12.28}$$

We now consider a crude of density 30 API. The ratios of densities of products and crude fed are shown in Table 12.3. We notice that ratios differ considerably. Table 12.4 shows the difference.

	API	Density $\rho_i^{(std)}$	Ratio $\left(\frac{\rho_i^{(std)}}{\rho_{crude}^{(std)}}\right)$
Crude	30	0.87616099	
Residue	10	1	1.141343
Gas–oil	20	0.9339934	1.066007
Diesel	35	0.84984985	0.96997
Kerosene	45	0.80169972	0.915014
Gasoline	55	0.75871314	0.865952
LPG	70	0.70223325	0.801489

Table 12.3 Densities of Crude and Topping Products

Measured Volumetric Flow (bbl/h)	Standard Deviation	Reconciled Volumetric Flow (bbl/h)	Measurement Converted to Mass Flow (lb/h)	Reconciled Mass Flows (lb/h)	Difference
5049.99	1.0%	5022.91	1493.74	1485.52	0.014%
1445.02	2.7%	1466.73	481.04	488.46	–0.038%
658.63	2.5%	667.33	206.02	208.82	–0.033%
365.69	2.0%	369.73	110.34	111.59	–0.028%
892.71	1.5%	900.07	258.48	260.66	–0.020%
1174.27	1.5%	1183.66	315.29	317.87	–0.020%
431.82	1.5%	435.38	97.31	98.13	–0.021%

Table 12.4 Comparison of Data Reconciliation Using Volumetric Flows and Mass Flows

Industrial Applications of Data Reconciliation and Gross Error Detection

There are several papers presenting data reconciliation applied to several systems. Most of these case studies correspond to the use of data reconciliation and gross error detection in steady-state systems.

One of the first applications was for refineries and/or crude distillation units (Yamamura et al., 1988; Swartz et al., 1989, Charpentier et al., 1991, Ravikumar et al., 1994; Voros et al., 1999; Li et al., 2001; Bhat and Saraf., 2004). Sánchez and Romagnoli (1996) present the details of the variable classification and data reconciliation of an olefin plant case and alkylation study. Islam et. al. (1994) and Weiss et al. (1996) present the case study of a pyrolisis reactor. In turn, Ganguly et al. (1993), Noorai (1994) and Noorai et al. (1994) presented the study of a distilation column. These three cases are also reviewed in Sánchez and Romagnoli (2000). Several papers were written on data reconciliation for metallurgical processes (Smith Ichiyen, 1973; Heraud et al., 1991), chemical reactors (Sheel and Crowe, 1969; Murthy, 1974; Madron et al., 1977), paper pulping processes (Stephenson and Shewchuck, 1986), steam-metering systems (Serth and Heenan, 1986), non-isothermal flash (McDonald and Howat, 1988), mineral-processing systems (Reid et al., 1982; Flament and Hodouin, 1985; Hodouin and Everell, 1991; Simpson et al., 1991; Makni and Hodouin, 1994; Singh et al., 2001; de Andrade Lima 2006), food processes (Meyer et al., 1993), parameter estimation in residue treatment plant (Krishnan et al., 1993), reactors (Ravikumar et al., 1994), hydrogen and sulfur plants (Chiari et al., 1997), NGL recovery/nitrogen rejection unit (Nair and Jordache, 1991), ethylene plants (Sanchez et al., 1992, 1996), ethylene dichloride plants (Natori et al., 1992), catalytic processes to produce ammonia, methanol and syngas (Christiansen et al., 1997), ammonia plant (Placido and Loureriro, 1998; Amand et al., 2001), extraction plants (Holly et al., 1989), hazardous waste plants (Rovaglio et al., 1994), ketone plants (Dempf and List, 1998), desalination plants (Bourouis et al., 1998), industrial utility plants (Lee et al., 1998), smelting furnaces regenerative heat exchangers and blast furnaces (Minet et al., 2001; Eksteen et al., 2002), paper deinking process (Brown et al., 2003), heat exchanger networks (Wang and Romagnoli, 2003), on-line estimation in reactors (Musch et al., 2004), sulfuric acid and alkylation plants (Ozyurt and Pike, 2004), polimerization reactors (Sirohi et al., 1996; Barbosa et al., 2000; Benqlilou et al., 2001), dynamic data reconciliation in large plants (Soderstrom et al., 2000), urban water networks (Ragot and Maquin, 2006), MTBE units (Al-Arfaj, 2006), coke ovens (Faber et al., 2006), coking plants (Hu and Shao, 2006), pepelenes (Reddy et al., 2006; Alves and Nacimento, 2007), and propylene reactors (Martinez Prata et al., 2009). Narasimhan and Jordache (2000) provide an additional list. Finally, Benqlilou et al. (2002) presented a flexible and open architecture for data reconciliation and parameter estimation.

Remark: Although there are several isolated studies on the use of dynamic data reconciliation in industry the fact is that with the exception of the Kalman filter there is little commercial software available.

References

Akbaryan, F. and P. R. Bishnoi, "Smooth Representation of Trends by a Wavelet-Based technique," *Comp. Chem. Eng.* **24**:1913–1943 (2000).

Almasy, G. A. and R. S. H. Mah, "Estimation of Measurement Error Variances from Process Data," *Ind. Eng. Chem. Process Des. Dev.* **23**:779 (1984).

Al-Arfaj, M. A., "Shortcut Data Reconciliation Technique: Development and Industrial Application," *AIChE J.* **52**:414–417 (2006).

Alves, R. M. B. and C. A. O. Nascimento, "Analysis and Detection of Outliers and Systematic Errors in Industrial Plant Data," *Chem. Eng. Comm.* 382–397 (2007).

Amand, Th., G. Heyen, and B. Kalitventzeff, "Plant Monitoring and Fault Detection: Synergy between Data Reconciliation and Principal Component Analysis," *Comp. Chem. Eng.* **25**:501–507 (2001).

Bagajewicz, M., "On the Probability Distribution and Reconciliation of Process Plant Data," *Comp. Chem. Eng.* **20**(6/7):813 (1996).

Bagajewicz, M. and C. Gonzales, "Is the Practice of Using Unsteady Data to Perform Steady State Reconciliation Correct?" *AIChE Spring Meeting,* Houston, Tex. (2001).

Bagajewicz, M. and Q. Jiang, "Comparison of Steady State and Integral Dynamic Data Reconciliation," *Comp. Chem. Eng.* **24**(11):2367–2518 (2000).

Bagajewicz, M. and E. Cabrera, "Data Reconciliation in Gas Pipeline Systems," *Ind. Eng. Chem. Res.* **42**(22):5596–5606 (2003).

Bakhtazad, A., A. Palazoglu, and J. A. Romagnoli, "Process Trend Analysis Using Wavelt-Based De-Noising," *Proc. IFAC Conf.,* Korea (1998).

Barbosa, V. P., M. R. M. Wolf, and R. Maciel Fo, "Development of Data Reconciliation for Dynamic Nonlinear System: Application to Polymerization Reactor," *Comp. Chem. Eng.* **24**:501–506 (2000).

Benqlilou, C. R., V. Tona, A. Espuña, and L. Puigjaner, "On-Line Application of Dynamic Data Reconciliation," *Proc. Pres'01, 4th Conference on Process Integration, Modeling and Optimisation for Energy Savings and Pollution Preventions,* Florence, Italy:403–406 (2001).

Benqlilou, C., M. Graelis, A. Espuña, and L. Puigjaner, "An Open Software Architecture for Steady-Stae Data Reconciliation and Parameter Estimation," *Comp. Aided Chem. Eng.* **10**:853–858 (2002).

Bhat, S. A. and D. N. Saraf, "Steady-State Identification, Gross Error Detection, and Data Reconciliation for Industrial Process Units," *Ind. Eng. Chem. Res.* **43**:4323–4336 (2004).

Binner, R., J. JHOwell, G. Janssens-Maenhout, D. Sellinschegg, and K. Zhao, "Practical Issues Relating to Tank Volume Determination," *Ind. Eng. Chem. Res.* **47**:1533–1545 (2008).

Bourouis, M., L. Pibouleau, P. Floquet, S. Domenech, D. M. K. Al-Goabaisis, "Simulation and Data Validation in Multistage Flash Desalination Plants," *Desalination* **115**:1–14 (1998).

Bristol, E. H. "Swinging Door Trending: Adaptive = Trend Recording?," *ISA Nat. Conf. Proc.*:749–753 (1990).

Brown, P. R. and R. R. Rhinehart, "Automated Steady-State Identification in Multivariable Systems," *Hydrocarbon Process.* (Sep.) (2000).

Brown, D., F. Maréchal, G. Heyen, and J. Paris, "Application of Data Reconciliation to the Simulation of System Closure Options in a Paper Deinking Process," *Computer Aided Chem. Eng.* **14**:1001–1006 (2003).

Cao, S. and R. R. Rhinehart, "An Efficient Method for On-line Identification of Steady State," *J. Proc. Cont.* **5**(6):363–374 (1995).

Cao, S. and R. R. Rhinehart, "Critical Values for a Steady-State Identifier," *J. Process Control* **7**:149 (1997).

Charpentier, V., L. J. Chang, G. M. Schwenzer, and M. C. Bardin, "An Online Data Reconciliation System for Crude and Vacuum Units," *Proc. NPRA Comp. Conf.,* Houston, Texas (1991).

Chen, J., A. Bandoni, and J. A. Romagnoli., "Robust Estimation of Measurement Error Variance/Covariance from Process Sampling Data," *Comp. Chem. Eng.* **21**(6):593–600 (1997).

Chiari, M., G. Bussari, M. G. Grottoli, and S. Pierucci, "On-line Data Reconciliation and Optimization: Refinery Applications," *Comp. Chem. Eng.* **21**:S1185–S1190 (1997).

Chmielewski, D. J., V. Manousiouthakis, B. Tilton, and B. Felix, "Loss Accounting and Estimation of Leaks and Instrument Biases Using Time-Series Data," *Ind. Eng. Chem. Res.* **39**:2336–2344 (2000).

Chmielewski, D. J. and T. Agrawal, "On the Scheduling of Leak Detection Personnel," *Comp. Chem. Eng.* **28**:417–424 (2004).

Christiansen, L. J., N. Bruniche-Olsen, J. M. Cartensen, and M. Schroeder, "Performance Evaluation of Catalytic Processes," *Comp. Chem. Eng.* **21** (Suppl.): S1179–S1184 (1997).

Crow, E. L., F. A. Davis, and M. W. Maxwell, *Statistics Manual,* Dover, New York, NY, p. 63 (1955).

Darouach, M. and M. Zasadzinski, "Data Reconciliation in Generalized Linear Dynamic Systems," *AIChE J.* **37**:193 (1991).

Darouach, M., R. Ragot, M. Zasadzinski, and G. Krzakala, "Maximum Likelihood Estimator of Measurement Error Variances in Data Reconciliation," *IFAC. AIPAC Symp.* **2**:135–139 (1989).

de Andrade Lima, L. R. P., "Nonlinear Data Reconciliation in Gold Processing Plants," *Minerals Eng.* **19**:938–951 (2006).

Dempf, D. and T. List, "On-line Data Reconciliation in Chemical Plants," *Comp. Chem. Eng.* **22** (Suppl.):S1023–1025 (1998).

Eksteen, J. J., S. J. Frank, and M. A. Reuter, "Dynamic Structures in Variance Based Data Reconciliation Adjustments for a Chromite Smelting Furnace," *Mineals. Eng.* **15**:931–943 (2002).

Faber, R., B. Li, P. Li, and G. Wozny, "Data Reconciliation for Real-Time Optimization of An Industrial Coke-Oven-Gas Purification Process," *Sim. Model. Prac. Theo.* **14**:1121–1134 (2006).

Flament, F. and Hodouin D., "BILMAT Computer Program for Material Balance of Mineral Processing Circuits," *SPOC Manual*, CANMET, Special Report SP85-31/E, Energy, Mines and Resources Canada, Ottawa, Ontario, 140(1985).

Flehmig, F., R. V. Watzdorf, and W. Marquardt, "Identification of Trends in Process Measurements Using the Wavelet Transform," *Comp. Chem. Eng.* **22** (Suppl.): S491–S496 (1998).

Ganguly, S., V. Sairam, and D. N. Saraf, "Nonlinear Parameter Estimation for Real-Time Analytical Distillation Models," *Ind. Eng. Chem. Res.* **32**:99–107(1993).

Halle, J. C. and H. L. Sellars, "Historical Data Recording for process Computers," *Chem. Eng. Prog.* (Nov.):38–43 (1981).

Heraud, N., D. Maquin, and J. Ragot, "Multilinear Balance Equilibration: Application to a Complex Metallurgical Process," *Int. J. Miner. Proc.* **7**:91–116 (1980).

Heyen, G., E. Maréchal, B. Kalitventzeff B. "Sensitivity Calculations and Variance Analysis in Process plant Measurement Reconciliation," *Comp. Chem. Eng.* **20**(Suppl.):539–544 (1996).

Hodouin, D. and M. D. Everell, "A Hierarchical Procedure for Adjustment and Material Balancing of Mineral Process Data," *Min. Metall. Proc.* **11**:197–204 (1991).

Holly, W. R., R. Cook, and C. M. Crowe, "Reconciliation of Mass Flowrate Measurements in a Chemical Extraction Plants," *The Canadian J. of Chem. Eng.* **67**:595–601 (1989).

Hu, M. and H. Shao, "Theory Analysis of Nonlinear Data Reconciliation and Application to a Coking Plant," *Ind. Eng. Chem. Res.* **45**:8973–8984 (2006).

Huber, P. J. *Robust Statistics,* Wiley, Hoboken, N.J. (1981, 2004).

Islam, K., G. Weiss, and J. A. Romagnoli, "Nonlinear Data Reconciliation for an Industrial Pyrolysis Reactor," *Comp. Chem. Eng.* **18**:S217–221 (1994).

Jiang, Q. and M. Bagajewicz, "An Integral Approach to Dynamic Data Reconciliation," *AIChE J.* **43**(10) (Oct.):2546 (1997).

Jiang, T., B. Chen, X. He, and P. Stuart, "Application of Steady-State Detection Method Based on Wavelet Transform," *Comp. Chem. Eng.* **27**:569–578 (2003).

Keller, J. Y., M. Zasadzinski, and M. Darouach, "Analytical Estimator of Measurement Error Variances in Data Reconciliation," *Comp. Chem. Eng.* **16**:185 (1992).

Kelly, J., "Practical Issues in the Mass Reconciliation of Large Plant-Wide Flowsheets," *AIChE Spring Meeting*, Houston, Tex. (1999).

Kelly, J., "The Necessity of Data Reconciliation: Some Practical Issues," *NPRA Comp. Conf.,* (Nov.), Chicago, Ill. (2000).

Krishnan, S., G. W. Barton, and J. D. Perkins, "Robust Parameter Estimation in Online Optimization—Part I. Application to an Industrial Process," *Comp. Chem. Eng.* **17**:663–669 (1993).

Lee, M. H., A. J. Lee, C. Han, K. S. Chang, "Hierarchical On-line Data Reconciliation and Optimization for an Industrial Utility Plant," *Comp. Chem. Eng.* **22** (Suppl.): S247–S254 (1998).

Li, B., B. Chen, J. Wang, and S. Cong. "Steady-State Online Data Reconciliation in a Crude Oil Distillation Unit," *Hydrocarbon Processing*, (Mar.) (2001).

Loar, J., "How Can Steady-State Conditions be Identified Automatically?" *Control for the Process Industries* **8**:62 (1994).

Mah, R. S. H., A. C. Tamhane, S. H. Tung, and A. N. Patel, "Process Trending with Piecewise Linear Smoothing," *Comp. Chem. Eng.* **19**:129–137 (1995).

Makni, S. and D. Hodouin, "Recursive Bilmat Algorithm: An On-Line Extension of Data Reconciliation Techniques for Steady-State Bilinear Material Balance," *Minerals Eng.* **7**:1179–1191 (1994).

Madron, F., V. Veverka, and V. Vanacek, "Statistical Analysis of Material Balance of a Chemical Reactor," *AIChE J.* **23**:482–486 (1977).

Martinez Prata, D., M. Schwaab, E. L. Limaa, and J. C. Pinto, "Nonlinear Dynamic Data Reconciliation and Parameter Estimation through Particle Swarm Optimization: Application for an Industrial Polypropylene Reactor," *Chem. Eng. Sci.* **64**:3953–3967 (2009).

Mc Donald, R. J. and C. S. Howat, "Data Reconciliation and Parameter Estimation in Plant Performance Analysis," *AIChE J.* **34**:1–8 (1988).

Meyer, M., H. Pingaud, and M. Enjalbert, "An Investigation in Food Process Simulation using Data Reconciliation," *Comp. Chem. Eng.* **17** (Suppl.):S257–S262 (1993).

Minet, M., G. Heyen, B. Kalitventzeff, J. Di Puma, and M. Malmendier, "Dynamic Data Reconciliation of Regenerative Heat Exchangers Coupled to a Blast Furnace," *Computer Aided Chem. Eng.* **9**:1053–1058 (2001).

Morad, K., W. Y. Svrcek, and I. McKay, "Robust Direct Approach for Calculating Measurement Error Covariance Matrix," *Comp. Chem. Eng.* **23**:889 (1999).

Murthy, A. K. S. "Material Balance around a Chemical Reactor II," *Ind. Eng. Chem. Proc. Des. Dev.* **13**:347 (1974).

Musch, H., T. List, D. Dempf, and G. Heyen,"On-line Estimation of Reactor Key Performance Indicators: An Industrial Case Study," *Computer Aided Chem. Eng.* **18**:247–252 (2004).

Nair, P. and C. Jordache, "Rigorous Data Reconciliation is Key to Optimal Operations," *Control for the Process Industries* **IV**(10):118–123. Chicago: Putnam (1991).

Narasimhan, S., R. S. H. Mah, A. C. Tamhane, J. W. Woodward, and J. C. Hale, "A Composite Statistical Test for Detecting Changes of Steady States," *AIChE J.* **32**(9):1409–1418 (1986).

Natori, Y., M. Ogawa, V. S. Verneuil, Jr, "Application of Data Reconciliation and Simulation to a Large Chemical Plant," *Proc. Large Chemical Plants. 8th Intern. Symp.*:103–113 (1992).

Noorai, A., "Implementation of Advanced Operational Techniques for an Experimental Distillation Column within a DCS Environment," PhD thesis. University of Sidney, Sydeny, Australia (1994).

Noorai, A., G. Barton, and J. A. Romagnoli, "Online Data Reconciliation, Simulation and Optimizing Control of a Pilot-Scale Distillation Column," *Proc. Process Systems Eng. Con. (PSE'94)* **2**:1265–1268 (1994).

Ozyurt, D. B. and R. W. Pike, "Theory and Practice of Data Reconciliation and Gross Error Detection for Chemical Processes," *Comp. Chem. Eng.* **28**:381–402 (2004).

Placido, J. and L. V. Loureiro, " Industrial Application of Data Reconciliation," *Comp. Chem. Eng.* **22** (Suppl.):S1035–S1038 (1998).

Ravikumar, V., S. R. Singh, M. O. Garg, and S. Narasimhan, "RAGE-A Software Tool for Data Reconciliation and Gross Error Detection," *Proc. Foundations of Computer-aided Process Operations (FOCAPO) Conf.*: pp. 429–436, Amsterdam, Elsevier, (1994).

Reid, K. J., K. A. Smith, V. R. Voller, and M. Cross, "A Survey of Material Balance Computer Packages in the Mineral Industry," *Proc. 17th Applications of Computers and Operations Research in the Mineral Industry Conf.*, New York, AIME (1982).

Rollins, D. K. and S. Devanathan, "Unbiased Estimation in Dynamic Data Reconciliation," *AIChE J.* **39**:8 (1993).

Rota, R., S. Frattini, S. Astori, and R. Paludetto, "Emissions from Fixed-Roofed Storage Tanks: Modeling and Experiments," *Ind. Eng. Chem. Res.* **40**:5847–5857 (2001).

Rovaglio, M., D. Manca, G. Biardi, G. Nini, M. Mariano, and R. Gani, "Data Reconciliation and Process Optimization for Hazardous Wastre Incineration Plants," *Proc. PSE'94*:613–621 (1994).

Sánchez, M., and J. Romagnoli, "Use of Orthogonal Transformations in Data Classification—Reconciliation," *Comput. Chem. Eng.* **20**:483–493 (1996).

Sánchez, M., A. Bandoni, and J. Romagnoli, "PLADAT—A Package for Process Variable Classification and Plant Data Reconcilitaion," *Comp. Chem. Eng.* **S16**:499–506 (1992).

Sánchez, M. and J. Romagnoli, *Data Processing and Reconciliation for Chemical Process Operations.* Academic Press, San Diego (2000).

Sánchez, M., G. Sentoni, S. Schbib, S. Tonelli, and J. Romagnoli, "Gross Measurements Error Detection/Identification for an Industrial Ethylene Reactor," *Comp. Chem. Eng.* **20** (Suppl: 2) S1559–S1564 (1996).

Serth, R. W. and W. A. Heenan, "Gross Error Detection and Data Reconciliation in Steammetering Systems," *AIChE J.* **32**:733–742 (1986).

Sheel, J. P. and C. M. Crowe, "Simulation and Optimization of an Existing Ethylbenzene Dehydrogenation Reactor," *Can. J. Chem. Eng.* **47**:183–187 (1969).

Simpson, D. E., V. R. Voller, and M. G. Everett, "An Efficient Algorithm for Mineral Processing Data Adjustment," *Int. J. Miner. Proc.* **31**:73–96 (1991).

Singh, S. R., N. K. Mittal, and P. K. Sen, "A Novel Data Reconciliation and Gross Error Detection Tool for the Mineral Processing Industry," *Minerals Eng.* **14**(7):809–814 (2001).

Sirohi, A. and K. Y. Choi, "Online Parameter Estimation in a Continuous Polymerization Process," *Ind. Eng. Chem. Res.* **35**:1332–1343 (1996).

Soderstrom, T. A., and T. F. Edgar, L. P. Russo, and R. E. Young, "Industrial Application of a Large-Scale Dynamic Data Reconciliation Strategy," *Ind. Eng. Chem. Res.* **39**:1683–1693 (2000).

Smith, H. W. and N. Ichiyen, "Computer Adjustment of Metallurgical Balances," *CIM Bull.*:97–100 (1973).

Sundarraman, A. and R. Srinivasan, "Monitoring Transitions in Chemical Plants Using Enhanced Trend Analysis," *Comp. Chem. Eng.* **27**:1455–1472 (2003).

Stephenson, G. R. and C. F. Shewchuck, "Reconciliation of Process Data with Process Simulation," *AIChE J.* 32:247–254 (1986).

Swartz, C. L. E., K. H. Pang, V. S. Verneuil, and D. A. Eastham, "Refinery Implementation of a Data Reconciliation Scheme," *Proc. ISA Intern. Conf.*, Oct., Philadelphia, (1989).

Ullrich, C., G. Heyen, and C. Gerkens, "Variance of Estimates in Dynamic Data Reconciliation," *Comp. Aided Chem. Eng.* **26**(Suppl.):S357–S362 (2009).

Voros, N. G., C. T. Kiranoudis, and D. Marinos-Kouris, "Greek Refinery uses Data Reconciliation to Calibrate Instruments," *J. Oil and Gas* **97**:33–36 (1999).

Weiss, G., K. Islam, and J. A. Romagnoli, "Data Reconciliation—An Industrial Case Study," *Comp. Chem. Eng.* **20**:1441–1449 (1996).

Yamamura, K., M. Nakajima, and H. Matsuyama, "Detection of Gross Errors in Process Data Using Mass and Energy Balances," *Int. Chem. Eng.* **28**:91–98 (1988).

Wang, D. and J. A. Romagnoli, "A Framework for Robust Data Reconciliation Based on a Generalized Objective Function," *Ind. Eng. Chem. Res.* **42**:3075–3084 (2003).

Zhou, J. and R. H. Luecke, "Estimation of the Covariances of the Process Noise and Measurement Noise for a Linear Discrete Dynamic System," *Comp. Chem. Eng.* **19**:187–192 (1995).

CHAPTER 13

Value of Control Strategies*

In the case of control systems, smart plant operation refers to the automated retuning of plant controllers based on updated information about disturbance characteristics with the overall objective of maintaining the highest plant profitability (Chap. 1).

The evolution of disturbance characteristics necessitates the ability to retune controllers. To enable such retuning, two technologies are needed:

- A method to characterize disturbances
- A *profit*-focused tuning method that is responsive to disturbance models

Moreover, the proposed notion of disturbance characterization goes beyond simply modeling the impact of measured disturbances on process outputs. The desired models need to make predictions about future disturbances based on measurements of the past.

In this chapter we review existing methods for the first need area (predictive disturbance modeling) and then propose new control system design methods that make full use of the disturbance model. The envisioned smart plant is one in which disturbance characteristics can be estimated online. Then this updated disturbance model is sent an automated scheme for controller retuning. The net result is a control system that can adapt, not just to changing disturbance measurements, but to changing disturbance characteristics. In addition, all retuning will be guided by the economics of the particular situation. Connections to well-established real-time optimization (RTO) methods will be offered.

Classic Control

Rather than review the classic theory of control system design (see, for example, Stephanopoulos, 1984; Seborg et al., 2003; Riggs and Karim, 2006; Romagnoli and Palazoglu, 2006), we will assume some

*This chapter was written by Donald J. Chmielewski.

knowledge of it and use the following example to illustrate the relation between disturbance characteristics and classic controller tuning. Following this example, and motivated by it, we will introduce the notion of economic penalty, which will be fundamental to our subsequent definition of control system value.

Example 13.1: Consider the surge tank depicted in Fig. 13.1. The depicted flow controller (or servo-loop) is intended to regulate flow from the pump. Specifically, this controller (FC) receives a set-point command, $q^{(sp)}$, and then modifies valve position until the measured flow (FT) is equal to the given set-point. These low-level loops (typically mass flow controllers) serve to simplify process modeling for the higher-level controllers concerned with the plant as a whole, its profitability, and its safety. That is, rather than model valve and pump characteristics, the high-level controller will give a set-point command and (in most cases) assume it is perfectly executed. Thus, from this point we will assume that $q = q^{(sp)}$ for all time, and that q can be selected as a manipulated variable (through its relation to $q^{(sp)}$).

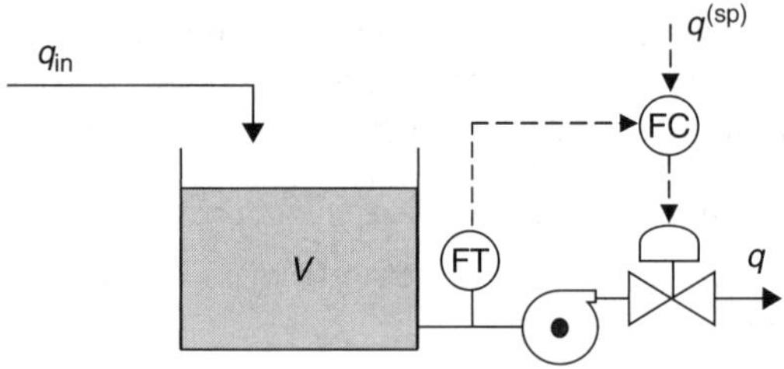

FIGURE 13.1 Surge tank with flow controller.

The higher-level controllers of interest typically operate with respect to a larger time scale. Consider, for example, the volume of liquid in the tank. Assume the inlet flow (q_{in}) is from a reactor that varies its throughput to achieve temperature regulation. The exit flow from the tank (q) is sent to a separation unit that demands little variation in its inlet flow. Obviously, the volume of liquid in the tank, V, should not exceed the total tank volume (which would cause an overflow), and the tank should not be allowed to run dry (which would damage the pump). Thus, the objective of the surge tank is to deliver a nearly constant downstream flow ($q = 30 \pm 1$ m^3/min), in the face of upstream variations ($q_{in} = 30 \pm 3$ m^3/min), while avoiding tank faults ($V = 10 \pm 10$ m^3).

We begin by assuming a constant exit flow ($q = 30$ for all time), and test the system under two arbitrarily selected inlet flow scenarios (q_{in} depicted in the plots). Fig. 13.2*a* depicts disturbance scenario (a) and indicates that all process objectives are met (the requirements of the downstream unit are satisfied ($29 < q < 31$) without violating the minimum and maximum hold-up constraints. The disturbance of scenario (b) has the same shape as that of (a), but the time period is doubled. Under this scenario, the tank is predicted to overflow. Thus, the success of this control policy (i.e., uncontrolled) is clearly dependent on the characteristics of the disturbance signal, q_{in}. While the magnitude characteristic, ±3 m^3/min, is clearly important, the disturbance duration (long or short period) is highlighted as a critical parameter. In other words, because the constraints have been violated, this control policy is clearly not acceptable and an alternative is needed.

Let us now consider the installation of a level controller (Fig. 13.3), where tank volume is the control variable (CV) and the manipulated variable (MV) is q. In this

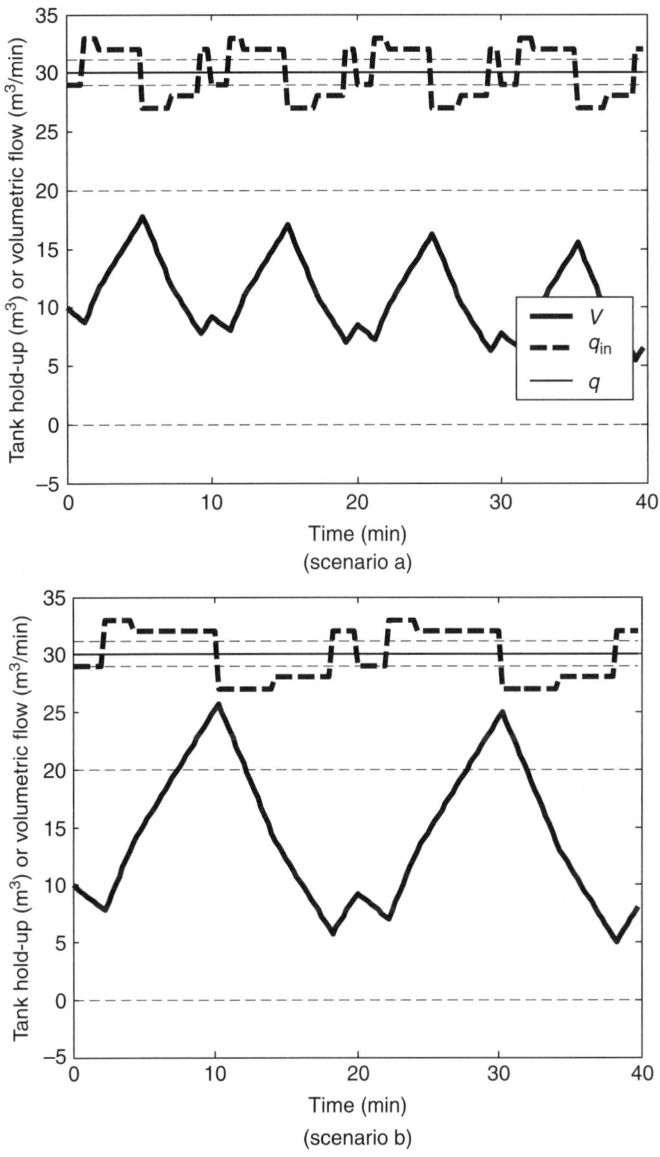

FIGURE 13.2 Surge tank simulations under uncontrolled policy.

cascade configuration, the level (or volume) controller is at a higher level than the flow controller (or equivalently LC is master to the slave FC). As we will see through this example, the issue of safety and profitability will quickly emerge.

For the moment assume the level controller is of the proportional integral (PI) variety. As an aid to PI tuning, it is suggested to develop a plant model. Thus, a volume balance yields

$$\dot{V} = q_{\text{in}} - q \tag{13.1}$$

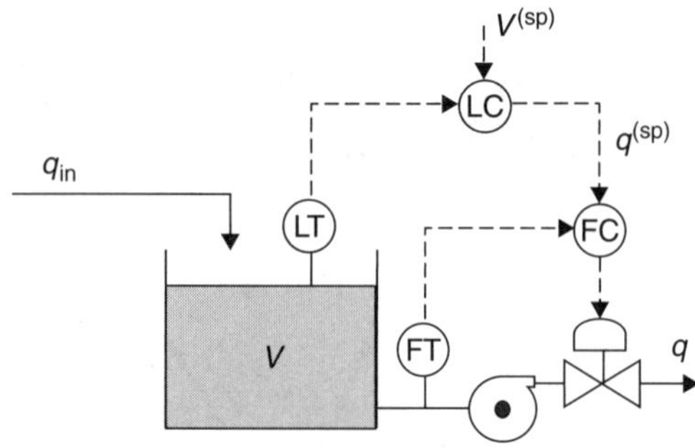

FIGURE 13.3 Surge tank with level controller.

If the nominal operating condition is $V^{nom} = V^{(sp)} = 10\ \text{m}^3$, $q_{in}^{\ nom} = 30\ \text{m}^3/\text{min}$, and $q^{nom} = 30\ \text{m}^3/\text{min}$, then one can define deviation variables as: $V' = V - V^{nom}$, $q'_{in} = q_{in} - q_{in}^{\ nom}$, and $q' = q - q^{nom}$, which generate the following plant model in deviation variables.

$$\dot{V}' = q'_{in} - q' \tag{13.2}$$

Taking the Laplace transform yields the following:

$$sV'(s) = q'_{in}(s) - q'(s) \tag{13.3}$$

Then the open-loop transfer function is

$$\dot{V}'(s) = G_p(s)q'(s) + G_d(s)q'_{in} \tag{13.4}$$

where the transfer function corresponding to the process is $G_p(s) = -1/s$ and that corresponding to the disturbance is $G_d(s) = 1/s$. The block diagram of the closed-loop system is depicted in Fig. 13.4, where the transfer function of the PI controller is

$$G_c(s) = K_c\left(1 + \frac{1}{\tau_I s}\right) \tag{13.5}$$

We now resort to commonly recommended tuning parameter values from the literature. For this process, Table 12.1 (entry E) of Seborg et al. (2003) suggests the following PI tuning relation:

$$K_c = -\frac{2}{\tau_c} \tag{13.6}$$

$$\tau_I = 2\tau_c \tag{13.7}$$

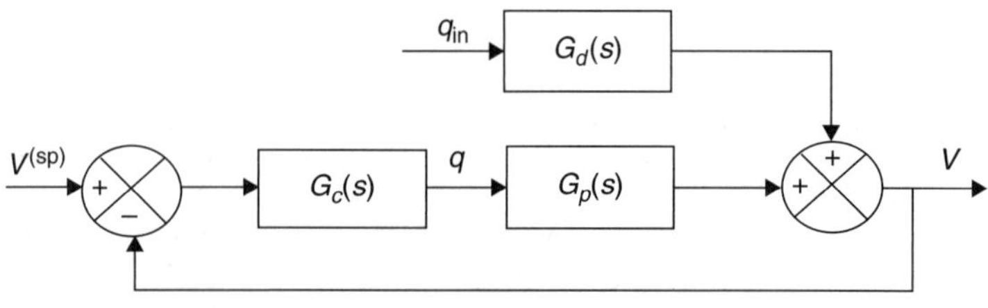

FIGURE 13.4 Block diagram of level controller.

where τ_c is the desired closed-loop time constant (to be selected by the user). This results in the following closed-loop transfer functions:

$$\frac{V'(s)}{q'_{\text{in}}(s)} = \frac{1}{s}\left(\frac{\tau_c s}{\tau_c s+1}\right)^2 \tag{13.8}$$

$$\frac{q'(s)}{q'_{\text{in}}(s)} = \frac{1}{s^2}\left(\frac{\tau_c s}{\tau_c s+1}\right)^2 \tag{13.9}$$

We now discuss different choices of τ_c.

- If we select $\tau_c = 1$ min and test the controller under the previous two disturbance scenarios (Fig. 13.5), we find very good response with respect to

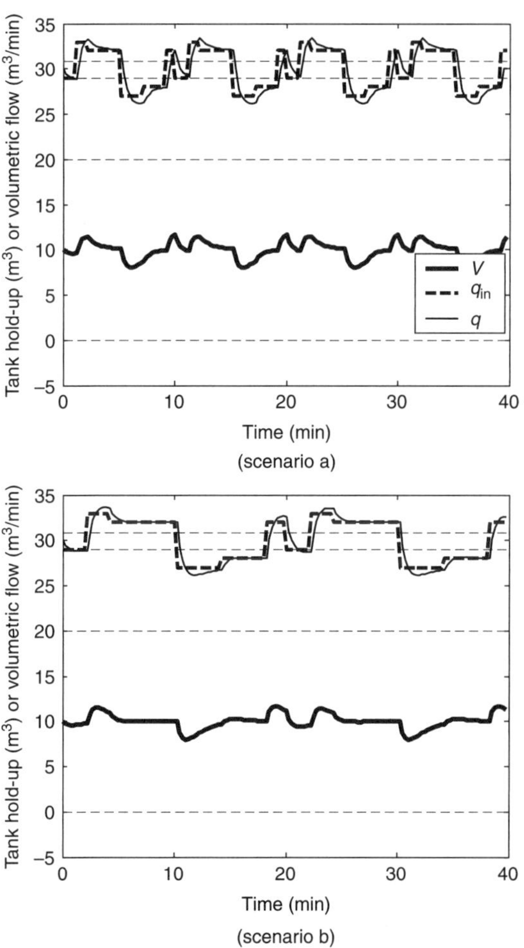

FIGURE 13.5 Surge tank simulations with PI control ($\tau_c = 1$).

tank volume, but the primary objective of reducing variation at the exit flow is completely lost.

- If, in turn, we select $\tau_c = 10$, then Fig. 13.6 shows tank hold-up variations within limit under disturbance (a). However, under the disturbance of scenario (b) the exit flow objective is violated.
- If we select $\tau_c = 50$, then Fig. 13.7 shows that the exit flow variation objective will be met, but at the expense of violating the volume constraints under scenario (b).

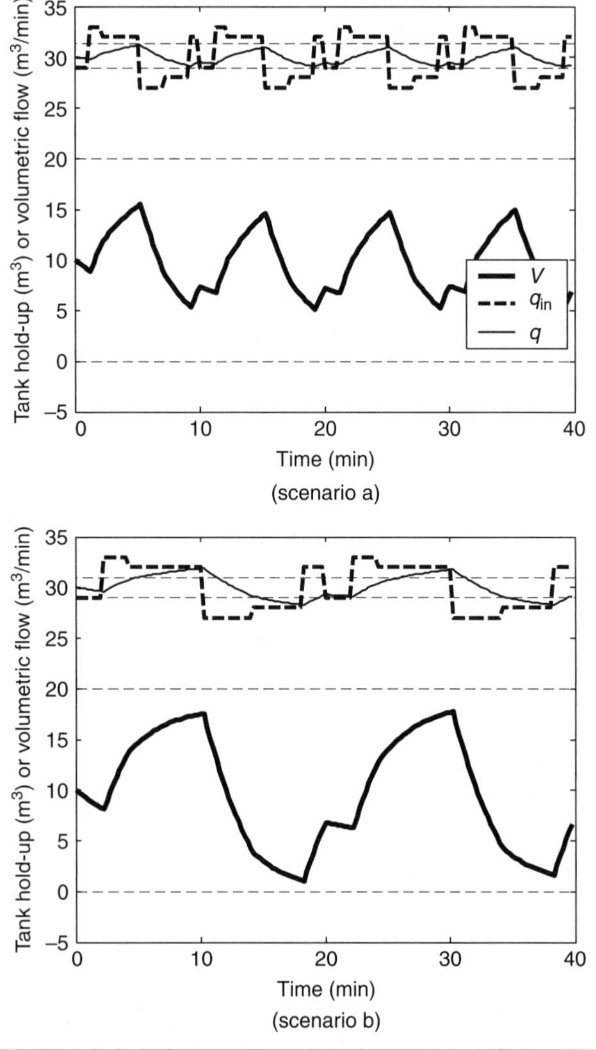

FIGURE 13.6 Surge tank simulations with PI control ($\tau_c = 10$).

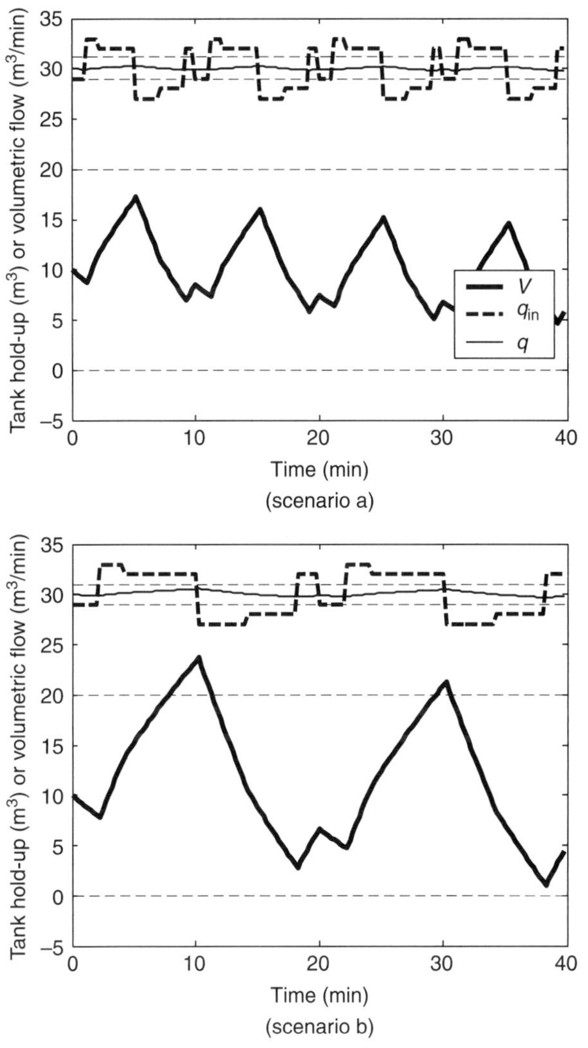

Figure 13.7 Surge tank simulations with PI control ($\tau_c = 50$).

The plots of Fig. 13.5 are those that one would typically find in a classic control textbook—the selected CV, tank volume, is being regulated at the expense of the MV, exit flow. Two extreme situations can be identified.

- The controller used in Fig. 13.5 completely ignores and violates the exit flow constraints.
- The other extreme is the constant exit flow (or uncontrolled) case (Fig. 13.2), where one would simply assume the tank is sufficiently large and the tank volume constraints can be ignored.

In most real systems, the situation rarely falls into one of these black and white scenarios, especially as plant designs become more and more economically focused. Thus, an objective of this chapter is to illustrate how to handle the gray areas of controller tuning, where the constraints of both the MV and CV are of concern. The controllers that produce the results of Figs. 13.6 and 13.7 are examples of trying to create this balance.

An additional point of Example 13.1 is that tuning decisions should be based on the disturbance scenarios one expects. For the two disturbance scenarios used in Example 13.1, all three candidates ($\tau_c = 1, 10$, or 50) fail at some point. Therefore one should select the least damaging failure. Since a tank overflow fault is likely more catastrophic than violating the exit flow bounds, $\tau_c = 10$ is likely the best choice.

During plant operation we expect the plant operator will take a central role. Imagine the following scenario:

- The operator begins a shift in which operation similar to Fig. 13.6*a* is observed.
- In this case all looks well and the operator decides to leave the tuning at $\tau_c = 10$.
- Half way through the shift, performance similar to Fig. 13.6*b* is observed.
- The operator must then decide if a tuning change (to meet the exit flow criteria) is desired.

Given the observed approach to tank volume limits, such a change to Fig. 13.7*b* is clearly ill advised and would likely have undesired consequences. These violations, if they occur, receive the name of parametric faults and will be discussed in Chap. 14.

Economic Perspective of Example 13.1: The operator's plight of being forced to violate exit flow constraints does, however, come with an economic penalty. To find this penalty, we must reconsider the downstream separation unit. Specifically, the unit was likely designed to accept an inlet flow of 30 m^3/min. While this flow can be increased, if it is increased beyond 31 m^3/min, then the end product will not meet purity specifications, and profit will be lost. If the flow is less than 30 m^3/min (e.g., 29 m^3/min), then end product purity will be good, but the production rate will be low, and again profit will be lost. Thus, the original specification, $q = 30 \pm 1$, is an attempt to reflect the actual objective but within a framework suitable for control system design.

Returning to the operator's plight, the appropriate action is to change the nominal throughput of the system, from 30 to 29 m^3/min. In that case (and under the disturbance of Fig. 13.6*b*), the exit flow would never exceed 31 m^3/min and push the product purity off specification. Thus, the resulting reduction in production rate (by 3%) is the actual economic penalty one would expect.

To quantify the economic penalty, one could employ the following simplified definition of value (i.e., change in profit):

$$\Delta P = K(q_{\text{actual}}^{\text{nom}} - q_{\text{design}}^{\text{nom}}) \tag{13.10}$$

where K indicates the value of the product (per m^3), $q_{\text{design}}^{\text{nom}}$ the nominal flow at the design condition, and $q_{\text{actual}}^{\text{nom}}$ the actual flow specified by the operator. Thus, if the actual nominal flow is equal to the design value, then there will be no change in profit. If the actual nominal flow is different, then process profit will change in the expected way (i.e., increased flow gives more profit and decreased flow gives less). Clearly, one could use other more complex functions that will account for inventory costs when there is overproduction or sales loss when there is underproduction.

Remark: In most textbooks on classic control, similar notions of control system profit are alluded to in the introductory chapters (see for example Riggs and Karim, 2006, Romagnoli and Palazoglu, 2006 or Edgar, 2004). However, the literature contains many efforts aimed at quantifying control system value (see the Control System Value section of this chapter for details and citations). It is also noted that our definition of smart plant operation (based on the notion of control system value and plant profit) is just one aspect of the problem. Other perspectives on the role of a control system in smart plant operations can be found in: Davis (2006), Christofides et al, (2007), Mhaskar et al, (2007), and Mahmood et al, (2008).

The conclusions of this section are

- Changes in disturbance characteristics will change the controller's ability to meet certain operational criteria.
- In some cases, adjustment of controller parameters (retuning) will recover the ability to meet operational criteria.
- In other cases, the only recourse is the modification of the nominal operating condition.
- Modification of the nominal operating condition will have an impact on plant profitability.

Model Predictive Control

It is well known that model predictive control (MPC) possesses a number of features not available from the classic methods. However, in this section we illustrate that most (if not all) of the themes of the previous section exist in the MPC case. Specifically, the relation between disturbance characteristics and controller tuning are nearly identical as is the subsequent impact on plant profitability.

Rather than review the general theory of MPC (see, for example, Garcia et al., 1989; Rawlings, 2000; Mayne et al., 2000; Qin and Badgwell, 2003; and Seborg et al., 2003), we will use the plant of Example 13.1 to illustrate the basic concepts behind MPC. Once the MPC controller for this plant has been developed, we will test its capabilities under disturbance scenarios similar to those of Example 13.1.

Consider again the surge tank depicted in Fig. 13.1, which has the following plant model in deviation variables. (We are dropping the *′ notation to indicate deviation variables, for example V' will be simply denoted as V. In the remainder of this section it is assumed that all variables are deviation variables.)

$$\dot{V} = q_{\text{in}} - q \tag{13.11}$$

Although MPC can be formulated with a continuous-time model [such as Eq. 13.11], practical implementation requires a discrete-time framework. To pursue the discrete-time approach, we employ a one-step explicit Euler method of model discretization (see Burl, 1998 for additional details on other discretization methods). Define $V_k = V(\Delta t\, k)$ to be a time sequence approximation of the function $V(t)$, where k is the discrete-time, unit less, time index. To recover "real" time, simply multiply k by the sample interval Δt. Similarly, define $q_k = q(\Delta t\, k)$ and $q^{(\text{in})}_k = q_{\text{in}}(\Delta t\, k)$, where it is assumed that $q(t)$ and $q_{\text{in}}(t)$ are constant over the interval $t = \Delta t\, k$ to $t = \Delta t\, (k+1)$. Then one can approximate Eq. (13.11) as

$$\frac{V_{k+1} - V_k}{\Delta t} \approx \dot{V} = q_k^{(\text{in})} - q_k \tag{13.12}$$

For notational convenience (and to be aligned with future notation) we will refer to the discrete-time model of Eq. (13.12) in the following form:

$$V_{k+1} = a_d V_k + b_d\, q_k + g_d\, q_k^{(\text{in})} \tag{13.13}$$

where $a_d = 1$, $b_d = -\Delta t$, and $g_d = \Delta t$, and we assume $\Delta t = 0.1$ min. The addition of coefficients will help in extending to a more general representation later on. In the simulations to follow, the model given by Eq. (13.13) is assumed to represent the actual plant.

We now introduce the notion of a prediction model. Assume we are at a time k, and would like to predict the tank volume at some other time i. Let $V_{i|k}$ be the predicted volume at time i, assuming we have information up to and including (or simply at) time k. Implicit in this definition is the assumption that $i > k$, otherwise $V_{i|k}$ is a tank volume from the present or past and is known to us (simply equal to V_k). If $i > k$, then $V_{i|k}$ will depend on the future values of q_k and $q^{(\text{in})}_k$. In the case of the manipulated variable, q_k, we define $q_{i|k}$ to be the proposed manipulation at time i given information at time k. Similarly, we define $q^{(\text{in})}_{i|k}$ to be the disturbance input we anticipate at time i given information at time k. Thus, our predictive model is

$$V_{i+1|k} = a_d V_{i|k} + b_d\, q_{i|k} + g_d\, q_{i|k}^{(\text{in})} \tag{13.14}$$

which we can use to predict the trajectory, V_i, via the sequence $V_{i|k}$. This prediction assumes that we are given the initial condition $V_{k|k} = V_k$ (the volume at the current time k), the anticipated disturbance sequence, $q^{(in)}_{i|k}$, and a proposed sequence of manipulations $q_{i|k}$. Disturbance anticipation is, in general, a difficult task, and is a main theme of a later section. For the moment we simply assume that a procedure exists to select values for the anticipated disturbance sequence. Selection of the manipulation sequence will be addressed next, but first a simple example to illustrate the predictive model is presented.

Example 13.2: Consider the following scenario: Assume we are at time $k = 2$ and the tank volume is known to be $V_2 = 1$, which defines $V_{2|2} = 1$. Additionally, we may anticipate the disturbance to be $q^{(in)}_{2|2} = 0$, $q^{(in)}_{3|2} = 1.0$ and $q^{(in)}_{i|2} = 0$ for $i > 3$. Similarly, a candidate manipulation sequence is $q_{2|2} = 6$ and $q^{(in)}_{i|2} = 0$ for $i > 2$. Then application of Eq. (13.14) yields

$$V_{3|2} = 1.0 - 0.6 + 0.0 = 0.4$$

$$V_{4|2} = 0.4 - 0.0 + 1.0 = 1.4$$

$$V_{5|2} = 1.4 - 0.0 + 0.0 = 1.4$$

and so on.

The basic idea behind model predictive control is to use the predictive model [Eq. (13.14)] to test a variety of possible manipulation sequences and select the one that best meets the performance objectives. For example, we can restrict the search to be such that the manipulation sequence satisfies $-1 \le q_{i|k} \le 1$, which would ensure the exit flow criterion of Example 13.1 is satisfied, at least in the prediction. Similarly, one can consider only those manipulation sequences that result in a volume sequence that satisfies $-10 \le V_{i|k} \le 10$.

Once a suitable manipulation sequence has been selected, the MPC will implement only the first element of the sequence (by applying to the process, q_k set equal to the first element of the selected manipulation sequence, $q_{i|k}$). Then, new information will be collected: measurements of V_{k+1} and $q^{(in)}_{k+1}$, the latter of which will generate a new sequence of anticipated disturbances $q^{(in)}_{i|k+1}$. Then the cycle repeats with a search for a new manipulation sequence, $q_{i|k+1}$, based on the updated information (at time k+1).

Unfortunately, there is usually more than one manipulation sequence that will satisfy the inequality constraints $-1 \le q_{i|k} \le 1$ and $-10 \le V_{i|k} \le 10$. To help select from the set of feasible manipulation sequences (i.e., those that satisfy the constraints $-1 \le q_{i|k} \le 1$ and $-10 \le V_{i|k} \le 10$), the predictive controller employs an additional measure of sequence quality. This quality (or objective) function typically indicates the amount of deviation each sequence has from their respective nominal values. (Since all variables are assumed to be in deviation form, the nominal value of each is zero.) Since the disturbance $q^{(in)}_{i|k}$ is outside the influence of the controller, it is usually left out of this objective function. The most commonly used objective function is the quadratic

$$\sum_{i=k}^{k+N-1} \{(V_{i|k})^2 + \gamma(q_{i|k})^2\} + \gamma_f\,(V_{k+N|k})^2 \tag{13.15}$$

where N is the prediction horizon and γ and γ_f are objective function weights. All three of these parameters must be positive and can be considered as tuning parameters, similar to τ_c of Example 13.1. Additional details on the motivation for the quadratic objective function and other objective function options are discussed by several authors (Garcia et al., 1989; Rawlings, 2000; Mayne et al., 2000; Qin and Badgwell, 2003; Seborg et al., 2003).

Thus, the search for the best manipulation sequence boils down to solving the following optimization problem (P1):

$$\text{(P1)} \quad \min_{\substack{V_{k|k}, V_{k+1|k}, \ldots, V_{k+N|k} \\ q_{k|k}, q_{k+1|k}, \ldots, q_{k+N|k}}} \sum_{i=k}^{k+N-1} \{(V_{i|k})^2 + \gamma(q_{i|k})^2\} + \gamma_f\,(V_{k+N|k})^2 \tag{13.16}$$

s.t.

$$V_{i+1|k} = a_d V_{i|k} + b_d\, q_{i|k} + g_d\, q^{(in)}_{i|k} \quad i = k \ldots k+N-1 \tag{13.17}$$

$$-1 \le q_{i|k} \le 1 \qquad i = k \ldots k+N-1 \tag{13.18}$$

$$-10 \le V_{i|k} \le 10 \quad i = k+1 \ldots k+N \tag{13.19}$$

$$V_{k|k} = V_k \tag{13.20}$$

$$q^{(in)}_{i|k} = q^{(in)}_k \qquad i = k+1 \ldots k+N \tag{13.21}$$

The first set of constraints [Eq. (13.17)] correspond to the prediction model and the next two sets [Eq. (13.18) and (13.19)] correspond to the process limitations. The fourth constraint [Eq. (13.20)] is the initial condition and the last set corresponds to the prediction of the disturbance sequence, which is set equal to the latest measurement of the disturbance, $q^{(in)}_k$.

Remark: While this type of disturbance anticipation [set equal to the latest measurement; Eq. (13.21)] is the most common, it is not the only option. For example, since this inlet flow is coming from an upstream unit, which is likely also controlled by MPC, the prediction model of the upstream unit could in some way be employed. The upcoming Example 13.3 will highlight the importance of the disturbance anticipation task, and motivates the subsequent section on disturbance modeling.

From an implementation standpoint, problem (P1) is solved numerically at every time step, k. As such, robustness of the search algorithm is of paramount importance. Unfortunately, for certain measured values of V_k and $q^{(in)}{}_k$ there may not exist a manipulation sequence, $q_{i|k}$, which satisfies the set of constraints. From a robustness standpoint, this sort of infeasible situation can be quite detrimental. Since the controller will not have a command to send to the actuator, the safety system will initiate an exception mode of operation due to the controller fault. Control loop faults will be covered in Chap. 14.

One way around this problem is to employ soft constraints on the constraints of lower priority (Zheng and Morari, 1995). As seen in Example 13.1, the constraint on tank volume is paramount and thus is considered hard. However, the exit flow constraints may be violated, although only as a last resort. Thus, one could replace the inequalities $-1 \le q_{i|k} \le 1$ with $-1 - s \le q_{i|k} \le 1 + s$, where s is termed the slack variable and is required to be positive or zero. To encourage this slack variable to be small, it is added to the objective function. Thus, the soft constrained MPC problem is

$$(\text{P1}') \quad \min_{\substack{V_{k|k}, V_{k+1|k}, \ldots, V_{k+N|k} \\ q_{k|k}, q_{k+1|k}, \ldots, q_{k+N|k}, s}} \left\{ \sum_{i=k}^{k+N-1} \{(V_{i|k})^2 + \gamma(q_{i|k})^2\} + \gamma_f (V_{k+N|k})^2 + \gamma_s s^2 \right\} \tag{13.22}$$

s.t.

$$V_{i+1|k} = a_d V_{i|k} + b_d\, q_{i|k} + g_d\, q^{(in)}_{i|k} \quad i = k \ldots k+N-1 \tag{13.23}$$

$$-1 - s \le q_{i|k} \le 1 + s \qquad i = k \ldots k+N-1 \tag{13.24}$$

$$-10 \le V_{i|k} \le 10 \qquad i = k+1 \ldots k+N \tag{13.25}$$

$$V_{k|k} = V_k \tag{13.26}$$

$$q^{(in)}_{i|k} = q^{(in)}_k \qquad i = k+1 \ldots k+N \tag{13.27}$$

$$s \ge 0 \tag{13.28}$$

The penalty parameters of the slack variable γ_s is set to a very large value (say 10^8). This penalty value will make the optimization problem prefer solutions that satisfy the constraints, that is, solutions with s small.

The following example will investigate the performance of this controller in the face of various disturbances and will serve to illustrate the potential benefit of disturbance characterization.

Example 13.3: Consider the MPC controller, defined by problem (P1′), applied to the tank described by Eq. (13.13) and depicted in Fig. 13.8. To illustrate the relationship between MPC and PI control, we begin with the unconstrained case, that is, assuming that the constraint Eqs. (13.24) and (13.25) ($-10 \le V_{i|k} \le 10$ and $-1 - s \le q_{i|k} \le 1 + s$) are removed from problem (P1′).

Figures 13.9 and 13.10 show responses similar to the PI controller of Example 13.1. In particular, aggressive tuning values encourage tight volume regulation (Fig. 13.9) while a more conservative tuning encourages flow regulation

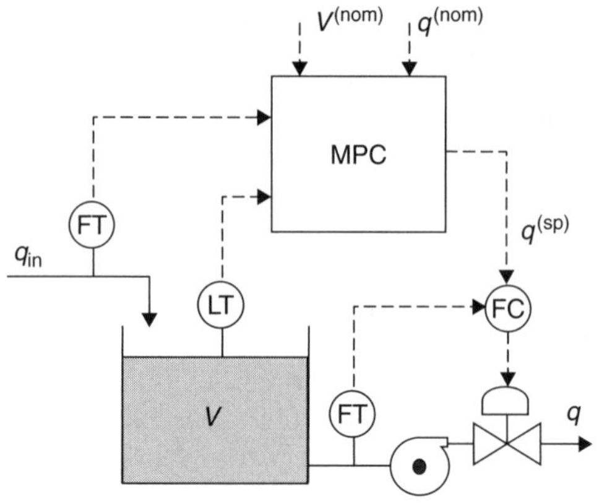

FIGURE 13.8 Surge tank with MPC controller.

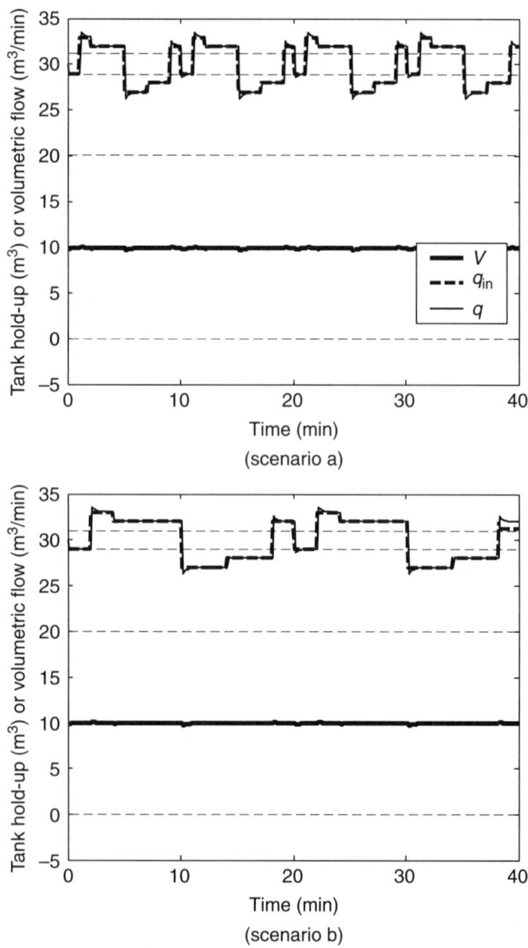

FIGURE 13.9 Surge tank simulations with unconstrained MPC [$\gamma = 0.1$, $\gamma_f = 4.16$, $N = 30$ (or 3 min)].

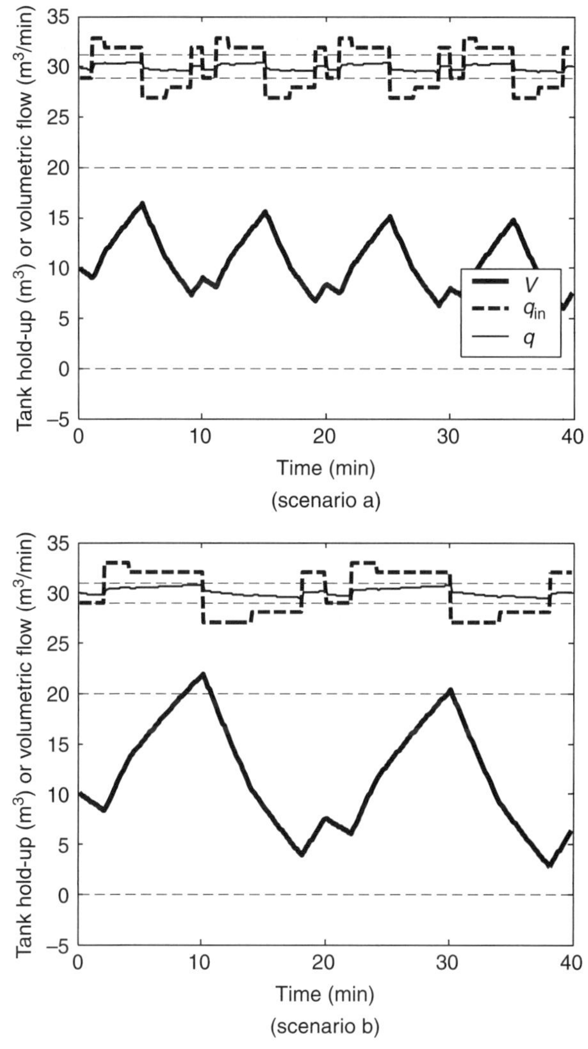

FIGURE 13.10 Surge tank simulations with unconstrained MPC [$\gamma = 500$, $\gamma_f = 224$, $N = 30$ (or 3 min)].

(Fig. 13.10) and may result in a volume overflow, depending on the disturbance characteristics.

If the constraint Eqs. (13.24) and (13.25) are reapplied to the controller (P1′), then Fig. 13.11*a* clearly indicates the influence of the constraints $-1 \le q_{i|k} \le 1$, and due to the aggressive tuning results in a bang-bang type controller, which is a controller that has two discrete values, a minimum and a maximum depending on the state of the process. In the current example $q_k = q^{\max} = 1$, if $V_k > 0$ and $q_k = q^{\min} = -1$, if $V_k < 0$. Figure 13.11*b* shows similar response, but also shows the action of the slack variables in that q_k is allowed to be greater than 1 when the

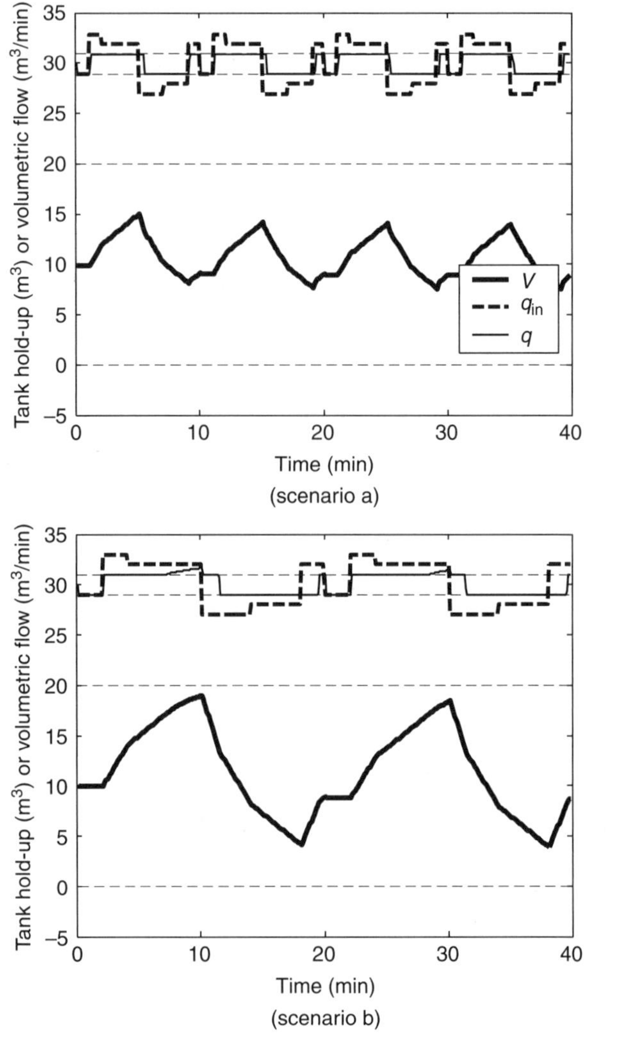

FIGURE 13.11 Surge tank simulations with constrained MPC [$\gamma = 0.1$, $\gamma_f = 4.16$, $N = 30$ (or 3 min)].

MPC model begins to predict that V_k is at risk of exceeding the bound of 20 (see q at times 8–10 min and 28–30 min).

The less aggressive tuning of Fig. 13.12 shows a more restrained response at q (as opposed to the bang-bang response). In fact the response seen in Fig. 13.12*a* is nearly identical to that of the unconstrained case (Fig. 13.10*a*). In Fig. 13.12*b*, however, we see again the impact of the slack variables. Thus, we conclude that under the tuning of Fig. 13.12, the constraints serve more as a backup feature (not needed in scenario a but exercised in scenario b), as opposed to essentially defining the control policy as was the case for Fig. 13.11.

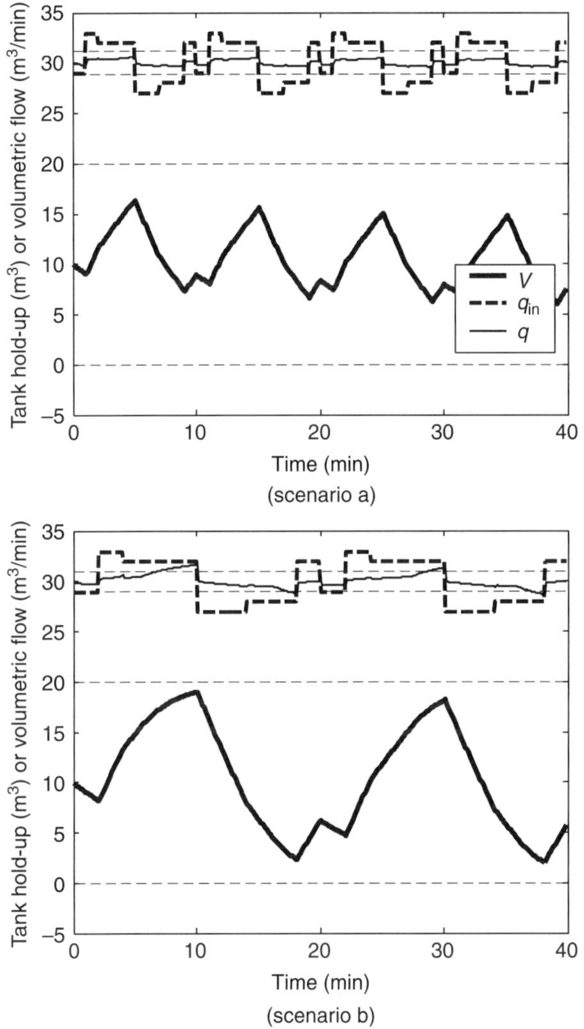

FIGURE 13.12 Surge tank simulations with constrained MPC [$\gamma = 500$, $\gamma_r = 224$, $N = 30$ (or 3 min)].

Figures 13.13 and 13.14 show the impact of reducing the size of the prediction horizon from 3 to 0.3 min. Of particular interest is Fig. 13.14*b*, where at about 8 min the controller seems to be caught off guard and must quickly activate the slack variables to prevent an overflow situation. This is in contrast to Fig. 13.12*b*, where at about 6 min the controller starts to increase exit flow in anticipation of a possible overflow situation. This clearly shows the predictive aspect of MPC.

Figures 13.15 and 13.16 show the impact of increasing prediction horizon to 6 min. In all four cases we see a premature activation of the slack variables.

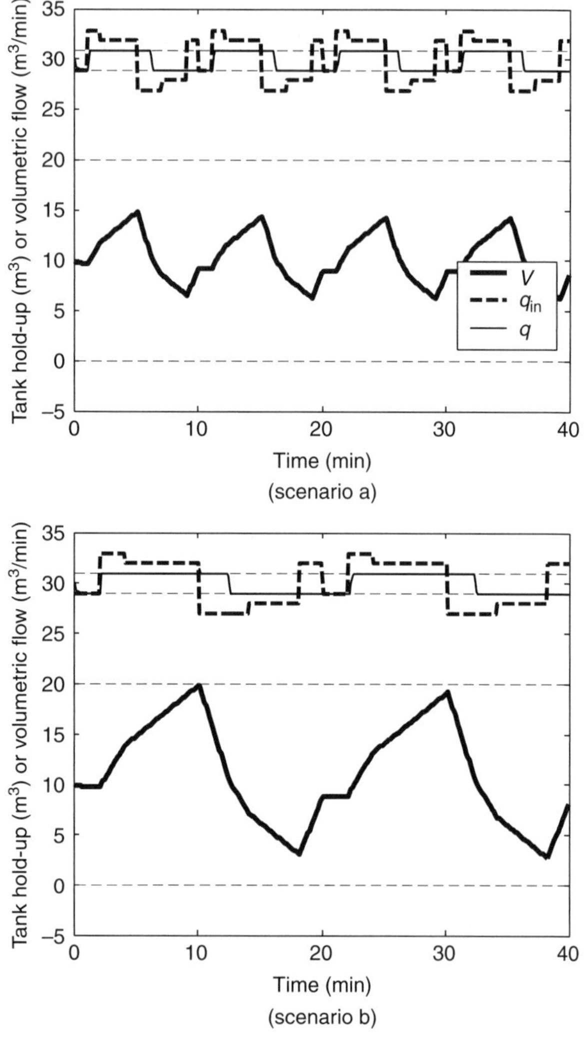

FIGURE 13.13 Surge tank simulations with constrained MPC and short prediction horizon [$\gamma = 0.1$, $\gamma_f = 4.16$, $N = 3$ (or 0.3 min)].

That is, q exceeds the bounds even though the volume appears to be far from its constraint. (In scenario a this occurs in the 1–2 min interval and in scenario b in the 3–6 min interval.) In this case, the predictive abilities of MPC seem to be overestimating the threat. Take, for example, scenario a. In the 1–2 min interval, the disturbance, $q^{(in)}_k$, is measured to be 3. Then in the MPC algorithm the anticipated disturbance, $q^{(in)}_{i|k}$, is set equal to 3 for the next 6 min. If this prediction turned out to be true (i.e., actual disturbance equal to 3 for 6 min), then activation of the slack variables would be warranted. However,

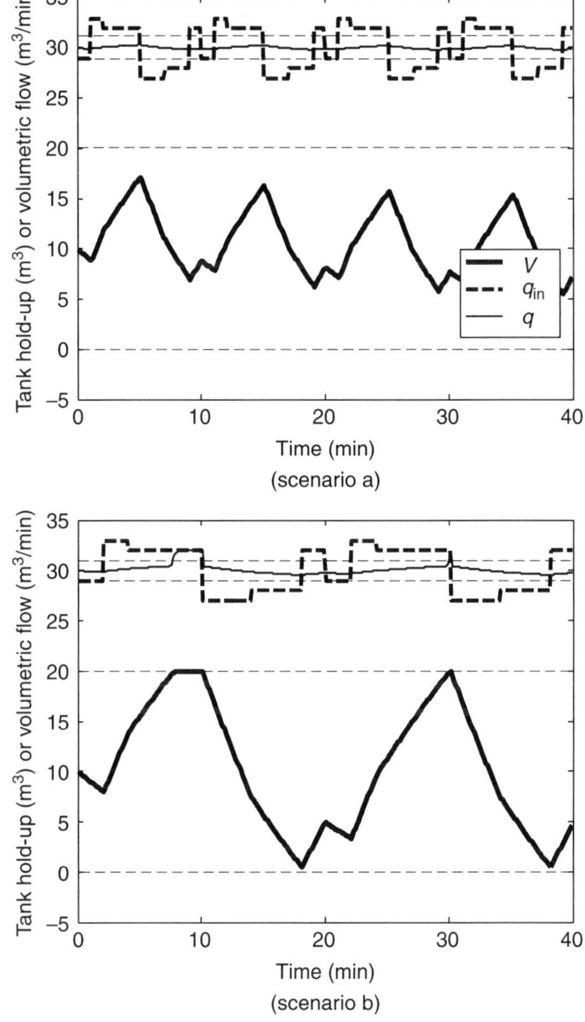

Figure 13.14 Surge tank simulations with constrained MPC and short prediction horizon [$\gamma = 500$, $\gamma_f = 224$, $N = 3$ (or 0.3 min)].

since the prediction was not true, such exuberant action was not warranted (the actual disturbance dropped down to 2 and eventually –3). Thus, the observed poor performance can be directly attributed to the implicit disturbance model employed by MPC. That is, the assumption $q^{(\text{in})}_{i|k} = q^{(\text{in})}_{k}$ (or the anticipated disturbance for the entire horizon is equal to the measured disturbance at time k) was a poor assumption for this case.

Clearly, one could argue that the selection of a 6-min prediction horizon was the source of the problem and the solution is to simply reduce the horizon size. However, this leads back to the question of controller tuning.

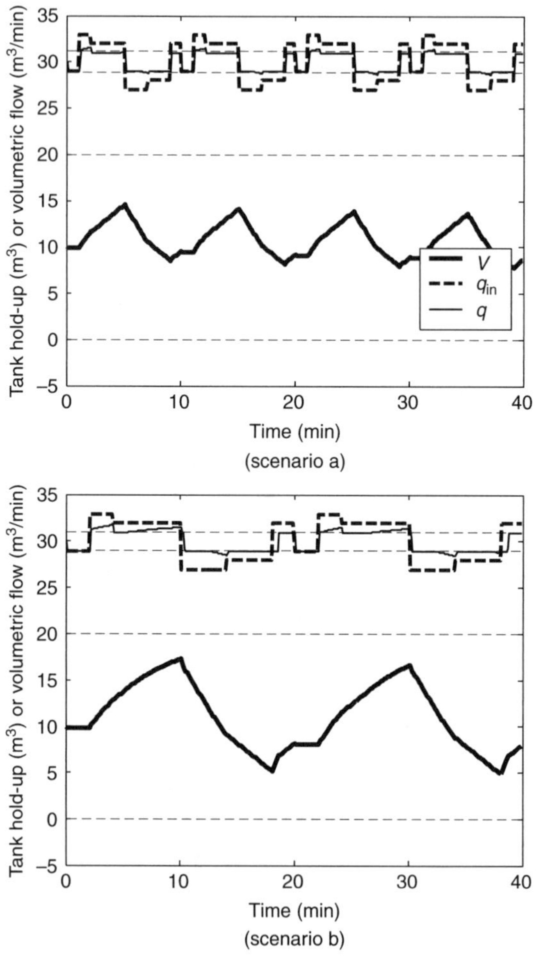

Figure 13.15 Surge tank simulations with constrained MPC and long prediction horizon [$\gamma = 0.1$, $\gamma_f = 4.16$, $N = 60$ (or 6 min)].

In practice, it is expected that those who regularly tune MPC controllers take into account the characteristics of the anticipated disturbances when selecting the prediction horizon. However, in the context of smart plant design and operation, the more attractive solution is a modified MPC algorithm that will automatically adjust (or retune) in the face of changing disturbance characteristics.

As an illustration of what may be possible, consider the case of perfect disturbance prediction. That is, assume the controller knows all future values of $q^{(in)}_k$. In this case, one would formulate MPC with the constraints $q^{(in)}_{i|k} = q^{(in)}_k$ replaced by $q^{(in)}_{i|k} = q^{(in)}_i$ (i.e., the anticipated is set equal to the actual). In Figs. 13.17 and 13.18, only Fig. 13.18*b* shows a slight constraint violation, which could be eliminated

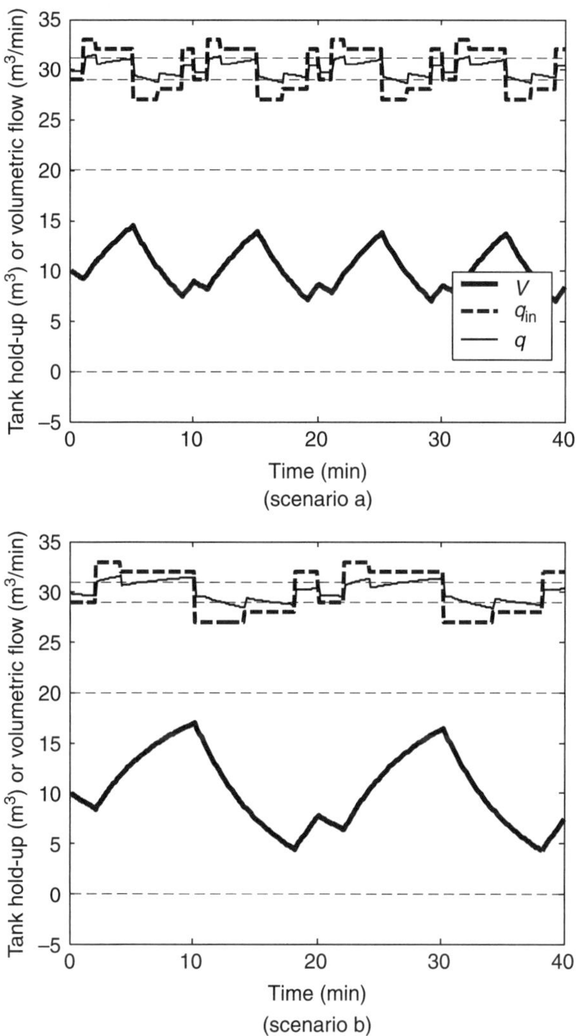

Figure 13.16 Surge tank simulations with constrained MPC and long prediction horizon [$\gamma = 500$, $\gamma_f = 224$, $N = 60$ (or 6 min)].

by increasing the prediction horizon. While this scenario of perfect disturbance prediction is somewhat unrealistic, it illustrates the desired responses, and how far the existing method is from that target.

Economic Perspective of Example 13.3: Now recall the economic penalty discussed earlier: If the actual variable q is expected to exceed 31 m^3/min, then the operator will need to reduce nominal flow and thus incur a loss of profit. This was summarized by the profit definition of Eq. (13.10).

In scenario a plots of Example 13.3 (the short duration disturbances), the aggressive tuning ($\gamma = 0.1$, $\gamma_f = 4.16$) resulted in q spanning the ± 1 window.

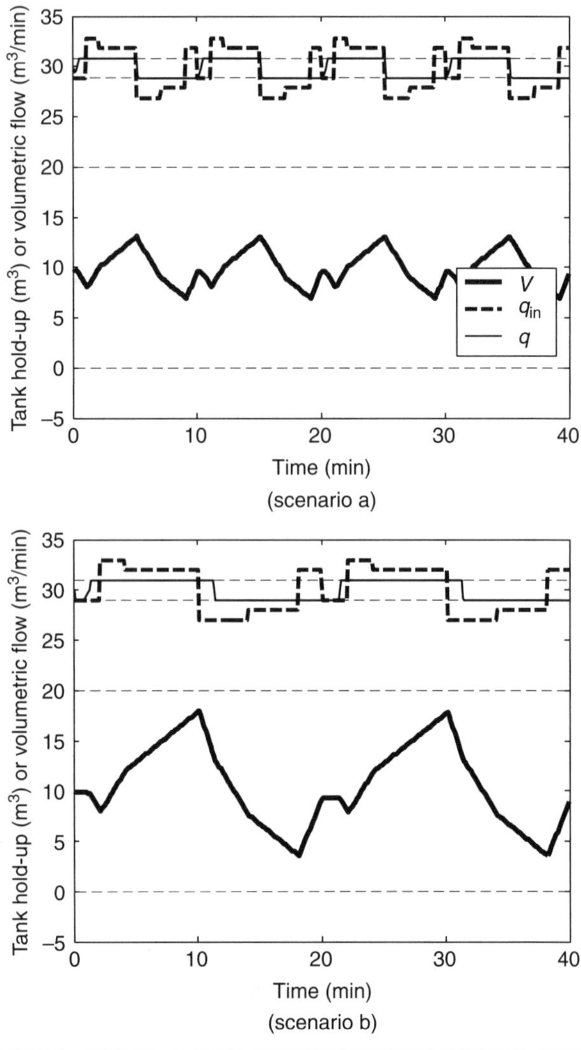

Figure 13.17 Surge tank simulations of constrained MPC with perfect disturbance prediction [$\gamma = 0.1$, $\gamma_f = 4.16$, $N = 30$ (or 3 min)].

In the less aggressive tuning ($\gamma = 500$, $\gamma_f = 224$) the span was much smaller, and suggests that throughput may be increased (see, for example, Fig. 13.12*a*).

In scenario b (the long-duration disturbances), the size of the disturbance was sufficient to cause the MPC policy to be dominated by constraint enforcement, causing the traditional tuning parameters, γ and γ_f to become much less important (i.e., each MPC controller was nearly identical in the face of these more influential disturbances). As illustrated in the figures, about the same level of throughput reduction would be required regardless of the tuning values. This suggests that a threshold exists, in the sense that for sufficiently small disturbances (scenario a) tuning can improve the situation, but for

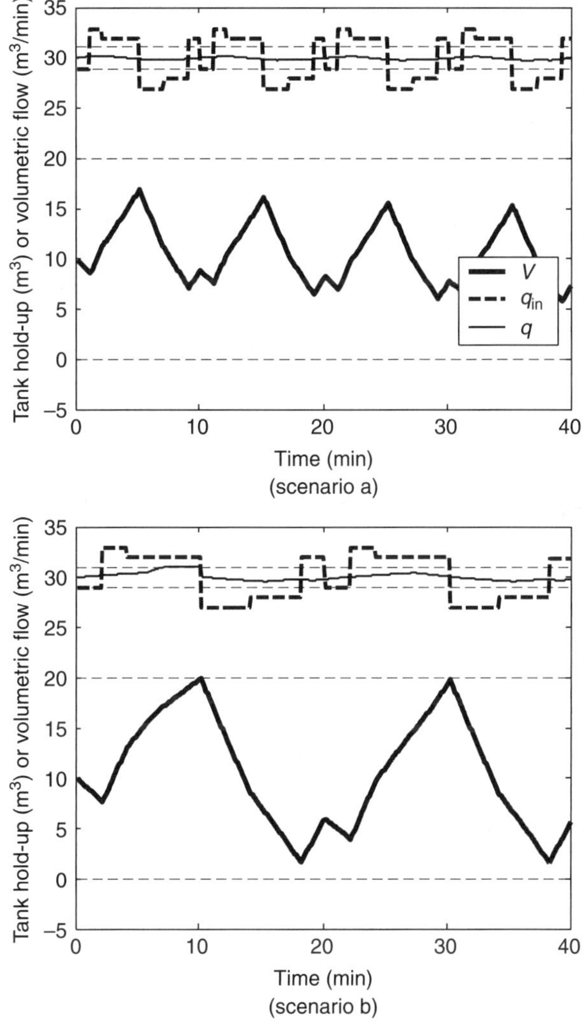

FIGURE 13.18 Surge tank simulations of constrained MPC with perfect disturbance prediction [$\gamma = 500$, $\gamma_f = 224$, $N = 30$ (or 3 min)].

sufficiently large disturbances (scenario b) nearly all tuning values will generate the same results.

Based on the above example, the following conclusions from the previous section are still valid:

- Changes in disturbance characteristics will change the controller's ability to meet certain operational criteria.
- In some cases, adjustment of controller parameters (retuning) will recover the ability to meet operational criteria.

- In other cases, the only recourse is the modification of the nominal operating condition.
- Modification of the nominal operating condition will have an impact on plant profitability.

We summarize by noting that the key calculation for smart plant operation is the required level of throughput reduction (or possibly throughput increase). As indicated above, the level of throughput change will depend on the amount of variation expected at q. It has also been illustrated that this variation at q is a function of the size and time-related characteristics of the disturbance $q^{(in)}$, as well as the controller tuning values. The next section will extend the scalar notions of the previous two sections to the generalized case.

The Hierarchy of the Modern Control Architecture

The modern chemical control architecture of Fig. 13.19 has been developed specifically to improve plant economics. (We omit here fault detection/resolution and all related features shown in a similar figure in Chap. 1.) Within this structure the hierarchal layers operate in master/slave relationships. The lowest level contains simple servo-loops. As illustrated in Example 13.3, the servo-loop set points are received from higher level controllers, typically MPC. The MPC receives information about the current state of the plant via a state estimator (to be discussed shortly). In addition, the MPC receives set-point commands from a real-time optimizer (RTO).

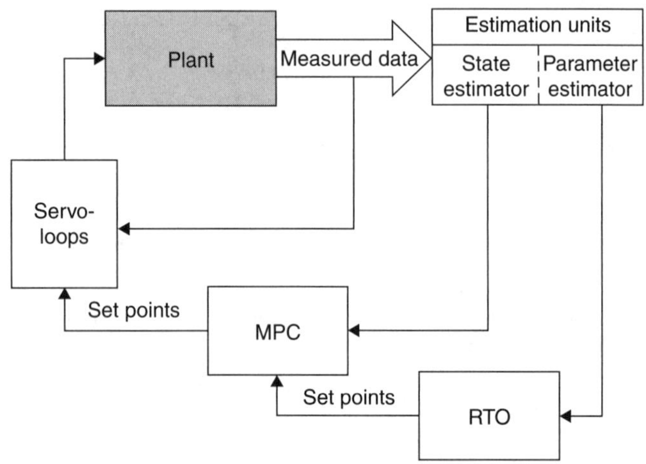

Figure 13.19 Modern control system architecture.

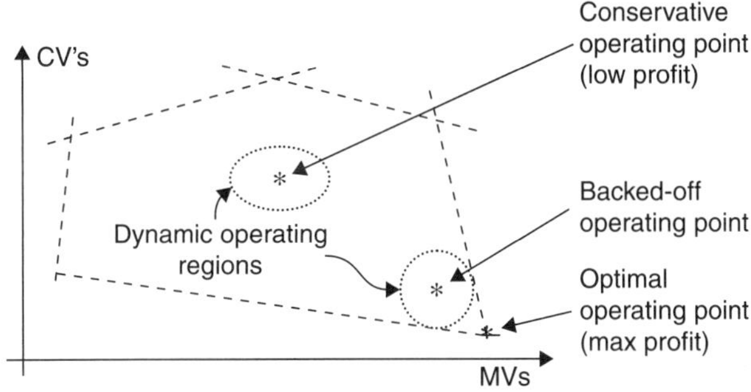

Figure 13.20 Synergy between RTO and MPC.

The RTO is responsible for selecting nominal (or steady-state) operating conditions such that profit is maximized. The RTO receives information about current model parameters from a parameter estimator (examples include heat exchanger fouling states, distillation column tray efficiency, etc.) In many cases this parameter estimator is very similar to the data reconciliation schemes presented in previous chapters.

As illustrated in the previous section, the key contribution of MPC is the ability to consider hardware and process limitations, such as maximum flowrates or minimum temperatures. The economic synergy between RTO and MPC is illustrated in Fig. 13.20. Traditionally, operating points were selected to be farthest from all constraints—at the center of the constraint set. This was to ensure the expected dynamic operating region (EDOR) would not extend beyond the constraint set, and thus minimize the likelihood of constraint violations. Unfortunately, the most profitable operating point, denoted as the optimal steady-state operating point (OSSOP), is typically near (if not at) the boundary of the constraint set (recall the throughput notions of the previous section). Thus, the benefit of MPC is that its constraint avoidance abilities enable operation within the most profitable regions of the constraint set—near the constraints. Clearly, one should select a back-off operating point (BOP) that is as close as possible to the OSSOP, while ensuring the EDOR does not extend beyond the constraint set.

Example 13.4: Consider again the process of Example 13.3. In the context of Fig. 13.20, the constraint set is defined by the maximum and minimum limits given for tank volume and exit flow. These constraints are indicated by the dashed lines of Fig. 13.21 and 13.22. Then, given the profit function of Eq. (13.10), the OSSOP is at the maximum value for exit flow (31 m^3/min) and any feasible tank volume (which we select for obvious reasons to be 10 m^3). Now consider the response of the MPC for the tuning values indicated in Fig. 13.12. The left plot of Fig. 13.21 depicts the EDOR corresponding to Fig. 13.12*a*, and clearly shows that an increase in profit is possible by moving the nominal throughput to about

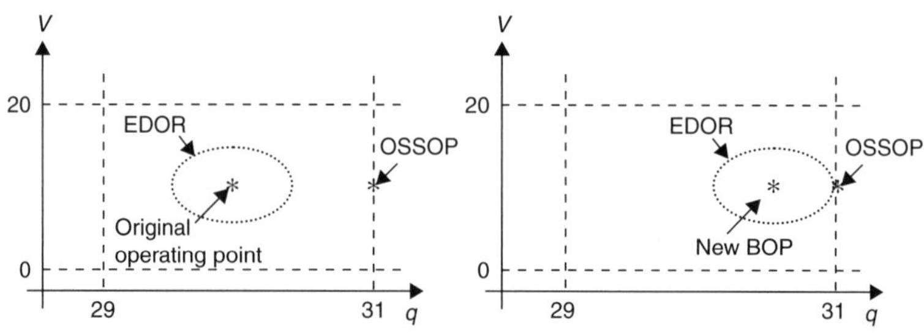

FIGURE 13.21 BOP selection for the scenario of Fig. 13.12*a*.

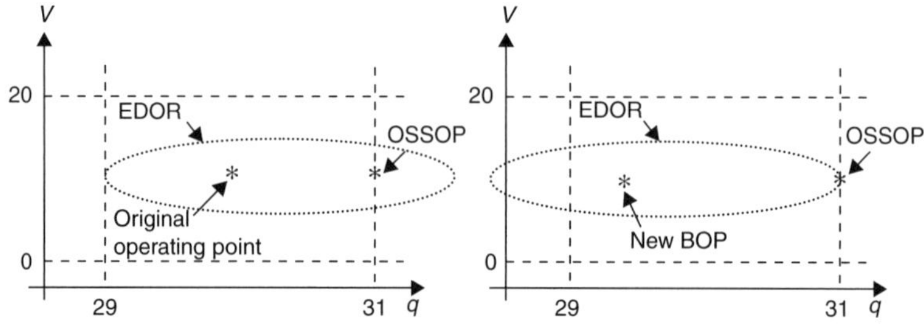

FIGURE 13.22 BOP selection for the scenario of Fig. 13.12*b*.

30.5 m^3/min. However, under disturbance scenario (b), Fig. 13.12*b* indicates a much larger EDOR. Thus, Fig. 13.22 shows the reduction in throughput that must be incurred due to this larger EDOR.

As indicated in the previous examples, the size of the EDOR will depend on the process dynamics, the controller (and its tuning values), and the disturbance characteristics. Thus, the outline for the next three sections is as follows. First, the state-space method of process modeling will be reviewed. Then an introduction to disturbance characterization for the stochastic class of disturbances is provided. Finally, we illustrate how such characterizations (of the process and disturbances) can be combined with a candidate controller to calculate the size and shape of the EDOR.

An important detail of the analysis tools we will cover next is that the complexity of the MPC controller will be replaced by the simplicity of a linear feedback array without constraints. While it may appear that such replacement will fail to capture the complex characteristics of MPC, it will be shown that, if selected appropriately, a simple linear array is sufficient to be used as a surrogate for MPC. Thus, in the next three sections we will simply assume an appropriate linear feedback array has been selected and thus can be used to characterize the EDOR.

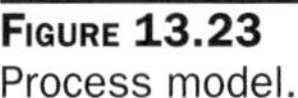

Figure 13.23 Process model.

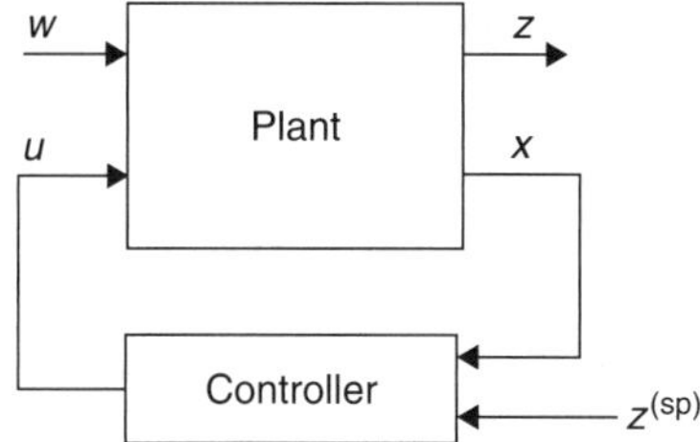

State-Space Process Modeling

In the MPC algorithm introduced previously, the process model was a scalar version of a state-space model. Since the remainder of the chapter will focus on state-space methods, we take this opportunity to introduce the generalized notation. A process with the following vector signals—state variable x, manipulated variable u, disturbance w, and performance output z—will be denoted as a state-space process:

$$\dot{x} = Ax + Bu + Gw \tag{13.29}$$

$$z = D_x x + D_u u + D_w w \tag{13.30}$$

Figure 13.23 illustrates the relation of the state-space process variables to that of the controller. Of particular note is the signal $z^{(sp)}$, which is the set-point command coming from the RTO. As indicated in the last section, the subsequent analysis will assume the controller is defined by a simple linear feedback array, L, which, as discussed earlier, will need to be selected appropriately. The resulting controller will then take a form similar to $u = Lx$. The influence of $z^{(sp)}$ on this controller will be discussed in a later section.

Example 13.5: Consider the pair of surge tanks depicted in Fig. 13.24. Similar to the previous examples the first objective is to deliver a constant exit flow, q_2, to the downstream unit, in the face of upstream variations at q_0. For each tank the volume of liquid should not exceed the total tank volume, and neither tank should be allowed to run dry.

A volume balance around each tank yields

$$\dot{V}_1 = q_0 - q_1 \qquad \dot{V}_2 = q_1 - q_2 \tag{13.31}$$

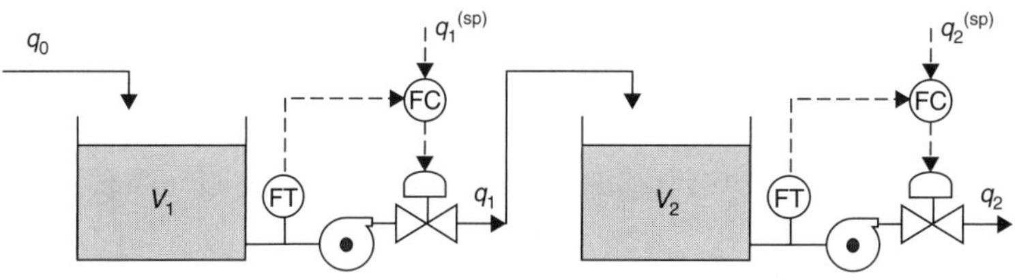

Figure 13.24 Surge tanks.

If the nominal operating condition is given to be $\{V_1^{nom}, V_2^{nom}, q_0^{nom}, q_1^{nom}, q_2^{nom}\}$, then one can define deviation variables as $V_i' = V_i - V_i^{nom}$ and $q_i' = q_i - q_i^{nom}$, which generates a deviation-based plant model.

$$\dot{V}_1' = q_0' - q_1' \qquad \dot{V}_2' = q_1' - q_2' \tag{13.32}$$

If one selects q_1 and q_2 as the manipulated variables and q_0 as the disturbance, then the following state space model is generated:

$$A = \begin{bmatrix} 0 & 0 \\ 0 & 0 \end{bmatrix} \qquad B = \begin{bmatrix} -1 & 0 \\ 1 & -1 \end{bmatrix} \qquad G = \begin{bmatrix} 1 \\ 0 \end{bmatrix} \tag{13.33}$$

where $x = [V_1' \; V_2']^T$, $u = [q_1' \; q_2']^T$, and $w = [q_0']$. Additionally, given the objectives of the process, it is reasonable to define the performance output as $z = [V_1' \; V_2' \; q_1' \; q_2']^T$, which gives

$$D_x = \begin{bmatrix} 1 & 0 \\ 0 & 1 \\ 0 & 0 \\ 0 & 0 \end{bmatrix} \qquad D_u = \begin{bmatrix} 0 & 0 \\ 0 & 0 \\ 1 & 0 \\ 0 & 1 \end{bmatrix} \qquad D_w = \begin{bmatrix} 0 \\ 0 \\ 0 \\ 0 \end{bmatrix} \tag{13.34}$$

As a first illustration of a state feedback controller ($u = Lx$), consider a pair of proportional controllers applied to the pair of surge tanks: $q_i' = K_{ci} V_i'$. In this case, the feedback array will take the following form:

$$L = \begin{bmatrix} l_{11} & l_{12} \\ l_{21} & l_{22} \end{bmatrix} \tag{13.35}$$

where $l_{11} = K_{c1}$, $l_{12} = 0$, $l_{21} = 0$, and $l_{22} = K_{c2}$.

As illustrated in Examples 13.1 through 13.4, the set of constraints used is central to evaluating controller performance. Within the state-space model, these constraints are specified through the performance signal, z: $z_i^{min} < z_i < z_i^{max}$. In the current example $z_1^{min} = V_1^{min} - V_1^{nom}$ and the other constants (z_i^{min} and z_i^{max}) are defined similarly.

Concerning modification of nominal operating conditions, one must take care to observe the steady-state process model: $0 = Ax^{nom} + Bu^{nom} + Gw^{nom}$. In the current example, these equality constraints indicate that all nominal flowrates must be equal, $q_0^{nom} = q_1^{nom} = q_2^{nom}$, and one has total freedom in selecting nominal tank hold-ups.

Remark: In the state-space version of MPC, measurements are typically used to construct an estimate of the system state, which is compared with the desired operating condition. In the previous section, perfect measurements were assumed, and thus the perfectly measured tank volume served as the state estimate. In the last example, perfect measurement (or full state information) was also assumed, resulting in the controller $u = Lx$. Perfect measurement is, as we saw in previous chapters, impossible, so state estimation is the alternative. A detailed introduction to state estimation and the partial state information (PSI) case will be provided later in the chapter.

Remark: As done in the MPC section, a continuous-time model can be converted to a discrete-time form.

$$x_{k+1} = A_d x_k + B_d u_k + G_d w_k \tag{13.36}$$

$$z_k = D_x x_k + D_u u_k + D_w w_k \tag{13.37}$$

where x_k, u_k, and w_k are defined identical to that of V_k, q_k, and $q^{(in)}{}_k$ of the previous section. If the Euler method is employed, then $A_d = I + \Delta tA$, $B_d = \Delta tB$, and $G_d = \Delta tG$. To achieve greater accuracy, it is recommended to use the sample and hold method described in Burl (1998) rather than the Euler method.

Remark: An alternative to the state-space model is an input-output model, potentially generated directly from step test data. These models usually take a discrete-time form. The output signal in time is collected into a time series $\bar{z} = [z_1\ z_2\ z_3 \ldots z_N]$, where the subscript is the time index. If the input signal is also collected into a vector, $\bar{u} = [u_1\ u_2\ u_3 \ldots u_M]$, then the time derivative of the step test data can be used to construct a matrix M_u indicating the relation between process inputs and outputs $\bar{z} = M_u\bar{u}$. The matrix M_u is typically denoted as the dynamic matrix, and (along with a disturbance version, M_w) forms the basis of the input-output version of MPC, typically denoted as dynamic matrix control (DMC). While DMC enjoys wide acceptance in the industrial community, the state-space version of MPC (introduced in the previous section) has been shown by the academic community as well as software vendors (Badgwell, 2008) to posses performance similar to DMC. Because of this, we do not cover DMC any further.

Disturbance Modeling

As suggested by Example 13.1, the disturbance is frequently assumed to be a sequence of step inputs. However, within a typical chemical process, disturbances of such structure (i.e., with abrupt changes) are infrequently observed. An alternative approach is to assume the disturbance is stochastic in nature.

In the discrete-time framework, a stochastic disturbance (or a stochastic process) is simply a sequence of random variables, where the index of the sequence is associated with time. Associated with a stochastic process is a sequence of statistics. If each random variable is gaussian (i.e., characterized by only the mean and variance), then the first step to characterizing a stochastic process is to identify the mean and variance sequence corresponding to the time indexed random variable sequence. Although not required, it is frequently assumed that the mean and variance sequence are constant over time. In this case of time invariant statistics, the stochastic process is called *stationary*.

Example 13.6: In Example 13.1, it is given that the inlet flow is anticipated to be $q_{in} = 30 \pm 3$ m^3/min. If this disturbance is assumed to be a gaussian stochastic process $q^{(in)}{}_k$, then it is clearly appropriate to assume the mean of $q^{(in)}{}_k$ is equal to 30 m^3/min at all times, k. Similarly, it is reasonable to assume that the variance of $q^{(in)}{}_k$ is equal to 9 m^6/min^2 for all time, which would yield a standard deviation of 3 m^3/min. Under these assumptions we expect that $q^{(in)}{}_k$ will be in the desired range (30 ± 3 m^3/min) 68% of the time. However, if we assume the standard deviation is 1.5 m^3/min (or variance equal to 9/4 m^6/min^2), then we would expect $q^{(in)}{}_k$ to be in the range 30 ± 3 m^3/min 95.5% of the time.

As illustrated in Example 13.1, the time-related structure of the disturbance will greatly influence how the disturbance impacts the outputs of the process. Recall how disturbance (b) caused larger variations at q and V as compared to disturbance (a), even though each disturbance had the same magnitude of variation. In disturbance (b) the longer duration of each step in the sequence suggested the following. If the disturbance is at a particular value at a particular time, then one expects the disturbance to be at that same value in the near future. However, as we go further into the future this expectation (of being at the same value) will diminish. Looking at the plot of disturbance (b), one could select "near future" to correspond to about 3 to 5 min, while for disturbance (a) the notion of near future would only be about 1 to 3 min. To capture similar (duration type) characteristics within the stochastic framework, we need to introduce the concept of autocorrelation. We do this next.

Let us return to a sequence of random variables, the collection of which forms a stochastic process. Denote this sequence as w_k, and define the mean and variance sequences as $\bar{w}_k = E[w_k]$ and $\Sigma_{w,k} = E[(w_k - \bar{w}_k)^2]$. Note that for the moment we assume each random variable, w_k, to be a scalar.

We now ask, how much each random variable in the sequence is correlated with a different random variable in the sequence. Such correlation is defined by the autocorrelation function (briefly introduced in Chap. 12).

$$R_{w,k,i} = E[(w_k - \bar{w}_k)(w_{k+i} - \bar{w}_{k+i})] \tag{13.38}$$

If w_k is a stationary process (statistics are time invariant), then the autocorrelation is a function only of the time difference between the random variables: In other words, $R_{w,k,i}$ is independent of k, and therefore one can drop the subindex k, and write $R_{w,k,i} = R_{w,i}$ for all time k.

Recall now the discussion of the disturbances of Example 13.1. Physical intuition suggested that we expect high correlation in the near future and lower correlation as we go further into the future. And, the distinction between the two disturbances was how quickly this correlation would diminish as we go further into the future. The autocorrelation function, $R_{w,k,i}$, exhibits identical behaviour. For a given k, the maximum is found at $i = 0$, $R_{w,k,0} = \Sigma_{w,k}$; that is, each random variable is most correlated with itself. As i increases, $R_{w,k,i}$ will decrease, and approach zero as i approaches infinity. In other words, as the time difference between two random variables increase, the correlation between the two will decrease to the limit of no correlation for very large time differences. If $R_{w,k,i}$ is different for two stochastic processes, then the time series characteristics of each process are different. Specifically, if the autocorrelation function decreases slowly, then it will have duration type characteristics similar to disturbance (b), while a fast decrease suggests disturbance (a).

In the limiting case of no correlation between any of the random variables in the sequence ($R_{w,k,i} = \Sigma_{w,k}$ if $i = 0$, and $R_{w,k,i} = 0$, otherwise), we denote the stochastic process as *white noise*. As we will see, the white noise process holds a special place in disturbance modeling. Specifically, it can be thought of as a blank canvas from which most other stochastic processes can be generated.

In the continuous-time framework, all of the above notions carry through, but with the obvious change in time index notation. Specifically, if the stochastic process is $w(t)$, then $\bar{w}(t) = E[w(t)]$, $\Sigma_w(t) = E[(w(t) - \bar{w}(t))^2]$, and

$$R_w(t,\tau) = E[(w(t) - \bar{w}(t))(w(t+\tau) - \bar{w}(t+\tau))] \tag{13.39}$$

The one exception concerns white, noise, which, to define precisely, is outside the current scope (see McGarty, 1973). However, for our subsequent purposes it will be sufficient to assume white noise is just as a stochastic process with autocorrelation $R_w(t,\tau) = \Sigma_w$ if $\tau = 0$, $R_w(t,\tau) = 0$, otherwise, which in essence says that the white noise has no correlation, as described above for the discrete case.

While the autocorrelation function provides an intuitively attractive description of disturbance characteristics, it is rarely used as a quantitative disturbance model. The more common approach is to model the stochastic disturbance as the output of a linear system driven by white noise.

The simplest disturbance model is that of a first-order system driven by white noise. Assume w_0 is a gaussian white noise process with zero-mean and variance $\Sigma_{w0} = 1$. Since the mean, variance, and autocorrelation is constant over time, w_0 is a stationary process. If this white noise input is used to drive the process defined by the following set of equations:

$$\dot{\omega} = a_\omega \omega + g_\omega w_0 \tag{13.40}$$

$$w = c_\omega \omega \tag{13.41}$$

then we obtain different disturbances as a function of the parameters a_ω, g_ω, and c_ω.

Example 13.7: Consider the white noise disturbance of Fig. 13.25. The disturbances shown in Fig. 13.26 can be generated using the following parameters:

Low correlation case: $a_\omega = -0.5$, $g_\omega = 1$, and $c_\omega = 3$

High correlation case: $a_\omega = -0.05$, $g_\omega = 0.1^{1/2}$, and $c_\omega = 3$

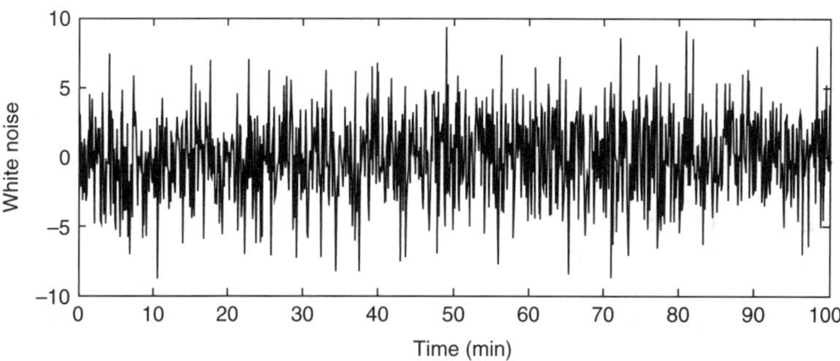

Figure 13.25 White noise disturbance.

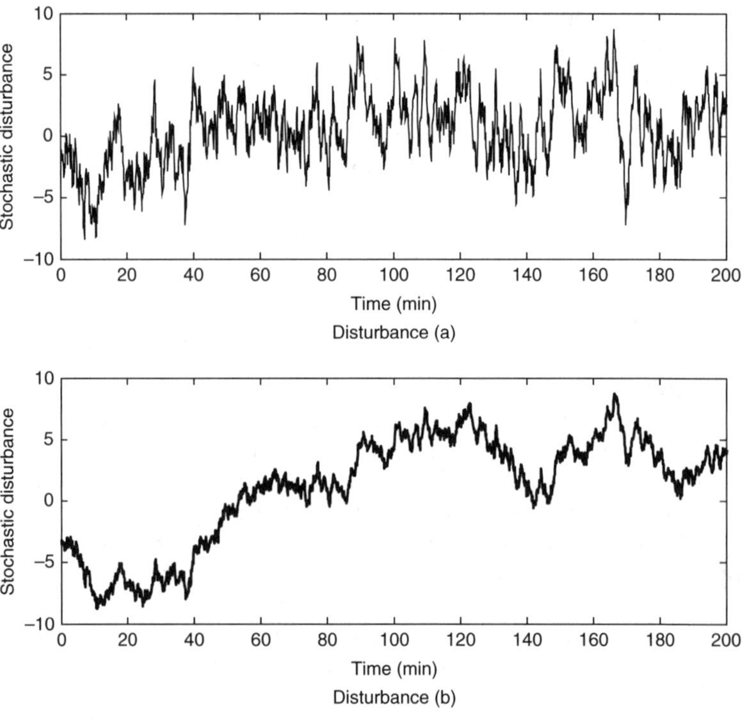

FIGURE 13.26 Stochastic disturbances: Disturbance (a) low correlation, disturbance (b) high correlation.

If $\bar{w}(0) = 0$ and $\Sigma_w(0) = 9$, then it can be shown that for both cases $\bar{w}(t) = 0$ and $\Sigma_w(t) = 9$ for all t [see the next section and Burl (1998) for details]. It can also be shown (Burl, 1998) that $R_w(t, \tau) = e^{a_\omega|\tau|}\Sigma_w(t)$, which further illustrates the high/low correlation nature that distinguishes each disturbance.

It is additionally noted that the plot of Fig. 13.26 was generated via the discrete-time equivalent of the process Eqs. (13.40) and (13.41). Specifically, the discrete-time model is

$$\omega_{\kappa+1} = a_{\omega d}\, \omega_\kappa + g_{\omega d}\, w_{0,\kappa} \tag{13.42}$$

$$w_\kappa = c_\omega\, \omega_\kappa \tag{13.43}$$

with $w_{0,\kappa}$ being a white noise sequence with zero mean and a variance equal to $\Sigma_{w0d} = \Sigma_{w0}/\Delta t$, where Δt is the sample time (assumed 0.1 min) and $a_{\omega d}$ and $g_{\omega d}$ are calculated by the Euler method.

Example 13.8: We now reconsider the surge tank pair of Example 13.5. Assume $q_0^{\text{nom}} = q_1^{\text{nom}} = q_2^{\text{nom}} = 50\ \text{m}^3/\text{min}$, $V_1^{\text{nom}} = V_2^{\text{nom}} = 150\ \text{m}^3$, and the controller is the pair of PI controllers with $K_{c1} = K_{c2} = 0.05\ \text{min}^{-1}$. Now apply the disturbance signals of Example 13.7 to the inlet flowrate [i.e., $q_0 = q_0^{\text{nom}} + w$, where w is generated by the system Eqs. (13.40) and (13.41)]. While both disturbance inputs have identical variance (standard deviation equal to 3, as shown in the first two plots of Fig. 13.27), the outputs of the process (plots 3–6 of Fig. 13.27) indicate that output

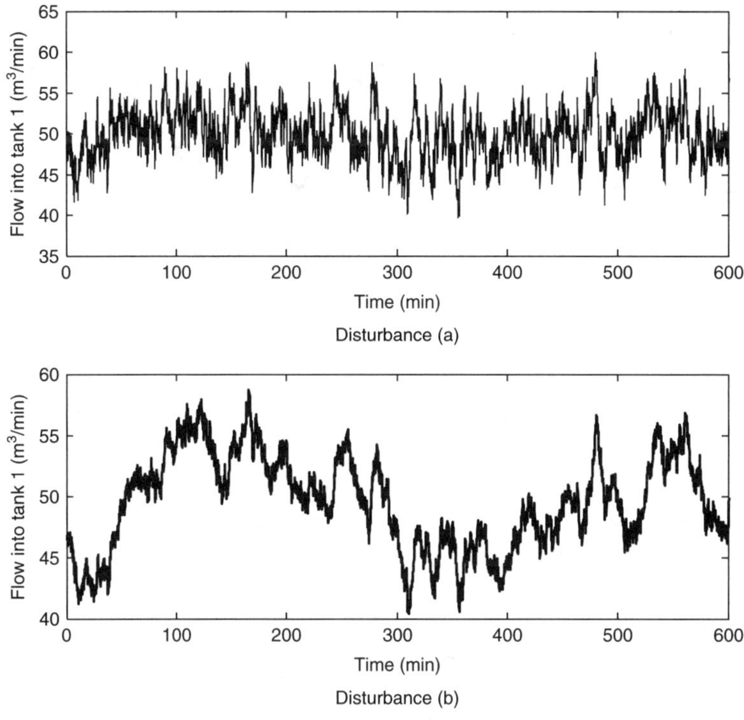

Figure 13.27 Response of surge tanks to different disturbances: Disturbance (a) thin line; Disturbance (b) thick line.

variance will depend on the amount of correlation present in the disturbance. This conclusion of output characteristics (i.e., size) being dependent on disturbance characteristics (i.e., size and correlation level) is nearly identical to the conclusions made in Examples 13.1 and 13.3. The only difference is the specific class of disturbances considered—sequence of steps or stochastic.

In the general case, the disturbance model takes a state-space form, similar to Eqs. (13.29) and (13.30).

$$\dot{\omega} = A_{\omega}\,\omega + G_{\omega} w_0 \tag{13.44}$$

$$w = C_{\omega}\,\omega \tag{13.45}$$

Thus, the disturbance modeling challenge is to determine the matrices A_{ω}, G_{ω}, and C_{ω} such that the characteristics of the generated disturbance signal are equal to that of the actual disturbances exciting the plant. Fortunately, a wealth of methods aimed at developing such models (from plant data) are available in the literature (see, for example, Bendat and Piersol, 1985; Soderstrom and Stoica, 1989; Ljung and Glad, 1994; Larimore, 1999; Zhu, 2001; and Odelson et al., 2006). Rather than discussing these techniques, the remainder of the chapter will assume that such modeling has been performed.

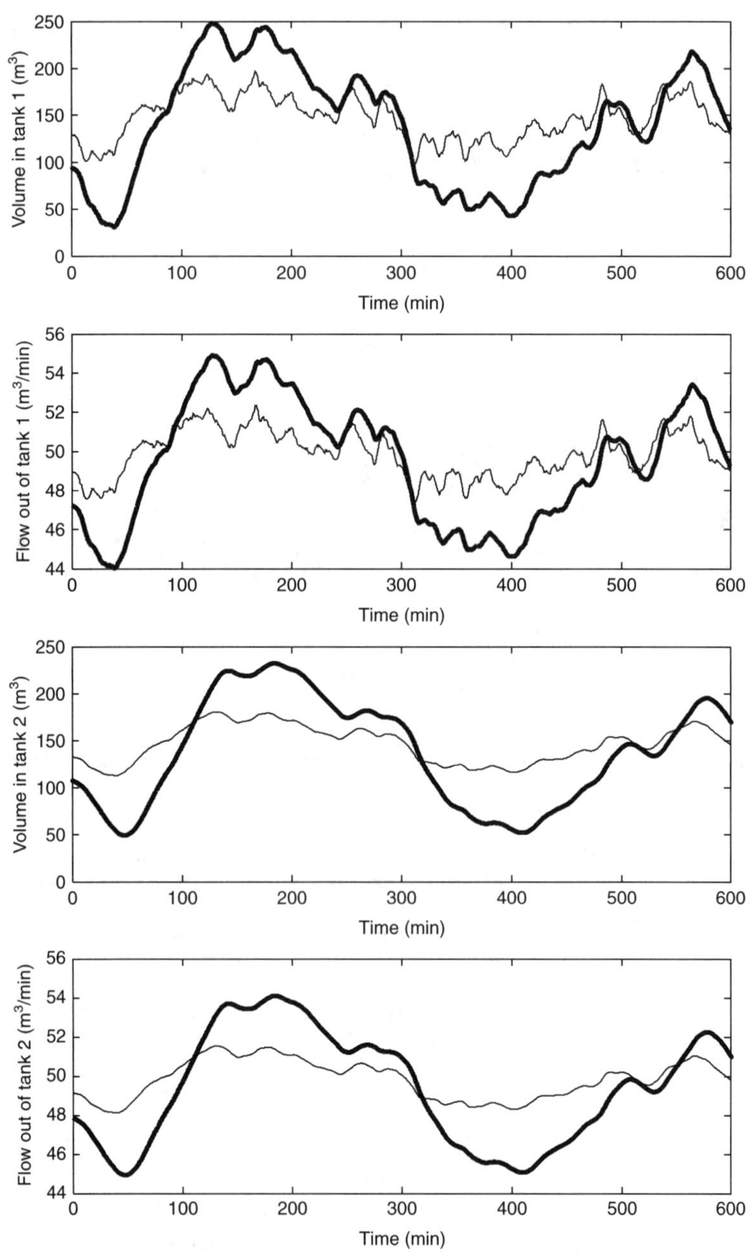

FIGURE 13.27 *(Continued)*

The next step is to formally augment the original process model to add the disturbance model. Simple algebraic manipulations of Eqs. (13.29) and (13.30) and (13.44) to (13.45) yield the following compound system:

$$\begin{bmatrix} \dot{x} \\ \dot{\omega} \end{bmatrix} = \begin{bmatrix} A & GC_\omega \\ 0 & A_\omega \end{bmatrix} \begin{bmatrix} x \\ \omega \end{bmatrix} + \begin{bmatrix} B \\ 0 \end{bmatrix} u + \begin{bmatrix} 0 \\ G_\omega \end{bmatrix} w_0 \tag{13.46}$$

$$z = \begin{bmatrix} D_x & D_w C_\omega \end{bmatrix} \begin{bmatrix} x \\ \omega \end{bmatrix} + D_u u \tag{13.47}$$

In the sequel we will refer to this compound system as

$$\dot{x}^{\text{cmp}} = A^{\text{cmp}} x^{\text{cmp}} + B^{\text{cmp}} u^{\text{cmp}} + G^{\text{cmp}} w^{\text{cmp}} \tag{13.48}$$

$$z = D_x^{\text{cmp}} x^{\text{cmp}} + D_u^{\text{cmp}} u^{\text{cmp}} + D_w^{\text{cmp}} w^{\text{cmp}} \tag{13.49}$$

where a compounded vector x^{cmp} is composed of x and ω. If in the discrete-time framework, the notation will be A_d^{cmp}, B_d^{cmp}, and G_d^{cmp}.

In the next section we illustrate how this model can be used to calculate the size and shape of the EDOR. In the context of Example 13.8, these methods will allow us to calculate, without simulation, the standard deviations observed in the plots of Fig. 13.27. These analytic methods will be a central component to the selection of an appropriate controller $u = Lx$.

Expected Dynamic Operating Region Characterization

As in the previous section we begin with the scalar discrete-time case. Consider the discrete-time process of Example 13.7.

$$\omega_{\kappa+1} = a_{\omega d}\, \omega_\kappa + g_{\omega d}\, w_{0,\kappa} \tag{13.50}$$

$$w_\kappa = c_\omega\, \omega_\kappa \tag{13.51}$$

The mean of ω_κ and w_κ are calculated as follows: Define $\bar{w}_{0,k} = E[w_{0,k}]$ for all k. Since $w_{0,\kappa}$ is assumed zero mean, $\bar{w}_{0,k} = 0$ for all k. Define the mean of ω_κ and w_κ similarly: $\bar{\omega}_k = E[\omega_k]$ and $\bar{w}_k = E[w_k]$. Then application of the expectation operator to Eqs. (13.50) and (13.51) yields

$$\bar{\omega}_{k+1} = a_{\omega d}\bar{\omega}_k \qquad \text{and} \qquad \bar{w}_k = c_\omega \bar{\omega}_k \tag{13.52}$$

If the original process is stable (in the scalar discrete-time case this is $|a_{\omega d}| < 1$), then the mean of both ω_κ and w_κ will approach zero at a rate equal to the time constant of the process. If the mean of the initial condition is assumed zero (i.e., $\bar{\omega}_0 = E[\omega_0] = 0$), then the mean of both ω_κ and w_κ will be zero for all k. To simplify the variance calculation we employ such assumption.

The variance of ω_κ and w_κ are calculated as follows: Since $w_{0,\kappa}$ is assumed stationary, its variance is constant with respect to time: $\Sigma_{w0d} = E[w_{0,k}{}^2]$. As in the previous section, the variance of ω_κ and w_κ are defined as

$$\Sigma_{\omega,k} = E[\omega_k^2] \tag{13.53}$$

$$\Sigma_{w,k} = E[(c_d\omega_k)^2] = c_\omega^2 \Sigma_{\omega,k} \tag{13.54}$$

The variance of $\omega_{\kappa+1}$ is now calculated as

$$\left.\begin{aligned} \Sigma_{\omega,k+1} &= E[(a_{\omega d}\omega_k + g_{\omega d} w_{0,k})^2] \\ &= a_{\omega d}^2 E[\omega_k^2] + 2a_{\omega d} g_{\omega d} E[\omega_k \omega_{0,k}] + g_{\omega d}^2 E[w_{0,k}^2] \\ &= a_{\omega d}^2 \Sigma_{\omega,k} + g_{\omega d}^2 \Sigma_{\omega 0d} \end{aligned}\right\} \tag{13.55}$$

where the independence between $w_{0,\kappa}$ and ω_κ causes the cross term, $E[\omega_k w_{0,k}]$, to be zero. If $|a_{\omega d}|<1$, then $\Sigma_\omega \triangleq \lim_{k\to\infty} \Sigma_{\omega,k}$ (the steady-state variance of ω_κ) will exist and can be calculated from

$$\Sigma_\omega = a_{\omega d}^2 \Sigma_\omega + g_{\omega d}^2 \Sigma_{\omega 0d} \tag{13.56}$$

Equation (13.56), usually denoted as the Lyapunov equation, is simply the limit of the following recursion: $\Sigma_{\omega,k+1} = a_{\omega d}^2 \Sigma_{\omega,k} + g_{\omega d}^2 \Sigma_{w0}$. Finally, the steady-state variance of w_κ is: $\Sigma_w = \lim_{k\to\infty} \Sigma_{w,k} = c_\omega^2 \Sigma_\omega$, where Σ_ω is given by Eq. (13.56).

Example 13.9: In the high correlation case of Example 13.8, the discrete-time model parameters are found to be $a_{\omega d} = 0.995$, $g_{\omega d} = 0.0316$, $c_\omega = 3$, and $\Sigma_{w0d} = 10$. Then by application of Eq. (13.56), one finds $\Sigma_\omega = 1$ and $\Sigma_w = 9$. Thus, the standard deviation of w_κ is calculated to be 3, which is the value observed in Fig. 13.27.

In the vector version of the discrete-time case, with a candidate controller $u_k = Lx_k$, the closed-loop process model is

$$x_{k+1} = (A_d + B_d L)x_k + G_d w_k \tag{13.57}$$

$$z_k = (D_x + D_u L)x_k + D_w w_k \tag{13.58}$$

Using a derivation similar to the scalar case, the steady-state covariance matrices (the vector generalization of variance) are given by

$$\Sigma_x = (A_d + B_d L)\Sigma_x (A + B_d L)^T + G_d \Sigma_w G_d^T \tag{13.59}$$

$$\Sigma_z = (D_x + D_u L)\Sigma_x (D_x + D_u L)^T + D_w \Sigma_w D_w^T \tag{13.60}$$

For the vector case recall that if $y = \Gamma x$, then the variance of y, Σ_y, is given by $\Sigma_y = \Gamma \Sigma_x \Gamma^T$, where Σ_x is the variance of the variable x [Chap. 4, Eq. (4.29)].

The interpretation of Σ_x is that each diagonal element provides the variance of the respective scalar sequence, $x^{(i)}_k$, and the off-diagonal terms indicate the cross-correlation between the various sequences within the vector x_k (similar to but not exactly the autocorrelation).

A similar interpretation of Σ_z exists with respect to z_k. However, in later sections we will make much greater use of the steady-state variance of each output signal, $z^{(i)}_k$. In fact, these steady-state variances will be the quantification of the EDOR. Thus, we define the following notion:

The size of the EDOR in the ith direction, ξ_i, is simply calculated as $\xi_i = \phi_i \Sigma_z \phi_i^T$, where ϕ_i is the ith row of an appropriately sized identity matrix. Thus ξ_i is defined as the steady-state variance of $z^{(i)}_k$.

In the continuous-time case, with a candidate controller $u = Lx$, the closed-loop system is

$$\dot{x} = (A+BL)x + Gw \tag{13.61}$$

$$z = (D_x + D_u L)x + D_w w \tag{13.62}$$

Similar to the discrete-time case, the steady-state covariance matrices of the closed-loop system are calculated with the following equation:

$$(A + BL)\Sigma_x + \Sigma_x(A + BL)^T + G\Sigma_w G^T = 0 \tag{13.63}$$

$$\Sigma_z = (D_x + D_u L)\Sigma_x (D_x + D_u L)^T + D_w \Sigma_w D_w^{\,T} \tag{13.64}$$

Then, the size of the EDOR in the i direction is calculated as $\xi_i = \phi_i \Sigma_z \phi_i^T$.

Example 13.10: Application of the scalar version of Eq. (13.63) to the continuous-time model of Example 13.8 yields

$$2a_\omega \Sigma_\omega + g_d^{\,2} \Sigma_{w0} = 0 \tag{13.65}$$

Thus, $\Sigma_\omega = -g_\omega^{\,2}\, \Sigma_{w0}/2a_\omega$ and $\Sigma_w = c_\omega^{\,2}\, \Sigma_\omega$. Then, application of the parameters given in Example 13.8 indicates that $\Sigma_\omega = 1$ and $\Sigma_w = 9$ for both cases. In general, if one selects

$$a_\omega = -1/\tau_\omega \qquad g_\omega = (2/\tau_\omega)^{1/2} \qquad \text{and} \qquad c_\omega = (\Sigma_w')^{1/2} \tag{13.66}$$

then, it is easily verified that the variance of w, Σ_w, is equal to Σ_w' for all values of τ_ω. Using these relations the parameters of Example 13.8 were generated from

Low correlation case: $\tau_\omega = 2$ and $\Sigma_w' = 9$

High correlation case: $\tau_\omega = 20$ and $\Sigma_w' = 9$

The utility of Eq. (13.66) is the construction of a simple two-parameter disturbance model, where the output variance is assigned by the parameter Σ_w' and the amount of correlation is related to the parameter τ_ω: $R_w(t, \tau_\omega) = e^{-1}\Sigma_w{}'$.

Example 13.11: To calculate the output variability depicted in Fig 13.27, the disturbance model of Example 13.8 will need to be augmented to the plant model of Example 13.5. The resulting compound system is

$$A^{\text{cmp}} = \begin{bmatrix} 0 & 0 & \sqrt{\Sigma'_w} \\ 0 & 0 & 0 \\ 0 & 0 & -\dfrac{1}{\tau_\omega} \end{bmatrix} \quad B^{\text{cmp}} = \begin{bmatrix} -1 & 0 \\ 1 & -1 \\ 0 & 0 \end{bmatrix} \quad G^{\text{cmp}} = \begin{bmatrix} 0 \\ 0 \\ \sqrt{\dfrac{2}{\tau_\omega}} \end{bmatrix} \tag{13.67}$$

$$L^{cmp} = \begin{bmatrix} K_{c1} & 0 & 0 \\ 0 & K_{c2} & 0 \end{bmatrix} \tag{13.68}$$

Then application of Eqs. (13.63) and (13.64) to the system Eqs. (13.67) and (13.68) yields the data of Table 13.1. These figures are equal to the standard deviations observed in the plots of Fig. 13.27.

	$\xi_1^{1/2}$	$\xi_2^{1/2}$	$\xi_3^{1/2}$	$\xi_4^{1/2}$
Low Correlation	18.1	13.4	0.905	0.668
High Correlation	42.4	36.7	2.12	1.84

Table 13.1 Standard Deviation of Outputs for Example 13.11

Remark: The calculation of Σ_x for the matrix case reduces to manipulating the matrix Eq. (13.63) into a set of linear equations. The function *"lyap"* within the Matlab control system toolbox was used to perform these manipulations and solve the linear equations for Example 13.11.

An important detail of the augmented process model concerns the availability of information. Specifically, the controller will not have direct access to the states internal to the disturbance model, ω. At best one will be able to measure the disturbance input to the plant, $w = C_\omega \omega$, but the internal states ω will likely be a nonphysical construct of the disturbance model and thus will be impossible to measure directly. The solution to this problem is to employ a state estimator.

The development of a state estimator requires the introduction of a measurement information structure [also known as partial state information (PSI)]. In this case, the system of interest is

$$\dot{x} = Ax + Bu + Gw \tag{13.69}$$

$$y = Cx + v \tag{13.70}$$

where Eq. (13.70) is the measurement equation. The term v is the measurement noise, assumed here to be gaussian white noise with zero mean and covariance Σ_v. If the original (unaugmented) process is used, then the state estimate is generated by the following Kalman filter (see Chap. 9), also known as the optimal state estimator.

$$\dot{\hat{x}} = A\hat{x} + Bu + K(y - C\hat{x}) \tag{13.71}$$

where $K = \Sigma_e C^T \Sigma_v^{-1}$ and Σ_e is the covariance matrix associated with the estimation error signal, $e = x - \hat{x}$, which is calculated as the solution of the following Riccati equation:

$$A\Sigma_e + \Sigma_e A^T + G\Sigma_w G^T - \Sigma_e C^T \Sigma_v^{-1} C\Sigma_e = 0 \tag{13.72}$$

If augmented with the disturbance model, then the PSI plant of interest is

$$\dot{x}^{\text{cmp}} = A^{\text{cmp}} x^{\text{cmp}} + B^{\text{cmp}} u^{\text{cmp}} + G^{\text{cmp}} w^{\text{cmp}} \tag{13.73}$$

$$y = C^{\text{cmp}} x^{\text{cmp}} + v \tag{13.74}$$

where A^{cmp}, B^{cmp}, and G^{cmp} are defined in the same way as those of the previous section and $C^{\text{cmp}} = [C^T\ 0]^T$. In this case, the design of the Kalman filter is a simple extension of the original (unaugmented) procedure.

To be able to implement a Kalman filter, certain linear algebraic criteria must be met. Specifically, the pair (A, C) must be detectable and (A, G) must be stabilizable (see Burl, 1998 for definitions). The following example extends the previous two tank example to meet these criteria.

Example 13.12: While the disturbance model of Example 13.11 is enough for basic covariance analysis, it cannot be used to design a Kalman filter due to insufficient excitement of the uncontrolled process [i.e., the pair (A, G) is not stabilizable]. To find additional excitement, we look more closely at the mass flow controllers regulating q_i. It is postulated that the actual flow will deviate from the set-point command: $q_i = q_i^{(\text{sp})} + w_i$. If we assume this deviation is zero-mean white noise with a standard deviation equal to 0.5% of the nominal flow, then the following disturbance augmented process model will result:

$$A = \begin{bmatrix} 0 & 0 & \sqrt{\Sigma'_w} \\ 0 & 0 & 0 \\ 0 & 0 & -\dfrac{1}{\tau_\omega} \end{bmatrix} \quad G = \begin{bmatrix} 0 & 1 & -1 & 0 \\ 0 & 0 & 1 & -1 \\ \sqrt{\dfrac{2}{\tau_\omega}} & 0 & 0 & 0 \end{bmatrix} \tag{13.75}$$

$$\Sigma_w = \begin{bmatrix} 1 & 0 & 0 & 0 \\ 0 & 0.25^2 & 0 & 0 \\ 0 & 0 & 0.25^2 & 0 \\ 0 & 0 & 0 & 0.25^2 \end{bmatrix} \tag{13.76}$$

Turning to the measurement equation, we assume that both tank volumes are measured as well as the flow inlet to tank 1. We assume the measurement noise to be zero-mean white noise with a standard deviation equal to 0.5% of the nominal hold-up/flow. Thus,

$$C = \begin{bmatrix} 1 & 0 & 0 \\ 0 & 1 & 0 \\ 0 & 0 & 1 \end{bmatrix} \quad \Sigma_v = \begin{bmatrix} 0.75^2 & 0 & 0 \\ 0 & 0.75^2 & 0 \\ 0 & 0 & 0.25^2 \end{bmatrix} \tag{13.77}$$

FIGURE 13.28 Partial state information process model.

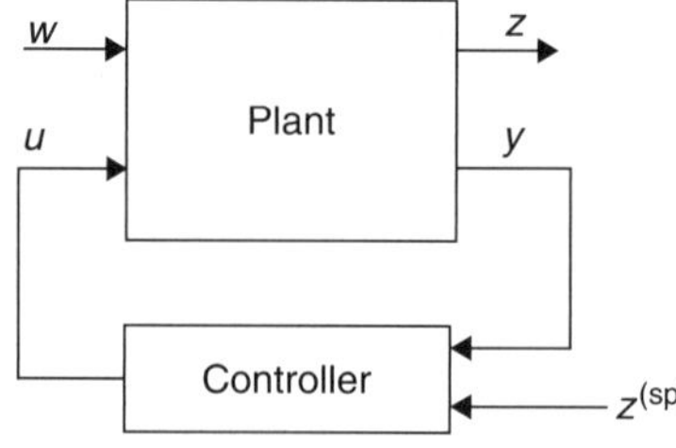

Then application of Eq. (13.72) to the case of $\tau_\omega = 20$ yields

$$\Sigma_e = \begin{bmatrix} 0.558 & -0.039 & 0.097 \\ -0.039 & 0.262 & 0.004 \\ 0.097 & 0.004 & 0.069 \end{bmatrix} \tag{13.78}$$

The objective of a state estimator is to make the (covariance of the) error signal small. The Kalman filter is termed optimal because it achieves the minimum error (i.e., the sum of the diagonal elements of Σ_e is minimized). As indicated in Fig. 13.28, the state of the state-space process is unavailable to the controller in the PSI case (i.e., $u = Lx$ is not possible). However, once the state estimator (which is driven by the measured signal, y) has been implemented, then it is possible for the controller to employ the state estimate, $\hat{x}$, in place of the state (i.e., $u = L\hat{x}$).

In the disturbance augmented case, where the disturbance is now assumed to be estimated, the controller will be

$$u = \begin{bmatrix} L_x & L_\omega \end{bmatrix} \begin{bmatrix} \hat{x} \\ \hat{\omega} \end{bmatrix} \tag{13.79}$$

In Example 13.11, the PI form of the controller implicitly assumed $L_\omega = 0$, and thus took no advantage of the disturbance estimate. In the more general case (where L_ω is not set to 0), the advantage is that the disturbance model will be used to make predictions about future values of the disturbance, based on the obtained estimates of ω. This additional information can then be utilized by the disturbance portion of the controller, L_ω. Recall the disturbance anticipation discussion at the end of Example 13.3. Disturbance estimation is the first step to employing the disturbance characterization within the controller. The next section on "Constrained Minimum Variance Control" is the second step.

In the PSI case, Eqs. (13.63) and (13.64) can no longer be used to calculate EDOR size. In Chmielewski and Manthanwar (2004), the following relations were developed to calculate the EDOR for the unaugmented PSI case:

$$(A+BL)\Sigma_{\hat{x}} + \Sigma_{\hat{x}}(A+BL)^T + \Sigma_e C^T \Sigma_v^{-1} C \Sigma_e = 0 \tag{13.80}$$

$$\Sigma_z = (D_x + D_u L)\Sigma_{\hat{x}}(D_x + D_u L)^T + D_x \Sigma_e D_x^T \tag{13.81}$$

where ξ_i is again calculated as $\phi_i \Sigma_z \phi_i^T$ and Σ_e is calculated via Eq. (13.72). The disturbance augmented case is again a simple extension of the unaugmented procedure.

Constrained Minimum Variance Control

From a controller design perspective (i.e., the selection of L), the system of Equations (13.63)–(13.64) can be of great utility. For a given disturbance characterization, Σ_w, one can evaluate the variance of each performance output as a function of the choice of L.

Example 13.13: Consider the single surge tank of Example 13.1: $\dot{V}' = q_0' - q_1'$, and assume q_0' is zero-mean white noise with $\Sigma_q = 9$. Then $a = 0$, $b = -1$, and $g = 1$, and we can define the performance as the state and the manipulated variable: $D_x = [1\ 0]^T$ and $D_u = [0\ 1]^T$. If the controller is $q_1' = LV'$, then by using the scalar version of Eq. (13.63), $0 = 2\Sigma_x(a + bL) + g^2\Sigma_w$, we find $\Sigma_x = 9/(2L)$. In turn, from Eq. (13.64), we find $\xi_1 = 9/(2L)$ and $\xi_2 = 9L/2$. Plots of ξ_1 and ξ_2 as a function of L (Fig. 13.29, left) indicate a general trend that we will see throughout this section. Specifically, efforts to decrease the variance of the state can only be achieved by increasing the variance of the manipulated variable. This can be seen more clearly in the parametric plot of ξ_1 vs. ξ_2 (Fig. 13.29, right).

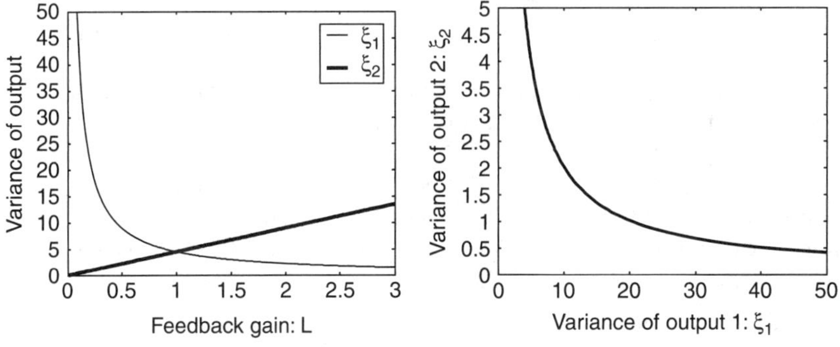

FIGURE. 13.29 Plots of output variance for Example 13.13.

The scalar notions of Example 13.13 extend quite naturally to the multidimensional case. Specifically, if we investigate the positive quadrant of the ξ_i space and ask if there exists a linear controller that will generate this set of variances, a Pareto frontier can be constructed as the boundary between controller existence and nonexistence. A Pareto frontier is a curve in the space of two or more objectives that define the points where one objective cannot be improved without worsening at least one of the others. Figure 13.30 shows this for the two-dimensional case. In the general case, the frontier will be a hypersurface in an n-dimensional space.

To characterize the Pareto frontier, one can define an optimization problem with an objective function constructed as the weighted sum of the ξ_i's. (In multicriteria optimization, this objective function is

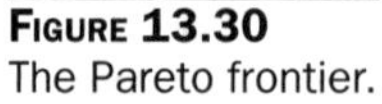

FIGURE 13.30 The Pareto frontier.

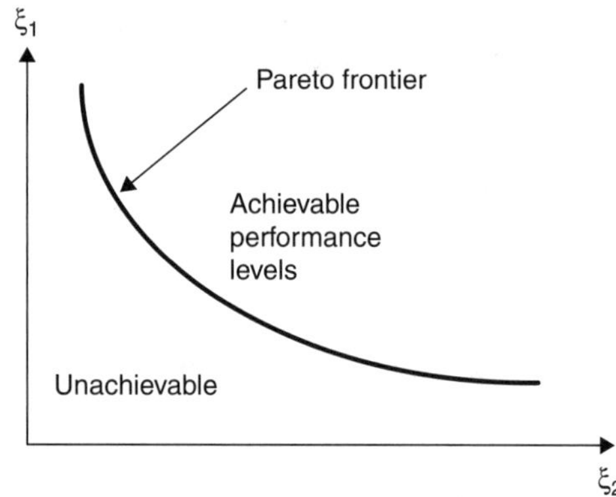

called a utility function.) Thus, the minimum variance control problem is defined as

$$\min_{\xi_i,\Sigma_z,\Sigma_x,L} \left\{ \sum_{i=1}^{n_z} d_i \xi_i \right\} \tag{13.82}$$

s.t.

$$(A + BL)\Sigma_x + \Sigma_x(A + BL)^T + G\Sigma_w G^T = 0 \tag{13.83}$$

$$\xi_i = \phi_i(D_x + D_u L)\Sigma_x(D_x + D_u L)^T \phi_i^T, \qquad i = 1 \ldots n_z \tag{13.84}$$

where d_i are weights intended to indicate the relative importance of each output signal. It is easily observed that the solution to Eq. (13.82) will lie on the Pareto frontier, and that the specific point of that solution is a function of the weights d_i. The Pareto interpretation is that all controllers generated by Eq. (13.82) are efficient in the sense that variance reduction in one direction can only be achieved by increasing the variance in another.

Example 13.14: Reconsider the pair of surge tanks of Example 13.5 with the disturbance model of Example 13.12; assume $\tau_\omega = 20$. Solutions to the minimum variance control problem [Eq. (13.82)] for various objective function weights are given in Table 13.2.

	$\xi_1^{1/2}$	$\xi_2^{1/2}$	$\xi_3^{1/2}$	$\xi_4^{1/2}$	Optimal Controller
$d_1 = d_2 =$ $d_3 = 1$ $d_4 = 1$	1.06	0.604	3.12	3.09	$L = \begin{bmatrix} 0.894 & -0.447 & 2.81 \\ 0.447 & 0.894 & 2.75 \end{bmatrix}$
$d_1 = d_2 =$ $d_3 = 1$ $d_4 = 10^4$	65.8	65.8	1.98	1.11	$L = \begin{bmatrix} 0.725 & -0.718 & 1.63 \\ 0.007 & 0.007 & 0.372 \end{bmatrix}$
$d_1 = d_2 =$ $d_3 = 1$ $d_4 = 10^8$	796	796	1.57	0.113	$L = \begin{bmatrix} 0.216 & -0.216 & 1.58 \\ 7\times10^{-5} & 7\times10^{-5} & 0.004 \end{bmatrix}$

TABLE 13.2 Solution to the Minimum Variance Control Problem for Example 13.14

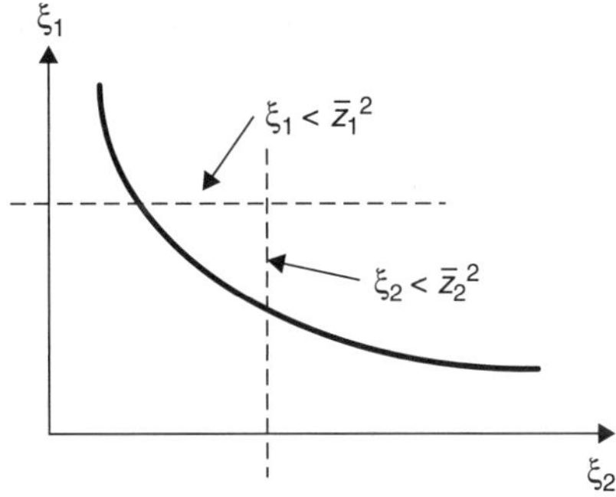

FIGURE 13.31 The constrained minimum variance control problem.

In many cases one would like to impose specific bounds on the variance of each output: $\xi_i < \bar{z}_i^2$. The result is the constrained minimum variance (CMV) control problem, defined as follows:

$$\min_{\xi_i, \Sigma_z, \Sigma_x, L} \left\{ \sum_{i=1}^{n_z} d_i \xi_i \right\} \tag{13.85}$$

s.t.

$$(A + BL)\Sigma_x + \Sigma_x(A + BL)^T + G\Sigma_w G^T = 0 \tag{13.86}$$

$$\xi_i = \phi_i(D_x + D_u L)\Sigma_x(D_x + D_u L)^T \phi_i^T, \qquad i = 1 \ldots n_z \tag{13.87}$$

$$\xi_i < \bar{z}_i^2 \qquad i = 1 \ldots n_z \tag{13.88}$$

In the context of Pareto efficiency the constrained problem is visualized in Fig. 13.31, where the feasible region is now a pie piece (above the frontier and less than the dashed lines) and again the solution will be found on the frontier. The important concept is that the imposition of variance bounds will override the intensions provided by the objective function weights, d_i. In fact, as shown in Chmielewski and Manthanwar (2004), these bounds can be thought of as modifying the weights so that a new unconstrained problem will have the same solution. For the control engineer, the constrained formulation and its reduced focus on weights is an advantage. Specifically, it is fairly easy to specify variability targets for most output signals, but the selection of objective function weights is, in most cases, a trial and error exercise.

Example 13.15: Reconsider the scenario of Example 13.14, with $d_1 = d_2 = d_3 = d_4 = 1$. If $\bar{z}_1 = \bar{z}_2 = 150$ and $\bar{z}_3 = 5$, then for various values of $\bar{z}_4$ the solution to the constrained minimum variance problem is presented in Table 13.3. The Pareto efficiency of the controller is clearly observed in the sense that ξ_1 and ξ_2 must increase as ξ_4 is forced to decrease.

	$\xi_1^{1/2}$	$\xi_2^{1/2}$	$\xi_3^{1/2}$	$\xi_4^{1/2}$	Optimal Controller
$\bar{z}_4 = 2.25$	14.3	14.2	2.64	2.249	$L = \begin{bmatrix} 0.726 & -0.687 & 2.12 \\ 0.039 & 0.041 & 1.34 \end{bmatrix}$
$\bar{z}_4 = 1.5$	40.6	40.6	2.20	1.499	$L = \begin{bmatrix} 0.712 & -0.700 & 1.77 \\ 0.013 & 0.014 & 0.640 \end{bmatrix}$
$\bar{z}_4 = 0.75$	109	109	1.79	0.7499	$L = \begin{bmatrix} 0.734 & -0.701 & 1.52 \\ 0.003 & 0.003 & 0.177 \end{bmatrix}$

TABLE 13.3 Solution to the Constrained Minimum Variance Control Problem for Example 13.15

While the Pareto interpretation is appealing with respect to optimization, the following EDOR visualization is more attractive from a process operations perspective. As indicated in Example 13.4, this visualization will be central to our subsequent definition of control system value. Consider the multidimensional elliptical region defined as z_o such that $z_o^T \Sigma_z^{-1} z_o < 1$, which indicates, generally, where one expects to find the trajectory $z(t)$. As indicated in Fig. 13.32, $\xi_i < \bar{z}_i^2$ defines a multidimensional constraint box, and the interpretation of Eq. (13.85) is that a feasible controller is one in which the resulting EDOR is contained in the constraint box. One can think of the EDOR as a multidimensional balloon that can be squeezed in one direction, but due to its Pareto efficiency must then expand in some other direction. However, this balloon squeezing can only occur if there exists a controller L such that Eqs. (13.63) and (13.64) are satisfied. Figure 13.33 provides the EDOR visualizations resulting from the controllers of Example 13.15.

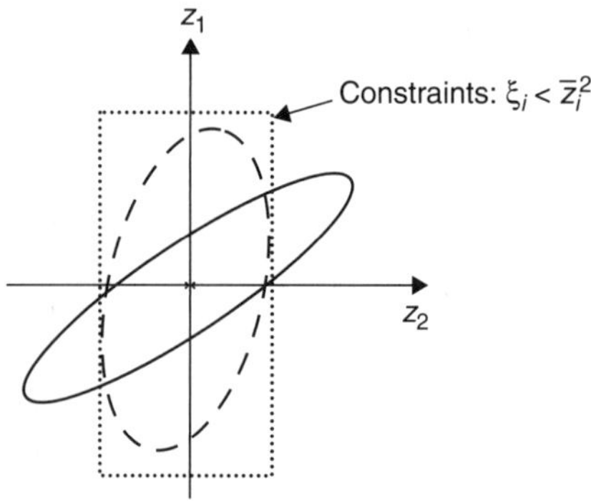

FIGURE 13.32 Feasible controllers for constrained minimum variance control problem (solid EDOR represents infeasible and dashed feasible).

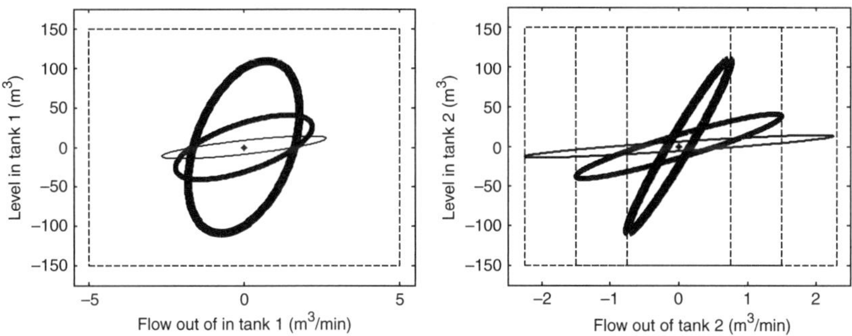

FIGURE 13.33 EDOR visualizations for Example 13.15.

Remark: The main computational concern with the CMV control problem, is that the variables Σ_x and L appear as nonlinear terms in the constraints. To alleviate this computational difficulty, the constraints of Eq. (13.85) can be converted to a convex form through the use of linear matrix inequalities (LMIs) (see Boyd et al., 1994 for an introduction to LMIs). We begin with the following theorem (Chmielewski and Manthanwar, 2004):

Theorem 13.1 There exists stabilizing $\Sigma_x \geq 0$, L and $\xi_i > 0$, $i = 1 \ldots n_z$ such that

$$(A+BL)\Sigma_x + \Sigma_x(A+BL)^T + G\Sigma_w G^T = 0$$

$$\xi_i = \phi_i(D_x + D_u L)\Sigma_x(D_x + D_u L)^T \phi_i^T, \qquad i = 1 \ldots n_z$$

$$\xi_i < \bar{z}_i^2 \qquad i = 1 \ldots n_z$$

if and only if there exists $X > 0$, L and $\xi_i > 0$, $i = 1 \ldots n_z$ such that

$$(AX+BY) + (AX+BY)^T + G\Sigma_w G^T < 0 \tag{13.89}$$

$$\begin{bmatrix} \xi_i & \phi_i(D_x X + D_u Y) \\ (D_x X + D_u Y)^T \phi_i^T & X \end{bmatrix} > 0 \qquad i = 1 \ldots n_z \tag{13.90}$$

$$\xi_i < \bar{z}_i^2, \qquad i = 1 \ldots n_z \tag{13.91}$$

Thus, Eq. (13.85) is exactly equivalent to the following convex optimization problem from which the global solution is readily obtained:

$$\min_{\xi_i, X, Y} \left\{ \sum_{i=1}^{n_z} d_i \xi_i \right\} \qquad \text{s.t.} \tag{13.92}$$

$$(AX+BY) + (AX+BY)^T + G\Sigma_w G^T < 0$$

$$\begin{bmatrix} \xi_i & \phi_i(D_x X + D_u Y) \\ (D_x X + D_u Y)^T \phi_i^T & X \end{bmatrix} > 0 \qquad i = 1 \ldots n_z$$

$$\text{and } \xi_i < \bar{z}_i^2 \qquad i = 1 \ldots n_z$$

If X^*, Y^* is the solution to Eq. (13.92), then the solution to Eq. (13.85) is reconstructed as $\Sigma_x = X^*$ and $L = Y^*(X^*)^{-1}$.

The PSI version of the constrained minimum variance control problem is

$$\min_{\xi_i,\Sigma_{\hat{x}},L}\left\{\sum_{i=1}^{n_z} d_i\xi_i\right\} \quad \text{s.t.} \tag{13.93}$$

$$(A+BL)\Sigma_{\hat{x}}+\Sigma_{\hat{x}}(A+BL)^T+\Sigma_c C^T\Sigma_v^{-1}C\Sigma_e=0$$

$$\xi_i=\phi_i(D_x+D_uL)\Sigma_{\hat{x}}(D_x+D_uL)^T\phi_i^T+\phi_iD_x\Sigma_eD_x^T\phi_i^T$$

$$\xi_I<\bar{z}_i^2,\qquad i=1\ldots n_z$$

where Σ_e is precalculated via Eq. (13.72). The following is a simple application of Theorem 13.1.

Theorem 13.2 There exists stabilizing $\Sigma_{\hat{x}}\geq 0$, L and $\xi_i>0$, $i=1\ldots n_z$ such that

$$(A+BL)\Sigma_{\hat{x}}+\Sigma_{\hat{x}}(A+BL)^T+\Sigma_e C^T\Sigma_v^{-1}C\Sigma_e=0$$

$$\xi_i=\phi_i(D_x+D_uL)\Sigma_{\hat{x}}(D_x+D_uL)^T\phi_i^T+\phi_iD_x\Sigma_eD_x^T\phi_i^T \qquad i=1\ldots n_2$$

$$\xi_i<\bar{z}_i^2 \qquad i=1\ldots n_z$$

if and only if there exists $X>0$, L and $\xi_i>0$, $i=1\ldots n_z$ such that

$$(AX+BY)+(AX+BY)^T+\Sigma_eC^T\Sigma_v^{-1}C\Sigma_e<0 \tag{13.94}$$

$$\begin{bmatrix}\xi_i-\phi_iD_x\Sigma_eD_x^T\phi_i^T & \phi_i(D_xX+D_uY)\\ (D_xX+D_uY)^T\phi_i^T & X\end{bmatrix}>0 \qquad i=1\ldots n_z \tag{13.95}$$

$$\xi_i<\bar{z}_i^2 \qquad i=1\ldots n_z \tag{13.96}$$

Thus, Eq. (13.93) is exactly equivalent to the following:

$$\min_{\xi_i,X,Y}\left\{\sum_{i=1}^{n_z} d_i\xi_i\right\} \quad \text{s.t.} \tag{13.97}$$

$$(AX+BY)+(AX+BY)^T+\Sigma_eC^T\Sigma_v^{-1}C\Sigma_e<0$$

$$\begin{bmatrix}\xi_i-\phi_iD_x\Sigma_eD_x^T\phi_i^T & \phi_i(D_xX+D_uY)\\ (D_xX+D_uY)^T\phi_i^T & X\end{bmatrix}>0 \qquad i=1\ldots n_z$$

$$\xi_i<\bar{z}_i^2 \qquad i=1\ldots n_z$$

If X^*, Y^* is the solution to Eq. (13.97), then the solution to Eq. (13.93) is reconstructed as $\Sigma_{\hat{x}}=X^*$ and $L=Y^*(X^*)^{-1}$.

Within the discrete-time framework, the CMV control problem is very similar; the objective function is identical and the constraints are due to discrete-time covariance analysis. The discrete-time versions of Theorems 13.1 and 13.2 are given in Chmielewski and Manthanwar (2004), Theorems 6.1 and A.1, respectively.

Connection between CMV Control and MPC

We now return to the question of replacing the complexity of MPC with a simple linear feedback array. In particular, we illustrate that the CMV problem appropriately selects a linear feedback, $u = Lx$, such that it can be used as a surrogate for MPC.

In Chmielewski and Manthanwar (2004), it is shown that both Eqs. (13.82) and (13.85) are equivalent to the linear quadratic gaussian (LQG) optimal control problem in the sense that Eq. (13.82) and (13.85) will generate a controller that is the solution to some LQG problem. In fact, for Eq. (13.82) one can directly specify the LQG problem as one with objective function weights equal to

$$Q = D_x^T D D_x \quad R = D_u^T D D_u \quad \text{and} \quad M = D_x^T D D_u \tag{13.98}$$

where $D = \sum_{i=1}^{n_z} d_i \phi_i^T \phi_i$. In this case [and with certainty equivalence connecting LQG with its deterministic counterpart the linear quadratic regulator (LQR)], the solution to Eq. (13.82) is $L = -R^{-1}(PB + M)^T$ where P is the positive definite solution to the algebraic Riccati equation:

$$A^T P + PA + Q - (PB + M)R^{-1}(PB + M)^T = 0 \tag{13.99}$$

In the case of Eq. (13.85), the story is similar in that there exists weights Q, R, and M for an LQR objective function such that solution via the Riccati equation will generate a controller identical to that produced by Eq. (13.85). However, the difference is that formulas similar to Eq. (13.98) do not exist. To obtain the LQR weights one would need the solution to Eq. (13.85) and then employ the following inverse optimality result (Chmielewski and Manthanwar, 2004). The utility of such weights will become evident shortly.

Theorem 13.3 If there exists $P > 0$ and $R > 0$ such that

$$\begin{bmatrix} L^T RL - A^T P - PA & -(L^T R + PB) \\ (L^T R + PB)^T & R \end{bmatrix} > 0 \tag{13.100}$$

then $Q = L^T RL - A^T P + PA$ and $M = -(L^T R + PB)$ will be such that

$$\begin{bmatrix} Q & M \\ M^T & R \end{bmatrix} > 0 \tag{13.101}$$

and P and L satisfy

$$A^T P + PA + Q - (P + M)R^{-1}(PB + M)^T = 0 \tag{13.102}$$

$$L = -R^{-1}(PB + M)^T \tag{13.103}$$

The discrete-time version of Theorem 13.3 can be found in Chmielewski and Manthanwar (2004).

Example 13.16: In the $d_1 = d_2 = d_3 = d_4 = 1$ case of Example 13.14, the solution to the minimum variance control problem is generated by the controller

$$L = \begin{bmatrix} 0.894 & -0.447 & 2.81 \\ 0.447 & 0.894 & 2.75 \end{bmatrix} \tag{13.104}$$

Using Eq. 13.98, it is found that

$$Q=\begin{bmatrix}1 & 0 & 0\\0 & 1 & 0\\0 & 0 & 0\end{bmatrix} \qquad R=\begin{bmatrix}1 & 0\\0 & 1\end{bmatrix} \qquad M=\begin{bmatrix}0 & 0\\0 & 0\\0 & 0\end{bmatrix} \tag{13.105}$$

Using Theorem 13.3,

$$Q=\begin{bmatrix}0.849 & -0.723 & -0.541\\-0.723 & 2.48 & 0.607\\-0.541 & 0.607 & 7.73\end{bmatrix} \tag{13.106}$$

$$R=\begin{bmatrix}1.75 & -1.09\\-1.09 & 1.58\end{bmatrix} \qquad M=\begin{bmatrix}-0.016 & 0.156\\-0.059 & -0.192\\-1.57 & -1.39\end{bmatrix} \tag{13.107}$$

However, in both cases, placement of Q, R, and M into Eq. (13.99) will recover the controller of Eq. (13.104). In the case of $\overline{z}_4 = 0.75$ of Example 13.15, then

$$Q=\begin{bmatrix}161 & -160 & -34.4\\-160 & 160 & 35.2\\-34.4 & 35.2 & 629\end{bmatrix} \tag{13.108}$$

$$R=\begin{bmatrix}299 & -81.8\\-81.8 & 444\end{bmatrix} \qquad M=\begin{bmatrix}20.2 & -60.8\\-20.8 & 58.8\\-395 & 28.8\end{bmatrix} \tag{13.109}$$

The connection between MPC and the CMV controller is twofold. First there are the constraints. If a time-domain bound $z_i(t) < \overline{z}_i$ is given, then the equivalent statistical constraint is $\xi_i < \overline{z}_i^2$. In this case, the CMV controller would give an 84% observance of the time domain bound (or 68%, if the bound is $|z_i(t)| < \overline{z}_i$). If greater confidence is desired, then one may enforce $\xi_i < (\overline{z}_i/\alpha_i)^2$, where $\alpha_i = 2$ will give a confidence of 98%. The net result is that the CMV controller is predisposed (or tuned) to avoid constraint violations.

Then there is the MPC objective function (which is identical to the LQR). As stated above, every CMV controller has a set of corresponding LQR objective function weights. Given that MPC is just LQR with constraints, it is easily concluded that an MPC implementation with no time-domain constraints and weights generated by CMV exactly regenerates the CMV controller (which as concluded in the previous paragraph is predisposed to avoid constraints). If we then reimpose the time domain constraints, the resulting MPC will possess an alignment between its objective function and its constraints. While this sort of alignment has important implications for MPC tuning, the more relevant conclusion is that the CMV controller can be used as a surrogate for the constraint enforcement ability of MPC. In essence, one can think of CMV as "pseudo-constrained MPC" in that constraints are not strictly enforced, but the tuning suggests the controller is aware of their existence and attempts to avoid them.

To summarize, consider the conceptual progression of the controllers just discussed. First, there is the LQG controller [which is essentially equal to Eq. (13.82), the minimum variance controller], where direct consideration of inequality constraints is not allowed and an arbitrary selection of objective function weights will result in arbitrary placement on the Pareto frontier. Then there is MPC in which constraint enforcement is central, but arbitrary selection of weights may result in misalignment between the objective and the constraints. Finally, there is the CMV controller which

- Guides placement on the Pareto frontier through the enforcement of statistical constraints
- Provides a means to tune MPC (creating alignment)
- Serves as an analytic surrogate for MPC in the subsequent analysis

Control System Value

As indicated in the control system hierarchy section, the value of a control system stems from its ability to operate within process constraints. More specifically, if the control system can mitigate the impact of disturbances and reduce the size of the EDOR, then one can expect increased profit by enabling process operation at a point closer to process limitations (see Fig. 13.34). In this section, we outline methods for selecting an operating point close to (or backed-off from) the OSSOP, while ensuring the EDOR is contained within the constraint set.

Remark: The notion of bringing economic motives to control system design is clearly no new concept. Consider for example the following from Morari et al. (1980) "... our main objective is to translate the economic objectives into process control objectives." Additional efforts to incorporate plant economics into control system design include: Miletic and Marlin (1998), Edgar (2004), Skogestad (2000, 2004), Aske et al. (2008), Engell, (2007), Bauer and Craig (2008) and Rawlings and Amrit (2009). The family of papers addressing Backed-off Operating Point (BOP) selection serves as a foundation to the methods advocated in this chapter. (Our definition of control system value is simply an extension of these important papers).

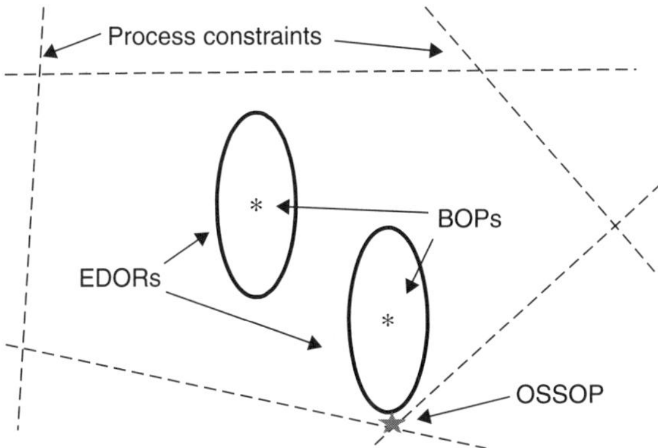

FIGURE 13.34 Economics of operating point selection.

Under the steady-state disturbance assumption, Bahri et al. (1996) consider an open-loop configuration and Contreras-Dordelly and Marlin (2000) assume perfect control. Similar steady-state versions of the BOP selection problem include; Loeblein and Perkins (1996, 1998); and Zhang and Forbes (2000). BOP selection methods based on step disturbance models include Bahri et al., (1995); Young et al., (1996); Figueroa et al., (1996); Figueroa, (2000), Figueroa and Desages, (2003); Arbiza et al., (2003); Biagiola et al., (2004); Young et al., (2004) and Soliman et al., (2008). BOP selection methods based on stochastic disturbance models include Loeblein and Perkins, (1999), Muske, (2003), Peng et al., (2005), Lee et al., (2008), Zhao et al., (2009) and Akande et al, (2009).

The following development of BOP selection is based on the stochastic disturbance perspective of the previous sections. Consider the original nonlinear process model.

$$\dot{s} = f(s,m,p) \tag{13.110}$$

$$q = h(s,m,p) \tag{13.111}$$

where s, m, p, and q are the state, manipulated, disturbance, and performance variables corresponding to the deviation variables x, u, w, and z. The RTO will use a steady-state version of this model to determine the OSSOP, based on a certain profit function $g(q)$ and operating constraints $q_i^{\min} < q_i < q_i^{\max}$. In much of the BOP selection literature, the nonlinear model is used directly. In other cases, linearization is used, which we analyze next.

The profit lost due to back-off can be approximated by linearizing the profit function

$$g(q^{\text{bop}}) \cong g(q^{\text{ossop}}) + g_q^{\,T} q' \tag{13.112}$$

where $q' = q^{\text{bop}} - q^{\text{ossop}}$ and the partial derivative g_q is evaluated at the OSSOP. Equation (13.112) is the generalization of Eq. (13.10), from the first section of the chapter. The set of available steady-state operating conditions is also approximated via linearization: $0 = As' + Bm' + Gp'$, $q' = D_x s' + D_u m' + D_w p'$, and $q_i'^{\min} < q_i' < q_i'^{\max}$, where A, B, G, D_x, D_u, D_w are the appropriate partials derivatives. The distance between the BOP and the constraints is given by $q_i' - q_i'^{\min}$ on one side and $q_i'^{\max} - q_i'$ on the other. If we now assume a stochastic disturbance perspective, as done in the earlier parts of the chapter, then this distance between the BOP and the constraints must be less than $\xi_i^{1/2}$ to ensure the EDOR is within the constraint set. Additionally, due to the assumption of zero-mean disturbances, $p' = 0$.

A conclusion of the previous literature review is that most BOP selection schemes assume the controller is known a priori, in which case the parameters ξ_i can be precalculated [using Eqs. (13.63) and (13.64)]. Under this assumption, the optimal BOP can be selected via the following linear program (Loeblein and Perkins, 1999):

$$\min_{s',m',q'} \left\{ g_q^T q \right\} \quad \text{s.t.} \tag{13.113}$$

$$0 = As' + Bm' \tag{13.114}$$

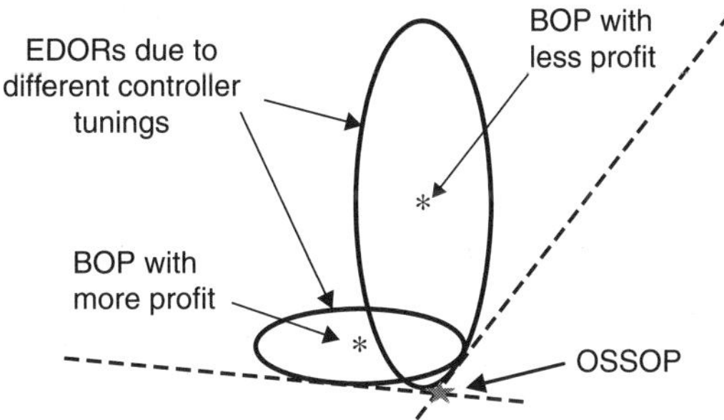

Figure 13.35 Simultaneous BOP selection and controller design.

$$q_i' = \phi_i (D_x s' + D_u m') \tag{13.115}$$

$$q_i'^{\min} < q_i' < q_i'^{\max} \tag{13.116}$$

$$\xi_i^{1/2} < q_i'^{\max} - q_i' \tag{13.117}$$

$$\xi_i^{1/2} < q_i' - q_i'^{\min} \tag{13.118}$$

An alternative is to design the controller while simultaneously selecting the BOP. The result (see Fig. 13.35) is the ability to move the BOP as well as change the shape of the EDOR (through modification of the controller). *This is a significant break from traditional back-off analysis.* Rather than use a given controller to determine back-off, the economic foundation of back-off analysis is used to guide controller design.

Under this assumption, the optimal BOP can be selected via the following nonlinear program (Peng et al., 2005).

$$\min_{\substack{s',m',q' \\ \xi_i, X>0, Y}} \left\{ g_q^T q \right\} \quad \text{s.t.} \tag{13.119}$$

$$0 = As' + Bm' \tag{13.120}$$

$$q_i' = \phi_i (D_x s' + D_u m') \tag{13.121}$$

$$q_i'^{\min} < q_i' < q_i'^{\max} \tag{13.122}$$

$$\xi_i^{1/2} < q_i'^{\max} - q_i' \tag{13.123}$$

$$\xi_i^{1/2} < q_i' - q_i'^{\min} \tag{13.124}$$

$$(AX + BY) + (AX + BY)^T + G\Sigma_w G^T < 0 \tag{13.125}$$

$$\begin{bmatrix} \xi_i & \phi_i (D_x X + D_u Y) \\ (D_x X + D_u Y)^T \phi_i^T & X \end{bmatrix} > 0 \tag{13.126}$$

Remark: The method of Zhao et al., (2008) is similar to Eq. (13.119). In addition to using an input-output model of the plant, the main difference is that the convex inequalities of Eq. (13.125) and (13.126) are replaced by a nonlinear equality constraint. This nonlinear equality constraint is based on the LQG benchmarking approach of Huang and Shah (1999).

The PSI version of Eq. (13.119) is achieved by replacing constraint Eqs. (13.125) and (13.126) with the following:

$$(AX+BY)+(AX+BY)^T+\Sigma_e C^T \Sigma_v^{-1} C\Sigma_e < 0 \tag{13.127}$$

$$\begin{bmatrix} \xi_i - \phi_i D_x \Sigma_e D_x^T \phi_i^T & \phi_i(D_x X + D_u Y) \\ (D_x X + D_u Y)^T \phi_i^T & X \end{bmatrix} > 0 \tag{13.128}$$

While the constraint Eqs. (13.125) and (13.126) [or Eqs. (13.127) and (13.128)] are nonlinear, they are convex and pose no computational challenge. However, in contrast to Eq. (13.113), the addition of ξ_i's as optimization variables causes the constraint Eqs. (13.123) and (13.124) to become reverse-convex and pose significant computational challenge. In Peng et al., (2005), a global solution procedure based on the branch and bound algorithm is provided. Using this method the following examples were solved in under a minute.

Example 13.17: We consider the mass spring damper of Fig. 13.36. While this is not strictly a chemical process, it is simple enough to illustrate procedural issues. The economic objective is to maximize the nominal value of r, given the constraint set $-1 = r_{\min} \le r \le r_{\max} = 1$ and $0 = f_{\min} \le f \le f_{\max} = 15$ in the face of a zero-mean white noise disturbance, w, with variance $\Sigma_w = 10$. The original system model is $\dot{r} = v$, $\dot{v} = -2v - 3r + f + w + 9.8$. The RTO solution is $r_{OSSOP} = 1$, $v_{OSSOP} = 0$, and $f_{OSSOP} = 12.8$. The steady-state model with respect to the OSSOP is $f' = 3r'$. The above BOP selection scenario is depicted in Fig. 13.37.

Then, the form of Eq. (13.119) is arrived at by defining

$$A = \begin{bmatrix} 0 & 1 \\ -3 & -2 \end{bmatrix} \quad B = \begin{bmatrix} 0 \\ 1 \end{bmatrix} \quad G = \begin{bmatrix} 0 \\ 1 \end{bmatrix} \tag{13.129}$$

$$D_x = \begin{bmatrix} 1 & 0 \\ 0 & 0 \end{bmatrix} \quad D_u = \begin{bmatrix} 0 \\ 1 \end{bmatrix} \quad D_w = \begin{bmatrix} 0 \\ 0 \end{bmatrix} \tag{13.130}$$

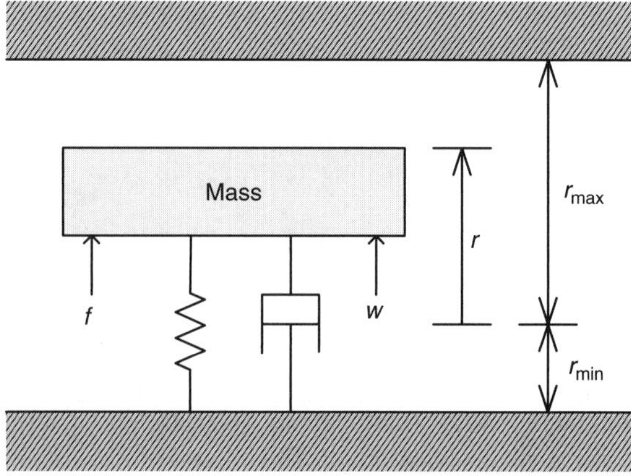

Figure 13.36 Mass-spring damper of Example 13.17.

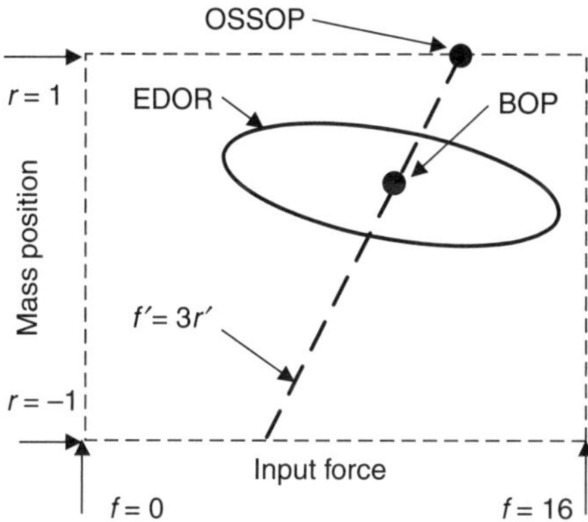

Figure 13.37 BOP selection scenario for Example 13.17.

The resulting optimization problem is

$$\min_{\substack{r',f',q_1',q_2' \\ \xi_1,\xi_2,X>0,Y}} \{-q_1\} \qquad \text{s.t.} \tag{13.131}$$

$$f' = 3r'$$
$$q_1' = r',\ q_2' = f'$$
$$-2 < q_1' < 0,\ -12.8 < q_2' < 2.2$$
$$\xi_1^{\ 1/2} < -q_1',\ \xi_1^{\ 1/2} < q_1' + 2$$
$$\xi_2^{\ 1/2} < 2.2 - q_2',\ \xi_2^{\ 1/2} < q_2' + 12.8$$
$$(AX + BY) + (AX + BY)^T + G\Sigma_w G^T < 0$$
$$\begin{bmatrix} \xi_1 & \phi_1 X \\ X^T \phi_1^{\ T} & X \end{bmatrix} > 0,\ \begin{bmatrix} \xi_2 & Y \\ Y^T & X \end{bmatrix} > 0$$

The solution to Eq. (13.131) is given in Fig. 13.38 (top) denoted as the FSI case. For the partial state information case of Fig. 13.38 (top), it is assumed that one velocity sensor is available (corrupted by white noise with zero mean and a standard deviation of 0.5 m/s).

In Fig. 13.38 (bottom) the impact of modifying the constraint set is illustrated. Case A is the same as the FSI case of Fig. 13.38 (top). Case B is the same as case A, but the max force is changed from 15 to 18. Case C is the same as case A, but the min force is changed from 0 to 9.5. These variations clearly indicate the complex relation between the EDOR, the steady-state model, the constraint set, and the economics of the process.

Example 13.18: Consider the furnace reactor process of Fig. 13.39, described by the following system matrices: $s = [T_F\ T_R\ C_{O2}\ C_{CO}]$, $m = [F_{\text{feed}}\ F_{\text{fuel}}\ P_v]$, $p = [T_0]$.

$$A = 1000^* \begin{bmatrix} -8 & 0 & 0 & 0 \\ 2 & -1.5 & 0 & 0 \\ 0 & 0 & -5 & 0 \\ 0 & 0 & 0 & -5 \end{bmatrix}$$

$$B = \begin{bmatrix} -75 & 75{,}000 & 0 \\ -25 & 0 & 0 \\ 0 & -8500 & 8.5^* 10^5 \\ 0 & 0 & -5^* 10^7 \end{bmatrix} \qquad G = \begin{bmatrix} 10{,}000 \\ 0 \\ 0 \\ 0 \end{bmatrix}$$

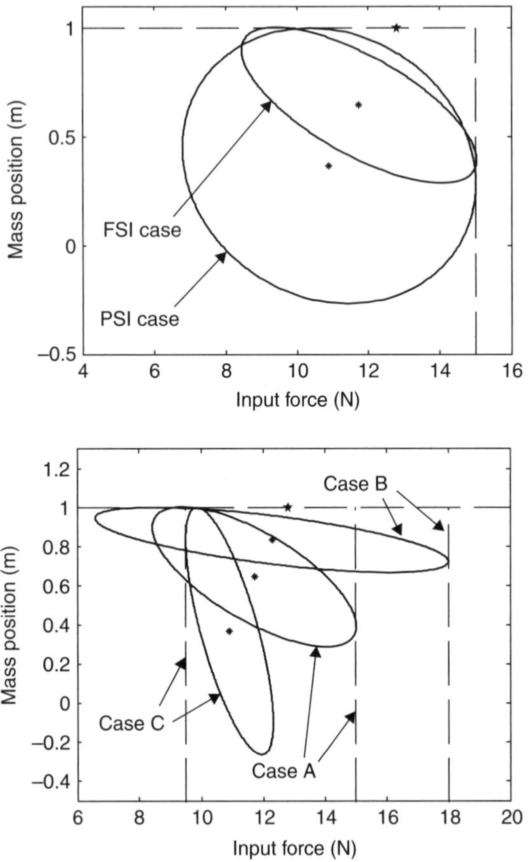

FIGURE 13.38 Solution to the BOP selection problem of Example 13.17.

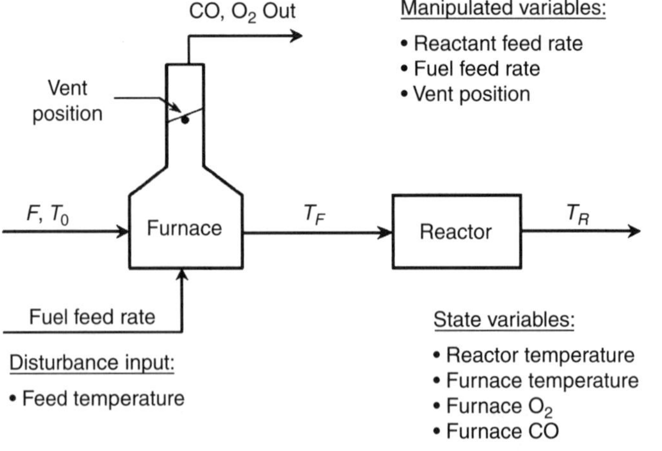

FIGURE 13.39 Process schematic for Example 13.18.

$D_x = [I_{4\times4} \,|\, 0]^T$; $D_u = [0 \,|\, I_{3\times3}]^T$, and $D_w = 0$. If $\Sigma_w = (0.13975)^2$ and the economic objective is $g_q^T = [0\ 0\ 0\ 0.01\ -10\ 30\ 0]$, then the optimal BOP selection is given in Fig. 13.40. (The constraint set is also indicated in Fig. 13.40. The PSI case assumes availability of a reactor temperature measurement, corrupted by white noise with zero mean and standard deviation of 0.1°F.) In this seven-dimensional problem, we clearly see that the EROR shape (and thus the LQG controller) is being modified so as to minimize the distance between the BOP and the OSSOP. We also note that reactant feed rate (of the center right plot) holds the greatest economic value, which causes the scheme to generate a controller (and EDOR) with zero back-off in that key direction.

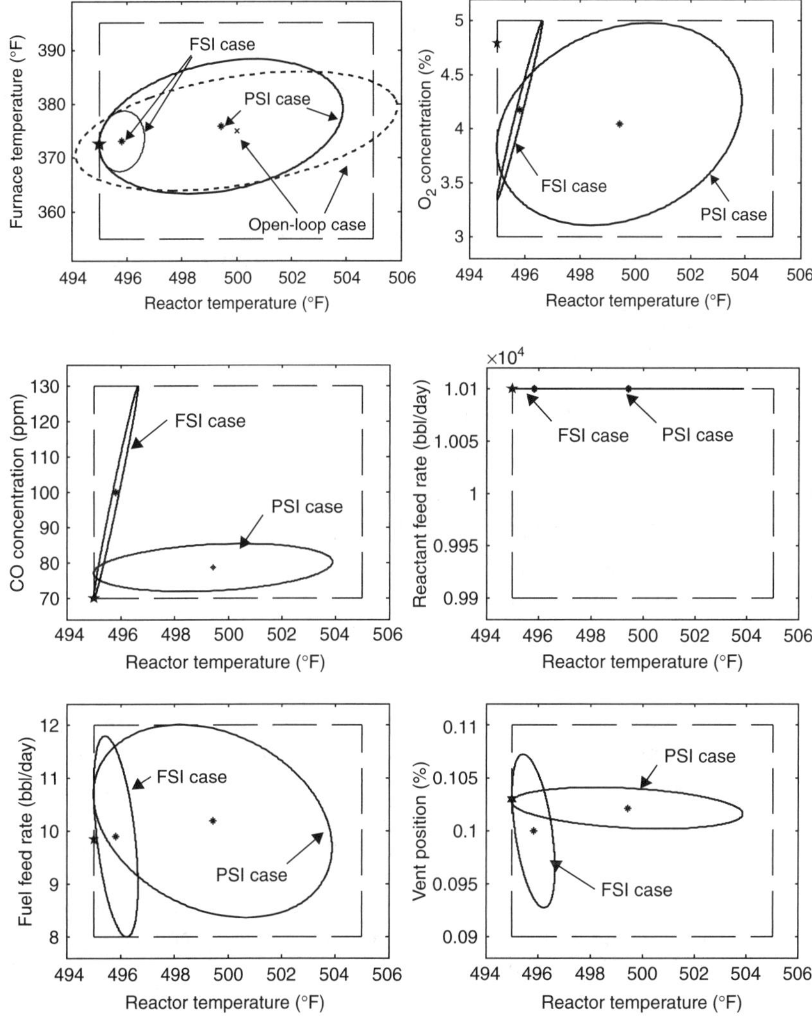

FIGURE 13.40 BOP selection solution for Example 13.18.

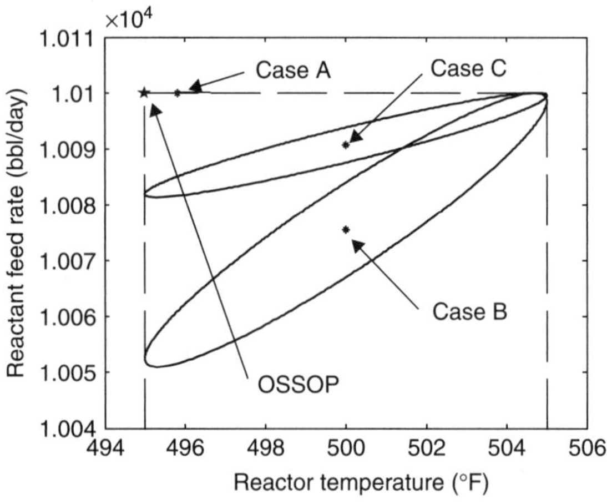

FIGURE 13.41 BOP selection solution for Example 13.18.

In Fig. 13.41, the impact of modifying the constraint set is illustrated. Case A is the same as the FSI case of Fig. 13.40. Case B is the same as case A, but the fuel feed bound is changed to 10 ± 0.25. Case C is the same as case A, but the O_2 concentration bound changed to from 3 to 4%. These modifications clearly indicate that the economics and constraint environment of the situation are having a direct influence on the tuning of the controller.

Impact of Process and Measurement Biases

The notion of bias is that the process disturbance (or equally measurement noise) is no longer assumed to be zero-mean. Implicit in this definition is the fact that the value of this nonzero mean is unknown, as it was discussed in several previous chapters. Otherwise, if it is known (directly or through gross error size estimation) one could simply incorporate this known nonzero mean into the disturbance nominal value, and then within the deviation variable model, the resulting deviation variable disturbance would again be zero-mean.

Additionally, the notion of bias suggests that this (unknown) nonzero mean disturbance is slowly varying, and in many cases is assumed to be a time-independent constant (see Chap. 2). Concerning the unknown aspect of this constant, it is usually assumed to be a random variable with known statistics (frequently a zero-mean gaussian with a given value for variance).

Consider, for example, a parameter estimation schemes used to update the RTO (see Fig. 13.19). In this case, the estimation error will be a potential bias, and the variance of this estimation error can be used to characterize the bias. The mean of the bias is the actual parameter estimate, which is subtracted out in the subsequent conversion to deviation variables. In many other cases (especially measurement bias), it may be difficult to identify a truly meaningful value for bias variance. However, if one employs a bias-detection algorithm, then (given statistics about the other parts of the process) one can calculate the maximum variance of an undetectable bias. This is covered in previous chapters in detail and therefore, in the following it is assumed that one can identify bias variance. Under the assumption of no bias-detection algorithm, the state estimator/control policy to be proposed below can serve to minimize the impact of biases (i.e., may substitute for bias detection and estimation).

Specifically, the undetectable biases will put additional uncertainty into the EDOR calculation and thus increase its size, which as described in the previous sections will serve to reduce the expected level of profit.

Unfortunately, the computational machinery described in the previous sections is not amenable to a nonstochastic description of bias. As such we will assume all biases are random processes but extremely slow varying. Recall the two parameter, scalar disturbance model of Example 13.10. The following is a trivial modification of that model (mostly to remove the unnecessary second equation):

$$\dot{\omega}_b = a_{\omega b}\, \omega_b + g_{\omega b}\, w_{0b} \tag{13.132}$$

where $a_{\omega b} = -1/\tau_{\omega b}$, $g_{\omega b} = (2\Sigma_{\omega b}/\tau_{\omega b})^{1/2}$, and $\Sigma_{w_{0b}} = 1$. It is easily verified that the variance of ω_b is equal to $\Sigma_{\omega b}$ for all values of $\tau_{\omega b}$. Thus, if one selects $\tau_{\omega b}$ to be a very large value (in comparison to the largest time constant of the original process/disturbance model), then ω_b will have the characteristics of an extremely slow varying random process. If, in addition, one selects $\Sigma_{\omega b}$ to be equal to the bias variance, then ω_b will be a good representation of the above notion of bias. For the case of multiple biases, Eq. (13.132) is generalized to

$$\dot{\omega}_b = A_{\omega b}\, \omega_b + G_{\omega b}\, w_{0b} \tag{13.133}$$

where $A_{\omega b}$ and $G_{\omega b}$ are diagonal matrices. The vector size of ω_b and w_{0b} is equal to the number of biases.

In the case of only process biases, the bias signals will enter into the original process/disturbance model in the following way:

$$\dot{x} = Ax + Bu + Gw + G_b\omega_b \tag{13.134}$$

$$y = Cx + v \tag{13.135}$$

$$z = D_x x + D_u u + D_w w \tag{13.136}$$

We can then augment this model with Eq. (13.133) to arrive at the following compound system:

$$\begin{bmatrix} \dot{x} \\ \dot{\omega}_b \end{bmatrix} = \begin{bmatrix} A & G_b \\ 0 & A_{\omega b} \end{bmatrix} \begin{bmatrix} x \\ \omega_b \end{bmatrix} + \begin{bmatrix} B \\ 0 \end{bmatrix} u + \begin{bmatrix} G & 0 \\ 0 & G_{\omega b} \end{bmatrix} \begin{bmatrix} w \\ w_{0b} \end{bmatrix} \tag{13.137}$$

$$y = \begin{bmatrix} C & 0 \end{bmatrix} \begin{bmatrix} x \\ \omega_b \end{bmatrix} + v \tag{13.138}$$

$$z = \begin{bmatrix} D_x & 0 \end{bmatrix} \begin{bmatrix} x \\ \omega_b \end{bmatrix} + D_x u \tag{13.139}$$

Given this augmented system, it is a simple matter to apply the PSI version of Eq. (13.119) to determine control system value in the presence of process biases. It is additionally noted that the state estimator portion of Eq. (13.119) will attempt to estimate ω_b. If the size of the bias is sufficient, then a reasonable estimate will be determined and the detectable portion will be compensated for in the resulting controller. If, however, the bias size is sufficiently small, or undetectable, (as would be the case if a bias detecting pre-processing unit was employed), then the bias will be estimated to be about zero and thus will be uncompensated. However, in both cases, the uncompensated portion will appear in the calculation of EDOR size and result in a larger EDOR, greater back-off, and thus lower profit, as illustrated in Fig. 13.42.

In the case of only measurement biases, the bias signals will enter into the original process/disturbance model in the following way:

$$\dot{x} = Ax + Bu + Gw \tag{13.140}$$

$$y = Cx + \nu + G_b\omega_b \tag{13.141}$$

$$z = D_x x + D_u u + D_w w \tag{13.142}$$

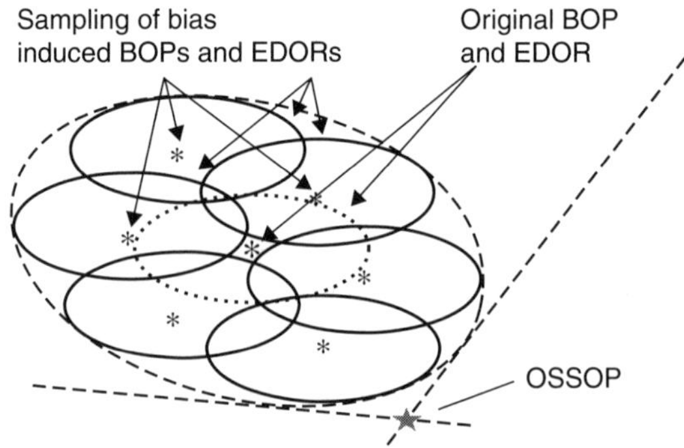

FIGURE 13.42 EDOR resulting from uncompensated biases.

where G_b is the identity matrix if all measurements are potentially biased. Augmentation with Eq. (13.133) results in the following compound system:

$$\begin{bmatrix} \dot{x} \\ \dot{\omega}_b \end{bmatrix} = \begin{bmatrix} A & 0 \\ 0 & A_{\omega b} \end{bmatrix} \begin{bmatrix} x \\ \omega_b \end{bmatrix} + \begin{bmatrix} B \\ 0 \end{bmatrix} u + \begin{bmatrix} G & 0 \\ 0 & G_{\omega b} \end{bmatrix} \begin{bmatrix} w \\ w_{0b} \end{bmatrix} \tag{13.143}$$

$$y = [C \quad G_b] \begin{bmatrix} x \\ \omega_b \end{bmatrix} + v \tag{13.144}$$

$$z = \begin{bmatrix} D_x & 0 \end{bmatrix} \begin{bmatrix} x \\ \omega_b \end{bmatrix} + D_x u \tag{13.145}$$

Once again, given this augmented system it is a simple matter to apply the PSI version of Eq. (13.119) to determine control system value in the presence of process biases. In this case of measurement bias, it is noted that the state estimator will find it difficult to distinguish between the true measurement and the bias of that measurement. Consider the case of $C = G_b = I$ and the first measurement $y_1 = x_1 + \omega_{b1} + v_1$. In this case, the linear algebraic argument of state observability suggest that x_1 and ω_{b1} cannot be distinguished. However, within the state estimator, which will consider the autocorrelation characteristics of x_1 and ω_{b1}, such distinction may be possible. The most relevant impact being that a solution to the Kalman filter Riccati equation is expected to exist, which is critical to the implementation of the PSI version of Eq. (13.119). Thus, if v_i is sufficiently small and the characteristics of x_i and ω_{bi} are sufficiently different, then the estimator may be able to distinguish between the two, and the controller will be able to partially compensate for the bias. However, if v_i is sufficiently large, then the bias will go uncompensated. In both cases, the uncompensated portion will again appear in the calculation of EDOR size and result in a larger EDOR, greater back-off, and thus lower profit.

Conclusions

In this chapter we have discussed two main concepts. The first is that an increased concern over plant economics has created an incentive for operation closer to the physical and safety limitations of the plant. As such, modern controller design methods must be aware of these limitations as well as the economic impact of not being able to satisfy all inequality constraints. In this case of an inability to meet all constraints, the design method must reselect the process operating condition such that all plant limitations can be observed. However, since there are numerous operating conditions that can achieve this goal, such a procedure should be guided by the economics of the plant so as to select the least costly alternative. In this chapter we have advocated RTO back-off as a measure of economic impact and CMV control as a design method to reflect these considerations within the controller.

The second main concept is the impact of disturbance characteristics on control system performance. Specifically, a change in disturbance characteristics (due to changes in the process or upstream units) will impact the controller's ability to satisfy the desired process limitations. Thus, a controller to be used within a smart plant (a smart controller) must be able to identify a change in disturbance characteristic and then redesign (or retune) itself such that physical and safety limitations continue to be observed and plant profit continues to be maximized.

Finally, this chapter assumes the instrumentation as already selected. To complete the discussion, Chap. 16 will present an example where the selection of instruments is performed to achieve an economical optimum of profit minus cost.

References

Akande, S., B. Huang, and K. H. Lee, "MPC Constraint Analysis-Bayesian Approach via a Continuous-Valued Profit Function," *Ind. Eng. Chem. Res.* **48**(8):3944–3954 (2009).

Arbiza, M., J. Bandoni, and J. Figueroa, "Use of Back-Off Computation in Multilevel MPC," *Lat. Am. App. Res.* **33**:251–256 (2003).

Aske, E. M. B., S. Strand, and S. Skogestad, "Coordinator MPC for Maximizing Plant Throughput," *Comp. Chem. Eng.* **32**(1–2):195–204 (2008).

Badgwell, T.A., *Personal communication* (2008).

Bahri, P., J. Bandoni, G. Barton, and J. Romagnoli, "Back-Off Calculations in Optimizing Control: A Dynamic Approach," *Comp. Chem. Eng.* **19**:S699–S708 (1995).

Bahri, P., J. Bandoni, and J. Romagnoli, "Effect of Disturbances in Optimizing Control: Steady-State Open-Loop Backoff Problem," *AIChE J.* **42**(4):983–994 (1996).

Bauer, M., and Craig I. K., "Economic Assessment of Advanced Process Control—A Survey and Framework," *J. Proc. Cont.* **18**(1):2–18 (2008).

Bendat, J. S., and A. G. Piersol, Random Data: Analysis and Measurement Procedures, 2d ed., John Wiley & Sons, New York (1985).

Biagiola, S., N. Galvez, and J. Figueroa, "Performance Criteria Based on Nonlinear Measures," *Comp. Chem. Eng.* **28**:1799–1808 (2004).

Boyd, S., L. El Ghaoui, E. Feron, and V. Balakrishnan, *Linear Matrix Inequalities in System Control Theory*, SIAM, Philadelphia (1994).

Burl, J., *Linear Optimal Control*, Addison-Wesley, Menlo Park, CA (1998).

Chmielewski, D. J., and A. Manthanwar, "On the Tuning of Predictive Controllers: Inverse Optimality and the Minimum Variance Covariance Constrained Control Problem," *Ind. Eng. Chem. Res.* **43**:7807–7814 (2004).

Christofides, P. D., J. F. Davis, N. H. El-Farra, D. Clark, K. R. D. Harris, and J. N. Gipson, "Smart Plant Operations: Vision, Progress and Challenges," *AIChE J.* **53**(11):2734-2741 (2007).

Contreras-Dordelly, J., and T. Marlin, "Control Design for Increased Profit," *Comp. Chem. Eng.* **24**:267–272 (2000).

Davis, J., *Workshop on Cyberinfrastructure (CI) in Chemical and Biological Process Systems: Impact and Directions. A University—Industry—CI Perspective*, September 25–26, Arlington, VA (2006). (available at: http://www.oit.ucla.edu/nsfci/)

Edgar, T., "Control and Operations: When do Controllability Equal Profitability?" *Comp. Chem. Eng.* **29**:41–49 (2004).

Engell, S., "Feedback Control for Optimal Process Operation," *J. Proc. Cont.* **17**(3):203-219 (2007).

Figueroa, J., "Economic Performance of Variable Structure Control: A Case Study," *Comp. Chem. Eng.* **24**:1821–1827 (2000).

Figueroa, J., and A. Desages, "Dynamic 'Back-Off' Analysis: Use of Piecewise Linear Approximations," *Optim. Cont. Appl. Meth.* **24**:103–120 (2003).

Figueroa, J., P. Bahri, J. Bandoni, and J. Romagnoli, "Economic Impact of Disturbances and Uncertain Parameters in Chemical Processes—A Dynamic Back-Off Analysis," *Comp. Chem. Eng.* **20**(4):453–461 (1996).

Garcia, C. E., D. M. Prett, and M. Morari, "Model Predictive Control: Theory and Practice," *Automatica* **25**(3):335–348 (1989).

Huang, B., and S. Shah, *Performance Assessment of Control Loops: Theory and Applications*, Springer, New York (1999).

Larimore, W. E., "Automated Multivariable System Identification and Industrial Applications," *Proceedings of the American Control Conference* 1148–1162 (1999).

Lee, K. H., B. Huang, and E. C. Tamayo, "Sensitivity Analysis for Selective Constraint and Variability Tuning in Performance Assessment of Industrial MPC," *Cont. Eng. Practice* **16**(10):1195–1215 (2008).

Ljung, L., and T. Glad, Modeling of Dynamic Systems, Prentice Hall, Englewood Cliffs, NJ (1994).

Loeblein, C., and J. Perkins, "Economic Analysis of Different Structures of On-Line Process Optimization Systems," *Comp. Chem. Eng.* **20**:S551–S556 (1996).

Loeblein, C., and J. Perkins, "Economic Analysis of Different Structures of On-Line Process Optimization Systems," *Comp. Chem. Eng.* **22**(9):1257–1269 (1998).

Loeblein, C., and J. Perkins, "Structural Design for On-Line Process Optimization: I. Dynamic Economics of MPC," *AIChE J.* **45**(5):1018–1029 (1999).

Mahmood, M., R. Gandhi, and P. Mhaskar, "Safe-Parking of Nonlinear Process Systems: Handling Uncertainty and Unavailability of Measurements," *Chem. Eng. Sci.* **63**(22):5434–5446 (2008).

Mayne, D. Q., J. B. Rawlings, C. V. Rao, and P. O. M. Scokaert, "Constrained Model Predicitve Control: Stability and Optimality," *Automatica* **36**(6):789–814 (2000).

McGarty, T., Stochastic Systems and State Estimation, John Wiley & Sons, New York (1973).

Mhaskar, P., A. Gani, C. McFall, P. D. Christofides, and J. F. Davis, "Fault-Tolerant Control of Nonlinear Process Systems Subject to Sensor Faults," *AIChE J.* **53**(3):654–668 (2007).

Miletic, I. P. and T. E. Marlin, "Results Diagnosis for Real-Time Process Operations Optimization," *Comp. Chem. Eng.* **22** (supp 1):S475–S482 (1998).

Morari, M., G. Stephanopoulos, and Y. Arkun, "Studies in the Synthesis of Control Structures for Chemical Processes, Part I: Promulgation of the Problem. Process Decomposition and the Classification of the Control Task. Analysis of the Optimizing Control Structures" *AIChE J.* **26**(2):220–232 (1980).

Muske, K., "Estimating the Economic Benefit from Improved Process Control," *Ind. Eng. Chem. Res.* **42**:4535–4544 (2003).

Odelson, B. J., R. M. R. Rajamani, and J. B. Rawlings, "A New Autocovariance Leastsquares Method for Estimating Noise Covariances," *Automatica* **42**(2): 303–308 (2006).

Peng, J-K, A. Manthanwar, and D. J. Chmielewski, "On the Tuning of Predictive Controllers: The Minimum Back-Off Operating Point Selection Problem," *Ind. Eng. Chem. Res.* **44**:7813–7822 (2005).

Qin, S. J. and T. A. Badgwell, "A Survey of Industrial Model Predictive Control Technology," *Cont. Eng. Pract.* **11**(7):733–764 (2003).

Rawlings, J. B., "Tutorial Overview of Model Predictive Control," *IEEE Cont. Syst. Mag.* **20**(3):38–52 (2000).

Rawlings, J. B. and R. Amrit, "Optimizing Process Economic Performance Using Model Predictive Control," *Lecture Notes in Control and Information Sciences* **384**:119–138 (2009).

Riggs, J. B. and M. N. Karim, *Chemical and Bio-Process Control,* 3d ed., Ferret Publishing Lubbock, Tex. (2006).

Romagnoli, J. A. and A. Palazoglu, *Introduction to Process Control*, Taylor & Francis, New York (2006).

Seborg, D. E, T. F. Edgar, and D. A. Mellichamp, *Process Dynamics and Control,* 2d ed., John Wiley & Sons, Hoboken, NJ (2003).

Skogestad, S., "Near-Optimal Operation by Self-Optimizing Control: From Process Control to Marathon Running and Business Systems," *Comp. Chem. Eng.* **29**(1):127–137 (2004).

Skogestad, S., "Self-Optimizing Control: The Missing Link Between Steady-State Optimization and Control," *Comp. Chem. Eng.* **24**(2–7):569–575 (2000).

Soderstrom, T. and P. Stoica, *System Identification*, Prentice Hall International, London, U.K. (1989).

Soliman, M., C. Swartz, and R. Baker, "A Mixed-Integer Formulation for Back-Off Under Constrained Predictive Control," *Comp. Chem. Eng.* **32**(10):2409–2419 (2008).

Stephanopoulos, G., *Chemical Process Control: An Introduction to Theory and Practice,* Prentice-Hall. Upper Saddle River, NJ (1984).

Young, J., C. Swartz, and R. Ross, "On the Effects of Constraints, Economics and Uncertain Disturbances on Dynamic Operability Assessment," *Comp. Chem. Eng.* **20**:S667–S682 (1996).

Young, J., R. Baker, and C. Swartz, "Input Saturation Effects in Optimizing Control—Inclusion within a Simultaneous Optimization Framework," *Comp. Chem. Eng.* **28**:1347–1360 (2004).

Zhang, Y. and J. Forbes, "Extended Design Cost: A Performance Criterion for Real-Time Optimization Systems," *Comp. Chem. Eng.* **24**(8):1829–1824 (2000).

Zhao, C., Y. Zhao, H. Su, and B. Huang, "Economic Performance Assessment of Advanced Process Control with LQG Benchmarking," *J. Proc. Contr.* **19**(14): 557–564 (2009).

Zheng, A. and M. Morari, "Stability of Model Predicitve Control with Mixed Constraints," *IEEE Transaction on Automatic Control* **40**(10):1818–1823 (1995).

Zhu, Y., *Multivariable System Identification for Process Control*, Elsevier Science Ltd, Oxford, U.K. (2001).

CHAPTER 14

Value of Parametric Fault Identification

Introduction

Process plants are subject to failures. These failures are of various types, all affecting production in one way or another. Aside from catastrophic accidents, some major failures require plant shutdown and other less serious equipment failures affect either the quality of the products (out of specification) or the product flows. All these affect the profitability of the plants.

In this chapter, we cover models that are suitable for fault detection and identification.

A process fault is a "degradation from normal operation conditions, and includes symptoms of a physical change (such as deviations in measured temperature or pressure) as well as the physical changes themselves (scaling, tube plugging, etc.) and deviations in parameters (such as a heat transfer coefficient)" (Wilcox and Himmelblau, 1994a).

Faults are manifested in deviations from normal conditions that propagate to a set of sensors (Fig. 14.1). In the case when these sensors are not faulty, the information can be used to determine whether there is a fault (detection) and what fault in particular it is. When the sensors are faulty, the detection and/or identification could be compromised.

A good alarm system should be able to filter all disturbances in the sensors (like voltage surges and other problems reviewed in Chap. 2) and have enough bias-free information (through redundant measurements) so that faults can be identified and appropriate corrective actions can be taken.

Fault Classification

In general, faults in a process can be classified as structural, parametric, or sensor faults. We now review briefly each one.

- *Structural Faults:* These faults are those related to equipment malfunction, wear and tear, etc. The response for these faults

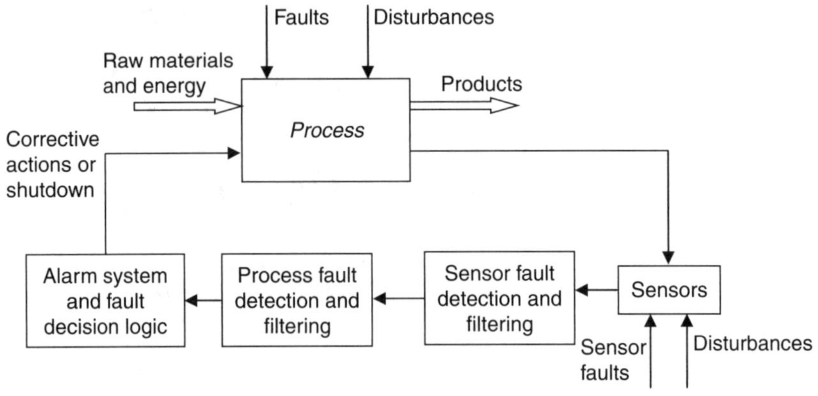

FIGURE 14.1 Fault diagnosis and alarm system.

is an appropriate level of maintenance, which is covered in Chap. 16.

- *Parametric Faults:* Examples of parametric faults include changes in flowrate, pressure, temperature, concentration, and so on. The deterioration of equipment functionality should also be included. Examples are heat exchanger fouling and catalyst deactivation.
- *Sensor Faults:* Of these, we recognize biases, which were discussed in Chap. 2, with their effect on parameter estimation and accuracy (Chaps. 10 and 11). There are several other possible sensor failure types such as freezing, missed values, etc.

In this chapter we only address the question of quantifying the value of a sensor network when parametric faults occur.

Fault Detection and Diagnosis Techniques

Fault detection and diagnosis has been addressed in several books (Himmelblau, 1978; Pau, 1981; Gertler, 1998). Several methods have been developed through the years. We note, however, that all of them assume that the necessary instrumentation to obtain the needed information is in place. These methods have been reviewed by Bagajewicz (2000) in detail. A more detailed review was done by Venkatasubramanian et al. (2003a, 2003b, 2003c). In general, fault diagnosis methods can be classified under three different categories. These are the qualitative, quantitative, and process history-based methods. We list some of them briefly next.

Quantitative methods:

- *Model-Based Approach*: These are quantitative models (usually linear) used for fault detection and isolation. They use dynamic models (Kalman filters, diagnostic observers, parity relations, parameters estimation, etc.) that help generate estimators of measured and unmeasured variables and parameters. Then, measurements and estimators are used to generate residuals, which are used for diagnostics (Gertler, 1988; Frank, 1990; and Patton, 1995) as well as the books by Himmelblau (1978), Patton et al., (1989), and Gertler (1998), among others.

Qualitative methods:

- *Fault Trees*: These are based on the identification of cause effect by tracing back the fault to the possible causes (Lapp and Powers, 1977; Aelion and Powers, 1993; Catino and Ungar, 1995), but are not useful for identification of the cause.
- *Rule-Based Approach*: It uses signed directed graphs to determine the positive or negative effect of a fault on a variable (Kramer and Palowitch, 1987; Chang and Yu, 1990; Mohindra and Clark, 1993).
- *Failure Propagation Networks*: These are based on the notion that the effect of the fault propagates through equipment. Digraphs with propagation times and failure probabilities are used (Kokawa and Shingai, 1982; Kokawa et al., 1983; and Qian, 1990).
- *Knowledge-Based Approach*: These are based on reasoning systems, expert systems, etc. (Kramer, 1987, Venkatasubramanian and Rich, 1988, Fathi et al., 1993).

Process history-based methods:

- *Neural Networks*: Based on layers of "neurons," nodes that are trained to process all the input information by adding all contributions from the previous layer and producing a new signal through a nonlinear transformation (Watanabe et al., 1989; Fan et al., 1993, Kavuri and Venkatasubramanian, 1993; Chen et al., 1998.)
- *Multivariate Statistical Methods*: These methods rely on principal component analysis (PCA) (Jackson, 1991), Inverse least squares, principal component regression, and partial least squares (PLS) (Wise and Gallagher, 1995; Kresta et al., 1991; Vinson and Ungar, 1994). Extensions of these two techniques have flourished in recent years: Multiblock PLS methods (McGregor et al., 1994), reconstruction procedures (Dunia and

Qin, 1998), neural networks embedded in PLS frameworks (Qin and McAvoy, 1992), nonlinear principal component analysis (PCA) based on neural networks (Dong and McAvoy, 1996), wavelet filtering followed by nonlinear PCA (Shao et al., 1998), PCA in conjuction with signed digraph (Vedam and Venkatasubramanian, 1998), and recursive PLS (Qin, 1998).

All the aforementioned partial list of methods refers to techniques that are capable of establishing a "signature" of some sort, which is very useful to perform the diagnosis. For example, statistical methods can use departures from normal conditions as long as they are properly trained. Model-based approaches can generate a set of estimators of parameters as well as measured and unmeasured variables, whose departure from normal operating conditions can be used as an indicator of a particular fault. Producing these "signatures" or residuals, as they are often called, is the basis of the analysis made in the rest of the chapter. The residuals can be generated through any one of the three types of fault diagnosis methods. The model-based approaches manipulate the analytical relationships between the variables to derive new equations that are affected by only a subset of faults. If none of these faults occur, then the derived equation (called residual) will show no deviation. If, on the other hand, the residual shows a deviation, one of the faults in the subset is implicated. If each subset contains only one fault then that means each residual is affected by only one fault and this condition is referred to as the diagonal relationship. Each residual can be monitored to detect and diagnose individual faults. Examples of the observer's approach for the development of the residuals can be found in Frank (1990). Examples of the so-called parity-based approaches for the development of residuals can be found in Gertler (1988).

The same idea of residuals can also be found in data-based approaches. PCA in combination with contribution charts can be used for fault diagnosis. The output nodes that denote various faulty conditions in a neural network can also be thought of as residuals. If such a residual can be developed for a particular fault based on a given sensor network, then the fault is observable/diagnosable.

In this chapter, we focus on a qualitative approach for observability analysis as well as a quantitative approach to determine the economic value of fault identification.

Fault Observability

Observability of a fault refers to the ability of a sensor network to detect its presence. Thus, a fault is observable if the sensor network can detect it regardless of its actual location.

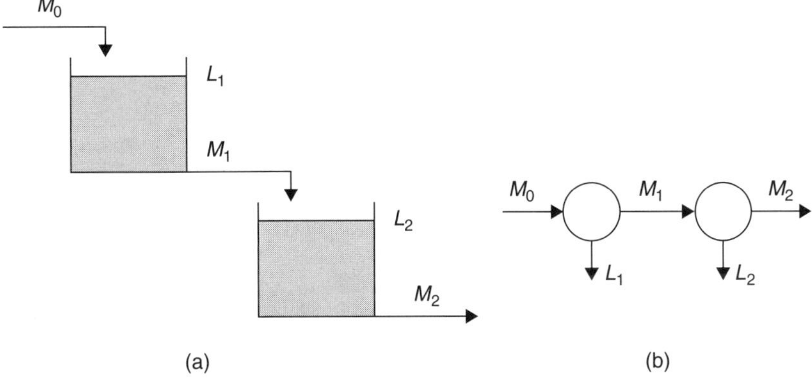

FIGURE 14.2 Two tanks in series.

We discuss observability in a qualitative form using digraphs. Consider the system representing two tanks in series presented in Fig. 14.2, which is taken from Wilcox and Himmelblau (1994a). Figure 14.2*b* shows a digraph, where the nodes are the units and the edges are the streams. To represent the level changes (increasing in the first tank and decreasing in the second), pseudo-streams are added.

Consider now another type of digraph where the nodes represent the streams of the process and the edges represent the influence that one exerts on the other. Such a digraph is shown in Fig. 14.3.

For simplicity, we assume that the control loops that manipulate flowrates M_1 and M_2 are set to maintain the flowrate constant (rather than the usual, which is maintaining the level of the preceding tank constant). In such case the directed graph loses its forward connections between levels and flows (Fig. 14.4).

Assume now that flowrate and level sensors are located in the system, S_{M0}, S_{M1}, S_{M2}, for the flowrates and S_{L1}, S_{L2} for levels. Then a graph can be made connecting variables to sensors (Fig. 14.5).

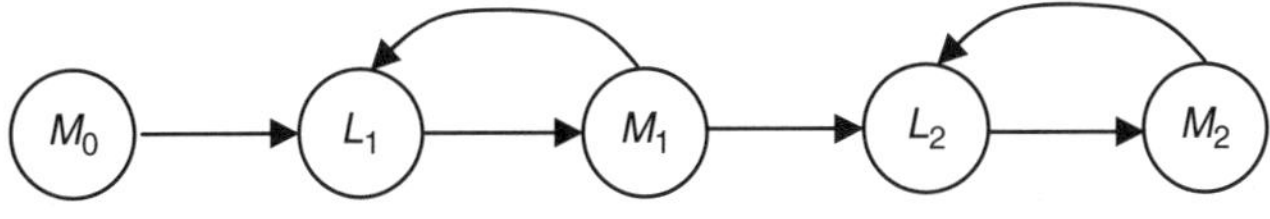

FIGURE 14.3 Directed digraph.

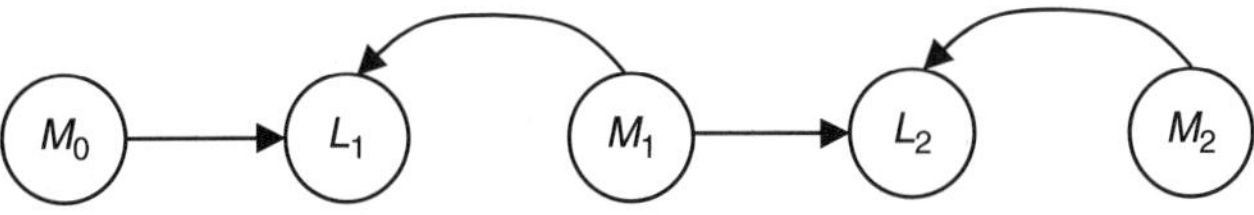

FIGURE 14.4 Directed digraph (no effect of level on flow).

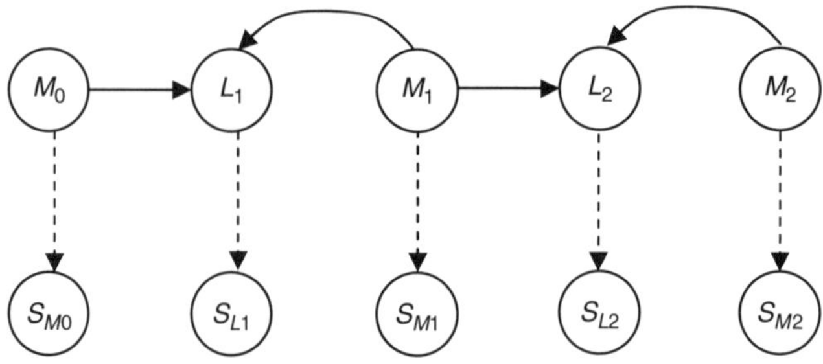

FIGURE 14.5 Directed digraph variable-sensor connectivity.

The digraph of Fig. 14.5 can be used to trace each abnormal condition to a sensor. We now add the faults: We consider for simplicity and without loss of generality that we only have changes in flowrate M_0 associated to failure of pumps/valves on the control loops, which we denote by F_{Vi} $(i = 1,2)$ and leaks in the tanks, which we denote F_{Li} $(i = 1,2)$. Thus, we modify our diagram to include such events (Fig. 14.6).

We notice that there is no specific fault associated to possible abnormal conditions in the inlet flow M_0. Abnormal conditions in this case are only abnormal flows and levels. One can now construct a graph connecting a fault to all sensors that will "see" their measurement changed. We call this the bipartite graph because the faults are in one row and the sensors in the other (Fig. 14.7). For example, an abnormal condition in the flowrate M_0 (indicated by F_{M0} in the bipartite graph) will be picked up by the flowrate sensor S_{M0} and the level sensor S_{L1}. The effect of this fault does not propagate.

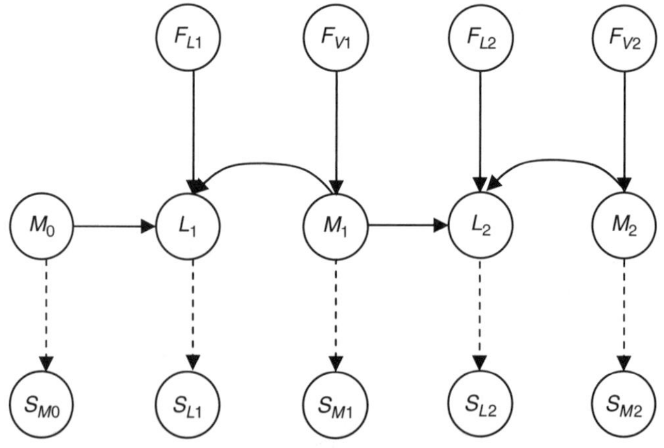

FIGURE 14.6 Directed digraph fault-variable-sensor connectivity.

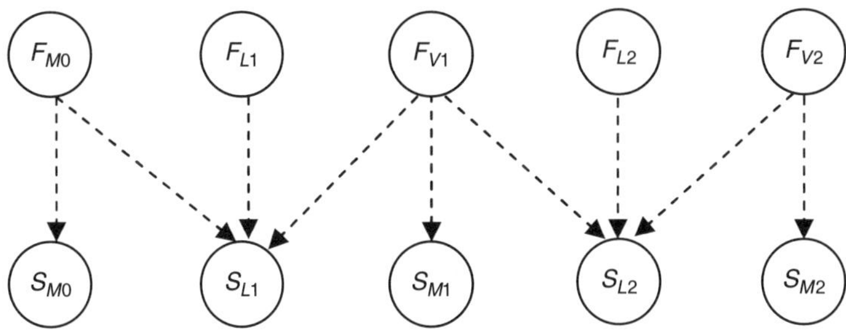

FIGURE 14.7 Bipartite graph.

Thus, a fault is observable if there is a connection from the fault to any sensor in the bipartite graph. In Fig. 14.6 all faults are observable. For example, if the level sensor S_{L1} is not installed, then the node disappears in Fig. 14.6 and fault F_{V1} becomes unobservable.

These digraphs can be extended to indicate that the effects of variables on variables, variables on sensors, and faults on variables have a positive or negative sign. Accordingly, these digraphs were extended to signed digraphs (SDG) (Iri et al., 1979; Umeda et al., 1980; Shiozaki et al., 1985; Wilcox and Himmelblau, 1994a, 1994b) and used to determine cause effect diagrams and ultimately bipartite graphs. A signed directed graph corresponding to Fig. 14.4 is shown in Fig. 14.8. For example, an increase in M_0 has a positive effect on level L_1, whereas an increase in M_1 has a negative effect on level L_1, and so on.

If we now consider the positive and negative effects of the faults, we get the digraph of Fig. 14.9. The corresponding bipartite graph is shown in Fig. 14.10.

In many instances, the sensors are omitted and the sign directed bipartite graph shows the connections from faults to variables only with the sensors assumed and, for simplicity, when a fault can have both signs it is represented as one node with both signs included (±). This is shown in Fig. 14.11, which should be read in such a manner that Fig. 14.10 is implied.

We finally have the influence of control loops, which by nature are cyclic. Consider the control loop of Fig. 14.12*a*. It is composed of a level sensor L_M, a controller C, and a control valve (we omit the actuator as a separate entity and consider it to be part of the valve). The corresponding DG is the one shown in Fig. 14.12*b*.

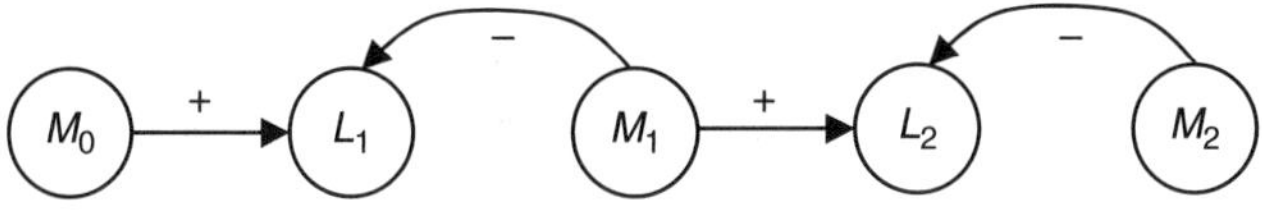

FIGURE 14.8 Signed directed digraph.

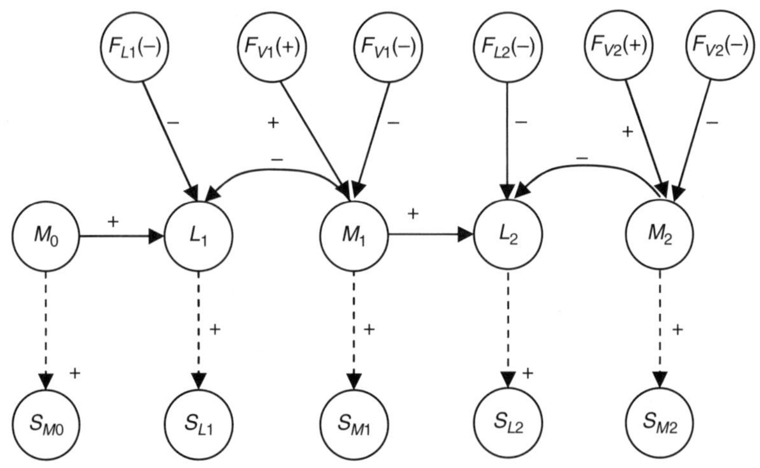

FIGURE 14.9 Directed signed digraph fault-variable-sensor connectivity.

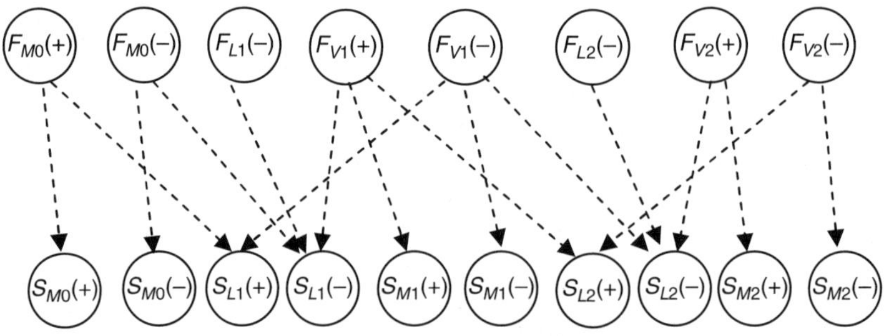

FIGURE 14.10 Bipartite graph for fault-variable-sensor connectivity.

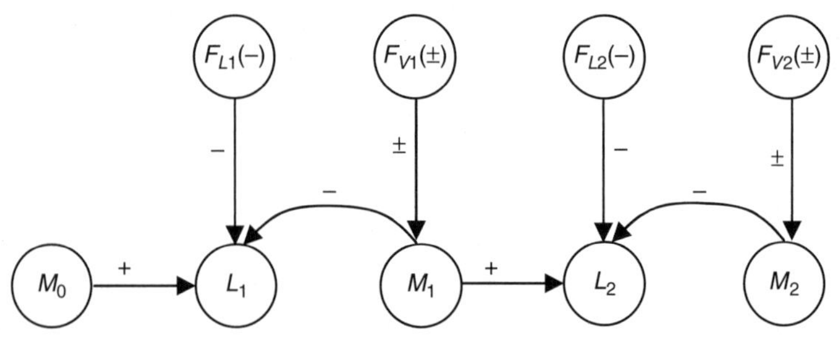

FIGURE 14.11 Fault to variable connectivity bipartite graph.

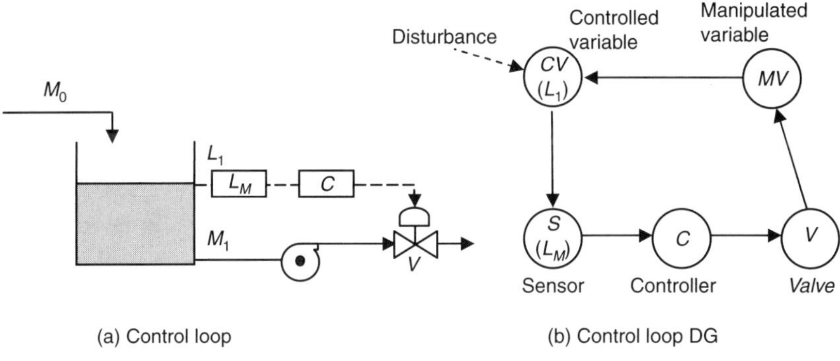

Figure 14.12 Level control loop.

A positive disturbance in the control loop (an increase in level, the controlled variable in this case) creates a change in the sensor, but the controller acts in such a way that it sends a signal to the valve, who lifts the stem (manipulated variable) and restores the level to its original value. This is called "perfect control" and the negative feedback loop created restores the variable to its original (intended) value. However, in doing so, it changes the value of the manipulated variable. If the manipulated variable cannot be changed anymore, like when the valve is already fully 100% open, then it is said that the loop "saturates." We omit discussing the SDG intricacies of control loops leading to perfect control and saturation in the presence of disturbances (see Bhushan and Rengaswamy, 2000 for further discussion).

Control loops can fail in many ways. The sensor may fail (be biased or produce no signal at all), the controller itself can fail, the valve can fail by failing to move the stem, or by the actuator failing. All these can lead to positive or negative deviations in the manipulated variable that ought to be analyzed. We present some more details of this in the example.

A more precise and general definition of fault observability is the following: Given a set of faults $\mathrm{F} = \{F_1, F_2, \ldots, F_n\}$, we define a binary vector fault indicator vector $\mathrm{f} = \{f_1, f_2, \ldots, f_n\}$ where the jth component of f is 1 if the jth fault has occurred and conversely, it is 0 if it has not occurred. Given a sensor network, $\mathrm{S} = \{S_1, S_2, \ldots, S_m\}$, the corresponding measurements $\mathrm{y} = \{y_{S_1}, y_{S_2}, \ldots, y_{S_m}\}$ and a fault diagnosis strategy, we define a "residual" vector, $\mathrm{r} = \{r_1, r_2, \ldots, r_n\}$ as the output of the diagnosis strategy. Then, a fault F_j is observable if $f_j \neq 0 \Rightarrow \mathrm{r} \neq 0$ (Narasimhan and Rengaswamy, 2007), regardless of the value that all the other elements of the vector f. The "residual" vector is in most cases, a measure of a deviation from normal conditions that can be obtained using the various methods presented above.

To put the above definition in context and to relate it to the above discussion, we now need to extend the bipartite graph to the use of

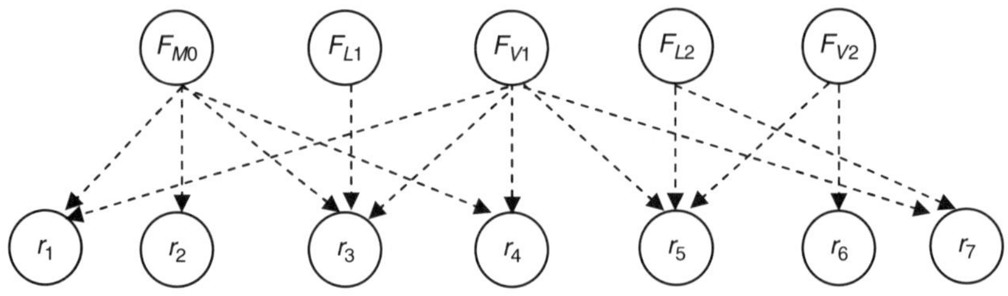

FIGURE 14.13 Bipartite fault-residual graph.

residuals. Because the sensor information is used to generate the residuals, we add them formally and generate a fault-residual bipartite graph. We omit discussing any particular residual generator, assuming one is present, and we show a generic case in Fig. 14.13.

The following step is to generate the fault-sensor (or faul-residual) maximum connectivity matrix (A), which contains the same information as the bipartite graph. In this matrix, the columns represent the faults and the rows represent the sensors. When $A_{ij} = 1$, then we say that there is a path from fault F_j to the potential sensor location S_i. For the case of Fig. 14.6, this matrix is the following:

$$A = \begin{array}{ccccc|l} F_{M0} & F_{L1} & F_{V1} & F_{L2} & F_{V2} & \\ \hline 1 & 0 & 0 & 0 & 0 & S_{M0} \\ 1 & 1 & 1 & 0 & 0 & S_{L1} \\ 0 & 0 & 1 & 0 & 0 & S_{M1} \\ 0 & 0 & 1 & 1 & 1 & S_{L2} \\ 0 & 0 & 0 & 0 & 1 & S_{M2} \end{array} \tag{14.1}$$

We now define a fault-sensor connectivity matrix $B(q) = [b_1, \ldots\ldots\ldots\ldots, b_n]$ using

$$b_j = a_j \otimes q \tag{14.2}$$

where b_j and a_j are the ith column of B and A, respectively, and $\otimes$ is the Hadamard product of two vectors, that is, $(a_j \otimes q) = a_{ji}\, q_i$. Thus, $A = B$ only when sensors are selected for all potential locations. For the particular case of our example,

$$B(q) = \begin{array}{ccccc|l} F_{M0} & F_{L1} & F_{V1} & F_{L2} & F_{V2} & \\ \hline q_{M0} & 0 & 0 & 0 & 0 & S_{M0} \\ q_{L1} & q_{L1} & q_{L1} & 0 & 0 & S_{L1} \\ 0 & 0 & q_{M1} & 0 & 0 & S_{M1} \\ 0 & 0 & q_{L2} & q_{L2} & q_{L2} & S_{L2} \\ 0 & 0 & 0 & 0 & q_{M2} & S_{M2} \end{array} \tag{14.3}$$

We now observe that the sum of all the elements of one column of B determines if the corresponding fault is observable or not, that is,

$$\text{If } \sum_{i=1}^{m} B_{ij}(q) = 0 \quad \text{then } F_j \text{ is unobservable} \tag{14.4}$$

If $\sum_{i=1}^{m} B_{ij} \geq 1$ then F_j is observable. This can be then summarized using the following formalism:

$$\text{If } \sum_{i \in M_1} b_k(q_i) \geq 1 \quad \text{then } F_k \text{ is observable} \tag{14.5}$$

Single Fault Resolution

The term resolution has been coined to refer to the ability of determining the exact process fault occurance, given the set of symptoms.

Consider two columns of the fault-sensor connectivity matrix. If both faults are observable, all one needs is these two faults not be connected to the same set of sensors. To guarantee that, at least one element of the two corresponding columns of B must be different. This is achieved by imposing the following condition:

$$\sum_{i \in M_1} [b_j(q) \otimes b_p(q)]_i \leq \text{Max}\left\{ \sum_{i \in M_1} [b_j(q)]_i, \quad \sum_{i \in M_1} [b_p(q)]_i \right\} - 1 \tag{14.6}$$

In the matrix shown in Eq. (14.1), faults F_1 and F_4 are observable and distinguishable for certain set of sensors. Indeed, $\sum_{i \in M_1} [b_1(q) \otimes b_2(q)]_i = q_2^2$ and $\text{Max}\left\{ \sum_{i \in M_1} [b_1(q)]_i, \sum_{i \in M_1} [b_4(q)]_i \right\} = \text{Max}\{q_1 + q_2, q_2 + q_3\}$. Thus, if $q_1 = 0$ and $q_3 = 0$, then inequality in Eq. (14.6) does not hold and the sensors cannot be distinguished (they would express both in sensor two only). In turn, if $q_2 = 0$, $q_1 = 1$, and $q_3 = 1$, then the inequality in Eq. (14.6) holds and the faults are distinguishable. They also are distinguishable, that is, the inequality in Eq. (14.6) holds if the three sensors are located ($q_1 = q_2 = q_3 = 1$). Now, consider faults F_1 and F_2. These two faults affect the same sensors and are indistinguishable, and there is no combination of sensors that can satisfy the inequality in Eq. (14.6). Indeed, in this case, $\sum_{i \in M_1} [b_1(q) \otimes b_2(q)]_i = q_1^2 + q_2^2$, and the sum of columns 1 and 2 is $q_1 + q_2$. Thus the right hand side of Eq. (14.6), that is, $q_1 + q_2 - 1$, which cannot be larger than the left hand side, no matter what values q_1 and q_2 assume.

Multiple Fault Resolution

Assume now that one wants to consider multiple faults occurring simultaneously and one wants to resolve them. Since multiple faults can have the same signature as single faults, one wants to resolve multiple faults from each other, regardless of their cardinality. Intuitively, two observable faults can be resolved if they produce "different" symptoms, the difference being appropriately defined. We now illustrate how this can be done for two faults.

Consider two faults taking place at the same time. They will have a signature corresponding to a vector that is the Boolean sum of the corresponding two columns of B, that is, their signature will be

$$d_{jk}(q) = [a_j \oplus a_k] \otimes q \tag{14.7}$$

Thus, we define the double fault-sensor connectivity matrix $D(q)$ through all column vectors the desired combinations of double faults $d_{jk}(q)$ that belong to a prespecified set. Resolution of these faults with respect to the others is now obtained through an expression similar to Eq. (14.6).

$$\sum_{i \in M_1} [d_{jk}(q) \otimes d_{pr}(q)]_i \leq \text{Max}\left\{\sum_{i \in M_1} [d_{jk}(q)]_i, \sum_{i \in M_1} [d_{pr}(q)]_i\right\} - 1 \quad \forall j,k,p,r \in M_{S2} \tag{14.8}$$

where (j, k) and (p, r) belong to M_{S2}, the set of all combinations of two faults that one wants to resolve. In addition $j \neq k \neq p \neq r$. In a similar fashion, one can distinguish double faults from single faults by writing

$$\sum_{i \in M_1} [d_{jk}(q) \otimes b_p(q)]_i \leq \text{Max}\left\{\sum_{i \in M_1} [d_{jk}(q)]_i, \sum_{i \in M_1} [b_p(q)]_i\right\} - 1 \quad \forall j,k \in M_{S2}, \forall p \in M_S \tag{14.9}$$

where $j \neq k \neq p$.

Alternative algorithmic methodologies to establish observability and resolvability were introduced by Bhushan and Rengaswamy (2000) and explicitly account for in the appendix of the article by Narasimhan and Rengaswamy (2007).

We end this section by introducing some notation. We consider first a set of faults $F = \{F_1, F_2, \ldots, F_n\}$ and we indicate by the singleton $\{F_j\}$ that a fault F_j is resolvable. When a subset of faults, say $F_{k1}, F_{k2}, \ldots F_{km}$ are unresolvable, then we use a set $\{F_{k1}, F_{k2}, \ldots, F_{km}\}$ to indicate this condition.

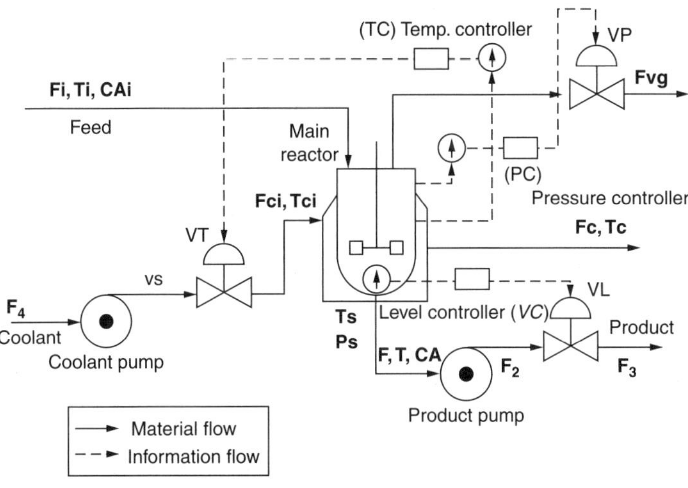

Figure 14.14 CSTR process schematic.

Example 14.1: We consider the CSTR example introduced in Example 8.5. We include in Fig. 14.14 the control loops. The CSTR involves an exothermic liquid-phase reaction: $A_{(l)} \rightarrow B_{(l)} + C_{(g)}$. The temperature controller controls the temperature of the reactor by manipulating the flowrate of the coolant flowing through the jacket. The level in the reactor is controlled by the level controlled by manipulating the outlet flowrate from the reactor. The pressure in the reactor is controlled by changing the vent gas flowrate. Both the reactor and the jacket are modeled using perfectly mixed-tank dynamics.

In Example 8.5 (Chap. 8), we presented model equations for steady state. The dynamic model equations are

Global mass balance:

$$F_i - F = \frac{dV}{dt} \tag{14.10}$$

Component mass balance (C_A):

$$\frac{F_i}{V} \cdot (C_{Ai} - C_A) - r_A = \frac{dC_A}{dt} \tag{14.11}$$

Assuming constant heat capacities and densities, the overall heat balance on the reactor is given by

$$\frac{F_i}{V}(T_i - T) + \frac{r_A(-\Delta H)}{\rho C_p} - \frac{UA(T - T_c)}{V\rho C_p} = \frac{dT}{dt} \tag{14.12}$$

Overall heat balance on the jacket:

$$\frac{F_c}{V_j}(T_{ci} - T_c) + \frac{UA(T - T_c)}{V_j \rho_j C_{pj}} = \frac{dT_c}{dt} \tag{14.13}$$

Gas phase balance:

$$r_A V - F_{vg} = \frac{dn}{dt} \tag{14.14}$$

The reaction rate is given by

$$r_A = C_d C_A k_0 e^{-E/RT} \tag{14.15}$$

The elemental mass balances in valves and pumps are (assuming no accumulation)

$$\left.\begin{aligned} F_3 - F_2 &= 0 \\ F_2 - F &= 0 \\ F_4 - F_c &= 0 \end{aligned}\right\} \tag{14.16}$$

Assuming ideal behavior, the pressure in the reactor is

$$PV_g = nRT \tag{14.17}$$

where V_g is the vapor space and is assumed constant.

Nominal values of all parameters and variables is given in Table 14.1.

Using the above equations (one can construct the DG diagram (Fig. 14.15). In this diagram, PfS, TfS, and VfS correspond to sensor fault nodes and VfP, VfT, and VfL to valve fault nodes. The bipartite graph for noncontrol faults is shown in Fig. 14.16 and the fault-sensor connectivity matrix for noncontrol loops is given in Table 14.2. Finally, the fault-sensor connectivity matrix is shown in Table 14.3.

Remarks:

- It seems from the DG diagram of the CSTR process (Fig. 14.15) that the sensor node T_c can observe the fault node T_{ci}. However, as Bhushan and Rengaswamy (2000) pointed out, because of the presence of the positive feedback loop between T and T_c, a disturbance in *Tci* affecting T_c will be compensated by the control mechanism so that both the controlled variable T and T_c return to their original steady-state values. Thus, T_c is not the sensor node that can observe the fault node T_{ci}.
- The fault node *CAi* cannot be differentiated from the fault node *Cd* using the qualitative fault detection method described above because they both have the same set of sensor nodes (in other words, they have the same symptom). The same thing is said for the two faults nodes *Ti* and *U*.

We consider the fault detection capability of the sensor network consisting of eight sensors obtained by Bhushan and Rengaswamy (2000). The eight sensors measuring {*CA* , *Fc*, *Tc*, *F*, *PC*, *TS*, *PS*, *VS*}. Sensors *TS*, *PS*, and *VS* are installed for control purposes and are also used for fault detection. The set of these eight sensors can observe all the fault nodes. Table 14.4 shows the set of single faults (noncontrol loop fault) and double faults (pair of noncontrol loop fault–control loop fault) that can be detected and identified (i.e., the fault positions are located) by this set of sensors.

Note: In Table 14.4, the fault node *CAi* can be equivalently replaced by the fault node *Cd*.

Notation	Variable	Nominal Value
V	Volume of liquid in the reactor	48 ft^3
C_A	Reactant concentration in the reactor	0.2345 lb-mol-A/ft^3
T	Reactor temperature	600°R
n	Number of moles in gas phase of reactor	28.3656 lb-mol-C
V_g	Volume of gas phase (constant)	16 ft^3
F_i	Inlet feed flowrate	40 ft^3/h
C_{Ai}	Inlet reactant concentration	0.5 lb-mol-A/ft^3
T_c	Jacket temperature	590.51°R
F_c	Coolant flowrate	56.626 ft^3/h
T_i	Inlet feed temperature	530°R
V_j	Volume of jacket	3.85 ft^3
k_0	Frequency factor	7.08 10^{10} h^{-1}
C_d	Catalyst activity	1
E	Activation energy	29,900 Btu/lb-mol
R	Gas constant	1.99 Btu/lb-mol°R
U	Heat transfer coefficient	150 Btu/h ft^2 °R
A	Heat transfer area	150 ft^2
T_{ci}	Inlet coolant temperature	530°R
ΔH	Heat of reaction	–30,000 Btu/lb-mol
C_p	Heat capacity (process side)	0.75 Btu/lb-mol°R
C_{pj}	Heat capacity (coolant side)	1.0 Btu/lb-m°R
ρ	Density of process mixture	50 lb-m/ft^3
ρ_j	Density of coolant	62.3 lb-m/ft^3
K_v, T_v	PI controller parameters for volume	1
K_t, T_t	PI controller parameters for temperature	4.3
K_p, T_p	PI controller parameters for pressure	0.5

Table 14.1 Nominal Values for CSTR Example 14.1

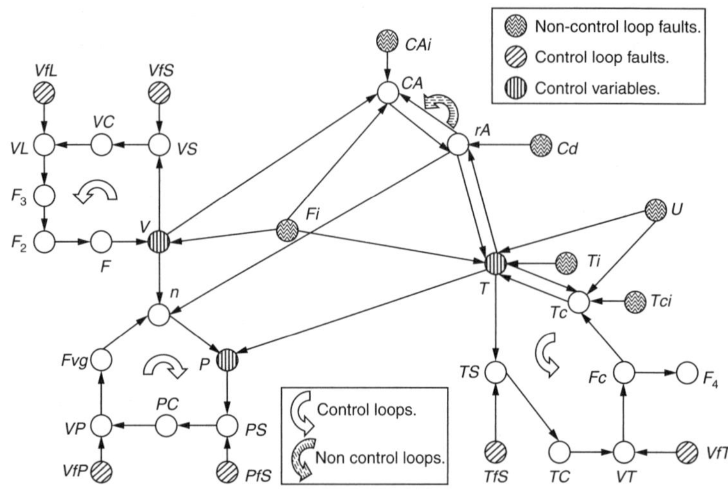

Figure 14.15 DG diagram of the CSTR.

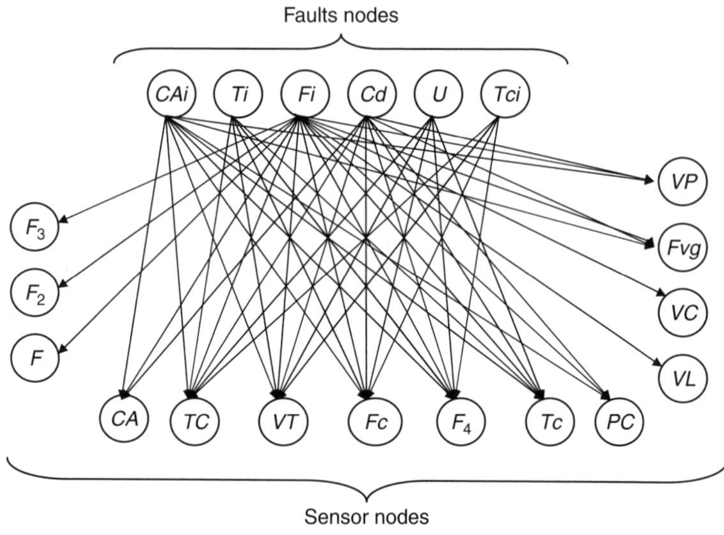

Figure 14.16 DG diagram for noncontrol loop faults.

Value of Fault Detection

The techniques presented above focus on fault observability and resolution. Both seem to be nonnegotiable objectives in the case of safety. Economics, however, may be an important issue, especially in parametric faults (deviation from normal states). Thus, one needs an

	CAi	Fi	Ti	Tci	Cd	U
C_A	x	x			x	
T_c	x	x	x		x	x
F_c	x	x	x	x	x	x
Fvg	x	x			x	
F		x				
F_2		x				
F_3		x				
F_4	x	x	x	x	x	x
TC	x	x	x	x	x	x
VT	x	x	x	x	x	x
N						
PC	x	x			x	
VP	x	x			x	
VC		x				
VL		x				

Table 14.2 Corresponding Nodes for Noncontrol Loop Faults

economic indicator to determine how valuable it is to be able to detect and resolve certain faults.

Despite its importance, not too many attempts have been made to assess the value obtained from the ability to detect and observe faults. One such attempt is based on computing the loss of profit that stems from nondetecting/resolving different parametric faults (Narasimhan and Rengaswamy, 2007). The idea is quite general, however; the actual determination of the value function required some assumptions, which are

- *Parametric Faults Only*: This includes abnormal events, upsets, disturbances (which may be measured or unmeasured, and will be considered abnormal if they exceed the normal range) that are relevant for diagnosis. They argue that there are certain deviations from normal states that are not necessarily adverse to profit and safety and they are usually tolerated. Since parametric faults are considered, it is possible to parameterize the respective quantity by a state variable. Each state variable x_i (the normal state denoted by μ_i) associated to a fault has lower and upper bounds (LB_N and UB_N) within which the process is assumed normal. Moreover, the variable deviations also have larger bounds that include normal and abnormal conditions (LB and UB).

	CA_i – VfL	CA_i – VfP	CA_i – VfT	T_i – VfL	T_i – VfP	T_i – VfT	F_i – VfL	F_i – VfP	F_i – VfT	T_{ci} – VfL	T_{ci} – VfP	T_{ci} – VfT	Cd – VfL	Cd – VfP	Cd – VfT	U – VfL	U – VfP	U – VfT
C_A	x	x	x			x	x	x	x			x	x	x	x			x
T_c	x	x	x		x	x	x	x	x			x	x	x	x	x	x	x
F_c	x	x		x	x		x	x		x	x		x	x		x	x	
Fvg	x	x	x			x	x	x	x			x	x	x	x			x
F								x	x									
F_2								x	x									
F_3								x	x									
F_4	x	x		x	x		x	x		x	x		x	x		x	x	
TC	x	x	x	x	x	x	x	x	x	x	x	x	x	x	x	x	x	x
VT	x	x		x	x		x	x		x	x		x	x		x	x	
N		x	x			x		x	x			x	x	x	x			x
PC	x	x	x			x	x	x	x			x	x	x	x			x
VP	x		x			x	x		x			x	x		x			x
VC							x	x	x									
VL								x	x									
TS			x			x			x			x			x			x
PS		x						x			x			x				
VS							x											

Table 14.3 Corresponding Nodes for Control Loop Faults (Valve Failures)

Single Fault Resolution (Noncontrol Loop Fault)	Double Fault Resolution (Noncontrol Loop Fault – Control Loop Fault)
CAi, Fi, Ti, Tci	*CAi – VfL, CAi – VfP, CAi – VfT, Ti – VfL, Tci – VfP, Ti – VfP, Fi – VfL, Fi – VfP, Fi – VfT*

Table 14.4 Fault Detection Capability of the Given Set of Sensors

- *Single Fault Occurrence*: Although diagnosis strategies based on this assumption are common in the literature (see, for example, the three parts 2003 review by Venkatasubramanian et al.), this is a strong assumption that should be tested. There is, however, some argument to be made from the probabilistic point of view. Narasimhan and Rengaswamy (2007) argue that the probability of two faults occurring simultaneously is very low. Monte Carlo simulations for sensor biases (see Chap. 11) confirm that the probability is low.
- *Probabilistic Description of Faults*: Each variable x_i associated to a fault f_i has an associated probability distribution $\varphi_i(x_i)$ (they use a truncated version), which is usually inferred from histograms or other historical data. They use a normal (Gaussian) truncated distribution. The distribution is truncated by the bounds (*LB* and *UB*) and normalized as follows

$$\varphi_i(x_i) = \frac{\phi(x_i,\mu_i,\sigma_i)}{\Phi(UB,\mu_i,\sigma_i) - \Phi(LB,\mu_i,\sigma_i)} \tag{14.18}$$

where $\phi(\bullet, 0, \sigma_i)$ is a normal distribution with variance σ^2_i and $\Phi(\bullet, 0, \sigma_i)$ its cumulative distribution, with the variance set by the condition

$$\int_{LB_N}^{UB_N} \varphi_i(x_i)dx_i = 1 - p_i \tag{14.19}$$

where p_i is the probability of the fault.

- *Profit Function*: The profit is a certain function of the deviation from normal values, that is, $c = c(\text{x})$. This profit function is defined over a certain period of time T. They denote c^* the profit under no deviation, that is, $c^* = c(\mu)$. The profit function, in a probabilistic sense, cannot increase with time. In other words, the expectation of profit decreases from period t to period $t + 1$, that is, $E[c(t + 1) \le E[c(t)]$.
- *Instant Correction of Detected Faults*: If the network identifies and resolves a fault, this fault is immediately corrected and the profit function is not affected. Undetected/unresolved faults are not corrected and as a result x is different from zero and the profit decreases.

- *Value*: Narasimhan and Rengaswamy (2007) identify value as the expectation of profit. This is calculated in two steps: In the first step, corresponding to the sensor network, the set of resolvable faults is determined. The value is then calculated by determining the expected value of the profit. Such expectation comes from the contributions of normal operation (no faults; V_0), resolvable faults (detected and corrected faults; V_r), unresolved faults (undetected and therefore uncorrected faults; V_{nr}), and "events" that actually contribute to improve profit (detected and corrected faults; V_1).

$$V = V_0 + V_r + V_{ur} + V_1 \tag{14.20}$$

The first two are equal to the normal profit c^* multiplied by the probability of the state and the fourth is ignored for simplicity. Finally, V_{nr} is given by

$$V_{nr} = \sum_i \int_{LB_N}^{UB_N} c(x_i)\varphi_i(x_i)dx_i \tag{14.21}$$

The integral has a summation over all probabilities of each fault separately, because the assumption of one fault at a time was made. Finally V_1 is ignored.

Because the profit function is monotone nonincreasing, value is assessed as the upper bound of profit and computed only in the first period.

- *Fault Detection*: A signed directed graph is used to identify the faults. These faults were classified as detectable and resolvable using their own algorithms. For the purpose of our analysis, the methodology outlined earlier can be used.

Example 14.2: Consider the CSTR example of Example 14.1(Fig. 14.2). Equations (14.10) through (14.17) were used for the CSTR model. The following data and results are taken from Narasimhan and Rengaswamy (2007). Nominal values for all parameters were shown above.

The following positive and negative disturbances are considered: Cai^+, Cai^-, Ti^+, Ti^-, Fi^+, Fi^-, Tci^+, Tci^-, corresponding to the concentration, temperature, and flowrate of the reacting inlet stream, and the inlet coolant temperature, respectively. Other faults considered are Cd^- and U^-, which correspond to catalyst deactivation and heat exchange area fouling.

Table 14.5 shows the lower and upper limits of operation (*LB* and *UB*) as well as the normal operation bounds of the quantities used to characterize the faults. The probability p_i for all faults was chosen to be 0.4. Table 14.6 depicts the set of variables that are affected by the fault events (control loops are assumed to be there by default). These variables have been determined using a sign directed graph (not shown).

Sensors can be placed in only 9 locations: *Ca*, *F*, *Fc*, *Tc*, *Fvg*, *Cai*, *Fi*, *Tci*, and *Ti*, resulting in 511 different combinations. Sensor costs are 1000 \$/yr for each sensor. The rest of the cost data are presented in Table 14.7. In this table, P_B represents the production rate of product B in lbmol/h.

Fault	Nominal Value	Normal Range		Abnormal Region	
		LB_N	UB_N	LB	UB
Cai^+ (lb-mol/ft^3)	0.5	0.5	0.55	0.55	0.65
Cai^-(lb-mol/ft^3)	0.5	0.45	0.5	0	0.45
Ti^+ (°R)	530	530	535	535	560
Ti^- (°R)	530	525	530	500	525
Fi^+ (ft^3/h)	40	40	45	45	100
Fi^- (ft^3/h)	40	35	40	0	35
Tci^+ (°R)	530	530	535	535	545
Tci^- (°R)	530	525	530	500	525
Cd^-	1	0.95	1	0.35	0.95
U^-	150	142.5	150	67.5	142.5

Table 14.5 Bounds on Variables for CSTR Example

Fault	Variables Affected by the Event
Cai^+	Ca^+, Fc^+, Tc^-, Fvg^+, Cai^+
Cai^-	Ca^-, Fc^-, Tc^+, Fvg^-, Cai^-
Ti^+	Fc^+, Tc^-, Ti^+
Ti^-	Fc^-, Tc^+, Ti^-
Fi^+	Ca^+, F^+, $Fc^\pm$, $Tc^\pm$, Fvg^+, Fi^+
Fi^-	Ca^-, F^-, $Fc^\pm$, $Tc^\pm$, Fvg^-, Fi^-
Tci^+	Fc^+, Tci^+
Tci^-	Fc^-, Tci^-
Cd^-	Ca^+, Fc^-, Tc^+, Fvg^-
U^-	Fc^+, Fc^-, Tc^-

Table 14.6 Variables Affected by Faults

Events Ti^- and Tci^- (corresponding to a decrease in inlet temperature of feed and coolant, respectively) lead to an increase in the operating profit, so their presence can be ignored. The sets of resolvable and unresolvable faults are shown in Table 14.8. Sensor network S_1 comprising [*Ca, F, Fc, Tc, Ti*] gives the best resolution properties as all faults can be resolved from each other. Sensor network S_2 comprising [*Cai*], that is, inlet concentration sensor can obviously observe and resolve only the deviations in inlet concentration. Likewise, sensor networks S_3 [*Fi:* inlet flowrate of reactant] and S_4 [*F:* outlet flowrate from reactor] can observe and resolve only deviations in inlet flowrates.

Revenue	Coolant Cost	Vapor Cost ($/ft^3)
$R = \begin{cases} 0.375\,P_B & \text{if } P_B \leq 10.61\,\text{mol/h} \\ -0.2433\,P_B + 6.56 & \text{if } P_B > 10.61\,\text{mol/h} \end{cases}$	0.015 ($/ft^3)	0.002255 ($/ft^3)

TABLE 14.7 Cost Data

Sensor Network	Sensors	Set of Resolvable Faults
S_1	[*Ca, F, Fc, Tc, Ti*]	{Cai^+}, {Cai^-}, {Ti^+}, {Ti^-}, {Fi^+}, {Fi^-}, {Tci^+}, {Tci^-}, {Cd^-}, {U^-}
S_2	[*Cai*]	{Cai^+}, {Cai^-}, {$Ti^\pm$, $Fi^\pm$, $Tci^\pm$, Cd^-, U^-}
S_3	[*Fi*]	{$Cai^\pm$, $Ti^\pm$, $Tci^\pm$, Cd^-, U^-}, {Fi^+}, {Fi^-}
S_4	[*F*]	{$Cai^\pm$, $Ti^\pm$, $Tci^\pm$, Cd^-,U^-}, {Fi^+}, {Fi^-}

TABLE 14.8 Fault Classsification as a Function of Sensor Networks

A naive expectation would be that since S_1 gives best resolution properties, it would be the most preferred sensor network from fault diagnosis perspectives that does not explicitly monetize the benefits of the diagnosis.

Table 14.9 computes the value (revenue minus coolant and vapor costs) assuming 8760 h of annual operation. While S_1 provides the largest value, S_2 [*Cai*] is better if the cost of sensors is subtracted, even though its resolution properties (resolving only inlet concentration deviations) are inferior to that of S_1. Sensor networks S_3 and S_4 are also preferred to S_1. The implication is that the benefits do not justify the extra expense incurred.

While the theory and procedures presented by Narasimhan and Rengaswamy (2007) are appealing, especially if low computing time is required, the simplifying assumptions have not been tested. In addition, it is unclear how tedious it would be to apply this methodology to large plants. One alternative, which Narasimhan and Rengaswamy (2007) disregarded as computationally too intensive, is to resort to Monte Carlo simulations over the horizon selected. Such technique is also used in Chap. 16 to assess the efficiency and cost of preventive maintenance.

Sensor Network	Sensors	Value	Sensor Cost	Net Value (Value-Cost)
S_1	[*Ca, F, Fc, Tc, Ti*]	27,453 $/yr	5,000 $/yr	22,453 $/yr
S_2	[*Cai*]	26,706 $/yr	1,000 $/yr	25,706 $/yr
S_3	[*Fi*]	26,417 $/yr	1,000 $/yr	25,417 $/yr
S_4	[*F*]	26,417 $/yr	1,000 $/yr	25,417 $/yr

TABLE 14.9 Sensor Network Value

References

Aelion, V. and G.J. Powers,"Risk Reduction of Operating Procedures and Process Flowsheets," *Ind. Eng. Chem. Res.* **32**:82–90 (1993).

Bagajewicz, M. *Process Plant Instrumentation. Design and Upgrade* (ISBN:1-56676-998-1), (Technomic Publishing Company) (http://www.techpub.com). Now CRC Press (http://www.crcpress.com) (2000).

Bhushan, M. and R. Rengaswamy, "Design of sensor network based on the signed directed graph of the process for efficient fault diagnosis," *Ind. Eng. Chem. Res.* **39**:999–1019 (2000).

Catino, C.A. and L.H. Ungar, "Model-based Approach to Automated Hazard Identification of Chemical Plants," *AIChE J.* **41**(1):97–109 (1995).

Chang, C.C. and C.C. Yu,"Online Fault Diagnosis using Signed Directed Graphs," *Ind. Eng. Chem. Res.* **29**:1290 (1990).

Chen, B.H., X.Z. Wang, and C. McGreavy, "Online Operational Support System for Fault Diagnosis in Process Plants," *Comp. Chem. Eng.* **22**(Suppl.):S973–S976 (1998).

Dong, D. and T.J. McAvoy, "Nonlinear principal component analysis based on principal curves and neural networks," *Comp. Chem. Eng.* **20**(1):65–78 (1996).

Dunia, R. and S.J. Qin, "A Unified Geometric Approach to Process and Sensor Fault Identification and Reconstruction: The Unidimensional Fault Case," *Comp. Chem. Eng.* **22**(7–8):927–943 (1998).

Fan, J.Y., M. Nikolau, and R.E. White, "An Approach to Fault Diagnosis of Processes via Neural Networks," *AIChE J.* **39**(1):82–88 (1993).

Fathi, Z., W.F. Ramirez, and J. Korbicz, "Analytical and Knowledge-based Redundancy for Fault Diagnosis in Process plants," *AIChE J.* **39**(1):42–56 (1993).

Frank, P.M, "Fault Diagnosis in Dynamic Systems using Analytical and Knowledge Based Redundancy. A Survey of some Results," *Automatica.* **26**(3):459–474 (1990).

Gertler, J.J, "A Survey of Model-based Failure Detection and Isolation in Complex Plants," *IEEE Cont. Syst. Mag.* **8**(6):3–11 (1988).

Gertler, J.J., *Fault Detection and Diagnosis in Engineering Systems.* Marcel Dekker, New York (1998).

Himmelblau D., *Fault Detection and Diagnosis in Chemical and Petrochemical Processes,* Elsevier, Amsterdam (1978).

Iri, M., K. Aoki, E. O'Shima, and H. Matsuyama., "An Algorithm for Diagnosis of System Failures in the Chemical Process," *Comp. Chem. Eng.* **3**:489–493 (1979).

Jackson, J.E., *A User's Guide to Principal Components,* Wiley, New York (1991).

Kavuri, S.N., and V. Venkatasubramanian., "Representing Bounded Fault Classes using Neural Networks with Ellipsoidal Activation Functions," *Comp. Chem. Eng.* **17**(2):139–163 (1993).

Kokawa, M. and S. Shingai., "Failure Propagation Simulation and Non-failure Path Search in Network Systems," *Automatica,* **18**:335 (1982).

Kokawa, M., S. Miyasaki, and S. Shingai, "Fault Location using Digraph and Inverse Direction Search with Application," *Automatica* **19**:729 (1983).

Kramer, M.A., "Malfunction Diagnosis using Quantitative Models with Non-boolean Reasoning in Expert Systems," *AIChE J.* **33**(1):130–140 (1987).

Kramer, M.A. and B.L. Palowitch, Jr., "A Rule-based Approach to Fault Diagnosis using the signed Directed Graph," *AIChE J.* **33**(7):1067–1078 (1987).

Kresta, J.V., J.F. McGregor, and T. Marlin., "Multivariate Statistical Monitoring of Process Operating Performance," *Can. J. Chem. Eng.* **69**:35–47 (1991).

Lapp, S.A. and G.J. Powers, "Computer-aided Synthesis of Fault Trees," *IEEE Trans. Reliab.* **2**:13 (1977).

McGregor, J.F., C. Jaeckle, C. Kiparissides, and M. Koutoudi, "Process Monitoring and Diagnosis by Multiblock PLS Methods," *AIChE J.* **40**(5):826–838 (1994).

Mohindra, S. and P.A. Clark, "A Distributed Fault Diagnosis Method based on Digraph Models: Steady-State Analysis," *Comp. Chem. Eng.* **17**(2):193–209 (1993).

Narasimhan, S. and R. Rengaswamy, "Quantification of Performance of Sensor Networks for Fault Diagnosis," *AIChE J.* **53**(4) April (2007).

Patton, R.J., "Robustness in Model-based Fault Diagnosis: The 1995 Situation," *Proceedings of the IFAC On-Line Fault Detection and Supervision in the Chemical Process Industries*, Newcastle upon Tyne, UK (1995).

Patton, R.J., P.M. Frank, and R.N. Clark, *Fault Diagnosis in Dynamic Systems: Theory and Application,* Prentice Hall, Englewood Cliffs, NJ (1989).

Pau, L.F., *Failure Diagnosis and Performance Monitoring*, Dekker, New York (1981).

Qian, D.Q., "An Improved Method for Fault Location of Chemical Plants," *Comp. Chem. Eng.* **14**(1): 41–48 (1990).

Qin, S.J., "Recursive PLS Algorithms for Adaptive Data Modeling," *Comp. Chem. Eng.* **22**(4/5): 503–514 (1998).

Qin, S.J., and T.J. McAvoy, "Nonlinear PLS Modeling using Neural Networks," *Comp.Chem. Eng.* **16**(4):379–391 (1992).

Shao, R., F. Jia, B. Martin, and A.J. Morris, "Fault Detection Using Wavelet Filtering and Non-linear Principal Component Analysis," *Proc. IFAC. Workshop on On-Line FaultDetection and Supervision in the Chemical Process Industries*, Lyon, France (1998).

Shiozaki, J., H. Matsuyama, E. O'Shima, and M. Iri, "An Improved Algorithm Fordiagnosis of System Failures in the Chemical Process," *Comp. Chem. Eng.* **9**:285 (1985).

Umeda, T., T. Kuriyama, E. O'Shima, and H. Matsuyama, "A Graphical Approach to Cause and Effect Analysis of Chemical Processing Systems," *Chem. Eng. Sci.* **35**:2379 (1980).

Vedam, H., and V. Venkatasubramanian, "Automated Interpretation of PCA-Based Process Monitoring and Fault Diagnosis Using Signed Digraphs," *Proc. IFAC. Workshop on On-Line Fault Detection and Supervision in the Chemical Process Industries*. Lyon, France (1998).

Venkatasubramanian, V., and S.H. Rich, "An Object Oriented Two-tier Aarchitecture for Integrating Ccompiled and Deep-level knowledge for process diagnosis," *Comp. Chem. Eng.* **12**(9/10): 903–921 (1988).

Venkatasubramanian, V., R. Rengaswamy, K. Yin, and S. Kavuri, "Review of Process Fault Detection Diagnosis—Part I: Quantitative Model Based Methods," *Comp. Chem. Eng.* **27**:239–311 (2003a).

Venkatasubramanian,V., R. Rengaswamy, and S. Kavuri, "Review of Process Fault Detection Diagnosis—Part II: Qualitative Models and Search strategies," *Comp. Chem. Eng.* **27**:313–326 (2003b).

Venkatasubramanian, V., R. Rengaswamy, S. Kavuri, and K. Yin, "Review of Process Fault Detection Diagnosis—Part III: Process History Models," *Comp. Chem. Eng.* **27**:313–326 (2003c).

Vinson, J.M., and L.H. Ungar, "Using PLS for Fault Analysis: A Case Study," *AIChE Ann. Meeting* (1994).

Watanabe, K., I. Matsura, M. Abe, M. Kubota, and D.M. Himmelblau, "Incipient Fault Diagnosis of Chemical Processes via Aartificial Neural Networks," *AIChE J.* **35**:1803 (1989).

Wilcox, N.A., and D.M. Himmelblau, "The Possible Cause and Effect Graphs (PCEG) Model for Fault Diagnosis-I. Methodology," *Comp. Chem. Eng.* **18**(2): 103–116 (1994a).

Wilcox, N.A., and D.M. Himmelblau, "The Possible Cause and Eeffect Graphs (PCEG) Model for Fault Diagnosis-II. Applications," *Comp. Chem. Eng.* **18**(2): 117–127 (1994b).

Wise, B.M., and N.B. Ghallagher, "The Process Chemometrics Approach to Process Monitoring and Fault Detection," *Proceedings of the IFAC On-Line Fault Detection and Supervision in the Chemical Process Industries*, Newcastle upon Tyne, UK (1995).

CHAPTER 15

Value of Instrumentation Upgrade—Monitoring and Faults Perspectives

Because data reconciliation, state estimation, advanced control, and advanced fault resolution methods are gaining acceptance in practice by process engineers and instrumentation engineers, the issue of instrument revamping becomes important, simply because existing instrumentation does not usually allow these methods to perform well.

Replacing an existing measurement device with a more accurate one is the most straightforward approach for increasing reconciled data accuracy for production accounting and quality control (Chap. 11), increased control efficacy through reduction of expected dynamic operating region (EDOR) and therefore more advantageous back-off operating point (BOP) (Chap. 13), and better fault resolution through more accurate residuals (Chap. 14). However, this is not the most economical option and it may not even be the answer if the level of software redundancy is low. New instrumentation in new places is likely to be needed. Because such measurements are costly and because their effect on accuracy of the final data vary depending on the number type, precision, and location of these sensors, the trade-off between the beneficial effect of increasing accuracy of data and the cost of the new sensors needs to be resolved through optimization.

The answer to those needs is that sensor networks should be able to handle gross errors effectively, and be precise enough for back-off control purposes and effective in providing appropriate data to detect parametric and structural failures. In summary, sensor networks together with the appropriate software (data reconciliation, Kalman filters, fault identification protocols, etc.) should provide precise and bias-free variable estimators so that the process plant can run as intended. Sensor networks that can perform this function are called robust.

Summarizing, sensor networks should feature the following attributes:

(a) *Provide Accurate Estimators of Key Variables:* The accuracy of key variables obtained through data reconciliation should be of a level that satisfies the needs of fault monitoring, control, production accounting (material balance), and quality control. This includes (as we saw in previous chapters)
   - *Detecting Biases and Gross Errors Efficiently:* The ability of a sensor network to effectively detect gross errors is connected to the redundancy of the system. The more measurements one adds, the larger is the network ability to detect gross errors.
   - *Detecting Leaks Efficiently and Being Capable of Differentiating them from Biases:* In other words, not be subject to equivalencies.

(b) *Retain Certain Degree of Accuracy when Gross Errors are Eliminated:* Once a process like serial elimination is used, the remaining system should guarantee certain accuracy for key important variables. This property is called *residual accuracy*. A similar concept was introduced for precision (see Bagajewicz, 2000).

(c) *Reduce Smearing:* When gross errors are not detected, the smearing effect should be low. We call this property resilience, and as we discussed in Chap. 11, it is now embedded in the concept of accuracy.

(d) *Be Reliable:* Instruments that break often are a frequent source of gross errors. Thus nonreliable instruments should be avoided. Reliability is now embedded in the concept of stochastic accuracy.

All the above properties are desirable, but they all require capital investment.

Thus, the problem of instrumentation network design and upgrade consists of determining the optimal set of measured variables and selecting the accuracy and reliability of the corresponding instruments.

This selection must be done to achieve certain performance targets: accuracy of key variables, fault observability, resolution of certain selected faults, and desired variance/accuracy for control purposes.

There are, however, two objectives to consider

- Minimize cost when the performance targets are known, or
- Maximize value when functions relating profit to the instrumentation network are available.

In this chapter we briefly cover the design of instrumentation networks for fault detection, parameter estimation, and monitoring in general. The next chapter covers instrumentation for control purposes.

Cost-Optimal Instrumentation Design

We first consider a motivating example of a design of an instrumentation network for monitoring purposes, where precision is the targeted performance metric.

Example 15.1: Consider for example the process flow diagram of Fig. 15.1.

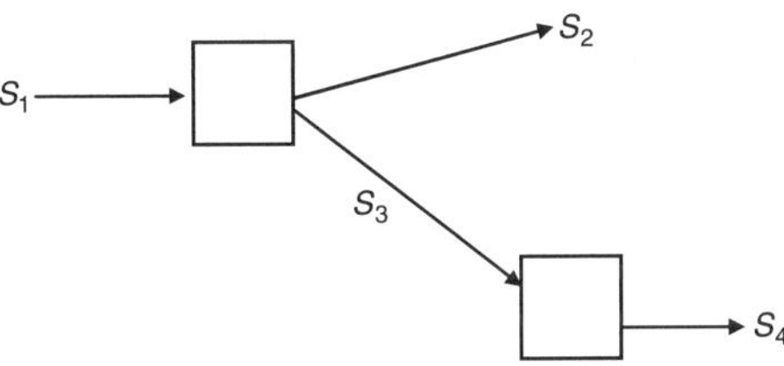

FIGURE 15.1 Flow sheet for Example 15.1.

Assume the flowrates are given by $F = (150.1, 52.3, 97.8, 97.8)$. Assume that the available instrumentation is that of Table 15.1, with costs (in arbitrary units) that depend on precision, but for simplicity not on flowrate. Assume now that the precision required in different streams is the one shown in Table 15.2.

Precision	Cost
1%	2,500
2%	1,500
3%	800

TABLE 15.1 Cost of Available Instrumentation

Stream	Precision Required
S_1	1.5%
S_4	2.0%

TABLE 15.2 Precision Required

One can solve this problem by simple inspection. However, for larger flow sheets, one needs to resort to some computational tools (Bagajewicz, 1997; 2000) as discussed below. Two globally optimal solutions of equal cost ($C = 3000$) are shown in Table 15.3.

Solution	S_1	S_2	S_3	S_4
A	–	2%	2%	–
B	–	2%	–	2%

TABLE 15.3 Solutions of the Precision Constrained Problem

The solution consists of a nonredundant network. In case B, measuring S_2 and S_4 makes S_1 observable with a precision that is lower than the required one. At the same time, S_4 is nonredundant and has the required precision.

Details of the Feasibility Calculations: We start with the expression of the variance of the estimators for redundant measurements (Chap. 4), the variance for nonredundant measurements, and the variance of the estimators for unobservable variables.

$$\hat{S}_R = [I - S_R E_R^T (E_R S_R E_R^T)^{-1} E_R] S_R \tag{15.1}$$

$$\hat{S}_{NR} = S_{NR} \tag{15.2}$$

$$\hat{S}_O = E_{RO} \hat{S}_R E_{RO}^T + E_{NRO} S_{NR} E_{NRO}^T \tag{15.3}$$

Consider the case of Table 15.3. Solution A, where the measurements are nonredundant. In this case, the canonical form of the incidence matrix is

$$\begin{matrix} & S_1 & S_4 & S_2 & S_3 \end{matrix}$$

$$C = \begin{bmatrix} 1 & 0 & -1 & -1 \\ 0 & 1 & 0 & -1 \end{bmatrix} \tag{15.4}$$

where

$$\begin{matrix} & S_2 & S_3 \end{matrix}$$

$$E_{NRO} = \begin{bmatrix} -1 & -1 \\ 0 & 1 \end{bmatrix} \tag{15.5}$$

$$\begin{matrix} & S_1 & S_4 \end{matrix}$$

$$E_O = \begin{bmatrix} 1 & 0 \\ 0 & 1 \end{bmatrix} \tag{15.6}$$

Note that there are no redundant measurements and therefore E_R does not exist. Then,

$$\hat{S}_{NR} = \begin{bmatrix} \sigma_2^2 & 0 \\ 0 & \sigma_3^2 \end{bmatrix} = \begin{bmatrix} (0.02 \times 52.3)^2 & 0 \\ 0 & (0.02 \times 97.8)^2 \end{bmatrix} \tag{15.7}$$

$$\hat{S}_O = E_{NRO} S_{NR} E_{NRO}^T = \begin{bmatrix} -1 & -1 \\ 0 & -1 \end{bmatrix} \begin{bmatrix} \sigma_2^2 & 0 \\ 0 & \sigma_3^2 \end{bmatrix} \begin{bmatrix} -1 & 0 \\ -1 & -1 \end{bmatrix} = \begin{bmatrix} -1 & -1 \\ 0 & -1 \end{bmatrix} \begin{bmatrix} -\sigma_2^2 & 0 \\ -\sigma_3^2 & -\sigma_3^2 \end{bmatrix} = \begin{bmatrix} \sigma_2^2 + \sigma_3^2 & \sigma_3^2 \\ \sigma_3^2 & \sigma_3^2 \end{bmatrix} \tag{15.8}$$

Thus, the diagonal of $\hat{S}_O$ gives the precision of the estimators for F_1 and F_4, that is,

$$\hat{\sigma}_1 = \sqrt{(0.02\times 52.3)^2 + (0.02\times 97.8)^2} = 2.218 \tag{15.9}$$

$$\hat{\sigma}_4 = \sigma_3 = 0.02\times 97.8 = 1.956 \tag{15.10}$$

Thus, percentage wise $\hat{\sigma}_1$ (%) = (100 × 2.218/150.1) = 1.48% < 2% and $\hat{\sigma}_4$ (%) = (100 × 1.956/97.8) = 2%.

Example 15.2: Assume now that the cost of the 3% sensors drops to 700. Then the optimal solution is no longer the one shown in Table 15.3. In this case two solutions of equal cost ($C = 2900$) and are shown in Table 15.4.

Solution	S_1	S_2	S_3	S_4
C	3%	3%	2%	–
D	3%	3%	–	2%

Table 15.4 Solutions of the Precision Constrained Problem (New Cost in 3% Instrument)

This solution is redundant and cheaper than the nonredundant given in Table 15.3.

Details of the Feasibility Calculations: In this case, for Solution C, we have

$$C = \begin{array}{c} \begin{matrix} S_4 & S_1 & S_2 & S_3 \end{matrix} \\ \begin{bmatrix} 0 & 1 & -1 & -1 \\ 1 & 0 & 0 & -1 \end{bmatrix} \end{array} \tag{15.11}$$

Now, rearranging and operating as in Chap. 3, we get

$$E_R = \begin{array}{c} \begin{matrix} S_1 & S_2 & S_3 \end{matrix} \\ \begin{bmatrix} 1 & -1 & -1 \end{bmatrix} \end{array} \tag{15.12}$$

$$E_{RO} = \begin{array}{c} \begin{matrix} S_1 & S_2 & S_3 \end{matrix} \\ \begin{bmatrix} 0 & 0 & -1 \end{bmatrix} \end{array} \tag{15.13}$$

To obtain the variance of the estimators of the redundant flowrates, we first calculate $(E_R S_R E_R^T)^{-1}$. Thus,

$$E_R S_R = \begin{bmatrix} 1 & -1 & -1 \end{bmatrix} \begin{bmatrix} \sigma_1^2 & 0 & 0 \\ 0 & \sigma_2^2 & 0 \\ 0 & 0 & \sigma_3^2 \end{bmatrix} = \begin{bmatrix} \sigma_1^2 & -\sigma_2^2 & -\sigma_3^2 \end{bmatrix} \tag{15.14}$$

$$E_R S_R E_R^T = \begin{bmatrix} 1 & -1 & -1 \end{bmatrix} \begin{bmatrix} \sigma_1^2 & 0 & 0 \\ 0 & \sigma_2^2 & 0 \\ 0 & 0 & \sigma_3^2 \end{bmatrix} \begin{bmatrix} 1 \\ -1 \\ -1 \end{bmatrix} = \begin{bmatrix} \sigma_1^2 & -\sigma_2^2 & -\sigma_3^2 \end{bmatrix} \begin{bmatrix} 1 \\ -1 \\ -1 \end{bmatrix} = \sigma_1^2 + \sigma_2^2 + \sigma_3^2 \tag{15.15}$$

and therefore,

$$(E_R S_R E_R^T)^{-1} = \frac{1}{\sigma_1^2 + \sigma_2^2 + \sigma_3^2} \tag{15.16}$$

$$(E_R S_R E_R^T)^{-1} E_R S_R = \frac{1}{\sigma_1^2 + \sigma_2^2 + \sigma_3^2}\begin{bmatrix}\sigma_1^2 & -\sigma_2^2 & -\sigma_3^2\end{bmatrix}$$

$$= \begin{bmatrix}\dfrac{\sigma_1^2}{\sigma_1^2 + \sigma_2^2 + \sigma_3^2} & \dfrac{-\sigma_2^2}{\sigma_1^2 + \sigma_2^2 + \sigma_3^2} & \dfrac{-\sigma_3^2}{\sigma_1^2 + \sigma_2^2 + \sigma_3^2}\end{bmatrix} \tag{15.17}$$

$$S_R E_R^T (E_R S_R E_R^T)^{-1} E_R S_R = \begin{bmatrix}\sigma_1^2 \\ -\sigma_2^2 \\ -\sigma_3^2\end{bmatrix}\begin{bmatrix}\dfrac{\sigma_1^2}{\sigma_1^2 + \sigma_2^2 + \sigma_3^2} & \dfrac{-\sigma_2^2}{\sigma_1^2 + \sigma_2^2 + \sigma_3^2} & \dfrac{-\sigma_3^2}{\sigma_1^2 + \sigma_2^2 + \sigma_3^2}\end{bmatrix} \tag{15.18}$$

$$S_R E_R^T (E_R S_R E_R^T)^{-1} E_R S_R = \begin{bmatrix}\dfrac{\sigma_1^4}{\sigma_1^2 + \sigma_2^2 + \sigma_3^2} & \dfrac{-\sigma_1^2\sigma_2^2}{\sigma_1^2 + \sigma_2^2 + \sigma_3^2} & \dfrac{-\sigma_1^2\sigma_3^2}{\sigma_1^2 + \sigma_2^2 + \sigma_3^2} \\ \dfrac{-\sigma_1^2\sigma_2^2}{\sigma_1^2 + \sigma_2^2 + \sigma_3^2} & \dfrac{\sigma_2^4}{\sigma_1^2 + \sigma_2^2 + \sigma_3^2} & \dfrac{\sigma_2^2\sigma_3^2}{\sigma_1^2 + \sigma_2^2 + \sigma_3^2} \\ \dfrac{-\sigma_1^2\sigma_3^2}{\sigma_1^2 + \sigma_2^2 + \sigma_3^2} & \dfrac{\sigma_2^2\sigma_3^2}{\sigma_1^2 + \sigma_2^2 + \sigma_3^2} & \dfrac{\sigma_3^4}{\sigma_1^2 + \sigma_2^2 + \sigma_3^2}\end{bmatrix} \tag{15.19}$$

Finally, $\hat{S}_R = [I - S_R E_R^T (E_R S_R E_R^T)^{-1} E_R] S_R$ renders

$$\hat{S}_R = \begin{bmatrix}\sigma_1^2 - \dfrac{\sigma_1^4}{\sigma_1^2 + \sigma_2^2 + \sigma_3^2} & \dfrac{\sigma_1^2\sigma_2^2}{\sigma_1^2 + \sigma_2^2 + \sigma_3^2} & \dfrac{\sigma_1^2\sigma_3^2}{\sigma_1^2 + \sigma_2^2 + \sigma_3^2} \\ \dfrac{\sigma_1^2\sigma_2^2}{\sigma_1^2 + \sigma_2^2 + \sigma_3^2} & \sigma_2^2 - \dfrac{\sigma_2^4}{\sigma_1^2 + \sigma_2^2 + \sigma_3^2} & \dfrac{-\sigma_2^2\sigma_3^2}{\sigma_1^2 + \sigma_2^2 + \sigma_3^2} \\ \dfrac{\sigma_1^2\sigma_3^2}{\sigma_1^2 + \sigma_2^2 + \sigma_3^2} & \dfrac{-\sigma_2^2\sigma_3^2}{\sigma_1^2 + \sigma_2^2 + \sigma_3^2} & \sigma_3^2 - \dfrac{\sigma_3^4}{\sigma_1^2 + \sigma_2^2 + \sigma_3^2}\end{bmatrix} \tag{15.20}$$

Thus, the diagonal of $\hat{S}_R$ gives

$$\hat{\sigma}_1^2 = \sigma_1^2\left[1 - \frac{\sigma_1^2}{\sigma_1^2 + \sigma_2^2 + \sigma_3^2}\right] \tag{15.21}$$

$$\hat{\sigma}_2^2 = \sigma_2^2\left[1 - \frac{\sigma_2^2}{\sigma_1^2 + \sigma_2^2 + \sigma_3^2}\right] \tag{15.22}$$

$$\hat{\sigma}_3^2 = \sigma_3^2\left[1 - \frac{\sigma_3^2}{\sigma_1^2 + \sigma_2^2 + \sigma_3^2}\right] \tag{15.23}$$

Thus, percentage wise $\hat{\sigma}_1 = (100 \times 2.218/150.1) = 1.48\% < 2\%$. In turn $\hat{S}_O = E_{\mathrm{RO}} \hat{S}_R E_{\mathrm{RO}}^T$ renders

$$\hat{S}_O = \begin{bmatrix}0 & 0 & -1\end{bmatrix}\begin{bmatrix}\sigma_1^2 - \dfrac{\sigma_1^4}{\sigma_1^2 + \sigma_2^2 + \sigma_3^2} & \dfrac{\sigma_1^2\sigma_2^2}{\sigma_1^2 + \sigma_2^2 + \sigma_3^2} & \dfrac{\sigma_1^2\sigma_3^2}{\sigma_1^2 + \sigma_2^2 + \sigma_3^2} \\ \dfrac{\sigma_1^2\sigma_2^2}{\sigma_1^2 + \sigma_2^2 + \sigma_3^2} & \sigma_2^2 - \dfrac{\sigma_2^4}{\sigma_1^2 + \sigma_2^2 + \sigma_3^2} & \dfrac{-\sigma_2^2\sigma_3^2}{\sigma_1^2 + \sigma_2^2 + \sigma_3^2} \\ \dfrac{\sigma_1^2\sigma_3^2}{\sigma_1^2 + \sigma_2^2 + \sigma_3^2} & \dfrac{-\sigma_2^2\sigma_3^2}{\sigma_1^2 + \sigma_2^2 + \sigma_3^2} & \sigma_3^2 - \dfrac{\sigma_3^4}{\sigma_1^2 + \sigma_2^2 + \sigma_3^2}\end{bmatrix}\begin{bmatrix}0 \\ 0 \\ -1\end{bmatrix} \tag{15.24}$$

$$\hat{S}_O = \begin{bmatrix} 0 & 0 & -1 \end{bmatrix} \begin{bmatrix} \dfrac{-\sigma_1^2\sigma_3^2}{\sigma_1^2+\sigma_2^2+\sigma_3^2} \\ \dfrac{\sigma_2^2\sigma_3^2}{\sigma_1^2+\sigma_2^2+\sigma_3^2} \\ -\sigma_3^2 + \dfrac{\sigma_3^4}{\sigma_1^2+\sigma_2^2+\sigma_3^2} \end{bmatrix} = \sigma_3^2 - \frac{\sigma_3^4}{\sigma_1^2+\sigma_2^2+\sigma_3^2} \tag{15.25}$$

Thus, $\hat{S}_O = \hat{\sigma}_4$ and therefore, percentage-wise, we have $\hat{\sigma}_4\,(\%) = (100 \times 2.218/150.1) = 1.48\% < 2\%$.

We conclude that efficient handling of gross errors requires redundant sensor networks, as we saw in previous chapters. Moreover, we expect the redundancy requirements to be higher if gross errors are to be effectively detected. We conclude that to be able to handle both precision and bias/gross error filtering, we need to use accuracy as a constraint. If we also want to add reliability, stochastic accuracy will have to be used instead.

Cost-Optimal Design

As in every design there is a representation of the variables: Assume that z is the vector of all mass flows and let q be a vector of binary variables defined by

$$q_i = \begin{cases} 1 & \text{if } z_i \text{ is measured} \\ 0 & \text{otherwise} \end{cases} \tag{15.26}$$

When other performance metrics are considered, like fault observability, fault resolution, variance for backed-off control, or accuracy for monitoring purposes, then these metrics will have to have targets defined. We therefore consider the following mathematical statement of the *minimum cost formulation* of the sensor network design/upgrade problem.

$$\left.\begin{array}{l} \text{Min sensor network cost}(q) \\ \text{s.t.} \\ \text{Performance metrics }(q) \le \text{performance targets} \end{array}\right\} \tag{15.27}$$

As stated above, the performance targets are accuracy, fault resolution, and variability in control.

Cost-Optimal Design for Precision or Accuracy

We now consider our first performance metrics: accuracy. Recall from Chap. 8 that a general nonlinear system of equations can be classified using reduction to canonical forms, matrix projection, or

Q-R factorization after one linearizes the system ($f(x) = 0$) around the design or operating point. The general form is

$$D\,x \cong b \tag{15.28}$$

where matrix D represents the Jacobian of $f(x)$ around x_0 and b is the corresponding constant. We also recall that the matrix D is partitioned in submatrices D_M and D_U, which are related to the vector of measured variables (x_M), unmeasured parameters, and state variables (x_U), respectively, such that

$$[D_M \quad D_U]\begin{bmatrix} x_M \\ x_U \end{bmatrix} = b \tag{15.29}$$

We can now say that in the context of instrumentation design, matrices D_M and D_U are a function of the choice of instruments, that is, $D_M = D_M(q)$ and $D_U = D_U(q)$. We also recall that

$$\hat{S}_O = H^{-1} \tag{15.30}$$

$$\hat{S}_R = S_R - S_R D_M^T (I - G^{-1} B H^{-1} B^T) G^{-1} D_M S_R \tag{15.31}$$

where $G = D_M S_R D_M^T$ and $H = D_U^T G^{-1} D_U$. Thus, the variance of the reconciled values depends on what is the set of sensors selected, that is, $\hat{\sigma}_i = \hat{\sigma}_i(q)$

$$\sigma_i(q) = \begin{cases} \sqrt{\left[\hat{S}_R(q)\right]_{ii}} & \text{if } z_i \text{ is measured and redundant} \\ \sqrt{\left[\hat{S}_O(q)\right]_{ii}} & \text{otherwise} \end{cases} \tag{15.32}$$

Finally, software accuracy of order one is given by

$$\tilde{a}_i^{MP(p,1)} = \sqrt{\left[\hat{S}_R\right]_{ii}} + Z_{\text{crit}}^{(p)} \operatorname*{Max}_{\forall s} \frac{\left[I - \left(\hat{S}_R D_M^T \left(D_M \hat{S}_R D_M^T\right)^{-1} D_M\right)\right]_{is}}{\sqrt{\left[D_M^T \left(D_M \hat{S}_R D_M^T\right)^{-1} D_M\right]_{ss}}} \tag{15.33}$$

and stochastic accuracy is given by

$$a_i(q) = \hat{\sigma}_i(q) + E[\delta_i(q)] \tag{15.34}$$

Consider now the definition of cost. If for each variable z_i there is only one potential measuring device with associated cost c_i, then the total cost is given by

$$C(q) = \sum_{\forall i} c_i\, q_i \tag{15.35}$$

Remark: In addition one can consider maintenance costs as part of the total cost and require fault detection capabilities in addition to accuracy. Thus, cost would be given by

$$\text{Cost} = \sum_{\forall i} d_i\, q_i + \text{maintenance cost}\,(q) \tag{15.36}$$

where d_i is some form of annualized cost of the instruments. In Chap. 17, we discuss the different ways in which maintenance cost can be assessed. In addition, we recognize that accuracy will also be a function of the frequency of maintenance. For simplicity, we omit incorporating the maintenance cost in the rest of this chapter.

When the performance targets are accuracy of certain key variables belonging to a specific set I_s, then the design of the sensor network is an optimization problem that can be written as follows:

$$\left.\begin{array}{ll} \text{Min} \ \sum_{\forall i} c_i q_i & \\ \text{s.t.} & \\ \quad a_k(q) \le a_{k,*} & \forall k \in I_s \\ \quad q_i \in \{0,1\} & \forall i \end{array}\right\} \tag{15.37}$$

where $a_{k,*}$ is the corresponding maximum value of accuracy tolerated for each variable in the set of variable of interest I_s.

Remark: If more than one device is being considered as potential candidate to be used in each variable measurement, the objective function requires the use of additional binary variables and additional constraints. We omit the details, which are described by Bagajewicz (1997). Thus, in the rest of this chapter we will consider that each location that is candidate for an instrument has one potential sensor to be installed.

Historical Note: Different approaches and driving forces have been used for the design of sensor networks. Vaclavek and Loucka (1976) used graph theory to guarantee variable observability. Kretsovalis and Mah (1987) used a combinatorial search based on the effect of the variance of measurements on the precision of key variables. Madron and Veverka (1992) proposed to classify measured and unmeasured variables of linear systems according to a preestablished criterion of "required" and "nonrequired." Unmeasured variables are later ordered from "hardly measured" to "easily measured." Madron proposed to use two objective functions: cost and overall precision of the system. By means of matrix decomposition and an elaborate column permutation procedure, suboptimal structures were found. Madron (1992) also presents details of this procedure based on graph theory. The concept of cost-edged graph is introduced and minimum spanning trees of these graphs are used to obtain minimum cost or optimal overall precision sensor networks. However, the method cannot target desired precision levels on individual variables. Ragot et al. (1992) presented a procedure that allows the identification of the set of sensors for which the system becomes observable. Luong et al. (1994) presented a method that provides solutions that feature minimal observability of those variables required for control and high degree of redundancy of variables. They use reliability as means of screening alternatives with equal cost. Maquin et al. (1994) proposed to obtain the location of sensors by inverting the expression that provides the variance of reconciled variables as a function of the variance of measurements. Ali and Narasimhan (1993) proposed to maximize reliability, which is based on sensor failure probability, observability of variables, and redundancy. While looking at all networks containing the minimum set of sensors to achieve observability, they propose a Max-Min problem using reliability as the objective function. Another graph-oriented procedure was proposed by Meyer et al. (1994) using cost as the objective function and providing solutions featuring networks containing the minimum set of sensors. Later, Ali and Narasimhan (1995) extended their previous work to redundant networks. Their algorithm uses graph theory to build networks with a specified number of sensors and maximum reliability. To address network reliability, Benqlilou et al. (2004) combined quantitative process knowledge and fault tree analysis for evaluating the reliability of process variable estimation. This reliability is used to assess the sensor network reliability, which, in turn, is used for the design and retrofit of the network. The solution is obtained using genetic algorithm (GA). Departing from graph theory and linear algebra approaches, Bagajewicz (1997) proposed a mixed integer nonlinear problem to obtain cost-optimal network structures for linear systems subject to constraints on precision, error detectability, and estimation availability. Simultaneously, Alhéritière et al. (1997) proposed to design a system that maximizes the precision of one variable, subject to cost constraints, which they tested in a refinery setting (Alhéritière et al. 1998). The method had several deficiencies (see Bagajewicz, 2000, Chap. 8). Finally, the connection between cost-optimal and maximum precision mathematical programming models was established by Bagajewicz and Sánchez (1999b).

Tree Search Procedure for the Cost-Optimal Formulation

Equation (15.38) is a mixed integer nonlinear problem (MINLP). Typical and most popular solution procedures for these problems require that the constraints be separable, that is, the constraints can be written as the sum of a nonlinear function of the continuous variables plus a second linear function of the binary variables only. This is not the case here. Moreover, each set of selected instruments gives rise of different matrices: $D_M(q)$ and $D_U(q)$. In each case these matrices have different dimensions and correspond to different variables. In fact, a classification step is needed. Thus, obtaining an explicit representation of matrices $\tilde{Q}_M(q)$ and $\tilde{Q}_U(q)$ seems difficult at first glance, especially because of the classification procedure that is required.

Remark: Bagajewicz and Cabrera (2002) proposed that all variables be considered as "measured" even when $qi = 0$. To be able to obtain the correct value of precision $\sigma_i(q)$, they proposed that the unmeasured variables be assigned a very large variance. Thus, they managed to get rid of the classification step, which is the major hurdle for obtaining explicit expressions of precision as a function of q.

They later manipulated the constraints so that a linear model could be obtained. This mixed integer linear programming (MILP) model was then solved using standard software (GAMS/CPLEX). Because they observed scaling problems among other difficulties, they abandoned pursuing this route. Chmielewski et al. (2002) also proposed to construct linear matrix inequalities and obtain a convex nonlinear problem. This model was only solved for small-scale problems, and its computational performance for larger cases has not been tested. Finally Kelly and Zyngier (2008) presented an alternative MILP formulation that exploited the properties of the Schur complement of the incidence matrix to determine observability without the need to deal with nonsparse matrices corresponding to certain matrix inverses that are found in Bagajewicz and Cabrera (2002). They are capable of solving the problem for minimum cost, subject to observability, redundancy, and precision constraints. It is unclear if this method can be extended to consider accuracy, but if this is possible, then all the existing problems that the tree searches exhibit, even after the latest modeling efforts (Gala and Bagajewicz, 2006a,b; Nguyen and Bagajewicz, 2008) could be overcome.

Although precision and software accuracy could in principle be explicitly written as a function of q, we saw in Chap. 10 that software accuracy does not seem to be a good estimator of the real accuracy, which is better obtained using stochastic accuracy. Because the use of stochastic accuracy requires using the Monte Carlo approach, we consider that pursuing mathematical programming formulations that would exploit the power of known solvers is at this time not warranted.

In view of the above difficulties implicit branch and prune tree enumeration algorithms are proposed. The first branch and prune algorithm was proposed by Bagajewicz (1997). In the next example we describe the basic aspects of this tree.

Example 15.3: Consider the tree depicted in Fig. 15.2, which is the tree corresponding to the flow sheet of Fig. 15.1.

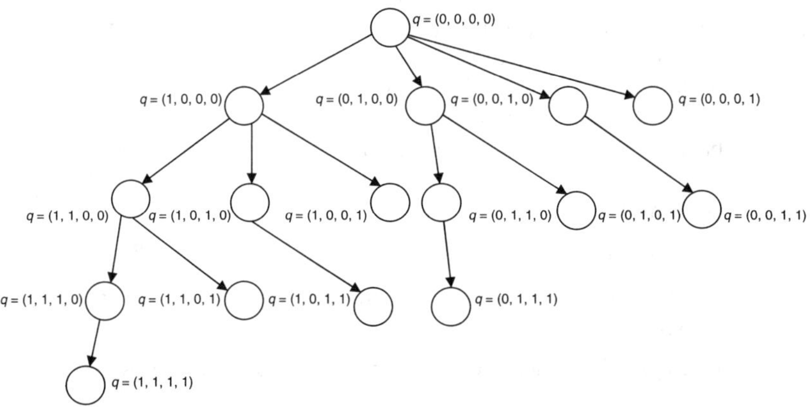

Figure 15.2 Tree of solutions.

This tree has some important properties.

1. The node $q = \{0, 0, 0, 0\ldots\}$ is trivially infeasible.
2. The tree is asymmetric. Indeed, the tree is more populated on the left than on the right because repeated nodes already enumerated are avoided.
3. As measurements are added to a branch, the nodes are infeasible first. This is because key variables will be unobservable or not enough precision has been achieved (low redundancy). Eventually, as measured variables are added, a node will become feasible. The node level at which feasibility is attained varies from branch to branch.
4. Cost increases from one level to the next if only one type of instrument is used. In such case, the first feasible node in each branch is the one with lowest cost in the tree it spans.
5. Finally, retrofit can be handled easily by setting the corresponding values of q to one.

Thus, the branch and prune algorithm is the following:

1. Start with a root node containing no variables being measured $q = \{0, 0, 0, 0,\ldots,0\}$. Consider that the upper bound of cost is the sum of the costs of all instruments plus a small tolerance, that is, $\mathrm{UB} = \sum_{\forall i} c_i + \varepsilon$. We state that the current best solution is $q^* = \{1, 1, 1, 1,\ldots,1\}$.
2. Verify if q^* is feasible, that is, if the performance metrics are below the targets. In the case of accuracy, we compute $a_k(q), \forall k \in I_s$. If the node is infeasible [$a_k(q^*) > a_{k,*}$, for some $k \in I_s$ in the case of accuracy performance metrics], then go to step 3. Otherwise, if the node is feasible go to step 4.
3. Use the branching criteria (discussed in the following section) to pick up the next node of the tree in the next level, that is, a node that contains the instruments of the current node, plus one additional instrument. If no node is left to be picked, go back to the previous level and use the branching criteria to pick up another node. If no nodes are left, then the algorithm is terminated.
4. Compute the cost of the node as $\sum_{\forall i} c_i q_i$. If $\sum_{\forall i} c_i q_i > \mathrm{UB}$, then go to step 3. This is one stopping criteria. Otherwise go to step 5.
5. If $\sum_{\forall i} c_i q_i < \mathrm{UB}$, then we update the upper bound, that is, we make the following assignment $\mathrm{UB} \leftarrow \sum_{\forall i} c_i q_i$, and we also

update the current best solution by making $q^* = q$. Regardless of the cost, we go to step 3.

Branching Criteria

Because what we are looking for is minimum cost, then one logical branching criteria is to pick the next sister node in the tree (one that has the same number of instruments) as the one that has the smallest cost. This means that one should pick the cheapest new instrument to add from the list of instruments that remain to be picked. Using this criterion helps identify cheap networks faster.

Thus, our branching criteria is to pick the node that is less expensive than all the alternatives. There are two pruning criteria.

- *Pruning Criteria 1:* Any node, feasible or not, which has a cost larger than the current upper bound UB, has no chance of generating a better solution, because any additional instrument will increase the cost. Thus the tree below this node is not explored. This pruning criteria is used in step 4.
- *Pruning Criteria 2:* Any feasible node has no chance of generating a better solution, because any additional instrument will increase the cost. Thus, the tree below this node is not explored. This pruning criteria is used in step 5.

Example 15.4: We illustrate the branch and prune tree enumeration procedure. Consider the process flow diagram of Fig. 15.1. To simplify the illustration, we consider that potential sensors have 2% precision. Costs are also 1500 for these sensors. We now illustrate the search in Fig. 15.3. Nodes are numbered in the sequence they are explored.

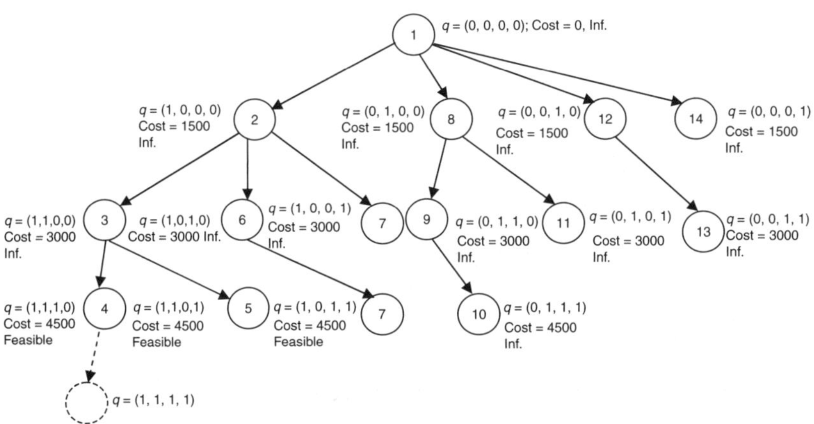

FIGURE 15.3 Illustration of the solution procedure.

Example 15.5: We consider the process introduced in Chap. 10, Example 10.2. We reproduce it in Fig. 15.4. It has seven streams and four nodes. We consider here that the performance metric is accurate.

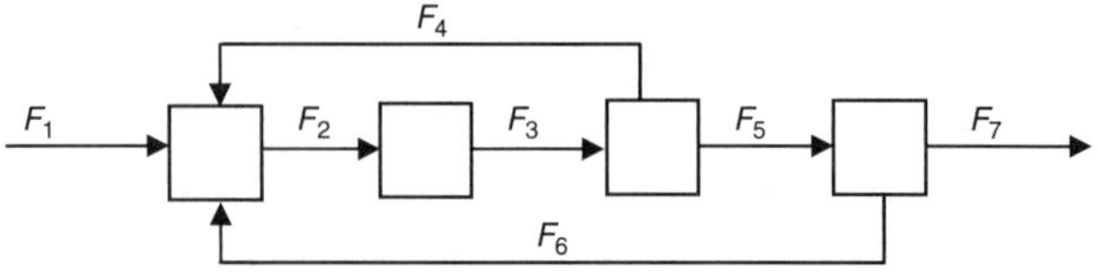

Figure 15.4 Flow sheet for Example 15.5.

Assume accuracy of 3% in stream 1 and 2% in stream 2 allowing only one gross error ($n_T = 1$) is requested. The sensor precisions, flowrates, and sensor costs are given Table 15.5.

Stream	Sensor Precision (σ_i)	Flow Rates (F_i)	Sensor Cost
S_1	1	100	20
S_2	0.447	140	25
S_3	1	140	20
S_4	1	20	15
S_5	1	120	20
S_6	1	20	10
S_7	1	100	30

Table 15.5 Sensor Precisions, Flowrates, and Sensor Costs of Example 15.5

The optimal solution, obtained using the branch and prune tree enumeration algorithm, indicates that all streams but S_7 and S_2 should be measured. The network cost is 85.

It is also of interest to study the way accuracy changes as sensors are added. This is shown in Table 15.6. This would correspond to the first branch of the enumeration tree. To do this, we consider the case of two gross errors in the system ($n_T = 2$). In this case, several equivalencies of gross errors can appear and therefore many combinations of errors lead to unbounded accuracy. Table 15.6 shows only the nodes of the tree that do not show unbounded results. Clearly, in this problem, one needs to measure all streams if accuracy is to be below 2.8 and 2.0 for streams S_1 and S_2, respectively. Only one node with six streams qualifies for a requirement of accuracy below 3.3 and 2.5 for streams S_1 and S_2, respectively. All other bounded nodes exhibit accuracies that are unacceptable.

Streams Measured	Cost	Accuracy (a_i) %
Any 2 measurements		S_1 = unbounded S_2 = unbounded
Any 3 measurements (1,2,5)	65	S_1 = 797.07 S_2 = 557.53
4 Measurements (1,2,3,5)	85	S_1 = 770.07 S_2 = 283.01
4 Measurements (1,2,5,7) or (1,3,5,7)	95	S_1 = 396.71 S_2 = 557.92
5 Measurements (1,2,3,6,7)	105	S_1 = 391.32 S_2 = 284.15
6 Measurements (1,2,3,4,5,7)	130	S_1 = 396.71 S_2 = 2.198
6 Measurements (1,2,3,4,6,7)	120	S_1 = 3.264 S_2 = 2.446
6 Measurements (1,2,3,5,6,7)	125	S_1 = 3.136 S_2 = 283.149
All measurements	140	S_1 = 2.748 S_2 = 1.98

Table 15.6 Accuracy and Cost as a Function of the Number of Sensors ($n_T = 2$)

Example 15.6: We consider the CSTR process of Example 8.5, which for completeness is shown in Fig. 15.5. We consider that the performance metrics here are precision only.

The variables of interest are $[F_i, C_{Ai}, C_A, T, T_i, T_c, F_c, T_{ci}, F_{vg}, F, F_2, F_3, F_4]$. The equations corresponding to this example were shown in Chap. 8. The nominal

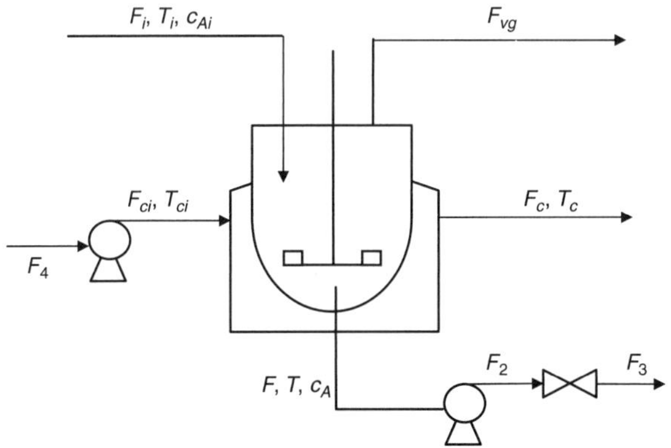

Figure 15.5 The CSTR problem.

operation conditions were given in Table 8.1. The linearized matrix D was also shown in Chap. 8. We reproduce it next.

$$
\begin{array}{c}
\begin{array}{ccccccccccccc}
F_i & c_{Ai} & c_A & T & T_i & T_c & F_c & T_{ci} & F_{vg} & F & F_2 & F_3 & F_4
\end{array}\\
D=\begin{pmatrix}
-0.00531 & -0.8333 & 1.7763 & 0.00923 & 0 & 0 & 0 & 0 & 0 & 0 & 0 & 0 & 0\\
1.4583 & 0 & -754.4 & 5.9503 & -0.8333 & -125 & 0 & 0 & 0 & 0 & 0 & 0 & 0\\
0 & 0 & 0 & -93.8067 & 0 & 108.5 & 15.7169 & -14.708 & 0 & 0 & 0 & 0 & 0\\
0 & 0 & -45.2612 & -0.443 & 0 & 0 & 0 & 0 & 1 & 0 & 0 & 0 & 0\\
1 & 0 & 0 & 0 & 0 & 0 & 0 & 0 & 0 & -1 & 0 & 0 & 0\\
0 & 0 & 0 & 0 & 0 & 0 & 0 & 0 & 0 & 0 & 1 & -1 & 0\\
0 & 0 & 0 & 0 & 0 & 0 & & 0 & 0 & 0 & -1 & 0 & 0\\
0 & 0 & 0 & 0 & 0 & 0 & -1 & 0 & 0 & 0 & 0 & 0 & 1
\end{pmatrix}
\end{array}
\tag{15.38}
$$

The costs of sensors that measure variables $V_1, V_2, \ldots, V_{13}$ are 100, 270, 300, 50, 55, 60, 105, 45, 85, 90, 95, 80, 82, respectively. The sensor precisions are 1% (for all sensors).

To illustrate the technique we consider two cases:

Case Study CSTR1: Precision thresholds of $\sigma_{CA} = 0.95\%$, $\sigma_T = 0.95\%$, and $\sigma_F = 0.95\%$ are required. The solution is to measure variables C_{Ai}, C_A, F_{vg}, *and* F_3; the minimum cost is 735.

Case Study CSTR2: Precision thresholds of $\sigma_{CA} = 1.5\%$, $\sigma_T = 1.5\%$, $\sigma_{Tci} = 1.5\%$, and $\sigma_F = 1.5\%$ are required. In addition residual precision thresholds of $\sigma_{CA\ (Residual)} = 2.5\%$, $\sigma_{T(Residual)} = 2.5\%$, $\sigma_{Tci(Residual)} = 2.5\%$ and $\sigma_{F(Residual)} = 2.5\%$ are required. All methods give the same solution, which is to measure variables C_{Ai}, C_A, T, T_i, T_{ci}, F, F_3, and F_4; the minimum cost is 972.

Remark: The performance of the above branch and prune tree searching strategy is not very good, and several different methods to improve it have been proposed:

- Gala and Bagajewicz (2006a, 2000b) proposed the use of cutsets and a decomposition procedure. Cutsets are special sets of streams that separate a connected flow sheet into two connected subflow sheets. Cutsets have the property that a material balance can be written using the elements of the set. The decomposition procedure they proposed allowed the reduction of computational time for large linear systems. They were able to reduce the design of an instrumentation network for the crude unit shown in Example 11.2 from more than 45 days to a few minutes.
- Nguyen and Bagajewicz (2008) proposed to use equation related sets—the analog of cutsets—to solve nonlinear cases. This method showed to be of no real advantage over the variable tree enumeration. Illustrations include the Tennessee Eastman problem.
- Nguyen and Bagajewicz (2008) also proposed to invert the tree, that is, to start with a node that has all measurements and keep removing them until the optimum is identified. The branching and pruning criteria were modified to accomodate to a decreasing cost.
- Branch and prune methods are also amenable to parallel computing.

Example 15.7: Consider the well-known challenge problem, the TE process, which was presented in detail in Chap. 10, Example 10.4. Three design cases are considered; they are described below.

Case Study TE1: This is a simple design case (low specification) with only six key variables (F_6, y_{A6}, y_{G6}, y_{H6},F_7, F_{10}). These six variables are required to be observable with precision thresholds of 2%.

Case Study TE2: This is a realistic design case with 17 key variables ($F_{6,}$ y_{A6}, y_{G6}, y_{H6}, F_7, y_{G7}, y_{H7}, y_{A9}, y_{G9}, y_{H9}, F_{11}, y_{G11}, y_{H11}, P_r, T_r, P_s, T_s). These variables are required to be observable with precision thresholds of 2%.

Case Study TE3: In this case, observability is required for *all* variables, except the following eight variables (y_{D9}, y_{E9}, y_{F9}, y_{G9}, y_{H9}, $F_{10,}$ y_{D10}, y_{E10}) (Thus, there are 47 – 8 = 39 key variables). Required precision thresholds are 1.6% for y_{G8}, y_{H8} and 1.5% for all other key variables. In addition, redundancy is also required for all 39 key variables except 6 key variables (y_{G8}, y_{H8}, P_r, T_r, P_s, T_s) with required residual precision thresholds of 4%.

For TE1, the optimal solution is to directly measure these six variables with sensors cost of 3200. Unlike the case study TE1 where only a few key variables are involved, case study TE2 targets many more key variables. None of the methods used by Nguyen and Bagajewicz (2008) was able to finish in a respectable time. The current best solution is to measure F_6, y_{A6}, y_{G6}, y_{H6}, F_7, y_{A7}, y_{A9}, y_{G9}, y_{H9}, y_{G10}, y_{H10}, $F_{11,}$ y_{H11}, P_r, T_r, P_s, and T_s. It has a cost of 9630. This solution contains only 14 of the 17 key variables: $F_{6,}$ y_{A6}, y_{G6}, y_{H6}, F_7, y_{A9}, y_{G9}, y_{H9}, F_{11}, y_{H11}, P_r, T_r, P_s, and T_s.

This example points out that the performance of the equations-based tree search method is still not satisfactory when solving realistic large-scale nonlinear sensor network problems. More computationally efficient methods for designing large-scale nonlinear sensor networks are needed.

Finally, case study TE3 is one that requires a lot of measurements, because the requirements are very strict. In this case, the best solution has all variables, except four (y_{E9}, F_{10}, y_{E10}, P) measured.

Remark: Genetic algorithms have been used to solve the minimum cost with precision constraints problem efficiently (Sen et al., 1998; Carnero et al., 2001a, b, 2005a, b; Heyen et al., 2002; Wongrat et al., 2005; El-Zonkoly, 2006). Lai et al. (2003) used genetic algorithms for maximizing availability, including maintenance and constraining life-cycle cost. Gerkens and Heyen (2005) combined the GA with parallel.

Remark: Chakraborty and Deglon (2008) developed a set of heuristics for the solution of the problem. They use the general principle of variance reduction through data reconciliation. They developed a set of generic design principles/heuristics for maximizing the variance reduction of process streams.

Remark: Benqlilou et al. (2003, 2005) also studied sensor placement for dynamic systems where the Kalman filter is used. In turn, Muske and Georgakis (2003) also presented a methodology where the system is represented by a state-space model and where the cost and the "process information" metric (a scalar metric based on the variance-covariance matrix, or the reliability) are explored in a form of a pareto optimal graph.

Remark: There are several articles studying the design of sensor networks using goals corresponding to normal monitoring operations. Aside from cost, different other objective functions such as precision (Madron and Veverka, 1992), reliability (Ali and Narasimhan, 1993, 1995; Bagajewicz and Sanchez, 2000a; Wang et al., 2007) or simply observability (Madron, 1992; Maquin et al., 1994; Luong et al., 1994; Bagajewicz and Sánchez, 1999a) were used. In the case of simultaneous reliability and cost Wailly et al. (2005, 2008) used symbolic calculus to make sensor placement using Groebner bases. Wang et al. (2002) proposed a methodology to locate sensors in order to improve PCA-based monitoring in processes.

Remark: Different techniques were also employed, such as graph theory (Madron, 1992; Ali and Narasimhan, 1993; 1995), mathematical programming (Bagajewicz, 1997; Bagajewicz and Cabrera, 2002).

Remark: Bagajewicz and Sánchez (1999b) showed in detail that there is a connection between the maximum precision and the minimum cost models. This connection states that the solution of one problem is the solution of the other. Bagajewicz and Sánchez (2000a) also showed that this same type of duality holds for the objectives of reliability and cost. In essence, it is sufficient to use one formulation.

Remark: The problem has also been extended to incorporate upgrade considerations (Bagajewicz and Sánchez, 2000b) and maintenance costs (Sánchez and Bagajewicz, 2000). Bagajewicz (2000) reviews all these methods and also discusses the applications to bilinear and fully nonlinear systems.

Remark: Recent work includes multiobjective optimization and two-dimensional pareto optimal graphs (Bagajewicz and Cabrera, 2003). This shows a departure from other multiobjective approaches (Viswanath and Narasimhan, 2001; Carnero et al., 2001a). Muske et al. (2003) also use Pareto optimal representations. Brown et al. (2005) also presented a multiobjective framework, which is solved with the aid of genetic algorithms.

Remark: Li and Chen (2002) studied the flexibility of sensor networks to adapt to conditions in which observability and redundancy may change. They used mathematical programming and graphical-theoretical approaches for analyzing the flexibility of a given sensor network, for designing a flexible sensor network, and for upgrading a sensor network to improve its flexibility.

Remark: There is a large literature of sensor positioning in units (reactors, distillation columns, etc.) and in general distributed systems (governed by PDE), which we have omitted in this book entirely because we concentrated on process plants. (See, for example, Colantuoni and Padmanabhan, 1977; Kumar and Seinfel, 1978; Alvarez et al., 1981; Waldraff et al., 1998; Wouwer et al., 2000; Singh and Hahn, 2005, 2006; Zamprogna et al., 2005; Venkateswarlu and Kumar, 2006; Brewer et al., 2007.) Other sensor location procedures for control purposes are also omitted here. Chapter 16 discusses some modern procedures mostly based on material presented in Chap. 13.

Cost-Optimal Design for Parametric Faults

As a consequence, the problem of designing an alarm system consists of determining the cost-optimal position of sensors, such that all process faults, single or multiple and simultaneous, can be detected and distinguished from instrument malfunction (biases). One of the first attempts to present a technique to locate sensors was done by Lambert (1977), who used fault trees based on failure probabilities. Failure probabilities are hard to assess, fault trees cannot handle cycles, and the construction of the tree is cumbersome for large-scale systems. Due to these limitations, the technique has not been developed further.

We turn to the models discussed in Chap. 14. The minimum cost model for fault observability and resolution is

Minimize {Total cost}
s.t.

- Desired observability of faults
- Desired level of resolution of faults

We recall from Chap. 14 that the maximum connectivity matrix (A) contains the same information as the bipartite graph. In this matrix, the columns represent the faults and the rows represent the sensors. When $A_{ij} = 1$, then we say that there is a path from fault F_j to the potential sensor location S_i. In that chapter we also defined the fault-sensor connectivity matrix $B(q) = [b_1, \ldots, b_n]$ using $b_j = a_j \otimes q$, where b_j and a_j are the ith column of B and A, respectively, and $\otimes$ is the Hadamard product of two vectors, that is, $(a_j \otimes q)_i = a_{ji} q_i$. Thus, we determined that when $\sum_{i \in M_1} b_k(q_i) \geq 1 \rightarrow F_k$ is observable [Eq. (14.5)].

This condition can then be used to establish observability. Because, $b_j = a_j \otimes q$, this condition can be written as $\sum_{i \in M_1} a_{ki} q_i \geq 1$.

We are now in a position to present our model for cost-optimal sensor network design for fault observability.

$$\left.\begin{aligned} &P_1 = \text{Min} \sum_{i \in M_1} c_i q_i \\ &\text{s.t.} \\ &\sum_{i \in M_1} a_{ki} q_i \geq 1 \quad \forall k \in M_O \\ &q_i \in \{0, 1\} \end{aligned}\right\} \tag{15.39}$$

where M_O is the set of faults that one wants to be able to observe.

We now recall Eq. (14.6) that described single fault resolution and reproduce it here.

$$\sum_{i \in M_1} [b_j(q) \otimes b_p(q)]_i \leq \text{Max}\left\{ \sum_{i \in M_1} [b_j(q)]_i \quad \sum_{i \in M_1} [b_p(q)]_i \right\} - 1 \tag{15.40}$$

We now use $b_j = a_j \otimes q$ to write

$$\left[b_j(q) \otimes b_p(q)\right]_i = a_{ji} a_{pi} q_i \tag{15.41}$$

which is easily verifiable. Thus, Eq. (15.38) can be written as follows:

$$\sum_{i\in M_1} a_{ji}\,a_{pi}\,q_i \le \text{Max}\left\{\sum_{i\in M_1} a_{ji}\,q_i,\ \sum_{i\in M_1} a_{pi}\,q_i\right\} - 1 \qquad \forall j,p \in M_S; j \ne p \qquad (15.42)$$

where M_S is the set of faults that one wants to be able to observe and resolve. Note that M_S should be included in M_O.

Similarly, taking into account that $d_{jk}(q) = (a_j \otimes a_k)_i\, q_i = a_{ji}\, a_{ki}\, q_i$, then the double fault resolution condition [Eq. (14.8)] is

$$\sum_{i\in M_1} [d_{jk}(q) \otimes d_{pr}(q)]_i \le \text{Max}\left\{\sum_{i\in M_1} [d_{jk}(q)]_i,\ \sum_{i\in M1} [d_{pr}(q)]_i\right\} - 1 \qquad \forall\, j,k,p,r \in M_{S2} \qquad (15.43)$$

where (j, k) and (p, r) $(j \ne k \ne p \ne r)$ belong to M_{S2} the set of all combinations of two faults that one wants to resolve, can be rewritten as follows:

$$\sum_{i\in M_1} a_{ji}a_{ki}a_{pi}a_{ri}q_i \le \text{Max}\left\{\sum_{i\in M_1} a_{ji}a_{ki}q_i,\ \sum_{i\in M_1} a_{pi}a_{ri}q_i\right\} \qquad \forall j,k,p,r \in M_{S2}; j \ne k \ne p \ne r \qquad (15.44)$$

Note, that M_{S2} should also be included in M_O.

The final model for observability, single and double fault resolution is

$$\left.\begin{aligned}
&\text{Min} \sum_{i\in M_1} c_i q_i \\
&\text{s.t.} \\
&\sum_{i\in M_1} a_{ki}\,q_i \ge 1 && \forall k \in M_O \\
&\sum_{i\in M_1} a_{ji}\,a_{pi}\,q_i \le C_{jp} - 1 && \forall j,p \in M_S; j \ne p \\
&C_{jp} = \text{Max}\left\{\sum_{i\in M_1} a_{ji}q_i,\ \sum_{i\in M_1} a_{pi}q_i\right\} && \forall j,p \in M_S; j \ne p \\
&\sum_{i\in M_1} a_{ji}a_{ki}a_{pi}a_{ri}q_i \le G_{jkpr} && \forall j,k,p,r \in M_{S2}; j \ne k \ne p \ne r \\
&G_{jnpr} = \text{Max}\left\{\sum_{i\in M_1} a_{ji}a_{ki}q_i\,,\ \sum_{i\in M_1} a_{pi}a_{ri}q_i\right\} && \forall j,k,p,r \in M_{S2}; j \ne k \ne p \ne r
\end{aligned}\right\} \qquad (15.45)$$

We recall that the max {•} operator can be linearized (Chap. 14). Thus the above models can be made linear.

The group of Rengaswamy has also studied sensor networks for fault detection (Raghuraj et al., 1999; Bhushan and Rengaswamy, 2000a,b; 2001; 2002a,b; Bhushan et al., 2003; 2008). In addition, Kotecha et al. (2007, 2008) investigated the use of constrained programming to obtain solutions to various sensor network design formulations. More recently Chen and Chang (2008) presented their own signed directed graph (SDG) method that is solved using integer programming.

Example 15.6: We consider again the CSTR case of Example 15.3 and we request single and double fault resolution in the following: single fault resolution (noncontrol loop fault) for the set $M_S = \{C_{Ai}, Fi, Ti\}$ and double fault resolution (noncontrol loop fault – control loop fault) for the set $M_{S2} = \{C_{Ai} - VfL, C_{Ai} - VfP, C_{Ai} - VfT, Ti - VfL, Ti - VfP, Fi - VfL, Fi - VfP, Fi - VfT, Tci - VfP, U - VfL\}$. Table 15.7 shows that an alternative arrangement of sensors can be obtained using the same cost for all measurements. A cross indicates that the corresponding node

	Case 1						Case 2	
Node Name	**Costs**	**Alternative Solutions (*q*)**				**Bhushan et al.**	**Costs**	***q***
CA	100	X				X	100	
TS	0	X	X	X	X	X	0	X
TC	100						100	
VT	100						100	
Fc	100	X	X	X	X	X	1	X
F_4	100						1	
Tc	100	X	X	X	X	X	1	X
N	100						—	
PS	0	X	X	X	X	X	0	X
PC	100			X		X	100	
VP	100				X		100	
Fvg	100		X				1	X
VS	0	X	X	X	X	X	0	X
VC	100						100	
VL	100						100	
F_3	100						1	
F_2	100						1	
F	100	X	X	X	X	X	1	X
Number of sensors		7				8		7

TABLE 15.7 Minimum Number of Sensors Solutions for the CSTR System

should be measured. These four solutions are different from the one obtained by Bhushan and Rengaswamy (2000), which is also shown in Table 15.7. The fault detection capability of these four solutions (indicated by the sets M_S and M_{S2}) is less than the capability of Bhushan and Rengaswamy's solution only in the single fault resolution criterion. Bhushan and Rengaswamy's solution can resolve single fault resolution for the set $\{C_{Ai}, Fi, Ti, Tci\}$ while the four solutions can resolve single fault resolution for the set $\{C_{Ai}, Fi, Ti\}$.

We also modified the problem costs (case 2). For this last case we consider that the cost of measuring concentration is too high and the node N (number of mol in gaseous phase) is unmeasurable, thus forbidding it. Furthermore, we assign high costs to the use of controllers and valves for detecting faults.

A few observations are appropriate:

- For this small example, we have found four alternative solutions that feature minimum number of sensors (seven).
- If costs are considered, less alternative solutions featuring the same cost are likely, but one could expect to identify suboptimal solutions with costs close to the optimal.
- Some nodes are essential and will be part of all solutions, no matter what cost is assigned to measure them. This is the case of T_c, F_C, and F. In the Bhushan and Rengaswamy work (2000) node, F_c is the node where the sensor should be placed to observe all the faults (except control loop failures) under the single fault assumption. For observability under the double faults fault assumption (control loop–noncontrol loop faults and noncontrol–noncontrol faults), node T_c has been selected to observe such pairs. Finally, node F is the node where the sensor should be placed for solving the single fault resolution problem.
- Those faults that affect the same set of nodes are not distinguished.

Remark: Chou et al. (1993) presented a method that explores faulty trees to determine the sensors of a power plant. Fijany and Vatan (2005) use an MILP formulation and a special branch and bound algorithm to determine sensor networks that deal with diagnosability of faults at minimum cost. Gerkens and Heyen (2008) proposed a genetic algorithm approach to solve the minimum cost sensor location problem for fault identification. Recently, Orantes et al. (2007, 2008) used the concept of classification using entropy together with simulations.

Integrated Cost-Optimal Design

It is quite clear that different sets of sensors are needed for different perspectives (control, faults, and monitoring). The issue, however, is that these set might intersect. Therefore, one needs to solve for minimum cost, but attending these different perspective needs simultaneously. Bagajewicz et al. (2004a) presented a simple example where control, monitoring, and fault identification are considered at the same time. Bagajewicz et al. (2004b) presented one approach where fault identification and precision are requested simultaneously. The example we present next is based on their work. More recently. Kotecha et al. (2008) proposed an MINLP multiobjective formulation that also uses constraint programming.

Example 15.7: We consider the CSTR case of Example 15.6. For each variable from the precision group (Fi, C_{Ai}, CA, T, Ti, Tc, Fc, Tci, F_{vg}, F, F_2, F_3, F_4) sensors of $\sigma_1 = 1\%$ are available at a cost of 100. We first consider requiring only precision in key variables ($\sigma_{CA} = 2\%$, $\sigma_T = 1\%$, and $\sigma_F = 2\%$). Consider also that for each node not considered in the precision problem, the cost of locating a sensor is 100 (as in case 1 of Example 15.6). The optimal set of sensors for this problem is shown in Table 15.8.

	Sensors for Fault Detection				**Sensors for Precision**												**Sensors for Fault Detection and Precision**
F_i					X	X	X										X
C_{Ai}					X			X	X	X							
CA	X				X	X	X	X	X	X	X	X	X	X	X	X	X
TS	X	X	X	X		X					X	X	X				X
T_i																	
Tc	X	X	X	X													X
Fc	X	X	X	X													X
Tc_i																	
F_{vg}		X					X							X	X	X	
F	X	X	X	X				X			X			X			X
F_2									X			X			X		
F_3										X			X			X	
F_4																	
TC																	
VT																	
N																	
PS	X	X	X	X													X
VS	X	X	X	X													X
PC			X														
VP																	
VC																	
VL																	

Table 15.8 Precision Required for CA, T, F (Precision Threshold $\sigma^*_{CA} \le 2\%, \sigma^*_T \le 1\%, \sigma^*_F \le 2\%$)

While seven sensors are enough for fault detection, the same number of sensors is enough for fault detection and the estimation of the desired variables, although the sensors are different. Note that the simple superposition of solutions for fault detection and precision requires 7 and 10 sensors as minimum and maximum, respectively.

Value-Optimal Instrumentation Design

Note that the performance target values are fixed constants and must be selected by the process engineer. This required commitment to target values has been the Achilles heel of all minimum cost formulations, because until recently no methodology or criterion has been ever suggested on what is actually a good target. As an alternative one can use value, as defined in Chaps. 11, 13, and 14. Thus, the following maximum value formulation, (Bagajewicz et al., 2004a) which allows the target values to float, is proposed:

$$\left.\begin{array}{l}\text{Max } \{V(q) = \text{profit [performance targets]} - \text{sensor network cost } (q)\} \\ \text{s.t.} \\ \text{Performance metrics}(q) \leq \text{performance targets}\end{array}\right\} \tag{15.46}$$

We assume that the profit function is monotonously increasing with the performance metrics. For example, we notice that downside expected financial loss is monotone with precision (the larger the variance, the larger the loss). Quite clearly it is also monotone with accuracy values. The same is true for fault resolution cases, that is, the larger the ability to resolve faults, the larger is the profit. Indeed, catching faults and implementing corrective actions reduces losses and therefore increases profit. Finally, in the case of control, it is quite clear that when variability is increased, profit is reduced.

If the performance targets are allowed to float, it is easy to see that the performance metrics will be equal to the performance targets. Indeed, if the performance metrics are smaller than the performance targets, then the real profit, which is based on the performance metrics will be smaller than the profit based on the performance targets. This is true only when the monotonicity of profit with performance metrics holds, as it does in our case. Thus, if the constraint in Eq. (15.46) is binding, that is, performance targets are equal to performance metrics, then the value formulation can be represented by the following unconstrained problem.

$$\text{Max } \{V(q) = \text{profit [performance metrics } (q)] - \text{sensor network cost}(q)\} \tag{15.47}$$

This is an important *paradigm shift* because it moves the responsibility of selecting targets off the engineer and places it on the optimization problem. Once the engineer has properly defined the profit function, the solution of the optimization problem will give direct justification for upgrading projects—specifically when the profit increase is greater than upgrade cost.

In the case of retrofits, the starting point is an existing network, and the cost is the cost of the new sensors, or the cost of relocating sensors, or the cost of the increased preventive maintenance. We consider the case of adding sensors only for simplicity, although this does not reduce the level of generality of the argument.

Let q_0 be the existing sensor network and let q be the upgraded network. The maximum value formulation needs to be formulated slightly differently. The cost is the cost of the new sensors (given by $q - q_0$), and the profit needs to be considered as the extra profit [ΔProfit (q, q_0)] generated by the addition of the new sensors. Then we solve:

$$\text{Max}\ \{V(q - q_0) = \Delta\text{profit}\ (q, q_0) - \text{sensor network cost}\ (q - q_0)\} \quad (15.48)$$

where

$$\Delta\text{profit}\ (q, q_0) = \text{profit [performance metrics } (q)] - \text{profit [performance metrics } (q_0)] \quad (15.49)$$

If we consider that the profit is zero when there is no instrument, that is, for $q_0 = 0$, then Eq. (15.46) is equivalent to Eq. (15.47).

Example 15.8: Consider the case of Example 15.5. The data for the example is given in Table 15.9. The key variable is flowrate of stream S_7 and the time horizon is 30 days.

Sensor	Precision	Cost
1	2%	55
2	2%	40
3	2%	60
4	2%	50
5	2%	45
6	2%	55
7	2%	60

Table 15.9 Data for Example 15.8

The results are obtained for two cases corresponding to two values of K_s parameter. It can be seen that the larger the K_s value, the more sensors need to be used to minimize the financial loss. Table 15.10 shows the results which were obtained by inspection.

K_s	50	80
Measured variables	1,7	1,5,6,7
Sensor costs	115	215
Objective value	603	860

Table 15.10 Results for Example 15.8

Example 15.9 Consider the flow sheet of Example 10.7. We solved this problem for key variables (product streams): 1, 9, 14 with K_s value = 25, 20, 20, respectively and a time horizon of 30 days. The costs of sensors are shown in Table 15.11.

Stream	Flow	Cost	Stream	Flow	Cost
1	140	19	13	10	12
2	20	17	14	10	12
3	130	13	15	90	17
4	40	12	16	100	19
5	10	25	17	5	17
6	45	10	18	135	18
7	15	7	19	45	17
8	10	6	20	30	15
9	10	5	21	80	15
10	100	13	22	10	13
11	80	17	23	5	13
12	40	13	24	45	13

TABLE 15.11 Costs for Streams of Example 15.8

The solution to this problem is to measure the following set: $\{S_1, S_2, S_3, S_4, S_6, S_7, S_8, S_9, S_{10}, S_{12}, S_{13}, S_{14}, S_{16}, S_{18}, S_{20}, S_{23}, S_{24}\}$, with objective value 495; computation time is 1 hour 16 min.

We also used a genetic algorithm obtaining $\{S_1, S_2, S_3, S_4, S_6, S_7, S_8, S_9, S_{10}, S_{12}, S_{13}, S_{14}, S_{15}, S_{16}, S_{20}, S_{21}, S_{22}, S_{23}, S_{24}\}$, with objective value of 501 (suboptimal); the computation time, however, is 19 min.

When value is considered, then the tree described in Fig. 15.2 looses one of its very important properties, monotonicity. Indeed, adding one instrument to an existing network is not guaranteed to increase its value because the increase in profit could be smaller than the cost of the sensor. Thus, the pruning criteria that we discussed in the case of minimum cost formulation does not apply anymore. In addition, all nodes are feasible. Thus the branch and prune algorithm of the minimum cost formulations needs to be reformulated.

The first observation one has to make is that the order in which sensors are added give rise to different values as the sensors are added. For example, consider the case of Fig. 15.1 and the corresponding FULL tree. To this date, this has not been fully investigated.

References

Alhéritière, C., N. Thornhill, S. Fraser, and M. Knight, "Evaluation of the Contribution of Refinery Process Data to Performance Measure," *AIChE Annual Meeting*, Los Angeles (1997).

Alhéritière, C., N. Thornhill, S. Fraser, and M. Knight, "Cost Benefit Analysis of Refinery Process Data:Case Study," *Comp. Chem. Eng*. **22** (Suppl.):S1031–S1034 (1998).

Ali, Y. and S. Narasimhan, "Sensor Network Design for Maximizing Reliability of Linear Processes," *AIChE J.* **39**(5) (1993).

Ali, Y. and S. Narasimhan, "Redundant Sensor Network Design for Linear Processes," *AIChE J.* **41**(10) (1995).

Ali, Y. and S. Narasimhan, "Sensor Network Design for Maximizing Reliability of Bilinear Processes," *AIChE J.* **42**:2563–2555 (1996).

Alvarez, J., J. A. Romagnoli, and G. Stephanopoulos, "Variable Measurements Structures for the Control of a Tubular Reactor," *Chem. Eng. Sci.* **36:**1695 (1981).

Angelini, R., C. A. Méndez, E. Musulin, and L. Puigjaner, "An Optimization Framework to Computer-Aided Design of Reliable Measurement Systems," *Comp. Chem. Eng.* **21**:1293–1298 (2006).

Bagajewicz, M., "Design and Retrofit of Sensor Networks in Process Plants," *AIChE J.* **43**(9):2300 (1997).

Bagajewicz, M., *Design and Upgrade of Process Plant Instrumentation*, *ISBN:1-56676-998-1*, Technomic Publishing Company (http://www.techpub.com) (2000).

Bagajewicz, M. and E. Cabrera, "A New MILP Formulation for Instrumentation Network Design and Upgrade," *AIChE J.* **48**(10):2271–2282 (2002).

Bagajewicz, M. and E. Cabrera, "Pareto Optimal Solutions Visualization Techniques for Multiobjective Design and Upgrade of Instrumentation Networks," *IECR.* **42**(21):5195–5203 (2003).

Bagajewicz, M. and M. Sánchez, "Design and Upgrade of Non-redundant and Redundant Linear Sensor Networks," *AIChE J.* **45**(9):1927–1939 (1999a).

Bagajewicz, M. and M. Sánchez, "Duality of Sensor Network Design Models for Parameter Estimation," *AIChE J.* **45**(3):661–664 (1999b).

Bagajewicz, M. and M. Sánchez, "Cost-Optimal Design of Reliable Sensor Networks," *Comp. Chem. Eng.* **23**(11/12):1757–1762 (2000a).

Bagajewicz, M. and M. Sánchez, "Reallocation and Upgrade of Instrumentation in Process Plants," *Comp. Chem. Eng.* **24**(8):1961–1980 (2000b).

Bagajewicz, M., D. Chmielewski, and R. Rengaswamy, "Integrated Process Sensor Network Design," *Proc. Sensor Topical Conference:AIChE Annual Meeting*. Austin, Texas (2004a).

Bagajewicz, M., A. Fuxman, and A. Uribe, "Instrumentation Network Design and Upgrade for Process Monitoring and Fault," *AIChE J.* **50**(8):1870–1880 (2004b).

Benqlilou, C., M. Graells, E. Musulin, and L. Puigjaner, "Design and Retrofit of Reliable Sensor Networks," *Ind. Eng. Chem. Res.* **43**(25):8026–8036 **(**2004**).**

Benqlilou, C., M. Bagajewicz, A. Espuña, and L. Puigjaner, "Sensor-Placement for Dynamic Processes," *Comp- Aided Chem. Eng.* **14**:371–376 (2003).

Benqlilou, C., E. Musulin, M. Bagajewicz, and L. Puigjaner, "Instrumentation Design based on Optimal Kalman Filtering," *J. Proc. Contr.* **15**:629–638 (2005).

Bhushan, M. and R. Rengaswamy, "Design of Sensor Network Based on the Signed Directed Graph of the Process for Efficient Fault Diagnosis," *Ind. Eng. Chem. Res.* **39** 999–1019 (2000a).

Bhushan, M. and R. Rengaswamy, "Design of Sensor Network Based on Various Fault Diagnostic Observability and Reliability Criteria," *Comp. Chem. Eng.* **24**:735–741 (2000b).

Bhushan, M. and R. Rengaswamy, "Comprehensive Design of a Sensor Network for Chemical Plants Based on Various Diagnosability and Reliability Criteria: I. Framework," *Ind. Eng. Chem. Res.* **41**:1826–1839 (2002a).

Bhushan, M. and R. Rengaswamy, "Comprehensive Design of a Sensor Network for Chemical Plants Based on Various Diagnosability and Reliability Criteria: II. Applications." *Ind. Eng. Chem. Res.* **41**:1840–1860 (2002b)

Bhushan, M., S. Narasimhan and R. Rengaswamy, "Sensor Network Reallocation and Upgrade for Efficient Fault Diagnosis," *Proc. Fourth International Conference on Foundations of Computer-Aided Process Operations*, Coral Springs, Florida: 443–446 (2003).

Bhushan, M., S. Narasimhan, and R. Rengaswamy, "Robust Sensor Network Design for Fault Diagnosis," *Comp. Chem. Eng.* **32**:1067–1084 (2008).

Brewer, J., Z. Huang, A. K. Singh, M. Misra, and J. Hahn, "Sensor Network Design via Observability Analysis and Principal Component Analysis," *Ind. Eng. Chem. Res.* 46:8026–8032 (2007).

Brown, D., F. Maréchal, G. Heyen, and J. Paris, "Application of Multiobjective Optimisation to Process Measurement System Design," *Comp-Aided Chem. Eng.* **20**(2):1153–1158 (2005).

Carnero, M., J. Hernández, M. Sánchez, and A. Bandoni, "Multiobjective Evolutionary Optimization in Sensor Network Design," *Proc. ENPROMER* **1**:325–330. Argentina (2001a).

Carnero, M., J. L. Hernández, M. Sánchez, and A. Bandoni, "An Evolutionary Approach for the Design of Non-redundant Sensor Networks," *Ind. Eng. Chem. Res.* **40**:5578–5584 (2001b).

Carnero, J., L. Hernández, and M. C. Sánchez, "Optimal Sensor Network Design and Upgrade Using Tabular Search," *Comp-Aided Chem. Eng.* **20**:1447–1452 (2005a).

Carnero, M., J. L. Hernández, M. Sánchez, and A. Bandoni, "On the Solution of the Instrumentation Selection Problem," *Ind. Eng. Chem. Res.* **44**(2):358–367 (2005b).

Chakraborty, A. and D. Deglon, "Development of a Heuristic Methodology for Precise Sensor Network Design," *Comp. Chem. Eng.* **32**(3):382–395 (2008).

Chen, J. Y. and C. Chang, "Development of an Optimal Sensor Placement Procedure Based on Fault Evolution Sequences," *Ind. Eng. Chem. Res.* **47**:7335–7346 (2008).

Chmielewski, D., T. Palmer, and V. Manousiouthakis, "On the Theory of Optimal Sensor Placement," *AIChE J.* **48**(5):1001–1012 (2002).

Chmielewski, D. and J-K Peng, "Covariance-Based Hardware Selection-Part I: Globally Optimal Actuator Selection," *IEEE Trans. Cont. Sys. Tech.* **14**(2):355–361 (2006).

Chou, H. P., J. N. Ning, and T. M. Tsai, "A Methodology for the Design and Analysis of Sensor Failure Detection Network," *Nuc. Technol.* **101**:101–109 (1993).

Colantuoni, G. and L. Padmanabhan, "Optimal Sensor Location for Tubular-Flow Reactor Systems," *Chem. Eng. Sci.* **32**:1035 (1977).

El-Zonkoly, A., "Optimal Meter Placement using Genetic Algorithm to Maintain Network Observability," *Expert Sys. Appl.* **31**:193–198 (2006).

Fijany, A. and F. Vatan, "A New Efficient Algorithm for Analyzing and Optimizing the System of Sensors," *Proc. IEEEAC Conference:Paper 1568.* March 2005.

Gala, M. and M. Bagajewicz, "Rigorous Methodology for the Design and Upgrade of Sensor Networks Using Cutsets," *Ind. Eng. Chem. Res.* **45**(21):6679–6686 (2006a).

Gala, M. and M. Bagajewicz, "Efficient Procedure for the Design and Upgrade of Sensor Networks using Cutsets and Rigorous Decomposition," *Ind. Eng. Chem. Res.* **45**(21):6687–6697 (2006b).

Gerkens, C. and G. Heyen, "Use of Parallel Computers in Rational Design of Redundant Sensor Networks," *Comp. Chem. Eng.* **29**(6):1379–1387 (2005).

Gerkens, C. and G. Heyen, "Sensor Placement for Fault Detection and Localization," *Comp-Aided Chem. Eng.* **25**:355–360 (2008).

Harris, T., J. F. Macgregor, and J. D. Wright, "Optimal Sensor Location with an Application to a Packed Bed Tubular Reactor," *AIChE J.* **26**:910 (1980).

Heyen, G., M. N. Dumont, and B. Kalitventzeff, "Computer-Aided Design of Redundant Sensor Networks," *Comp-Aided Chem. Eng.* **10**:685–690 (2002).

Kelly, J. D. and D. Zyngier, "A New and Improved MILP Formulation to Optimize Observability, Redundancy and Precision for Sensor Network Problems," *AIChE J.* **54**:1282–1291 (2008).

Kotecha, P. R., M. Bhushan, and R. D. Gudi, "Constraint Programming Based Robust Sensor Network Design," *Ind. Eng. Chem. Res.* **46**:5985–5999 (2007).

Kotecha, P. R., M. Bhushan, and R. D. Gudi, "Design of Robust, Reliable Sensor Networks Using Constraint Programming," *Comp. Chem. Eng.* **32**:2030–2049 (2008).

Kretsovalis, A. and R. S. H. Mah, "Effect of Redundancy on Estimation Accuracy in Process Data Reconciliation,"*CES* **42**:2115 (1987).

Kumar, S.and S. H. Seinfel, "Optimal Location of Measurements in Tubular Reactors," *Chem. Eng. Sci.* **33**:1507 (1978).

Lambert, H. E., "Fault Trees for Locating Sensors in Process Systems," *CEP*. (Aug.):81–85 (1977).

Lai, C., C. Chang, C. Ko, and C. Chen, "Optimal Sensor Placement and Maintenance Strategies for Mass-Flow Networks," *Ind. Eng. Chem. Res.* **42:** 4366–4375 (2003).

Li, B. and B. Chen, "Study on Flexibility of Sensor Network for Linear Processes," *Comp. Chem. Eng.* **26**(10):1363–1368 (2002).

Luong, M., D. Maquin, C. T. Huynh, and J. Ragot, "Observability, Redundancy, Reliability and Integrated Design of Measurement Systems,"*Proc. 2nd IFAC Symposium on Intelligent Components and Instrument Control Applications.* Budapest (1994).

Madron, F., *Process Plant Performance*, Ellis Horwood, Chichester, England (1992).

Madron, F. and V. Veverka, "Optimal Selection of Measuring Points in Complex Plants by Linear Models," *AIChE J.* **38**(2):227 (1992).

Maquin, D., M. Luong, and J. Ragot, "Observability Analysis and Sensor Placement," *Proc. SAFE PROCESS '94 IFAC/IMACS. Symposium on Fault Detection, Supervision, and Safety for Technical Process, June 13–15*, Espoo, Finland (1994).

Meyer, M. J., M. Le Lann, B. Koehret, and M. Enjalbert, "Optimal Selection of Sensor Location on a Complex Plant Using a Graph Oriented Approach," *Comp. Chem. Eng.* **18** (Suppl.):S535–S540 (1994).

Muske, K. R. and C. Georgakis, "Optimal Measurement System Design for Chemical Processes," *AIChE J.* **49**:1488–1494 (2003).

Nguyen, D. and M. Bagajewicz, "Design of Nonlinear Sensor Networks for Process Plants." *Ind. Eng. Chem. Res.* **47**(15):5529–5542 (2008).

Orantes, A., T. Kempowsky, M.-V. Le Lann, L. Prat, S. Elgue, C. Gourdon, and M. Cabassu, "Selection of Sensors by a New Methodology Coupling a Classification Technique and Entropy Criteria," *Chem. Eng. Res. Design*, **85**: 825–838 (2007).

Orantes, A., T. Kempowsky, M.-V. Le Lann, and J. Aguilar-Martin, "A New Support Methodology for the Placement of Sensors Used for Fault Detection and Diagnosis," *Chemical Engineering and Processing: Process Intensification* **47**:330–348 (2008).

Rao, R., M. Bhushan, and R. Rengaswamy, "Locating Sensors in Complex Chemical Plants Based on Fault Diagnostic Observability Criteria," *AIChE J.* **45**:310–322 (1999).

Ragot, J., D. Maquin, and G. Bloch, "Sensor Positioning for Processes Described by Bilinear Equations," *Revue Diagnostic et Surete de Fonctionnement.* **2**(2):115–132 (1992).

Sánchez, M. and M. Bagajewicz, "On the Impact of Corrective Maintenance in the Design of Sensor Networks," *Ind. Eng. Chem. Res.* **39**(4):977–981 (2000).

Raghuraj, R., M. Bhushan, and R. Rengaswamy, "Locating Sensors in Complex Chemical Plants Based on Fault Diagnosis Observability Criteria," *AIChE J.* **45**:310–322 (1999).

Sen, S., S. Narasimhan, and K. Deb, "Sensor Network Design of Linear Processes using Genetic Algorithms," *Comp. Chem. Eng.* **22**(3):385–390 (1998).

Singh, A. K. and J. Hahn., "Determining Optimal Sensor Locations for State and Parameter Estimation for Stable Nonlinear Systems," *Ind. Eng. Chem. Res.* **44**:5645 (2005).

Singh, A. K. and J. Hahn, "Sensor Location for Stable Nonlinear Dynamic Systems: Multiple Sensor Case," *Ind. Eng. Chem. Res.* **45**:3615–3623 (2006).

Vaclavek, V. and M. Loucka, "Selection of Measurements Necessary to Achieve Multicomponent Mass Balances in Chemical Plants," *CES* **31**:1199–1205 (1976).

Vaclavek, V., "Studies on Systems Engineering-III. Optimal Choice of the Balance Measurements in Complicated Chemical Engineering Systems," *CES* **24**:947–955 (1969).

Venkateswarlu, C. and B. J. Kumar, "Composition Estimation of Multicomponent Reactive Batch Distillation with Optimal Sensor Configuration," *Chem. Eng. Sci.* **61**:5560–5574 (2006).

Viswanath, A. and S. Narasimhan, "Multiobective Sensor Network Design Using Genetic Algorithms," *Proc. 4th IFAC Workshop on On-Line Fault Detection & Supervision in the Chemical Process Industries, June 8–9*, Seoul, Korea (2001).

Wailly, O. and N. Héraud, "Cost-Optimal Design of Reliable Sensor Networks Extended to Multilinear System," *Comp. Chem. Eng.* **29**:1083–1087 (2005).

Wailly, O., N. Héraud, and O. Malasse, "Design of Instrumentation in Process Plants Using Groëbner Bases," *Comp. Chem. Eng.* **32**:2179–2188 (2008).

Waldraff, W., D. Dochain, S. Bourrel, and A. Magnus, "On the Use of Observability Measures for Sensor Location in Tubular Reactor," *J. Proc. Contr.* **8**:497 (1998).

Wang, X., G. Rong, and J. Li, "A New Approach to Design Reliable General Sensor Network on the Basis of Graph Theory," *Ind. Eng. Chem. Res.* **46**(8):2520–2525 (2007).

Wang, H., Z. Song, and H. Wang, "Statistical Process Monitoring Using Improved PCA with Optimized Sensor Locations," *J. Proc Contr.* **12**:735–744 (2002).

Wongrat, W., T. Srinophakun, and P. Srinophakun, "Modified Genetic Algorithm for Nonlinear Data Reconciliation," *Comp. Chem. Eng.* **29**:1059–1067 (2005).

Wouwer, A. V., N. Point, S. Porteman, and M. Remy, "An Approach to the Selection of Optimal Sensor Locations in Distributed Parameter Systems," *J. Proc. Contr.* **10**:291 (2000).

Zamprogna, E., M. Barolo, and D. E. Seborg, "Optimal Selection of Soft Sensor Inputs for Batch Distillation Columns Using Principal Component Analysis," *J Proc. Contr.* **15**:39–52 (2005).

CHAPTER 16

Value of Instrumentation Upgrade—Control Perspective*

The topic of measurement device selection for control system applications has a long and distinguished history. The subject was first addressed by Kushner (1964), Meier et al. (1967) and Athans (1972), where focus was on measurement time selection for communication systems. Johnson (1969) and Muller and Weber (1972) both focused on the subject of selecting a sensor array to maximize the observability of a dynamic process. Yu and Seinfeld (1973) proposed a scheme for the selection of measurement locations within a dynamic distributed parameter system using the objective of minimizing estimation error variance. [Similar approaches can be found in Harris et al. (1980), as well as in the citations of the survey paper by Kubrusly and Malebranche (1985).] Other early efforts from the chemical process control literature include Mellefont and Sargent (1978), Morari and O'Dowd (1980), Morari and Stephanopoulos (1980) and Romagnoli et al. (1981). In the context of control structure (CV and MV) selection, an enormous body of literature has been created; see for example Morari et al. (1980), Hovd and Skogestad (1993), Narraway and Perkins (1994), Lee et al. (1995), Loeblein and Perkins (1996), Cao and Rossiter (1997), Wisnewski and Doyle (1998), Heath et al. (2000), and Van de Wal and de Jager (2001). Recent contributions not in the context of control structure selection include; Van de Wouwer et al. (2000) Faulds and King (2000), Antoniades and Christofides (2001), Chmielewski et al. (2002), Muske and Georgakis (2002), Alonso et al. (2004), Peng and Chmielewski (2005, 2006), Singh and Hahn (2006), and Armaou and

*This chapter was written by Donald J. Chmielewski.

Demetriou (2006). The objective of the current chapter is to simply connect the profit focused control system design notions (of chapter 13) to the topic of sensor selection.

Before describing the impact of instrumentation on control system value, we must review the pertinent aspects of Chap. 13. In particular, one must start by assuming that the controller is subject to a partial state information (PSI) structure (in the alternative case of a full state information structure, all state variables are assumed to be measured perfectly). In the PSI case, the dynamic system of interest is

$$\dot{x} = Ax + Bu + Gw \tag{16.1}$$

$$y = Cx + v \tag{16.2}$$

where Eq. (16.2) is the measurement equation. Of particular interest is the term v, which is the measurement noise, assumed to be gaussian white noise with zero mean and covariance Σ_v. Within the PSI scenario, the controller is typically a linear feedback of the state estimate, $u = L\hat{x}$, where $\hat{x}$ is generated by the Kalman filter as follows:

$$\dot{\hat{x}} = A\hat{x} + Bu + K(y - C\hat{x}) \tag{16.3}$$

The Kalman filter gain is $K = \Sigma_e C^T \Sigma_v^{-1}$, where Σ_e is the covariance matrix associated with the estimation error signal, $e = x - \hat{x}$. The estimation error covariance matrix is calculated as the solution of the Riccati equation.

$$A\Sigma_e + \Sigma_e A^T + G\Sigma_w G^T - \Sigma_e C^T \Sigma_v^{-1} C\Sigma_e = 0 \tag{16.4}$$

In Chap. 13, the following equations were suggested to calculate the expected dynamic operating region (EDOR) in the PSI case.

$$(A + BL)\Sigma_{\hat{x}} + \Sigma_{\hat{x}}(A + BL)^T + \Sigma_e C^T \Sigma_v^{-1} C\Sigma_e = 0 \tag{16.5}$$

$$\Sigma_z = (D_x + D_u L)\Sigma_{\hat{x}}(D_x + D_u L)^T + D_x \Sigma_e D_x^T \tag{16.6}$$

In this case, ξ_i is calculated as $\phi_i \Sigma_z \phi_i^T$ and Σ_e is defined by Eq. (16.4). Then the PSI version of the constrained minimum variance control problem was stated as

$$\min_{\xi_i, \Sigma_{\hat{x}}, L} \left\{ \sum_{i=1}^{n_z} d_i \xi_i \right\} \quad \text{s.t.} \tag{16.7}$$

$$(A + BL)\Sigma_{\hat{x}} + \Sigma_{\hat{x}}(A + BL)^T + \Sigma_e C^T \Sigma_v^{-1} C\Sigma_e = 0$$

$$\xi_i = \phi_i (D_x + D_u L)\Sigma_{\hat{x}}(D_x + D_u L)^T \phi_i^T + \phi_i D_x \Sigma_e D_x^T \phi_i^T$$

and $\xi_i < \bar{z}_i^2,\ i = 1 \ldots n_z$

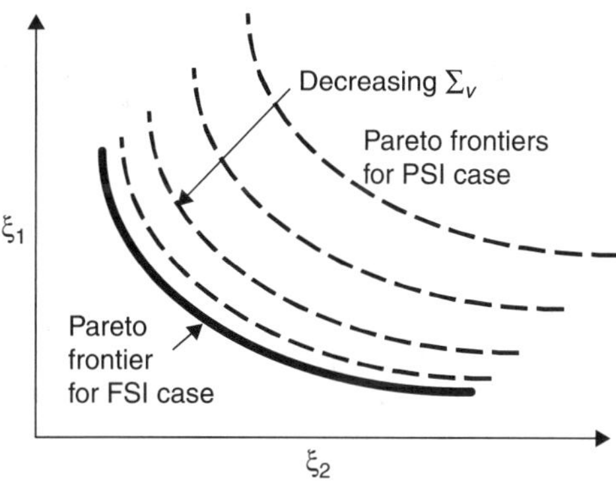

FIGURE 16.1 The Pareto frontiers as a function of instrumentation network quality.

Given Eq. (16.7), one can calculate a Pareto frontier for each instrumentation configuration of interest. In general, the result will be a shifting of the frontier based on the quality of instruments (Fig. 16.1). Specifically, higher-quality sensors (i.e., small measurement error, Σ_v) will result in better state estimates (i.e., smaller estimation error variance, Σ_e) and thus better control of the process (i.e., smaller values within Σ_z). In the limit of a high-quality instrumentation network, the Pareto frontier will approach that of the FSI case.

The impact of an instrumentation network on the value of a control system is depicted in Fig. 16.2. Specifically, a higher-quality instrumentation network will result in a smaller EDOR, and thus the ability to move the backed-off operating point (BOP) closer to the optimal steady-state operating point (OSSOP). However, in the instrumentation network design problem (to be defined below), the improvement in plant operating profit will be offset by the purchase and maintenance costs of the sensor network.

In the subsequent design problem, it will be convenient to have a sensor representation that smoothly transitions from sensor presence to sensor absence. Begin by constructing the C matrix under

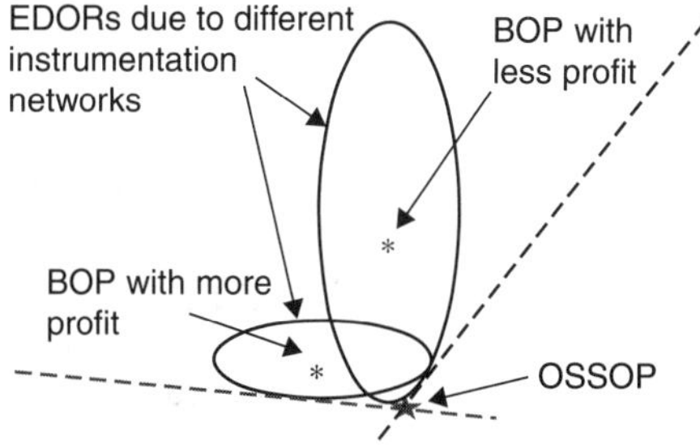

FIGURE 16.2 BOP selection with using different instrumentation networks.

the assumption that all possible sensors have been included in the network. Additionally, assume that the measurement noise term of each of these sensors is independent of the others. In this case, Σ_v will be diagonal and Σ_v^{-1} can be defined as $\mathrm{diag}\{\beta_i / \bar{\sigma}_{v_i}^2\}$, where $\bar{\sigma}_{v_i}^2$ is the actual precision of sensor i and β_i is a decision variable indicating the presence/absence of sensor i. Specifically, if $\beta_i \to 0$, then the ith element of Σ_v will become large, indicating that the ith sensor is extremely noisy. In this case, the Kalman filter will ignore this noisy sensor and make it appear to be absent in the construction of $\hat{x}$. If $\beta_i \to 1$, then the ith element of Σ_v will equal $\bar{\sigma}_{v_i}^2$ and indicate that sensor i is present with the appropriate precision.

An implication of the above sensor representation is the fact that Σ_e will be a function of the β_i optimization variables. As such, the resulting BOP selection scheme will need to include Eq. (16.4), with Σ_e as an optimization variable. In addition, Σ_e appears in a nonlinear term of Eq. (16.5). In this case of Σ_e being one of the optimization variables, Theorem 13.2 will not result in a convex formulation for CMV controller design. To achieve a convex formulation, we will need to begin with the following alternative formulation of Eq. (16.7), the equivalence of which is described in Peng and Chmielewski (2006):

$$\underset{\xi_i,\Sigma_x\geq 0,\Sigma_e\geq 0,L}{\mathrm{Min}} \left\{\sum_{i=1}^{n_z} d_i\xi_i\right\} \tag{16.8}$$

s.t.

$$A\Sigma_x + \Sigma_x A^T + BL(\Sigma_x - \Sigma_e) + (\Sigma_x - \Sigma_e)L^T B^T + G\Sigma_w G^T = 0 \tag{16.9}$$

$$A\Sigma_e + \Sigma_e A^T - \Sigma_e C^T \Sigma_v^{-1} C\Sigma_e + G\Sigma_w G^T = 0 \tag{16.10}$$

$$\xi_i = \phi_i\left[(D_x + D_u L)(\Sigma_x - \Sigma_e)(D_x + D_u L)^T + D_x\Sigma_e D_x^T\right]\phi_i^T \tag{16.11}$$

$$\xi_i < \bar{z}_i^2, \qquad i = 1 \ldots n_z \tag{16.12}$$

Then the following can be used to convert the constraints of the above problem into a convex form.

Theorem 16.1 There exists stabilizing $\Sigma_x \geq 0$, $\Sigma_e \geq 0$, L, and ξ_i, such that

$$A\Sigma_x + \Sigma_x A^T + G\Sigma_w G^T + BL(\Sigma_x - \Sigma_e) + (\Sigma_x - \Sigma_e)L^T B^T = 0 \tag{16.13}$$

$$A\Sigma_e + \Sigma_e A^T - \Sigma_e C^T \Sigma_v^{-1} C\Sigma_e + G\Sigma_w G^T = 0 \tag{16.14}$$

$$\xi_i = \phi_i\left[(D_x + D_u L)(\Sigma_x - \Sigma_e)(D_x + D_u L)^T + D_x\Sigma_e D_x^T\right]\phi_i^T \tag{16.15}$$

$$\xi_i < \bar{z}_i^2, \quad i = 1 \ldots n_z \tag{16.16}$$

if and only if there exists $X > 0$, $Z > 0$, Y, and ξ_i, such that

$$(AX + BY) + (AX + BY)^T + G\Sigma_w G^T < 0 \tag{16.17}$$

$$\begin{bmatrix} -ZA - A^T Z + C^T \Sigma_v^{-1} C & ZG^T \\ GZ & \Sigma_w^{-1} \end{bmatrix} > 0 \tag{16.18}$$

$$\begin{bmatrix} \xi_i & \phi_i (D_x X + D_u Y) & \phi_i D_x \\ (D_x X + D_u Y)^T \phi_i^T & X & I \\ D_x^T \phi_i^T & I & Z \end{bmatrix} > 0 \tag{16.19}$$

$$\xi_i < \bar{z}_i^2, \quad i = 1 \ldots n_z \tag{16.20}$$

Q.E.D.

A direct extension of our previous sensor selection formulation (Peng and Chmielewski, 2006) would result in a mixed integer convex program, due to the integer aspects of β_i. Unfortunately, this mixed integer formulation does not combine well with the reverse-convex aspects of the BOP selection problem. (Such a scheme could be implemented, but the branch-and-bound routine would be unnecessarily complicated.) To create uniformity in the nonconvex constraints, we suggest the following method of approximating integer constraints with reverse-convex inequalities. First, introduce a new set of variables $\beta_{c,i}$ and $\beta_{p,i}$. These will serve to split β_i into a cost aspect, $\beta_{c,i}$, and a precision aspect, $\beta_{p,i}$. That is, $\beta_{c,i}$ will appear in the objective function to reflect the cost of sensor i, and $\beta_{p,i}$ will appear in the Σ_v^{-1} function to reflect the precision of sensor i. The two βs are then connected by the inequality $\beta_{p,i} \le h(\beta_{c,i})$, where

$$h(x) = \begin{cases} m_0 x & \text{if} \quad x < x_0 \\ m_1 x + b_1 & \text{if} \quad x \ge x_0 \end{cases} \tag{16.21}$$

where $m_0 = y_0/x_0$, $m_1 = (1 - y_0)/(1 - x_0)$, and $b_1 = (y_0 - x_0)/(1 - x_0)$. Selecting $x_0 = 0.85$ and $y_0 = 0.1$ yields the function h of Fig. 16.3. (In the example to follow we used $x_0 = 0.9999$ and $y_0 = 0.0001$.) This reverse-convex inequality mimics the integer constraint by only allowing the precision aspect to be large (i.e., $\beta_{p,i} > y_0$) if the cost aspect is also large (i.e., if $\beta_{c,i} > x_0$).

Combining all of the above concepts, one finally arrives at the following simultaneous sensor, controller, and BOP selection design problem.

$$\min_{\substack{s', m', q', \xi_i, \\ X>0, Z>0, Y \\ \beta_{c,i}, \beta_{p,i}}} \left\{ g_q^T q + g_c^T \beta_c \right\} \tag{16.22}$$

s.t.

$$0 = As' + Bm' + Gp' \tag{16.23}$$

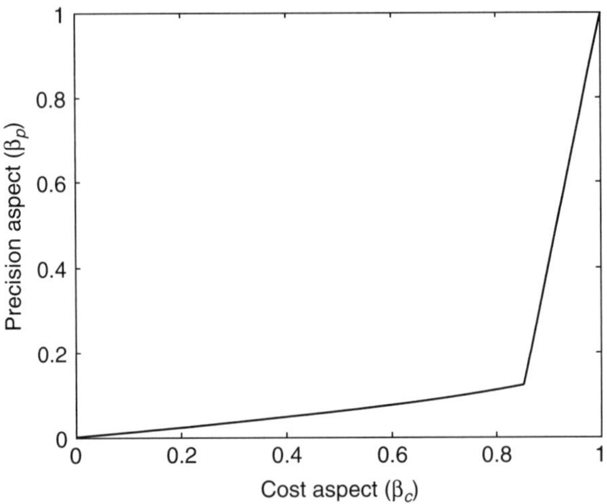

FIGURE 16.3 Reverse-convex constraint approximating integer constraint.

$$q_i' = \phi_i\,(D_x s' + D_u m' + Gp'),\ q_i'^{\min} < q_i' < q_i'^{\max} \tag{16.24}$$

$$\xi_i^{1/2} < q_i'^{\max} - q_i',\ \xi_i^{1/2} < q_i' - q_i'^{\min} \tag{16.25}$$

$$(AX + BY) + (AX + BY)^T + G\Sigma_w G^T < 0 \tag{16.26}$$

$$\begin{bmatrix} -ZA - A^T Z + C^T \Sigma_v^{-1} C & ZG^T \\ GZ & \Sigma_w^{-1} \end{bmatrix} > 0 \tag{16.27}$$

$$\begin{bmatrix} \xi_i & \phi_i\,(D_x X + D_u Y) & \phi_i D_x \\ (D_x X + D_u Y)^T \phi_i^T & X & I \\ D_x^T \phi_i^T & I & Z \end{bmatrix} > 0 \tag{16.28}$$

$$0 \le \beta_{c,i} \le 1,\quad 0 \le \beta_{p,i} \le 1 \quad \beta_{p,i} \le h(\beta_{c,i}) \quad i = 1 \ldots n_\beta \tag{16.29}$$

$$\Sigma_v^{-1} = \mathrm{diag}\left\{\beta_{p,i} / \bar{\sigma}_{v_i}^2\right\} \tag{16.30}$$

The elements of g_c are the time normalized cost of each sensor. If X^*, Y^*, W^* are the solution to the above problem, then a controller achieving this performance can be reconstructed as $L = Y^*(X^* - W^{*-1})^{-1}$. In this case, $\Sigma_e = W^{*-1}$ and $\Sigma_x = X^*$.

Example 16.1: Consider the reactor of Fig. 14.14 (Example 14.1). Process-modeling equations and nominal operating conditions can be found in Bhushan and Rengaswamy (2000). In the current example, the three PI controllers depicted in Fig. 14.14 will be replaced by the PSI controller described above. The resulting nonlinear model has five states $s = [C_A \; T \; T_c \; V \; P]^T$, three manipulated variables $m = [F_c \; F \; F_{vg}]^T$, and two disturbances $p = [F_i \; C_{Ai}]^T$. In addition, we define the performance output simply as $q = [s^T \; m^T \; p^T]^T$. The set of possible steady-state operating conditions is defined by the following equality constraints. $0 = As + Bm + Gp$, where A, B, and G are partial derivatives of the nonlinear model evaluated at the nominal conditions. Combining the above equalities with a set of upper and lower inequality bounds on each variable (see the dashed line of Fig. 16.4) defines the set of feasible steady-state operating points.

The profit function is

$$g(C_A, F_c, F, F_{vg}) = M_{\text{an}}[\alpha_1(C_{Ai} - C_A)F - \alpha_2 F_c - \alpha_3 F_{vg}]$$

where α_1 = \$0.375/mol B, α_2 = \$ 0.015/ft^3 of cooling water, α_3 = \$ 0.00225/ft^3 of vapor pumped, and M_{an} = 8760 h/yr. The linearized profit function is $g \cong g(q^{\text{nom}}) + g_q^{\,T} q'$, where $g_q^{\,T}$ is the partial derivative of the profit function evaluated at the nominal conditions. This along with the set of feasible steady-state operating points is used to determine OSSOP, indicated by the * points in Fig. 16.4, which has a profit \$47,370/yr. This solution represents the amount of profit one would yield if zero cost, perfect sensors were available and no disturbances acted on the system. It should be noted that the nominal value of C_{Ai} was assumed constant in the steady-state portion of the procedure. This is in contrast to F_i, which was allowed to be a free variable in the selection of the OSSOP.

Returning to the system dynamics, it was assumed that the disturbances, F_i and C_{Ai} have standard deviations equal to 0.1 and 0.01, respectively. It was additionally assumed that the existing sensor network consisted of four sensors, at C_A, T, V, and P, each with a precision of 2%. If we solve the BOP problem using this existing network, we find the expected profit to be \$28,970/yr. The BOP and EDOR resulting from the existing sensor network are indicated by the triangle points of Fig. 16.4 (and case 10 of Table 16.1).

Then, we assumed that new 1% precision sensors are available at each state (i.e., at C_A, T, T_c, V, and P). However, if a new sensor is placed, then the old one must be removed. The placement of a new sensor will have an annualized cost of \$1000/yr, and there will be no annualized cost due to leaving an old sensor in place. If we now place a new sensor at all five locations and apply the BOP selection method, we find that the profit will increase to \$37,060/yr, due to our ability to move the BOP closer to the OSSOP. This solution is indicated by the square points of Fig. 16.4. If we then subtract this amount from the profit of the existing network, we find that the value of the five sensor configuration is \$8090/yr, and the increase in profit (value minus sensor cost) is \$3080/yr (see case 8 of Table 16.1).

Application of a branch-and-bound, global search scheme to Eq. (16.22) reveals that replacement of sensors at C_A and P will yield the greatest increase in profit (see the circle points of Fig. 16.4 and case 1 of Table 16.1). The EDRs with triangles correspond to the existing network (configuration 10), the EDORs with squares correspond to the full upgrade (configuration 8), and the EDORs with a circle point correspond to the optimal upgrade (configuration 1). The profits and corresponding values as well as costs for other configurations can be found in Table 16.1.

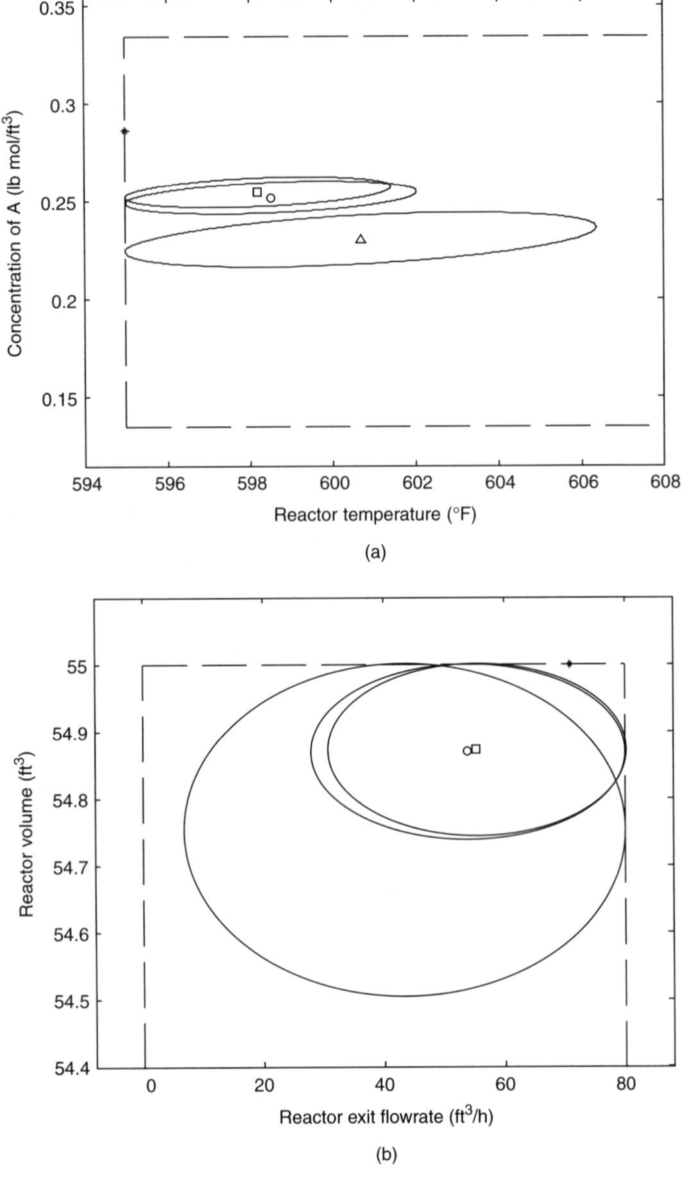

Figure 16.4 EDORs for Example 16.1: (a) Concentration vs. reactor temperature, (b) reactor volume vs. reactor exit flowrate, (c) jacket flowrate vs. jacket temperature, (d) exit flowrate vs. reactor pressure.

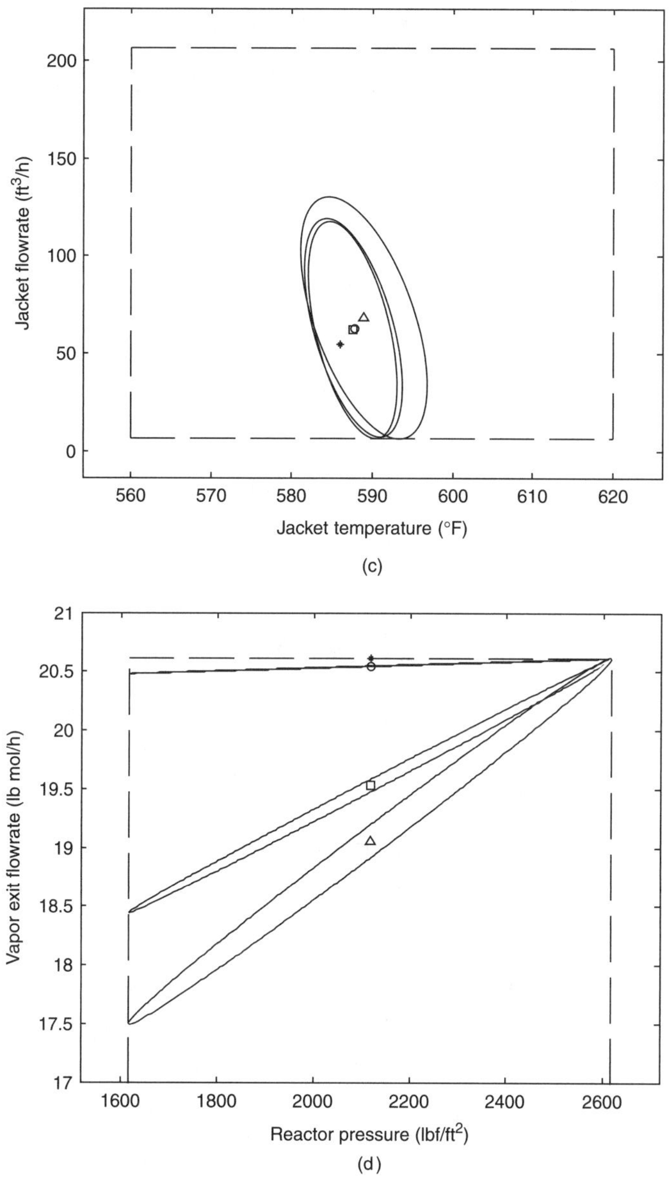

FIGURE 16.4 *(Continued)*

No	New Sensors	Profit ($/yr)	Value ($/yr)	Sensor Costs ($/yr)	Value-Sensor Costs ($/yr)
1	C_A, P	36,030	7,060	2,000	5,060
2	C_A, T_c, P	36,600	7,630	3,000	4,630
3	C_A, T, P	36,590	7,610	3,000	4,620
4	C_A, V, P	36,060	7,090	3,000	4,080
5	C_A	33,840	4,870	1,000	3,870
6	P	33,470	4,500	1,000	3,500
7	T, P	34,390	5,420	2,000	3,420
8	C_A, T, T_c, V, P	37,060	8,090	5,000	3,080
9	T, T_c, V, P	35,120	6,150	4,000	2,140
10	None	28,970	0	0	0

Table 16.1 Profits and Values of Upgrade Configurations

References

Athans, M., "On the Determination of Optimal Costly Measurement Strategies for Linear Stochastic Systems," *Automatica* **8**:397–412 (1972).

Alonso, A. A., I. G. Kevrekidis, J. R. Banga, and C. E. Frouzakis, "Optimal Sensor Location and Reduced Order Observer Design for Distributed Process Systems," *Comp. Chem. Eng*. **28**(1–2):27–35 (2004).

Antoniades, C. and P. D. Christofides, "Integrating Nonlinear Output Feedback Control and Optimal Actuator/Sensor Placement for Transport-Reaction Processes," *Chem. Eng. Sci*. **56**:4517–4535 (2001).

Armaou, A. and M. A. Demetriou, "Optimal Actuator/Sensor Placement for Linear Parabolic PDEs Using Spatial H2 Norm," *Chem. Eng. Sci*. **61**(22):7351–7367 (2006).

Bhushan, M. and R. Rengaswamy, "Design of Sensor Network Based on the Signed Directed Graph of the Process for Efficient Fault Diagnosis," *Ind. Eng. Chem. Res*. **39**(4):999–1019 (2000).

Cao, Y. and D. Rossiter, "An Input Pre-Screening Technique for Control Structure Selection," *Comp. Chem. Eng*. **21**(6):563–569 (1997).

Chmielewski, D. J., T. Palmer and V. I. Manousiouthakis, "On the Theory of Optimal Sensor Placement," *AIChE J*. **48**(5):1001–1012 (2002).

Faulds, A. L. and B. B. King, "Sensor Location in Feedback Control of Partial Differential Equation Systems," *Proc. IEE Conf. Cont. App*. 536–541 (2000).

Harris, T. J., J. F. Macgregor, and J. D. Wright, "Optimal Sensor Location with an Application to a Packed Bed Tubular Reactor," *AIChE J*. **26**(6):910–916 (1980).

Heath, J. A., I. K. Kookos, and J. D. Perkins, "Process Control Structure Selection Based on Economics," *AIChE J*. **46**(10):1998–2016 (2000).

Hovd, M. and S. Skogestad, "Procedure for Regulatory Control Structure Selection with Application to the FCC Process," *AIChE J*. **39**(12):1938–1953 (1993).

Johnson, C. D, "Optimization of a Certain Quality of Controllability and Observability for Linear Dynamical System," *J. Basic Eng., ASME Trans.*, Series D:228–238 (1969).

Kubrusly, C. S. and H. Malebranche, "Sensors and Controllers Location in Distributed Systems—A Survey," *Automatica* **21**:117–128 (1985).

Kushner H. J. "On the Optimal Timing of Observations for Linear Control Systems with Unknown Initial States," *IEEE Trans. Aut. Cont*. **9**:144–150 (1964).

Lee, J. H., R. D. Braatz, M. Morari, and A. Packard, "Screening Tools for Robust Control Structure Selection," *Automatica* **31**(2):229–235 (1995).

Loeblein, C. and J. Perkins, "Economic Analysis of Different Structures of On-Line Process Optimization Systems," *Comp. Chem. Eng.* **20**:S551–S556 (1996).

Meier, L., J. Peschon, and R. M. Dressler, "Optimal Control of Measurement Subsystems," *IEEE Trans. Aut. Cont.* **12**(5):528–536 (1967).

Mellefont, D. and R. Sargent, "Selection of Measurements for Optimal Feedback Control," *Ind. Eng. Chem. Process Des. Dev.* **17**:549 (1978).

Morari, M. and M. J. O'Dowd, "Optimal Sensor Location in the Presence of Nonstationary Noise," *Automatica* **16**:463 (1980).

Morari, M. and G. Stephanopoulos, "Optimal Selection of Secondary Measurements within the Framework of State Estimation in the Presence of Persistent Unknown Disturbances," *AIChE J.* **26:**247 (1980).

Morari, M., G. Stephanopoulos, and Y. Arkun, "Studies in the Synthesis of Control Structures for Chemical Processes, Part I: Promulgation of the Problem. Process Decomposition and the Classification of the Control Task. Analysis of the Optimizing Control Structures" *AIChE J.* **26**(2):220–232 (1980).

Muller, P. C. and H. I. Weber, "Analysis and Optimization of Certain Qualities of Controllability and Observability for Linear Dynamical Systems," *Automatica* **8**:237–246 (1972).

Muske, K. R. and C. Georgakis, "A Methodology for Optimal Sensor Selection in Chemical Processes," *Proc. Am. Cont. Conf.* 4274–4278 (2002).

Narraway, L. and J. Perkins, "Selection of Process Control Structure Based on Economics," *Comp. Chem. Eng.* **18**:511–515 (1994).

Peng, J. K. and D. J. Chmielewski, "Optimal Sensor Network Design Using the Minimally Backed-Off Operating Point Notion of Profit," *Proc. Am. Cont. Conf.* 220–224 (2005).

Peng, J.K. and D. J. Chmielewski, "Covariance Based Hardware Selection-Part II: Equivalence Results for the Sensor, Actuator and Simultaneous Selection Problems," *IEEE Trans. Cont. Sys. Tech.* **14**(2):362–368 (2006).

Romagnoli, J., J. Alvarez, and G. Stephanopolus, "Variable Measurement Structures for Process Control," *Int. J. Control* **33**:269 (1981).

Singh, A. K. and J. Hahn., "Sensor Location for Stable Nonlinear Dynamic Systems: Multiple Sensor Case," *Ind. Eng. Chem. Res.* **45**(10):3615–3623 (2006).

Van de Wal, M. and B. de Jager, "Review of Methods for Input/Output Selection," *Automatica* **37**(4): 487–510 (2001).

Van de Wouwer, A., N. Point, S. Porteman, and M. Remy, "An Approach to the Selection of Optimal Sensor Locations in Distributed Parameter Systems," *J. Proc. Cont.* **10**:291 (2000).

Wisnewski, P. A. and F. J. Doyle, "Control Structure Selection and Model Predictive Control of the Weyerhaeuser Digester Problem," *J. Proc. Cont.* **8**(5–6):487–495 (1998).

Yu, T. K. and J. H. Seinfeld, "Observability and Optimal Measurement Location in Linear Distributed Parameter Systems," *Int. J. Control* **18**(4):785–799 (1973).

CHAPTER 17

Structural Faults and Value of Maintenance*

In Chap. 14, we covered the detection and resolution of parametric faults, but structural faults were left aside. As it was discussed there, structural faults refer to minor and major equipment malfunction, which needs to be detected and corrected. In this chapter, we will assume that structural faults can be detected and resolved, which can be done using the same methods as those mentioned in Chap. 14. We will also assume that the proper set of alarms is in place. Thus, we will concentrate on the maintenance actions that will correct the malfunction of equipment and/or prevent them to take place in the first place.

Structural failure in process plants can be classified as follows:

- Slow deterioration or so-called wear and tear. Examples of these are wear and tear of rotating equipment (pumps, compressors, turbines) or fouling of heat exchangers.
- Major malfunction leading to equipment shutdown, either suddenly or slowly. When there is no spare equipment to switch operations to, the plant is totally or partially shut down. Examples of these malfunctions are motor failure in main process compressors or rupture in process vessels.

To fix failed equipment, corrective maintenance (or repair) is used, whereas to prevent deterioration or major malfunction, preventive maintenance is used. All these activities are usually centralized and coordinated, although a recent tendency is seen on trying to compartmentalize certain groups. One example of this compartmentalization is the cleaning of fouled heat exchanger networks, for which

*This chapter was written by DuyQuang Nguyen and Miguel Bagajewicz.

there are special optimization techniques (Lavaja and Bagajewicz, 2004; 2005a, b). Another example is instrumentation maintenance to maintain accuracy, for which special crews sometimes exist. This is covered in a later chapter in detail.

The economic impact of maintenance activities is sizable: Not counting scheduled outages, typical refineries experience about 10 days downtime per year due to equipment failures with an estimated economic loss of $20,000 to $30,000 per hour (Tan and Kramer, 1997). Dhillon (2002) also estimated that

1. The typical size of a plant maintenance group in a manufacturing organization varied from 5 to 10% of the total operating force.
2. Over $300 billion are spent on plant maintenance and operations by U.S. industry each year.
3. Approximately 80% of this 300 billion is spent to *correct* the chronic failure of machines, systems, and human errors.
4. The annual cost of maintenance as a fraction of total operating budget can go up to 40 to 50% for the mining industry (Murthy et al., 2002) and 20 to 30% for the chemical industry (Tan and Kramer, 1997).
5. The elimination of these chronic failures through effective maintenance can reduce the cost between 40 and 60%.

Maintenance

Maintenance can be defined as all actions appropriate for retaining an item/part/equipment in, or restoring it to, a given condition (Dhillon, 2002). More specifically, maintenance is used to repair broken equipment, preserve equipment conditions, and prevent their failure, which ultimately reduces production loss and downtime as well as the environmental and the associated safety hazards.

Maintenance practices and strategies have been shifted from reactive mode (focusing on repairing failed equipments) to proactive mode (focusing on preventing failures and keeping systems functioning). Reliability centered maintenance (RCM), a modern maintenance practice developed in 1960s and 1970s, is an effective tool for managing risk and safety of engineering systems with demonstrated successful applications in the aerospace industry and the power generating industry. Total productive maintenance (TPM), another successful story of modern maintenance management philosophies with widespread application in the manufacturing industries, was developed in 1960s and was reported to remarkably increase productivity of the manufacturing plants.

There is a large collection of textbooks that discuss current maintenance practices and strategies. In addition, a large number of Computerized Maintenance Management Systems (CMMS) software packages devoted to help the users manage/organize the maintenance activities are available on the market (over 360 software packages are listed in the Web site www.plant-maintenance.com). These packages are excellent databases that help track repair orders and maintain appropriate book-keeping.

The question of how to use maintenance effectively and how much maintenance resources are enough arises naturally and is known as the maintenance optimization problem. To seek an appropriate schedule for corrective and preventive actions, a maintenance model is utilized to optimize decision variables in maintenance planning like preventive maintenance (PM) frequency, spare parts inventory policy, labor workforce size, etc. Unfortunately, the maintenance optimization problem is usually not discussed in textbooks nor included as a feature in the commercial CMMS software packages.

In this chapter, we provide general description of models that are applicable to chemical process plants. In the next chapter, we will discuss optimization and in Chap. 18, we focus on instrumentation.

Maintenance Policies

There are two major objectives for maintenance.

1. *Managing Risk and Safety:* A comprehensive risk management program involves both the software and the hardware factors. The software factors are the safety culture that requires every plant employee to comply with safety regulations and the use of various hazard evaluation techniques. Some of these techniques are
   - *Hazard and Operability Analysis (HAZOP):* This is the most popular technique for hazard evaluation (Center for Chemical Process Safety, 2008). By systematic and careful analysis of process or operation, the HAZOP team lists potential causes and consequences of process deviations as well as existing safeguards protecting against the deviation. The HAZOP technique requires detailed information regarding the design and operation of process.
   - *Failure Modes and Effects Analysis (FEMA):* This is another popular technique and is an important component in modern maintenance programs like reliability centered maintenance (Smith and Hinchcliffe, 2004). The FEMA tabulates failure modes (how equipment fails) and their effects on a system or plant.

- *Other Techniques: Fault tree analysis* is a popular technique that identifies and displays graphically various causes of a particular accident or system failure, and *event tree analysis*, which analyzes and shows graphically various outcomes of an accident initiated by an equipment failure or human error, and *cause–consequence analysis*, which is a blend of fault tree and event tree analysis.

The above techniques are useful for failure observability and resolution and can be compared to directed graphs, sign directed graphs, and bipartite graphs.

In turn, the hardware factors are the use of safety instrumented system to safeguard the process such as pressure relief valves, flame detectors, etc., and the use of maintenance to preserve equipment condition and improve system reliability.

2. *Minimizing Economic Losses:* Maintenance reduces downtime in a plant associated to major failures. However, not all equipment failures lead to plant shutdown. Some failures deteriorate plant performance and most have an associated lost revenue on top of the cost of repair. In addition, there are costs associated with the replacement of pieces of equipment during preventive maintenance.

Modern maintenance management philosophy suggests that it should focus on the whole system rather than on individual equipment/components, and preventive maintenance frequency needs to be optimized.

Plant maintenance policies can be classified into two main types.

1. *Corrective Maintenance (CM) or Equipment Repair:* This type of maintenance deals with fixing already malfunctioning equipment.
2. *Preventive Maintenance (PM):* Three main factors are considered when planning/scheduling PM activities for a specific equipment.
   (i) Equipment Importance: This is related to the role of equipment in the system, its reliability measure (MTBF), its failure history, the regularity of use, availability of backup copy, etc.
   (ii) Economics: The difference between the gain (the reduction in repair cost and economic loss) and the cost incurred.
   (iii) Availability of needed resources (labor with appropriate skills and spare parts) in the plant.

Other types of maintenance policies, not usually used in the industry, are repair limit policy (estimated repair cost is compared with a predetermined limit to decide whether to repair or replace the unit), age replacement, and block replacement policy. These policies were described in Wang and Pham (2006).

Preventive maintenance (PM) is in turn divided into two types.

- *Condition-Driven PM:* It is usually called predictive maintenance. With this type of maintenance, maintenance personnel monitor (online or periodically) equipment's condition to detect in advance any failure symptom of the equipment, then perform planned repair for the failure-prone equipment to avoid downtime. Popular predictive maintenance techniques are vibration monitoring and analysis, lubricating oil analysis, visual inspection, and others.
- *Time-Driven PM:* In this case, maintenance actions are scheduled at predetermined (i.e., fixed) time intervals based on some criteria (the three factors mentioned above). Various versions of time-driven preventive maintenance (PM) policy have been proposed and were summarized in Wang and Pham (2006). They are
    - *Age-Dependent PM:* The PM times are based on the age of the unit.
    - *Periodic PM:* A unit is preventively maintained at fixed time kT ($k = 1, 2, \ldots$), where T is the PM interval, independent of the failure history or age of the unit.
    - *Sequential PM:* A unit is preventively maintained at *unequal* time intervals. Usually, the time intervals become shorter and shorter.
    - *Failure Limit Policy:* PM is performed only when the failure rate or other reliability measures of a unit reach a predetermined level.

Periodic PM is the most commonly discussed in textbooks while age-dependent PM is probably the most common maintenance policy used in maintenance optimization research (Wang and Pham, 2006). Indeed, it makes more sense to plan PM based on the age of the unit connected to some deterioration rate. However, the difficulty with the quantification of failure rates of units as function of time hinders the application of age-dependent PM in industry. Nevertheless, in what follows we cover imperfect maintenance to some extent.

For multiunit systems with dependence between units, group maintenance or opportunistic maintenance policies have been proposed.

1. *Group Maintenance:* This policy is suitable for system consisting of groups of identical units. Cost saving results from the reduction in maintenance set-up cost.

2. *Opportunistic Maintenance:* Performing PM to a unit whenever an opportunity is open (in addition to the regularly scheduled PM) results in higher number of PM actions on the unit. The result is improved equipment reliability at the expense of higher maintenance cost.

We now present the basic concepts on reliability that will be later used in the assessment of the value of maintenance.

Reliability, Failure Rate, and Mean Time to Failure

We now review some basic definitions and then describe some of the reliability functions.

Definition *Service availability* $(A_i^s(t))$ *is the probability of the equipment i in a normal state at time t, given that as good as new at time t* = 0.

Definition *Average availability in a time interval* [0, *t*] $(A_i^{\mathrm{av}}(t))$ *is the expected fraction of time within the interval* [0, *t*] *that the system is able to operate normally.*

The average unavailability $[1-A_i^{\mathrm{av}}(t)]$, the expected fraction of downtime, can be associated with the economic loss due to downtime within the time interval.

The service availability $A_i^s(t)$ and average availability $A_i^{\mathrm{av}}(t)$ are called point availability and interval availability, respectively, in other literature sources (Birolini, 2007; Nakagawa, 2005). The average availability is calculated using the cumulative value of $A_i^s(t)$ at discrete points in time as follows:

$$A_i^{\mathrm{av}}(t)=\frac{1}{t}\int_0^t A_i^s(u)du \tag{17.1}$$

Clearly, if the service/point availability is constant, then the average availability will also be constant and equal to the service/point availability. Also, for small times they can be shown to be asymptotically equal using L'Hospital rule for the limit $A_i^{av}(t)$ as $t \to 0$.

Definition *Service reliability* $R_i^s(t)$ *at time t, is the probability that the equipment i is in the normal state in the interval* (0, *t*), *given that it was on the normal state at time t* = 0.

Because availability is a property at time *t*, regardless of history, and reliability is related to the interval (0, *t*), then

$$A_i^s(t) \geq R_i^s(t) \tag{17.2}$$

When no maintenance is performed, availability and reliability are equal. A typical reliability curve $R_i^s(t)$ is shown in Fig. 17.1. It is also called *survival distribution*. It is characterized by a relatively small

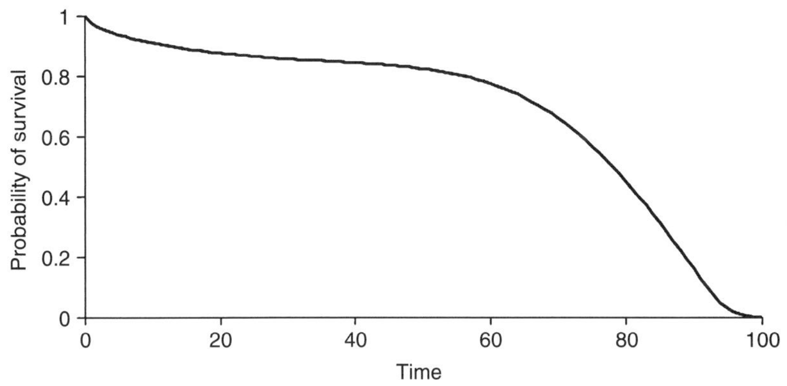

Figure 17.1 Reliability or survival distribution.

sharp decline at the beginning (0 to 20 in the figure), a stabilization period (20 to 55 in the figure), and a declining last period. These curves are constructed by simply starting with a large population of devices and monitoring the amount of devices surviving at each time.

Definition *Failure Density $f_i(t)$ at time t is the derivative of the survival distribution.*

$$f_i(t) = -\frac{dR_i^s(t)}{dt} \tag{17.3}$$

As it is shown in Fig. 17.2, there is a period of early failures, followed by a fairly constant failure rate period and finally by a wear-out failure period. The maximum at the end is attributed to the depletion of the sample.

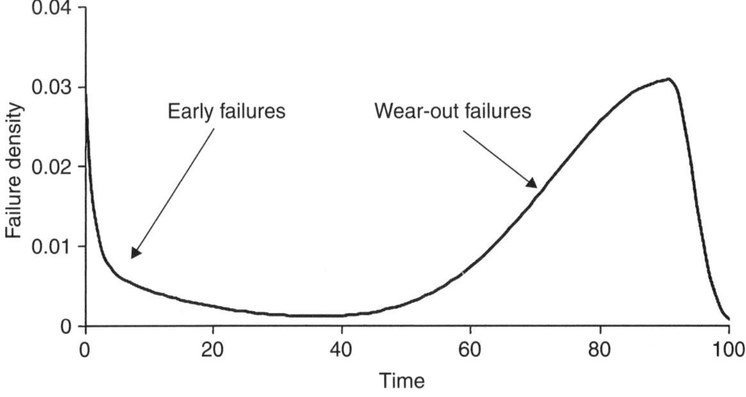

Figure 17.2 Failure density function.

However, one can relate this absolute rate to the existing sensors in the normal state. Thus, failure rate is defined as follows:

Definition *The failure rate $fr_i(t)$ at time t is the probability that the sensor fails per unit time at time t, given that it has survived until time t.*

This rate is expressed in conditional terms by requesting that the sensors be at normal state in the whole interval (0, *t*). In other words, when looked in terms of the thought experiment of a population of sensors at normal state at time zero, the failure rate is none other than the portion of sensors that fail calculated as a fraction of those that have survived. Indeed,

$$fr_i(t) = \frac{f_i(t)}{R_i^s(t)} \tag{17.4}$$

Since reliability is a function of time, a number representative of the reliability that is independent of time is the mean time to failure (MTTF). This is defined as follows:

$$\text{MTTF} = \int_0^\infty R_i^s(t)dt \tag{17.5}$$

For example, for a unit or equipment that has a failure intensity of two sensors/year, the mean time to failure is half a year. The mean time between failures (MTBF) is more commonly used in the industry. It is the (average) time interval between two consecutive downtimes (failure times); thus MTBF is the sum of the MTTF plus the mean time to repair (MTTR). Since MTTR is usually negligibly small, MTBF ≈ MTTF and they can be used interchangeably. Table 17.1 shows some reliability data for equipment. These values vary in the literature. We used values from Roup (1999) and Mannan (2005).

Equipment		Failure Rate (Failures per 10^6 h) Best	Average
Axial compressors		5.0	12.5
Centrifugal compressors	Clean service	7.6	14.3
	Fouling service	25.0	30.9
Reciprocating compressors	Lubricated	76.0	240.0
	Nonlube	114.0	340.0
	Labyrinth piston	23.0	56.0
Screw compressors	Dry	11.4	22.8
	Liquid injected	38.0	76.0

Table 17.1 Reliability Values for Different Equipment

Equipment		Failure Rate (Failures per 10^6 h) Best	Average
Electric motors	Induction <500 kW	9.5	14.2
	Induction >500 kW	5.7	9.5
Steam turbines	Condensing	16.2	22.8
	Back pressure	7.6	11.4
Centrifugal pumps	Process	17.5	33.0
	Utility service	17.8	33.5
Positive displacement pumps	Metering	33.0	73.0
	Screw	11.4	22.8
Gas turbines		19.0	22.8
Mixers		11.4	22.8
Heat exchangers		1	40
Process pressure vessel			0.308
Process storage vessel			0.205
Fired heaters			46
High temperature vessel			0.845
Low temperature vessel			0.171
		Range	
Pneumatic controllers		1–800	
Control valves		3–80	
Motorized valves		1–60	
Solenoid valves		1–25	
Relief Valves		1–10	
Checked Valves		0.8–10	
Butterfly Valves		1–30	
Flow sensors		0.1–10	
Level sensors		3–80	
Pressure indicators		1–9	
Temperature indicators		0.3–7	
Boilers-condensers		0.3–90	

Table 17.1 *(Continued)*

Failure Density Distributions

There are many types of distribution functions that have been used in reliability engineering: exponential, Weibull, normal distribution, Rayleigh, gamma, lognormal, binomial, etc. The three commonly used distributions are shown in Fig. 17.3. We now describe these distributions in detail.

Exponential Distribution

This distribution is described by

$$f_i(t) = \lambda_i e^{-\lambda_i t} \tag{17.6}$$

where λ_i is the distribution parameter. It is often used because it requires only one parameter. However, it applies well to certain systems: Smith and Hinchcliffe (2004) pointed out that for aerospace, electrical, and digital systems, the failure distribution is indeed exponential. The reliability function, hazard rate, and the MTTF corresponding to the exponential distributions can be shown to be

$$R_i^s(t) = e^{-\lambda_i t} \tag{17.7}$$

$$fr_i(t) = f_i(t) / R_i^s(t) = \lambda_i \tag{17.8}$$

$$\mathrm{MTTF} = 1/\lambda_i \tag{17.9}$$

We note that the failure rate $fr_i(t)$ and the MTTF (or MTBF) are constant (independent of time).

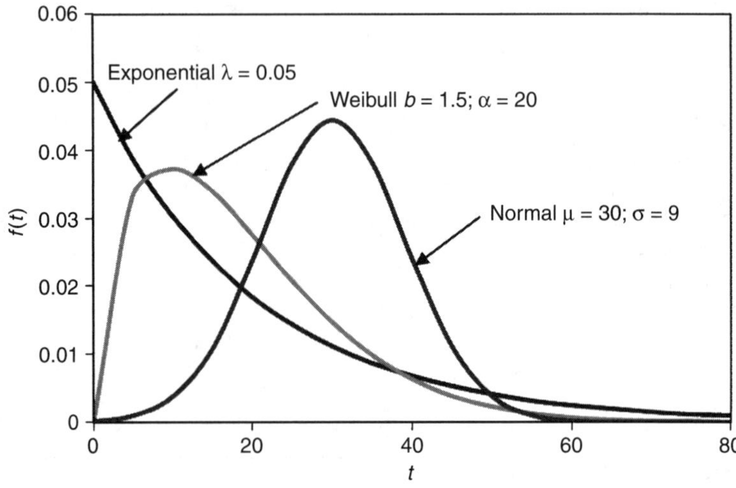

FIGURE 17.3 Distributions.

Weibull Distribution

This distribution is given by

$$f_i(t) = \frac{b_i t^{b_i - 1} e^{-(t/\alpha_i)^{b_i}}}{\alpha_i^{b_i}} \qquad \alpha_i > 0, b_i > 0 \tag{17.10}$$

where b_i and α_i are the shape and scale parameters, respectively. The Weibull distribution is known to be able to represent a wide range of distribution curves. The Weibull distribution with $b_i > 1$ is usually used to model failure-free time of components subjected to fatigue and/or wear-out (e.g., mechanical components). It can also be used to model other types of degradation processes. The scale parameter α_i, called the characteristic life, is defined as the time at which 63.2% of the population will have failed (i.e., the value of the cumulative failure distribution function at time $t = \alpha_i$ is 0.632, independent of the value of the shape parameter b_i).

The shape factor b_i determines the shape of the distribution. If $b_i > 1$, the failure rate increases monotonically; if $b_i < 1$, the failure rate decreases monotonically. When $b_i = 1$, the Weibull reduces to the exponential distribution (i.e., the failure rate is constant) and when $b_i = 2$, the Weibull becomes the Rayleigh distribution. When $b_i = 3.5$, it is similar to the normal distribution. The reliability function, failure rate, and MTTF corresponding to the Weibull distributions can be shown to be

$$R_i^s(t) = e^{-(t/\alpha_i)^{b_i}} \tag{17.11}$$

$$fr_i(t) = f_i(t)/R_i^s(t) = b_i t^{b_i - 1}/\alpha_i^{b_i} \tag{17.12}$$

$$\text{MTTF} = \alpha_i \Gamma(1 + 1/b_i) \tag{17.13}$$

where Γ is the complete gamma function, defined by $\Gamma(z) = \int_0^\infty x^{z-1} e^{-x} dx (z > 0)$. The MTTF is again constant.

Normal Distribution

This is well known and is defined by

$$f_i(t) = \frac{1}{\sigma_i \sqrt{2\pi}} \exp\left[-\frac{(t - \mu_i)^2}{2\sigma_i^2}\right] \tag{17.14}$$

where μ_i and σ_i are the mean and standard deviations of the distribution, respectively. This distribution is the best, well-known, two-parameter distribution with a wide range of applications. In reliability engineering, this distribution is used to model strengths of materials, percentages of defects in a population of products, etc. Failures described by normal distributions have failure rates increasing with time, and hence, these distributions can be used to model long-term wear or age-related failure phenomena. The reliability function, failure rate, and MTTF corresponding to the normal distributions can be shown to be

$$R_i^s(t) = \frac{1}{2} erfc\left(\frac{t-\mu_i}{\sqrt{2}\sigma_i}\right) \tag{17.15}$$

$$fr_i(t) = f_i(t)/R_i^s(t) = \frac{1}{\sigma_i}\sqrt{\frac{2}{\pi}} \exp\left[-\frac{(t-\mu_i)^2}{2\sigma_i^2}\right] / erfc\left(\frac{t-\mu_i}{\sqrt{2}\sigma_i}\right) \tag{17.16}$$

$$\text{MTTF} = \mu_i \tag{17.17}$$

where *erfc* is the complementary error function, defined by $erfc(x) = \frac{2}{\sqrt{\pi}}\int_x^{\infty} e^{-\xi^2} d\xi$.

Maintenance Models

A *maintenance model* is a mathematical model used to assess the effect of maintenance actions on reliability and economic performance as well as maintenance costs. Most of the existing maintenance models are analytical models obtained by using stochastic processes that have theoretical foundation based on probability theory. We will look at renewal processes (Wang and Pham, 1996) and Markov processes (Chan and Asgarpoor, 2006). They have some derivatives (semi-Markov, quasi-renewal).

Among nonanalytical models, stochastic simulation techniques stand out. We will look in detail at Monte Carlo simulation (Tan and Kramer, 1997) with some extensions proposed by Nguyen et al. (2008). Another one, which we will not cover here in detail, is the discrete-event production-oriented simulation (Charles et al., 2003).

Many other models can be found in the extensive review of published maintenance models performed by Wang and Pham (2006).

Of all existing models we pick the most popular and promising, namely the renewal, Markov, and Monte Carlo simulation approaches to highlight their ability to assess the economic cost and losses associated to maintenance decisions. Specifically, for each of these processes, we want to determine

- Availability as a function of time.
- How to compute economic losses and repair costs and the corresponding expressions/procedures to do it.
- Decision variables that influence the associated economics. This will allow us to optimize the system by manipulating their value.

In all three cases, we will assume that the failure density distribution is known and used as data.

Renewal Processes

An ordinary renewal process is a sequence of events, the intervals between which are independent and identically distributed (IID) random variables (Wang and Pham, 2006). The events we refer to in this context are the equipment failures. Thus, a renewal process is the number of failure of equipment within a time period with perfect repairs at failed times (perfect repair renews failed equipment instantaneously).

We recall from statistics that independent processes (in this case the renewal points) are those in which the probability of one event does not depend on the occurrence of other events. A renewal process with the inter-arrival times τ_k and the renewal points S_k at times $t_k(S_k)$ is shown in Fig. 17.4. The renewal times are random.

We now try to determine the failure distribution of each event, not only the first one.

We start with some nomenclature: The failure distribution of the nth inter-arrival time τ_n is $f_{i,n}(t)$. In turn, the probability that the nth event occurs at time t is given by the distribution $\hat{f}_{i,n}(t)$ [i.e., $\hat{f}_{i,n}(t)$ is the distribution of the nth renewal point $t_n(S_n)$]. Note that the *ordinary* renewal process requires that all the distributions $f_{i,n}(t)$ are equal, that is, $f_{i,n}(\tau_n) = f_{i,n-1}(\tau_{n-1}) = \cdots = f_{i,1}(\tau_1)$. Strictly speaking, one should be writing $\tau_{i,n}$ instead of simply τ_n because these time intervals refer to the equipment i. However, for simplicity of presentation, we use the latter. This is true for the case where equipment in continuous operation is presumably renewed at each failure in a negligible amount of time by a new statistically identical item. In other words, we consider that the repair is "as-good-as-new" (AGAN). Examples of IID sequences are the spins of a roulette, the sequence of dice rolls or coin flips, etc.

Example 17.1: For an ordinary renewal process where the failure distribution is given by the exponential distribution, one has $f_{i,n}(t) = \lambda_i e^{-\lambda_i t}$. In other words, every time that there is another failure, the probability function of the next failure is given by the same exponential distribution. Because all distributions are identical, then we can consider the repair is "as-good-as new" (AGAN).

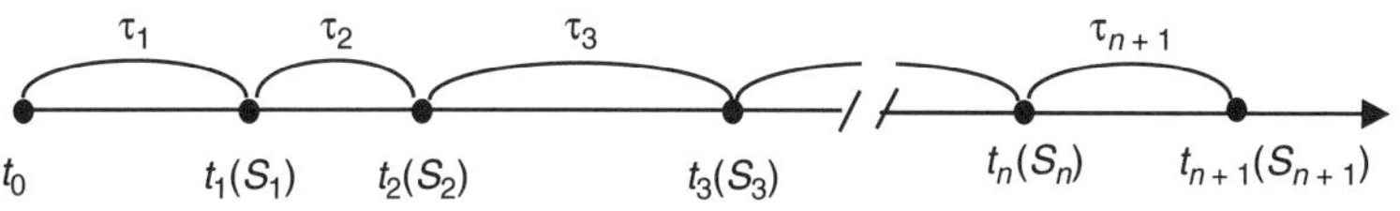

FIGURE 17.4 Illustration of renewal process.

We now attempt to obtain the probabilities $\hat{f}_{i,n}(t)$. The first failure time t_1 is characterized by a failure distribution $f_{i,1}(t)$. Because this is the first failure, it is easy to see that this distribution is equal to the equipment failure rate, that is,

$$\hat{f}_{i,1}(t) = f_{i,1}(t) \tag{17.18}$$

The second failure time t_2 is characterized by failure distribution $\hat{f}_{i,2}(t)$, which is obtained as follows: Assume that the first failure took place at time t_1. The probability of a second event at time $t > t_1$, once the first event has taken place at fixed time t_1, is given by $f_{i,2}(t-t_1)$. Because the events are independent, the probability is equal to the product of both probabilities $\hat{f}_{i,1}(t_1)\ f_{i,2}(t-t_1)$. According to probability theory, if X and Y are independent random variables with respective density functions h and g, then the density function of $X \cup Y$ is the convolution of f and g, that is, $(h^*g)(t) = \int_0^t h(\xi)g(t-\xi)dx = \int_0^t g(\xi)h(t-\xi)dx$, if $h(t)$ and $g(t)$ are defined for $t > 0$. This is equivalent to considering the integration of $\hat{f}_{i,1}(t-t_1)\ f_{i,1}(t_1)$ over all possible positive values of t_1. Thus,

$$\hat{f}_{i,2}(t) = \int_0^t \hat{f}_{i,1}(t-\xi)f_{i,2}(\xi)d\xi \tag{17.19}$$

Similarly, the cumulative distribution $\hat{F}_{i,2}(t)$ of S_2 is given by the integral from zero to t of Eq. (17.19).

$$\hat{F}_{i,2}(t) = \int_0^t \hat{F}_{i,1}(t-\xi)f_{i,2}(\xi)d\xi \tag{17.20}$$

We generalize this result to obtain $\hat{f}_{i,n}(t)$ recursively as follows:

$$\hat{f}_{i,1}(t) = f_{i,1}(t) \tag{17.21}$$

$$\hat{f}_{i,n}(t) = \int_0^t \hat{f}_{i,n-1}(t-\xi)f_{i,n}(\xi)d\xi \tag{17.22}$$

Similarly, $\hat{F}_{i,n}(t)$ is obtained using

$$\hat{F}_{i,1}(t) = F_{i,1}(t) \tag{17.23}$$

$$\hat{F}_{i,n}(t) = \int_0^t \hat{F}_{i,n-1}(t-\xi)f_{i,n}(\xi)d\xi \tag{17.24}$$

Example 17.2: Consider an ordinary renewal process where $f_{i,n}(t)$ is the exponential distribution, that is, $f_{i,n}(t) = \lambda_i e^{-\lambda_i t}$. Then,

$$\hat{f}_{i,2}(t) = \int_0^t \hat{f}_{i,1}(t-\xi)f_{i,2}(\xi)d\xi = \int_0^t \lambda_i e^{-\lambda_i(t-\xi)}\lambda_i e^{-\lambda_i\xi}d\xi = t\lambda_i^2 e^{-\lambda_i t} \tag{17.25}$$

$$\hat{f}_{i,3}(t) = \int_0^t \hat{f}_{i,2}(\xi)f_{i,3}(t-\xi)d\xi = \int_0^t \xi\lambda_i^2 e^{-\lambda_i\xi}\lambda e^{-\lambda_i(t-\xi)}d\xi = \frac{t^2\lambda_i^3}{2!}e^{-\lambda_i t} \tag{17.26}$$

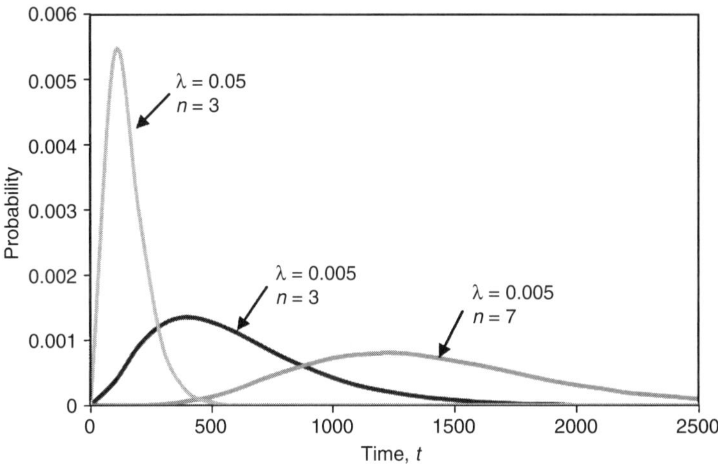

FIGURE 17.5 Probability density of ordinary renewal processes with exponentially distributed failure.

and in general

$$\hat{f}_{i,n}(t) = \int_0^t \hat{f}_{i,n-1}(\xi) f_{i,n}(t-\xi)d\xi = \frac{t^{n-1}\lambda_i^n}{(n-1)!}e^{-\lambda_i t} \tag{17.27}$$

which is known as the Erlang distribution with parameters λ_i and n. The cumulative distribution is obtained by using integration by parts.

$$\hat{F}_{i,n}(t) = 1 - \sum_{j=0}^{n-1} \frac{(\lambda_i t)^j e^{-\lambda_i t}}{j!} \tag{17.28}$$

Sketches of Erlang probability distribution for some values of λ_i and n are shown in Fig. 17.5.

As one would expect, the very first failures have larger probability at small time t while subsequent failures have larger probabilities at longer time. In other words, the distribution $\hat{f}_{i,n}(t)$ shifts to the right as n increases. If the failure rate λ_i increases, there is more chance for failures to occur earlier; thus the distribution $\hat{f}_{i,n}(t)$ shifts to the left as failure rate λ_i increases.

Expected Number of Repairs Because cost of repair is directly proportional to the number of repairs, we discuss how this is calculated next: To do it, we first write the relationship between expected number of repairs $\text{NR}_i(t)$ and the probabilities of having exactly n repairs ($n = 1, 2, 3, 4\ldots$) in the interval $[0, t]$ $P(N_i(t) = n)$. This relationship is

$$\text{NR}_i(t) = E[N_i(t)] = \sum_{n=1}^{\infty} n \times P\{N_i(t) = n\} \tag{17.29}$$

where $N_i(t)$ is the number of repairs in $[0, t]$. The relationship is actually the definition of the expectation of several discrete events.

To obtain the probability that of *exactly* n events occurring by time $t(P\{N_i(t)=n\})$, we subtract the probability of *at least* $(n+1)$ events occurring by time $t(P\{t_{n+1}\le t\})$, from the probability that there are *at least* n events $(P\{t_n \le t\})$, that is,

$$P\{N_i(t)=n\} = P\{t_n \le t\} - P\{t_{n+1}\le t\} = \hat{F}_{i,n}(t) - \hat{F}_{i,n+1}(t) \qquad n=1,2,\ldots \tag{17.30}$$

Therefore, the expected number of repairs $NR_i(t)$ is

$$\mathrm{NR}_i(t) = E[N_i(t)] = \sum_{n=1}^{\infty} n \times P\{N_i(t)=n\} = \sum_{k=1}^{\infty} k \times \left(\hat{F}_{i,k}(t) - \hat{F}_{i,k+1}(t)\right) = \sum_{k=1}^{\infty} \hat{F}_{i,k}(t) \tag{17.31}$$

We now determine the number of repairs for ordinary and nonordinary renewal processes. For ordinary renewal processes, we can rewrite Eq. (17.31) as follows:

$$\mathrm{NR}_i(t) = \sum_{k=1}^{\infty} \hat{F}_{i,k}(t) = \hat{F}_{i,1}(t) + \sum_{k=2}^{\infty} \hat{F}_{i,k}(t) = \hat{F}_{i,1}(t) + \sum_{k=1}^{\infty} \hat{F}_{i,k+1}(t) \tag{17.32}$$

But, using Eq. (17.23)

$$\mathrm{NR}_i(t) = \hat{F}_{i,1}(t) + \sum_{k=1}^{\infty} \hat{F}_{i,k+1}(t) = \hat{F}_{i,1}(t) + \sum_{k=1}^{\infty} \int_0^t \hat{F}_{i,k}(t-\xi) f_{i,k+1}(\xi)d\xi \tag{17.33}$$

which can be rewritten as

$$\mathrm{NR}_i(t) = \hat{F}_{i,1}(t) + \int_0^t \sum_{k=1}^{\infty} \hat{F}_{i,k}(t-\xi) f_{i,k+1}(\xi)d\xi \tag{17.34}$$

For ordinary renewal processes, $f_{i,1}(\xi) = f_{i,2}(\xi) = \cdots = f_{i,k}(\xi) = f_{i,k+1}(\xi)$, and therefore,

$$\mathrm{NR}_i(t) = \hat{F}_{i,1}(t) + \int_0^t \sum_{k=1}^{\infty} \hat{F}_{i,k}(t-\xi) f_i\,(\xi)d\xi \tag{17.35}$$

But, according to Eq. (17.33), $\sum_{k=1}^{\infty} \hat{F}_{i,k}(t-\xi) = NR_i(t-\xi)$; we also have $\hat{F}_{i,1}(t) = F_{i,1}(t)$. Therefore we get

$$\mathrm{NR}_i(t) = F_{i,1}(t) + \int_0^t \mathrm{NR}_i\,(t-\xi) f_i\,(\xi)d\xi \tag{17.36}$$

which is called the renewal equation. A Laplace transform of the renewal equation followed with a small manipulation yields

$$\mathrm{NR}_i(s) = \frac{F_{i,1}(s)}{1-f_i(s)} = \frac{F_i(s)}{1-sF_i(s)} \tag{17.37}$$

where the well-known property of the transform of a derivative function was used in the last step.

Two important results concerning asymptotic behavior (the behavior as time approaches infinity) of renewal processes are presented next without proof (Birolini, 2007).

1. *Elementary Renewal Theorem* The expected value of the number of failures per unit time $[\mathrm{NR}_i(t)/t]$ approaches asymptotically the expected failure rate, which is the reciprocal of the MTTF

$$\lim_{t\to\infty}\frac{\mathrm{NR}_i(t)}{t}=\frac{1}{\mathrm{MTTF}_i} \tag{17.38}$$

2. *Blackwell's Theorem* For long time (i.e., when $t \to \infty$) a renewal process experiences one renewal (failure) per time interval of duration equal to MTTF, and the expected number of failures in any interval is the duration of the interval divided by the MTTF. In other words, a renewal process "settles down" when $t \to \infty$. Thus, for a positive finite value of a,

$$\lim_{t\to\infty}[\mathrm{NR}_i(t+a)-\mathrm{NR}_i(t)]=\frac{a}{\mathrm{MTTF}_i} \tag{17.39}$$

Average Availability Because the repair time is negligible and repair takes place immediately after the equipment fails, then there is no downtime and $A_i^s(t)=A_i^{\mathrm{av}}(t)=1$.

Decision Variables This model does not have a decision variable.

Economic Impact There are two components of the economics: costs and losses.

- Maintenance Costs. In this model, only corrective maintenance is performed, and therefore, the cost of each repair is given by the associated spare parts and labor cost. Thus, the costs are given by the cost of repair (labor + parts) times the expected number of repairs.
- Economic Losses. Each failure results in some deterioration of the process performance, or even shut down. Although the assumption is that the repair is instantaneous, one can still assume that some associated losses exist and compute them.

Thus, the expected overall maintenance cost (MC) per unit time is

$$\mathrm{MC}_i=(C_{i,\mathrm{cm}}^l+C_{i,\mathrm{cm}}^p+L_{i,f})\frac{\mathrm{NR}_i(t^*)}{t^*} \tag{17.40}$$

where $\mathrm{NR}_i(t^*)$ is the expected number of repairs in a selected period of time t^*, $C_{i,\mathrm{cm}}^l$ the labor cost per repair in corrective maintenance, $C_{i,\mathrm{cm}}^p$ the spare parts cost per each repair, and $L_{i,f}$ the economic losses associated to the failure.

Example 17.3: For the exponential distribution the cumulative distribution is $F_i(t) = 1 - e^{-\lambda_i t}$, and therefore $F_i(s) = 1/s - 1/(s + \lambda_i)$. Thus,

$$\mathrm{NR}_i(s) = \frac{\lambda_i}{s^2} \tag{17.41}$$

$$\mathrm{NR}_i(t) = \lambda_i t = \frac{t}{\mathrm{MTTF}_i} \tag{17.42}$$

In other words, for the exponential distribution, the expected number of repairs within a period of time [0, *t*] is simply the failure rate (which is constant) multiplied by the time duration *t*. Birolini (2007) gives details of the analytical forms for other distributions, which are either complicated or unavailable (an approximation is used for such cases).

Finally, it is clear that the rate of failure is always constant, that is, $\mathrm{NR}_i(t)/t = \lambda_i = 1/\mathrm{MTTF}_i$. In addition, Blackwell's theorem result holds for all *t*, that is,

$$[\mathrm{NR}_i(t+a) - \mathrm{NR}_i(t)] = \left[\frac{(t+a)}{\mathrm{MTTF}_i} - \frac{t}{\mathrm{MTTF}_i}\right] = \frac{a}{\mathrm{MTTF}_i} \tag{17.43}$$

Delayed (Modified) Renewal Process

This is a process in which the probability distribution of the first event is different from the rest, which are all equal (Medhi, 2000). That is, $f_{i,n}(\tau_n) = f_{i,n-1}(\tau_{n-1}) = \cdots = f_{i,2}(\tau_2) \neq f_{i,1}(\tau_{n-1})$. Such a process is also called "general" renewal process (Medhi, 2000). It is easy to see that the only change here will be the resulting expected number of repairs. As in the ordinary renewal case, there are no decision variables, the availability is also constant (because the repairs are instantaneous) and the economics is given by the same expression. We now show how the expected number of repairs is obtained.

We return to Eq. (17.34) and note that $f_{i,k+1}$ is used inside the summation. Thus, if all the distributions, except the first, are equal, Eq. (17.37) becomes

$$\mathrm{NR}_i(s) = \frac{\hat{F}_{i,1}(s)}{1 - f_{i,2}(s)} = \frac{F_{i,1}(s)}{1 - sF_{i,2}(s)} \tag{17.44}$$

Example 17.4: Consider again the case of exponential failure distributions where there is some deterioration of the equipment after the first failure. Then $\lambda_{i,1} < \lambda_{i,2} = \cdots = \lambda_{i,n}$; that is, after the first repair, the probability of failure is larger than for the first for the same τ. In this case, we have

$$\hat{F}_{i,1}(s) = \frac{\lambda_{i,1}}{s(s + \lambda_{i,1})} \tag{17.45}$$

$$1 - f_{i,2}(s) = 1 - \frac{\lambda_{i,2}}{(s + \lambda_{i,2})} = \frac{s}{(s + \lambda_{i,2})} \tag{17.46}$$

Thus,

$$\mathrm{NR}_i(s) = \frac{\hat{F}_{i,1}(s)}{1 - f_{i,2}(s)} = \frac{\lambda_{i,1}}{s(s + \lambda_{i,1})} \frac{(s + \lambda_{i,2})}{s} = \frac{\lambda_{i,1}(s + \lambda_{i,2})}{s^2(s + \lambda_{i,1})} \tag{17.47}$$

The inverse Laplace transformation of $\mathrm{NR}_i(s)$ is

$$\mathrm{NR}_i(t) = \lambda_{i,2}t + \left(1 - \frac{\lambda_{i,2}}{\lambda_{i,1}}\right)(1 - e^{-\lambda_{i,1}t}) \tag{17.48}$$

It is clear that if $\lambda_{i,1} = \lambda_{i,n}$ we get the ordinary renewal processes, then Eq. (17.48) reduces to Eq. (17.42). Plots of the expected number of failures $NR(t)$ for the two cases: ordinary ($\lambda_{i,1} = \lambda_{i,n} = 0.005$, $n \geq 2$) and modified renewal processes ($\lambda_{i,1} = 0.005 < \lambda_{i,n} = 0.01$ and 0.02) are given in Fig. 17.6, from which we extract the following conclusions:

(i) The dependence of NR(t) on time t for the case $\lambda_{i,1} \neq \lambda_{i,2}$ slightly deviates from the linear relationship of the case $\lambda_{i,1} = \lambda_{i,2}$.

(ii) For small t, the failure behavior is described by the initial failure rate $\lambda_{i,1}$ (all three lines, which have the same $\lambda_{i,1}$, merge into one at small t) $\dfrac{d\mathrm{NR}_i(t)}{dt} = \lambda_{i,n} + \left(1 - \dfrac{\lambda_{i,n}}{\lambda_{i,1}}\right)\lambda_{i,1}e^{-\lambda_{i,1}t} \to \lambda_{i,1}$ for small t.

(iii) For large t, failure behavior of the unit is described by the subsequent failure rate $\lambda_{i,n}$: $\dfrac{d\mathrm{NR}_i(t)}{dt} = \lambda_{i,n} + \left(1 - \dfrac{\lambda_{i,n}}{\lambda_{i,1}}\right)\lambda_{i,1}e^{-\lambda_{i,1}t} \to \lambda_{i,n}$ at large t.

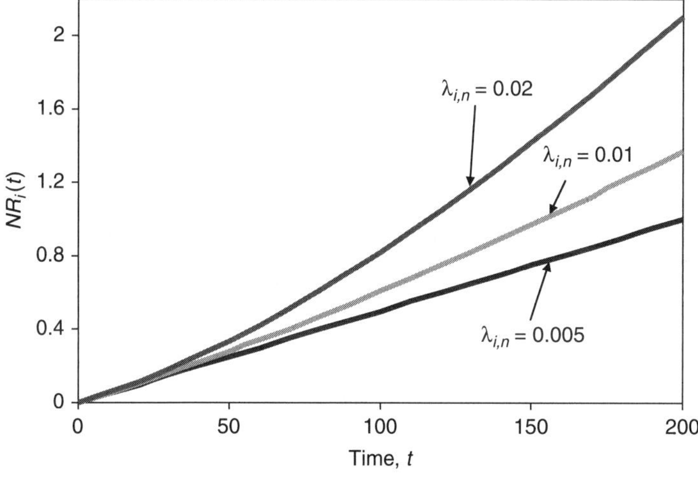

FIGURE 17.6 Number of failures $\mathrm{NR}_i(t)$ vs. time t for different failure rates ($\lambda_{i,1} = 0.005$ for all cases).

Remark: The case where the distributions $f_{i,n}(\tau_n)$ are not equal: $f_{i,n}(\tau_n) \neq f_{i,n-1}(\tau_{n-1}) \neq \cdots \neq f_{i,1}(\tau_1)$ cannot be solved analytically because Eq. (17.33) cannot be simplified when the distributions are different and simulation seems to be one alternative. The nonidentical distributions are usually a result of imperfect maintenance. Thus, the reliability of the equipment at renewal point $t_n(S_n)$ is different from the reliability at previous renewal points $t_k(S_k)$, $k < n$, usually smaller, and hence the failure times are not independent and identically distributed anymore.

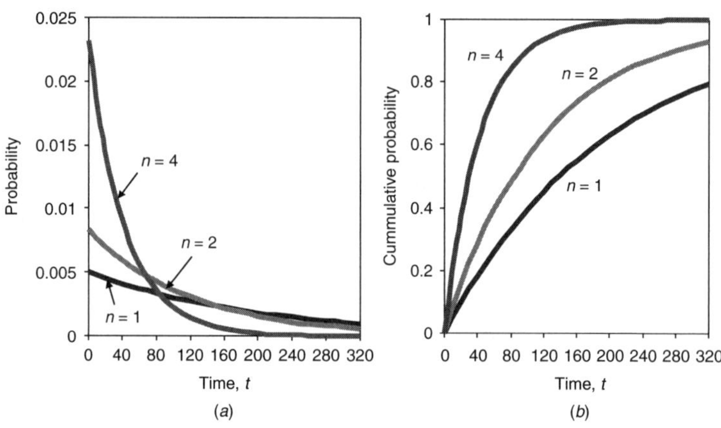

FIGURE 17.7 (*a*) Probability and (*b*) cumulative probability of failure-free times for quasi-renewal processes.

One particular case of this is the quasi-renewal process (Wang and Pham, 2006). In quasi-renewal processes, the distributions $f_{i,n}(\tau_n)$ of the inter-arrival points τ_n are still independent, but the distributions are not equal, but related to each other as follows: $f_{i,n}(t) = \alpha^{1-n} f_{i,1}(\alpha^{1-n}t)$. With $0 < \alpha < 1$, they model certain type of imperfect maintenance. Indeed, in imperfect maintenance, the subsequent failure-free times are expected to be shorter than the initial one; in other words, the distributions $f_{i,n}(t)$ shift to the left as n increases. This is illustrated next for the case where the distribution $f_{i,1}(t)$ is exponential, that is, $f_{i,1}(t) = \lambda_i e^{-\lambda_i t}$. Using the above relationship, the distribution of nth failure-free time is $f_{i,n}(t) = \alpha^{1-n}\lambda_i e^{-\lambda_i \alpha^{1-n} t}$, which is also exponential with increased failure rate $\lambda_{i,n} = \alpha^{1-n}\lambda_i > \lambda_i$. Plots of the distributions $f_{i,n}(t)$ and the associated cumulative distributions for the case $\lambda_i = 0.005$ and $\alpha = 0.6$ are given in Fig. 17.7.

Figure 17.7 reveals that the subsequent failure-free times ($n > 1$) are expected to occur sooner than the initial one ($n = 1$); in other words, the time between failures become shorter as time elapses. This result is due to the fact that the imperfect maintenance restores the equipment condition to some degree but not to the level of "as good as new"; hence reliability measures (e.g., failure rate) deteriorate after each failure. Because each subsequent failure distribution has some dependence on the previous, one can say that the system has some "memory."

All the above described renewal processes, especially those that do not have analytical solution, can be easily simulated by a Monte Carlo simulation: failure times τ_k are consecutively sampled in the same manner, as depicted in Fig. 17.4, using the distribution function of failure time $f_{i,n}(t)$ until the end of time period is reached; the number of failures within that time period can then be counted. This was covered in detail in Chap. 10 when stochastic accuracy was discussed and will also be discussed later in this chapter.

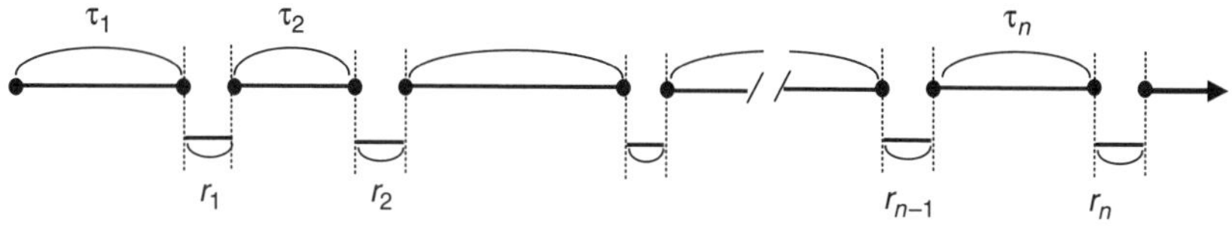

Figure 17.8 Alternating renewal process.

Alternating Renewal Process (Renewal Processes with Repair Time)
The renewal process, which is associated with an assumption of negligible maintenance time, does not account for downtime (i.e., repair time). This is unrealistic, especially when repair times are long.

We now extend the renewal process to consider the equipment repair, which occurs right after the failure event. If both inter-arrival times (τ_n) and repair times (r_n) are random variables, each one is independent and identically distributed according to its own distribution, then a random sequence of the pair (τ_n, r_n) is said to form an alternating renewal process. The process is illustrated in Fig. 17.8.

Expected Number of Repairs: It can be considered that the alternating renewal process (τ_n, r_n) is an ordinary renewal process with IID variables $\{z_n = \tau_n + r_n\}$ whose density and cumulative probability distributions are $h_i(t)$ and $H_i(t)$, respectively. Let $g_i(r)$ be the probability distribution of the repair time r_n. Then, the distribution of the random sequence $z_n = \tau_n + r_n$ is $h_i(\mathrm{t}) = (f_i^* g_i)(t)$ [i.e., h_i is the convolution of f_i and $g_i : h_i(t) = \int_0^t f_i(x) g_i(t-x)dx$]. The reason for this is that the two distributions are independent. The expected number of renewal points of z_n (which is also the expected number of failures) is therefore given by

$$\mathrm{NR}_i(t) = H_{i,1}(t) + \int_0^t \mathrm{NR}_i(t-\xi) h_i(\xi) d\xi \tag{17.49}$$

as in Eq. (17.36). Taking the Laplace transform and operating renders an expression identical to Eq. (17.37), which may or may not have an analytical inverse transform.

Average Availability: When downtime or a "down state" is considered, the availability can be readily evaluated. The point availability at time t, $A_i^s(t)$ is obtained as follows: We start with the same nomenclature as in previous section: $F_i(t) = F_{i,1}(t) = \cdots = F_{i,n}(t)$ is cumulative distribution of nth failure times τ_n; the nth renewal point in the alternative renewal processes is $t_n = \sum_{k=1}^{n} z_k = \sum_{k=1}^{n} (\tau_k + r_k)$, and the associated probability distribution and cumulative distributions are $\hat{f}_{i,n}^*(t)$ and $\hat{F}_{i,n}^*(t)$. These distributions can be obtained from the distribution of z_n, $h_i(t)$, in the same way as in the ordinary renewal process [Eqs. (17.18) through (17.24)]. The failure-free time intervals of equipment are

$[t_n, t_n + \tau_{n+1}]$ (see Fig. 17.8). Consider now the probability of the equipment working properly within a time interval $[t, t + x]$. We start with the following trivial relation:

$$Pr\{\text{item up in } [t,t+x]\} = \sum_{n=0}^{\infty} P(t_n < t,\ t+x < t_n + \tau_{n+1}) \tag{17.50}$$

which ties the desired probability to the probability of $[t, t + x]$ being in one failure-free time interval. We now expand this relation as follows:

$$\sum_{n=0}^{\infty} P(t_n < t,\ t+x < t_n + \tau_{n+1}) = P(0 < t, t+x < \tau_1) + \sum_{n=1}^{\infty} P(t_n < t,\ t+x < t_n + \tau_{n+1}) \tag{17.51}$$

However, the following straightforward relations hold:

$$P(0 < t, t+x < \tau_1) = P(\tau_1 > t+x) = 1 - F_i(t+x) \tag{17.52}$$

$$P(t_n < t, t+x < t_n + \tau_{n+1}) = P(t+x < t_n + \tau_{n+1} | t_n < t) \tag{17.53}$$

Recall that the distribution of the nth renewal point t_n is $\hat{f}^*_{i,n}(t)$; thus we write

$$P(t+x < t_n + \tau_{n+1} | t_n < t) = \int_0^t P(t+x < t_n + \tau_{n+1} | t_n = \xi)\hat{f}^*_{i,n}(\xi)d\xi \tag{17.54}$$

Finally, $\hat{f}^*_{i,n}(\xi)d\xi = d\hat{F}^*_{i,n}(\xi)$. Therefore, we write

$$Pr\{\text{item up in } [t,t+x]\} = 1 - F_i(t+x) + \sum_{n=1}^{\infty} \int_0^t P(t+x < \xi + \tau_{n+1}) d\hat{F}^*_{i,n}(\xi) \tag{17.55}$$

The distributions of $\tau_1, \ldots, \tau_n$ are $F_i(t) = F_{i,1}(t) = \cdots = F_{i,n}(t)$. Thus, $P(t+x < \xi + \tau_{n+1}) = P(t + x - \xi < \tau_{n+1}) = [1 - F_i(t+x-\xi)]$. Thus,

$$Pr\{\text{item up in } [t,t+x]\} = 1 - F_i(t+x) + \sum_{n=1}^{\infty} \int_0^t [1 - F_i(t+x-\xi)] d\hat{F}^*_{i,n}(\xi) \tag{17.56}$$

But $\sum_{n=1}^{\infty} \int_0^t [1 - F_i(t+x-\xi)] d\hat{F}^*_{i,n}(\xi) = \int_0^t [1 - F_i(t+x-\xi)] \sum_{n=1}^{\infty} d\hat{F}^*_{i,n}(\xi)$, and from Eq. (17.31) we also have $\sum_{k=1}^{\infty} d\hat{F}^*_{i,k}(\xi) = d\mathrm{NR}_i(\xi)$. Thus,

$$Pr\{\text{item up in } [t,t+x]\} = 1 - F_i(t+x) + \int_0^t [1 - F_i(t+x-\xi)] d\mathrm{NR}_i(\xi) \tag{17.57}$$

Let the renewal density function be $nr_i(t) = d\mathrm{NR}_i(t)/dt$. Then,

$$Pr\{\text{item up in } [t, t+x]\} = 1 - F_i(t+x) + \int_0^t [1 - F_i(t+x-\xi)] nr_i(\xi) d\xi \tag{17.58}$$

Finally, setting $x = 0$, we obtain the point availability

$$A_i^s(t) = 1 - F_i(t) + \int_0^t [1 - F_i(t-\xi)] nr_i(\xi) d\xi \tag{17.59}$$

A Laplace transform of Eq. (17.59) gives

$$A_i^s(s) = R_i(s) + R_i(s) nr_i(s) \tag{17.60}$$

where $R_i(s)$ is the Laplace form of the reliability function $R_i(t) = [1 - F_i(t)]$.

The two types of availability have the same asymptotic value, which is given by (Gertsbakh, 2000; Birolini, 2007)

$$\lim_{t\to\infty} A_i^{\mathrm{av}}(t) = \lim_{t\to\infty} A_i^s(t) = \frac{\mathrm{MTTF}}{\mathrm{MTTF} + \mathrm{MTTR}} \tag{17.61}$$

where MTTF is the mean time to failure ($= \mathrm{E}[\tau_n]$) and MTTR is the mean time to repair ($= \mathrm{E}[r_n]$). This result is the same as the one obtained using a Markov model (shown in the next section). Note also that the case of constant repair time is a special case of alternating renewal processes.

Example 17.5: Consider the case where both failure times τ_n and repair times r_n follow exponential distributions: $f_i(t) = f_{i,n}(t) = \lambda_i e^{-\lambda_i t}$, $g_i(t) = g_{i,n}(t) = \mu_i e^{-\mu_i t}$. The expected values of τ_n and r_n in this case are $\mathrm{E}[\tau_n]) = \mathrm{MTTF} = 1/\lambda_i$, $\mathrm{E}[r_n] = \mathrm{MTTR} = 1/\mu_i$.

A Laplace transform of $h_i(t)$ is

$$L\{h_i(t)\} = L\{(h_i \times g_i)(t)\} = h_i(s) = f_i(s) g_i(s) = \frac{\lambda_i}{s+\lambda_i} \frac{\mu_i}{s+\mu_i} \tag{17.62}$$

The Laplace transform of the expected number of failures, $\mathrm{NR}_i(s)$, is found from Eq. (17.37) with $H_i(s) = h_i(s)/s$ in place of $F_i(s)$.

$$\mathrm{NR}_i(s) = \frac{H_i(s)}{1 - h_i(s)} = \frac{1}{s^2} \frac{\lambda_i \mu_i}{s + \lambda_i + \mu_i} \tag{17.63}$$

Therefore, the Laplace transform of the renewal density is

$$nr_i(s) = s\mathrm{NR}_i(s) = \frac{1}{s} \frac{\lambda_i \mu_i}{s + \lambda_i + \mu_i} = \frac{\lambda_i}{s}\left[1 - \frac{s + \lambda_i}{s + \lambda_i + \mu_i}\right] \tag{17.64}$$

In turn, the Laplace transform of $R_i(t) = [1 - F_i(t)] = e^{-\lambda_i t}$ is $R_i(s) = 1/(s + \lambda_i)$. Substituting $R_i(s)$ and $nr_i(s)$ into Eq. (17.54) results in

$$A_i^s(s) = \frac{1}{s+\lambda_i} + \frac{1}{s+\lambda_i} \frac{\lambda_i}{s} - \frac{\lambda_i}{s} \frac{1}{s + \lambda_i + \mu_i} \tag{17.65}$$

The inverse Laplace transform yields the point availability

$$A_i^s(t) = \frac{\mu_i}{\lambda_i + \mu_i} + \frac{\lambda_i}{\lambda_i + \mu_i} e^{-(\lambda_i + \mu_i)t} \tag{17.66}$$

The same result is obtained if Markov processes theory is used (shown in next section). The asymptotic value of $A_i^s(t)$ is

$$\lim_{t\to\infty} A_i^s(t) = \frac{\mu_i}{\lambda_i + \mu_i} = \frac{\text{MTTF}}{\text{MTTF} + \text{MTTR}} \tag{17.67}$$

which confirms the result shown in Eq. (17.61).

Decision Variables As in the ordinary renewal processes model, none.

Economics (Cost and Losses) Unlike in the ordinary renewal process, the loss is now associated with downtime (i.e., unavailability) instead of the number of failures. The total cost is then

$$\text{MC}_i = (C_{i,\text{cm}}^l + C_{i,\text{cm}}^p)\frac{\text{NR}_i(t^*)}{t^*} + L_{UA} \times [1 - A_i^{\text{av}}(t^*)] \tag{17.68}$$

where L_{UA} is the loss rate ($ per unit time) if the system is unavailable.

Renewal Process Model with Preventive Maintenance

Consider that periodic preventive maintenance is used with time interval = T, that is, preventive maintenance times = T, $2T$, $3T$, etc. Here, T is the decision variable, and assume that both corrective and preventive maintenance are perfect leaving the equipment "as good as new" (all renewals are ordinary).

Figure 17.9 shows preventive maintenance at times T, $2T$, ..., kT. Within a preventive maintenance cycle $(0 < t \le T, T < t \le 2T, \ldots)$ failures may happen, which need to be immediately corrected by CM. Therefore, all failures within a preventive maintenance cycle clearly constitute a renewal process.

Expected Number of Repairs This is given by Eq. (17.49) or Eq. (17.36) applied to each interval of length T.

Average Availability Let t_{ER} and t_{PM} be the repair time and PM time, respectively, and let us assume they are constant. Also, let T_U and T_D be the cumulative uptime and downtime within a PM cycle, respectively. Then,

$$A_i^{\text{av}}(T) = \frac{T_U}{T_U + T_D} \tag{17.69}$$

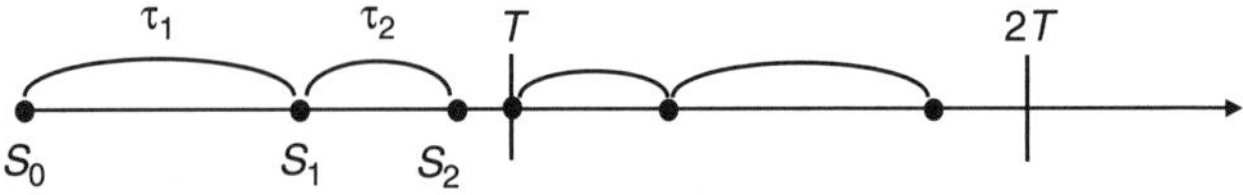

Figure 17.9 Illustration of renewal process—based PM model.

The downtime is the cumulative repair time $[\mathrm{NR}_i(T)^* t_{\mathrm{ER}}]$ plus the PM time (t_{PM}) and the uptime is given by $T_U = T - T_D$. Therefore,

$$A_i^{\mathrm{av}}(T) = \frac{T_U}{T_U + \mathrm{NR}_i(T) \times t_{\mathrm{ER}} + t_{\mathrm{PM}}} = \frac{T - \mathrm{NR}_i(T) \times t_{\mathrm{ER}} - t_{\mathrm{PM}}}{T} \quad (17.70)$$

Assuming $T \approx T_U$ (i.e., downtime is very small as compared to T, which is usually the case), one gets,

$$A_i^{\mathrm{av}}(T) = \frac{1}{1 + \dfrac{\mathrm{NR}_i(T)}{T} \times t_{\mathrm{ER}} + \dfrac{t_{\mathrm{PM}}}{T}} \quad (17.71)$$

When T decreases, the downtime due to repair $[t_{\mathrm{ER}} \times \mathrm{NR}_i(T)/T]$ expectedly decreases while the downtime due to PM (t_{PM}/T) increases.

Remark: For the case where the repair time and the PM time are randomly distributed Nakagawa (2005) was able to derive a more general expression for the average availability:

$$A_i^{\mathrm{av}}(T) = \frac{1}{1 + \dfrac{t_{\mathrm{ER}} F_i(T) + t_{\mathrm{PM}}[1 - F_i(T)]}{\int_0^T [1 - F_i(t)]dt}} \quad (17.72)$$

where $F(t)$ is the cumulative distribution of failure: $F_i(T) = \int_0^T f_i(t)dt$.

Decision Variables: The PM time interval T.

Economics: It can be seen that within a preventive maintenance cycle.

- The number of preventive maintenance actions is 1 and the PM cost is $C_{i,\mathrm{pm}}^l + C_{i,\mathrm{pm}}^p$.
- The expected number of failures is $\mathrm{NR}_i(T)$ (renewal function), and therefore the CM cost is $(C_{i,\mathrm{cm}}^l + C_{i,\mathrm{cm}}^p)\mathrm{NR}_i(T)$.
- The economic loss is $L_{UA} \times [1 - A_i^{\mathrm{av}}(T)]$.

Thus, the expected cost plus loss rate (cost + loss per unit time) is given by

$$\mathrm{MC}_i = \left\{ (C_{i,\mathrm{cm}}^l + C_{i,\mathrm{cm}}^p) \frac{\mathrm{NR}_i(T)}{T} + \frac{(C_{i,\mathrm{pm}}^l + C_{i,\mathrm{pm}}^p)}{T} \right\} + L_{UA} \times [1 - A_i^{\mathrm{av}}(T)] \quad (17.73)$$

Example 17.6: Consider a simple periodic PM renewal process with a failure distribution following an exponential distribution $f_{i,n}(t) = \lambda_i e^{-\lambda_i t}$ with failure rate $\lambda_i = 0.01$; the repair time and time to do PM are $t_{\mathrm{ER}} = t_{\mathrm{PM}} = 1$ (day), which are in this case negligibly small when compared with MTTF. The PM time interval $T = 200$ (days) and the cost parameters are $C_c = C_{i,\mathrm{cm}}^l + C_{i,\mathrm{cm}}^p = 10{,}000$,

$C_p = C^l_{i,\text{pm}} + C^p_{i,\text{pm}} = 500$, L_{UA} = 1000 (\$/day). The expected number of failures is given by Eq. (17.42): $\text{NR}_i(t) = \lambda_i t$ and the average availability is given by Eq. (17.71) with $\text{NR}_i(t) = \lambda_i t$. Thus,

$$A_i^{\text{av}}(T) = \frac{1}{1 + \lambda_i t_{\text{ER}} + \dfrac{t_{\text{PM}}}{T}} \tag{17.74}$$

which gives $A_i^{\text{av}}(T) = 0.9852$. In turn the total cost plus loss given by Eq. (17.73) is

$$\text{MC}_i = C_c \lambda_i + C_p/T + L_{UA} \times [1 - A_i^{\text{av}}(T)] \tag{17.75}$$

rendering $\text{MC}_i = 117.3$ (\$/day).

Markov Processes

As the renewal process, the *Markov process* is formally defined as a stochastic process whose behavior can be determined independently of its past; that is it has no memory.

In maintenance engineering, a Markov model is represented graphically by nodes (or states) and arcs (or transitions between the states). The states include failure-free (normal) state, failed state, in between (degraded state but not yet failed) or being taken down for preventive maintenance. The transitions between states are characterized by transition rates or transition probabilities. In this regard, Markov models are richer than renewal models, which can only model two states.

Consider a sequence of random variables $\{X_n\}$. The Markov property asserts that the distribution of X_{n+1} depends only on the current state $X_n = i_n$, not on the whole history. Formally, the process $\{X_n\}$ is called a Markov process if, for each n and every $i_0, \ldots, i_n$ and $j \in N$, the probability of transition from current state $X_n = i_n$ to state $X_{n+1} = j$ is

$$P_{i_n,j}^{n,n+1} = P[X_{n+1} = j | X_0 = i_0, \ldots, X_n = i_n] = P[X_{n+1} = j | X_n = i_n] \tag{17.76}$$

Homogeneous Markov processes are processes whose transition probabilities/transition rates are time independent (i.e., they are constants $P_{i,j}^{n,n+1} = P_{i,j}^{0,1}$). Most of Markov models in literature are assumed to be homogeneous.

The Markov processes can be classified depending on whether the time domain is discrete or continuous. The discrete Markov process (in time and space dimension) is often referred to as Markov chain.

Continuous Time Markov Model

The time-continuous and homogeneous Markov processes are fully defined by the initial values of probabilities of states and the rates of change between states represented graphically by nodes (or states) and arcs (or transitions between the states), as shown in Fig. 17.10.

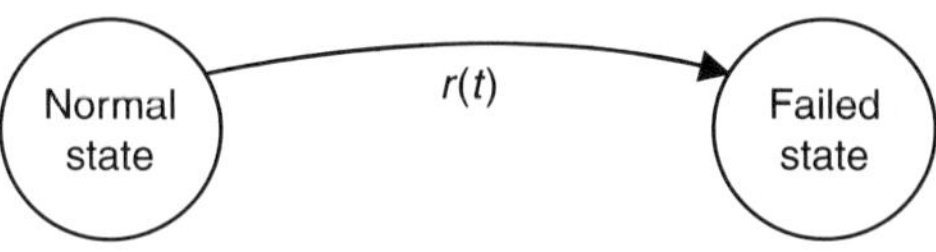

FIGURE 17.10 State transition.

The basis for Markovian models is that the rate of change (derivative) of the probability that equipment i is in the normal state $P_{i,N}(t)$, at time t is given by

$$\frac{dP_{i,N}(t)}{dt} = \text{inflow to normal state} - \text{outflow from normal state} \quad (17.77)$$

If there is no repair, then there is no inflow to the state from the failed state. There is, however, outflow from the normal state, that is, the failures. The outflow, in this case, is given by the rate of failure times the probability of failure. Thus,

$$\frac{dP_{i,N}(t)}{dt} = -\lambda_i(t)P_{i,N}(t) \quad (17.78)$$

A similar equation can be written for the failed state.

$$\frac{dP_{i,F}(t)}{dt} = \lambda_i(t)P_{i,N}(t) \quad (17.79)$$

where the inflow is the rate of change of the normal state. Most of Markov models in literature are assumed to be homogeneous, that is, λ is not a function of time.

We also recognize that $P_{i,N}(t) = A_i^s(t)$, which in this case is also equal to the reliability $A_i^s(t) = R_i^s(t)$. Thus, using $R_i^s(0) = 0$, the solution of Eq. (17.79) is

$$R_i^s(t) = e^{-\int_0^t \lambda_i(t)dt} \quad (17.80)$$

In the case of constant failure rate, the reliability and the availability of a system without repairs becomes

$$R_i^s(t) = A_i^s(t) = e^{-\lambda_i t} \quad (17.81)$$

which is the exponential distribution.

Average Availability: The average availability is easily derived as follows:

$$A_i^{\text{av}}(t) = \frac{1}{t}\int_0^t A_i^s(u)du = \frac{1}{t}\int_0^t e^{-\lambda_i u}du = \frac{1}{t\lambda_i}[1 - e^{-\lambda_i t}] \quad (17.82)$$

We note that both the service availability $A_i^s(t)$ and average availability $A_i^{\text{av}}(t)$ approach zero when time goes to infinity. This is because there is no repair.

When the failure rate is constant, the assumption is that the failure rate is proportional to the number of equipment in the normal state. It is, however, an optimistic assumption. Assuming that the failure is only proportional to the number of units in the normal state is assuming that there is no deterioration of their parts through time. In addition, the burn-in period is not considered.

Let us now consider the case where there are repairs. In this case $A_i^s(t) \neq R_i^s(t)$, so we write

$$\frac{dA_i^s(t)}{dt} = \mu_i(t)A_i^f(t) - \lambda_i(t)A_i^s(t) \tag{17.83}$$

where $\mu_i(t)$ is the rate of repair. Thus, the inflow of availability is the rate of repair times the probability of sensors being in the failed state $A_i^f(t)$. However, $A_i^s(t) + A_i^f(t) = 1$, which renders

$$\frac{dA_i^s(t)}{dt} = \mu_i(t)[1 - A_i^s(t)] - \lambda_i(t)A_i^s(t) \tag{17.84}$$

In the case of repair rate and failure rate being constant (the usual assumption), and assuming $A_i^s(0) = 1$, one can write

$$A_i^s(t) = \left(1 - \frac{\lambda_i}{\lambda_i + \mu_i}\right) + \left(\frac{\lambda_i}{\lambda_i + \mu_i}\right)e^{-(\lambda_i + \mu_i)t} \tag{17.85}$$

For a long period of time, the availability reaches a constant value, that is,

$$\lim_{t\to\infty} A_i^s(t) = \left(1 - \frac{\lambda_i}{\lambda_i + \mu_i}\right) = \frac{\mu_i}{\lambda_i + \mu_i} \tag{17.86}$$

Thus, if the repair rate is much larger than the failure rate, $(\mu_i >> \lambda_i)$, then $\lim_{t\to\infty} A_i^s(t) \approx 1$. This means that the equipment will always be available. We know that resource limitations prevent this situation from happening. Conversely, if the repair rate is much smaller than the failure rate, $(\mu_i << \lambda_i)$, then $\lim_{t\to\infty} A_i^s(t) \approx 0$, which means that all equipment will be in the failed state.

The asymptotic value of $A_i^{av}(t)$ (obtained as $t \to \infty$) is the same as the asymptotic value of $A_i^s(t)$, which is $(\mu_i)/(\lambda_i + \mu_i)$.

Expected Number of Repairs: The number of repairs per unit time is given by $\mu_i(t)[1 - A_i^s(t)]$. Thus, we can integrate over a certain period of time to get

$$\mathrm{NR}_i(T) = \mu_i \int_0^T [1 - A_i^s(t)]dt \tag{17.87}$$

which results in

$$\mathrm{NR}_i(T) = \frac{\lambda_i \mu_i}{\lambda_i + \mu_i}\left[T - \left(\frac{1}{\lambda_i + \mu_i}\right)(1 - e^{-(\lambda_i + \mu_i)T})\right] \tag{17.88}$$

Thus, if the repair rate is much larger than the failure rate, ($\mu_i >> \lambda_i$), and for relatively small T, then $NR_i(T) \to \lambda_i T$; for large T, we get $NR_i(T) = \lambda_i[T - 1/\mu_i] = \lambda_i T$, because $\lambda_i/\mu_i = 0$ when $\mu_i >> \lambda_i$. Thus if $\mu_i >> \lambda_i$ we have $NR_i(T) \to \lambda_i T$, that is, the number of repairs within time interval [0, T] tend to be equal to the failure rate times the duration of the interval. Conversely, by doing the same analysis if the repair rate is much smaller than the failure rate ($\mu_i << \lambda_i$), we obtain $NR_i(T) \to \mu_i T$, that is, the number of repairs is given by the repair rate. If repair rate and failure rate are at the same magnitude, the dependence of $NR_i(T)$ on time T also takes the shape of a linear relationship. It can also be seen that the Markov processes and renewal processes give the same result for the expected number of failures $NR_i(T)$: under the condition of exponentially distributed failure (underlying assumption of Markov processes), the renewal processes theory leads to $NR_i(T) = \lambda_i T$ (as shown in Example 17.2) while under the condition of negligible repair time (underlying assumption of renewal processes), that is, repair rate is very large $\mu_i >> \lambda_i$, then Markov processes theory leads to $NR_i(T) = \lambda_i T$.

Decision Variables There is no decision making involved in this Markov model for corrective maintenance.

Economics The expected cost plus loss rate within a time interval [0, T] is given by

$$MC_i = \left\{(C^l_{i,\text{cm}} + C^p_{i,\text{cm}})\frac{NR_i(T)}{T}\right\} + L_{UA} \times [1 - A^{\text{av}}_i(T) \qquad (17.89)$$

where the expected number of failures $NR_i(T)$ and the average availability $A^{\text{av}}_i(t)$ are given by Eq. (17.88).

Example 17.7: Consider that $\lambda_i = 0.005$. Figure 17.11 shows the dependence of availabilities $A^s_i(t)$ (solid lines) and $A^{\text{av}}_i(t)$ (dotted lines) on time t at different values of repair rate μ_i.

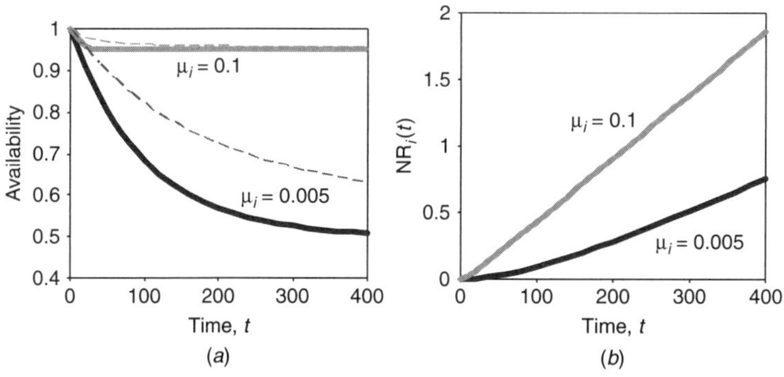

Figure 17.11 (a) $A^s_i(t)$ and $A^{\text{av}}_i(t)$. (b) Number of failures $NR_i(t)$ as a function of the repair rate μ_i.

We note that the larger the value of repair rate μ_i, the sooner the availability $A_i^s(t)$ reaches its steady-state value as can be inferred from Eq. (17.85). Now if the same cost parameters in Example 17.5 are used $C_c = C_{i,\text{cm}}^l + C_{i,\text{cm}}^p =$ 10,000, L_{UA} = 1000 ($/day) and assuming $\mu_i = 0.1$ and time $t = 400$, then the expected cost + loss rate $MC_i(t)$ is 93.78 (the average availability is 0.9527).

Remark: The nonhomogeneous Markov processes received little attention in literature. There are two main reasons for this: (i) nonhomogeneous Markov processes are much more difficult to solve analytically and (ii) their range of application is limited; they are used in the field of epidemiology and learning theory (Medhi, 2000).

Remark: Although homogeneous Markov processes are confined to exponential distributions only while renewal processes can consider any kind of distributions, the Markov processes can consider many states (failed, normal, or in between, etc.) while renewal processes are limited to two states only: up state (normal) and down state (failed). The Markov processes use transition rate or transition probability to describe evolution of the system while renewal processes use distributions of the events.

Corrective and Preventive Maintenance: We now look into the case of corrective and preventive maintenance. We assume that at any given time, a unit or component is either functioning normally, is failed, or is in preventive maintenance. Assume further that one transition occurs in a sufficiently small time interval and that the possibility of two or more transitions is negligible. Also the repairs and preventive maintenance tasks restore the component to a condition as good as new. The component is in the normal state at time $t = 0$. The process is represented in Figure 17.12.

We first introduce the preventive maintenance service availability $A_i^p(t)$. Therefore:

$$A_i^s(t) + A_i^f(t) + A_i^p(t) = 1 \tag{17.90}$$

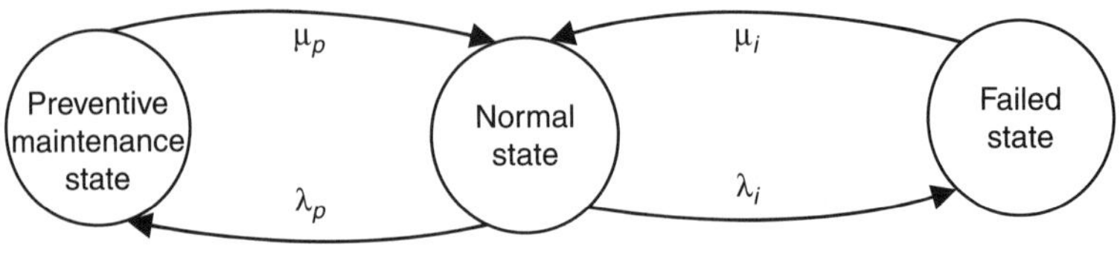

FIGURE 17.12 Markov representation of preventive and corrective maintenance.

We now need two equations: one for $A_i^s(t)$ and the other for $A_i^p(t)$

$$\frac{dA_i^S(t)}{dt} = \mu_i(t)A_i^f(t) - \lambda_i(t)A_i^s(t) + \mu_p A_i^p(t) - \lambda_p A_i^s(t) \tag{17.91}$$

$$\frac{dA_i^p(t)}{dt} = \lambda_p A_i^s(t) - \mu_p A_i^p(t) \tag{17.92}$$

Solving the system of three equations, we obtain (Dhillon, 2002)

$$A_i^S(t) = \frac{\mu_i \times \mu_p}{m_1 m_2} + \left[\frac{(m_1+\mu_i)(m_1+\mu_p)}{m_1(m_1-m_2)}\right]e^{m_1 t} - \left[\frac{(m_2+\mu_i)(m_2+\mu_p)}{m_2(m_1-m_2)}\right]e^{m_2 t} \tag{17.93}$$

$$A_i^f(t) = \frac{\lambda_i \times \mu_p}{m_1 m_2} + \left[\frac{\lambda_i(m_1+\mu_p)}{m_1(m_1-m_2)}\right]e^{m_1 t} - \left[\frac{\lambda_i(m_2+\mu_p)}{m_2(m_1-m_2)}\right]e^{m_2 t} \tag{17.94}$$

$$A_i^P(t) = \frac{\lambda_p \times \mu_i}{m_1 m_2} + \left[\frac{\lambda_p(m_1+\mu_i)}{m_1(m_1-m_2)}\right]e^{m_1 t} - \left[\frac{\lambda_p(m_2+\mu_i)}{m_2(m_1-m_2)}\right]e^{m_2 t} \tag{17.95}$$

where

$$m_1, m_2 = \frac{-B \pm \sqrt{B^2 - 4(\mu_p \times \mu_i + \lambda_p \times \mu_i + \mu_p \times \lambda_i)}}{2} \tag{17.96}$$

$$B = \mu_p + \mu_i + \lambda_p + \lambda_i \tag{17.97}$$

We note that $m_1 + m_2 = -B$ and $m_1 m_2 = \mu_p \times \mu_i + \lambda_p \times \mu_i + \mu_p \times \lambda_i$

Average Availability: The average availability is given by

$$A_i^{\mathrm{av}}(t) = \frac{\mu_i \times \mu_p}{m_1 m_2} + \left[\frac{(m_1+\mu_i)(m_1+\mu_p)}{m_1^2(m_1-m_2)}\right]\frac{[e^{m_1 T}-1]}{t} - \left[\frac{(m_2+\mu_i)(m_2+\mu_p)}{m_2^2(m_1-m_2)}\right]\frac{[e^{m_2 T}-1]}{t} \tag{17.98}$$

Expected Number of Repairs: The expected number of failures $\mathrm{NR}_i(T)$ within a time interval [0, T] for this case is given by

$$\mathrm{NR}_i(T) = \mu_i \int_0^T A_i^f(t)dt \tag{17.99}$$

that is,

$$\mathrm{NR}_i(T) = \mu_i T\frac{\lambda_i \mu_p}{m_1 m_2} + \left[\frac{\mu_i \lambda_i (m_1+\mu_p)}{m_1^2(m_1-m_2)}\right][e^{m_1 T}-1] - \left[\frac{\mu_i \lambda_i (m_2+\mu_p)}{m_2^2(m_1-m_2)}\right][e^{m_2 T}-1] \tag{17.100}$$

Decision Variables: In the Markovian PM model, λ_i, μ_i, μ_p are parameters. They are the reciprocals of the mean time to failure, mean time to repair, and mean time to do PM, respectively. The only decision variable is λ_p, which is the reciprocal of the mean time to take the unit off-line for doing PM (i.e., the PM time interval T).

Economics: We note that PM actions are done only when the unit is taken off-line for PM, that is, the PM cost rate is given by $C_p\lambda_p$. Thus, the expected cost + loss rate is then given by

$$\mathrm{MC}_i(T) = \left[\frac{C_c \mathrm{NR}_i(T)}{T} + C_p\lambda_p + L_{UA} \times [1 - A_i^{\mathrm{av}}(T)]\right] \tag{17.101}$$

Example 17.8: Consider the case: $\lambda_p = \lambda_i = 0.01$, $\mu_i = 0.1$ and μ_p is in the range [0.05, 0.5]. Figure 17.13 shows the dependence of $A_i^s(t)$ (solid lines) and $A_i^{\mathrm{av}}(t)$ (dotted lines) and the number of failures $\mathrm{NR}_i(t)$ on time t for different values of PM repair rate μ_p.

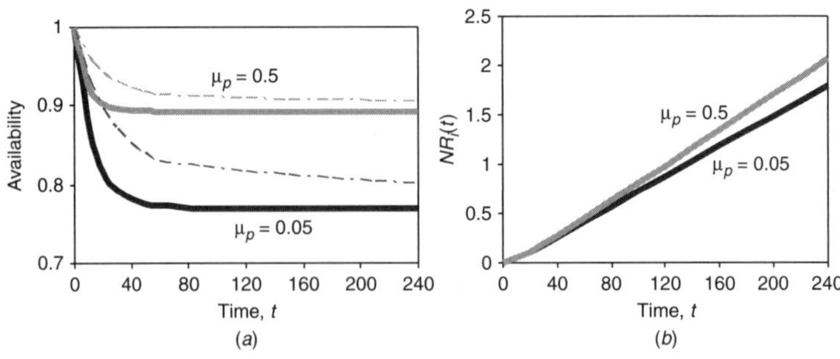

Figure 17.13 (a) $A_i^s(t)$ and $A_i^{\mathrm{av}}(t)$. (b) Number of failures $\mathrm{NR}_i(t)$ as a function of repair rate μ_p.

It can be seen from Fig. 17.13 and inferred from Eq. (17.93) that the availability $A_i^s(t)$ considerably increases when the PM repair rate μ_p increases (or shorter time to do PM) while the number of failures $\mathrm{NR}_i(t)$ is only slightly affected by the PM rate μ_p [because the number of failures $\mathrm{NR}_i(t)$ is only *indirectly* related to the PM rate μ_p].

The maintenance cost + loss rate [given by Eq. (17.101)] is 442.4 with the following cost parameters $C_c = C_{i,\mathrm{cm}}^l + C_{i,\mathrm{cm}}^p = 10{,}000$, $C_p = C_{i,\mathrm{pm}}^l + C_{i,\mathrm{pm}}^p = 500$, $L_{UA} = 1000$ (\$/day); the PM rate μ_p is 0.5 and the PM time interval $T = 200$. The average availability $A_i^{\mathrm{av}}(t)$ corresponding to these parameters is 0.8929.

Remark: When many states are considered, a homogeneous time-continuous Markov model can be written as a system of linear differential equations $dA(t)/dt = \boldsymbol{R} \times \boldsymbol{A}(t)$, where the matrix $\boldsymbol{R}$ (with constant coefficients) characterizes the transition between states and $\boldsymbol{A}(t)$ is the state probability vector whose elements are the probabilities of each state at time t.

Discrete Time Markov Models

A discrete-time Markov model is defined by a set of one-step transition probabilities, P_{ij} with $i,j = 1,2,\ldots n$ representing the probability to go from the state i to the state j in one step. For a system with n states, the state transition probabilities are given in a matrix, called *stochastic matrix of transition probabilities* **P**. Obviously, we have $P_{ij} \geq 0,\ \sum_{j=1}^{n} P_{ij} = 1$. Recall that when the Markov process is homogeneous then the probabilities are constant and not dependent on time.

Let $S_1, S_2, \ldots, S_n$ represent the possible states of the system and let Δt denote the time interval. Then the probability of state S_j at time $(t + \Delta t)$ is the sum of all possible transitions from other states (including the state S_j), that is:

$$p_{S_j}(t+\Delta) = p_{S_1}(t)P_{1j} + p_{S_2}(t)P_{2j} \ldots + p_{S_n}(t)P_{nj} = \sum_{i=1}^{n} p_{S_i}(t)P_{ij} \qquad (17.102)$$

Each term in the above sum is the combined probability that the system is in state S_i multiplied by the probability that the transition from state S_i to S_j takes place. Let $E(t)$ denote the state probability vector at a certain time t whose elements are the probabilities of each state at time t, that is $E(t) = [p_{S_1}(t), p_{S_2}(t), \ldots, p_{S_n}(t)]^T$, where $\sum_{i=1}^{n} p_{S_i}(t) = 1$. Equation (17.102) can be generalized to give:

$$E(t+\Delta t) = P^T \times E(t) \qquad (17.103)$$

where P^T is the transpose matrix of P. Thus, the discrete-time Markov processes are fully defined by the initial state $E(0)$ and the matrix of transition probabilities $P = ||P_{ij}||$. The steady state probabilities can be found assuming $E(t+\Delta t) = E(t)$ and solving the following system of algebraic equations:

$$E(t) = P^T \times E(t) \qquad (17.104)$$

We note that P is not the identity matrix, yet a solution exists. A simple Markov model for periodic PM is presented next.

Remark: The discrete Markov process is just the continuous one expressed in the form of an Euler step. The continuous Markov processes are preferred for time homogeneous cases while the discrete Markov processes are used when:

i) The rates of state transition like failure rates are constant over a short period of time but they change with time (that is, the Markov process is nonhomogeneous). This is the case when imperfect maintenance is considered

ii) Cases involving maintenance actions at discrete points in time are considered, especially when these actions occur at unequal time intervals like the nonperiodic PM actions in age-dependent PM policy.

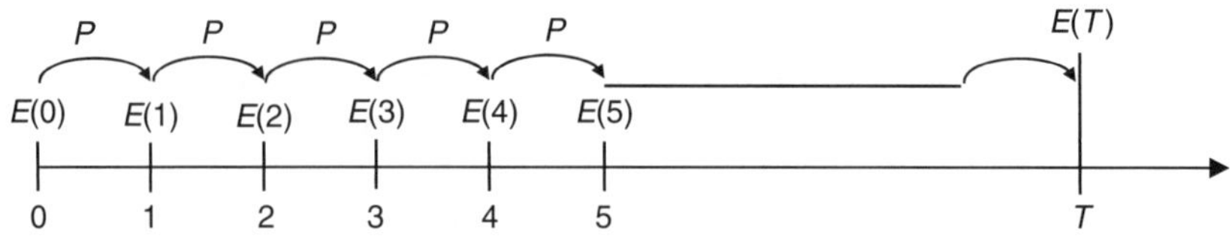

Figure 17.14 Discrete-time Markov model for periodic PM.

Discrete-time Markov model for periodic PM In this model, equipment are instantaneously repaired at failure to an as good as new state and preventively maintained at predetermined times (T, $2T$, ..., kT), the PM time interval T is the decision variable. Figure 17.14 depicts the process (the time interval, Δt, is 1 in Fig. 17.14).

Within a preventive maintenance cycle $[0, T]$, it is assumed that the equipment state is binary: either functioning or failed, these two states are denoted as S_1, S_2, respectively. Because PM restores equipment to the as good as new (AGAN) condition, then $\boldsymbol{E}(T) = \boldsymbol{E}(0)$.

The discrete Markov model is most valuable when PM is imperfect. There exist several methods to model imperfect maintenance (Wang and Pham, 2006), the improvement factor method is described here. In this method, imperfect PM restores equipment to a condition between as good as old AGAO (minimal maintenance) and AGAN (perfect maintenance). A simple model, where the deterioration is independent of time, but steady, one would write that the initial probabilities of failure at the end of period T, after the repair has occurred is $E(T) = M^T \times E(0)$, where M is not the identity matrix.

Let N be the number of time intervals within a PM cycle $[0,T]$. Then we obtain the following results:

Availability: Evaluation of the average availability is straightforward: it is the average value of state probabilities of the normal state S_1

$$A_i^{\text{av}}(T) = \frac{\sum_{k=1}^{N} P_{S_1}(k)}{N} \tag{17.105}$$

Expected Number of Repairs: The expected number of repairs in time interval k is $P_{S_2}(k) \times P_{21} = P_{S_2}(k)$. Thus the expected number of repairs is:

$$\text{NR}_i(T) = \sum_{k=1}^{N} P_{S_2}(k) \tag{17.106}$$

Decision Variable: The PM time interval (or PM cycle time) T

Economics: Costs (per maintenance action) of CM and PM are C_c and C_p, respectively (these costs include the labor and spare part cost: $C_c = C^l_{i,\text{cm}} + C^p_{i,\text{cm}}$; $C_p = C^l_{i,\text{pm}} + C^p_{i,\text{pm}}$).

The repair cost is CM cost per CM action C_c multiplied by the expected number of repairs, that is, $C_c \sum_{k=1}^{N} P_{S_2}(k)$. The PM cost within the PM cycle $[0,T]$ is simply C_p because PM is done only one time within a PM cycle. Thus, the total cost plus loss is

$$\mathrm{MC}_i(T) = \frac{C_c \sum_{k=1}^{N} P_{S_2}(k) + C_p}{T} + L_{UA} \times [1 - A_i^{\mathrm{av}}(T)] \tag{17.107}$$

which is minimum for certain T. We will discuss this later in the next section.

Remark: In the discrete-time Markov model that is associated with one-step transitions, it is understood that the repair of the failed equipment needs to be completed within the length of a time interval (Δt), so ideally the time interval Δt is at least equal to the repair time.

Illustrated examples of discrete-time Markov processes models for maintenance problems can be found in Márquez (2007).

Remark: The discrete-time Markov models offer more flexibility in modeling maintenance activities over the simple time-continuous Markov models, yet they are not flexible enough to represent realistic/complicated systems. Indeed, in discrete-time Markov processes, the system is "sitting" in its currents state (e.g. failed state) exactly one unit of time interval before it makes a transition to new state. We call the time the system stays in its current state the transition time (it is more commonly known as the stay/sojourn time). Examples of transition time are failure-free time and repair time. Thus, it can be seen that the cases where real transition times are random variables cannot be appropriately modeled by Markov models, the so-called *semi-Markov processes* (SMP) are used for such cases instead. Full treatment of discrete Markov processes and SMP is beyond the scope of this book. Interested readers are referred to the textbooks of Gertsbakh (1977), Márquez (2007) and Barlow and Proschan (1965) for more detail of these Markov processes and their applications in maintenance engineering.

Remark: A SMP is a Markov process with random transition times. Figure 17.15 depicts trajectory of a SMP which stays in state S_i a random time τ_{ij} before it jumps to state S_j.

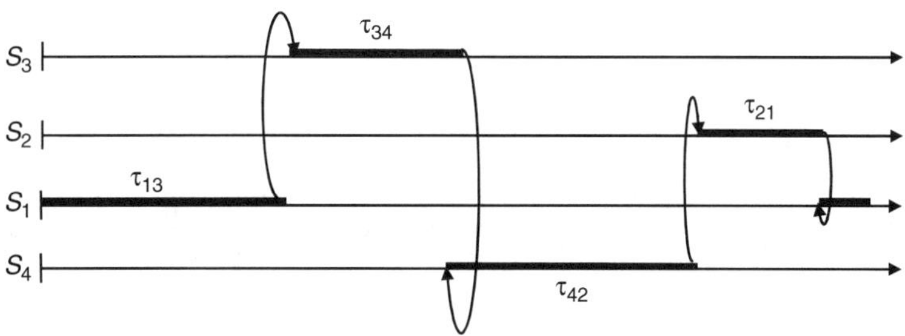

Figure 17.15 Semi-Markov processes.

The SMP are defined by two input parameters:

- The matrix of state transition probabilities $\boldsymbol{P}$
- The distributions $F_{ij}(t)$ of transition times τ_{ij}, called transition distributions and defined by

$$F_{ij}(t) = P\{\tau_{ij} \le t \mid \text{transition } i \to j\} \tag{17.108}$$

In the SMP, if we let $N_j(t)$ be the number of times that system is in state S_j within the time period $[0, t]$ (it is a random variable), the stochastic process $\{N_1(t), N_2(t), \ldots, N_m(t)\}$ is called a *Markov renewal process*. The Markov renewal process is also a useful maintenance modeling technique. It was described in Nakagawa (2005).

The SMP is the most powerful Markov processes-based modeling tool because it offers decision making regarding the type of maintenance actions in the process. More specifically, when the system is in a given state, several maintenance actions (MA) will be possible and the best MA is to be selected. For example, when the system is operating failure-free, the decision-maker can choose among several options: (i) do nothing; (ii) perform a preplanned PM action; (iii) replace the equipment by a new one if its age reaches a critical value.

The optimal strategy (optimal MA for each state and at every step) can then be determined by minimizing the expected costs (or maximizing the rewards). The SMP-based problems are usually solved by using dynamic programming, but other techniques such as genetics algorithms have also been used. More details and illustrated examples of SMP models can be found in Gertsbakh (1977) and Márquez (2007).

Remark: Even though Markov processes theory is a powerful modeling tool, it still cannot capture the dynamic nature of maintenance activities such as delays in some of the maintenance tasks in the waiting list due to unavailability of labor resource at the current time. Clearly, Markov processes are associated with time-independent rate of change (time-continuous processes), time-independent probabilities of one-step state transition (discrete-time processes) or random transition times (semi-Markov processes); hence the labor resource-dependent and spare parts-dependent properties of maintenance time cannot be taken into account by Markov processes. We consider these models our inheritance from the modeling effort that makes up for the lack of computer power, which forced engineers to seek simplifying assumptions. These days, the complexity of the constraints, the multiplicity of decisions that analytical and semi-analytical models could not handle is easily captured by using simulation methods. The most popular simulation method, the Monte Carlo method, is presented next.

Monte Carlo Simulation

The well-known Monte Carlo simulation method can sample a random event (like the failure time) according to the corresponding probability distribution. Upon the occurrence of each event, the type of maintenance actions (as good as new, as good as old or in between), can be also sampled. Based on the samplings, the reliability measures of the system, the maintenance cost/economic loss incurred in the process, etc., can be readily evaluated. "Monte Carlo simulation method is now recognized as playing an important role in system reliability, availability and *MTTF* (*MTBF*) assessment and optimal maintenance of large-scale complex networks" (Wang and Pham, 2000).

Monte Carlo simulations can easily handle

- The interaction between resources (labor, spare parts).
- The condition-based preventive maintenance.
- The time delays that are dependent on the state of other variables, etc.

We present now a Monte Carlo simulation–based maintenance model developed by Nguyen et al. (2008). The model incorporates three practical issues that have not been considered in previous work:

i) Different failure modes of equipment,
ii) Ranking of equipment for repair scheduling, according to the consequences of failure,
iii) A constraint of resource availability (including labor and spare parts) on maintenance activities as the basis for the optimization of labor work force size and spare inventory level.

Different failure modes of equipment and the relative importance of different equipments for repair purpose are addressed as follows:

- Failure modes of equipment: Equipment may have different failure modes involving different parts of the equipment. It can fail because of deterioration of mechanic parts (possible consequence is complete failure that requires parts replacement) or electronic parts malfunction (partial failure that can be repaired). Different failure modes need different repair costs and repair times and induce different economic losses.
- The sampling of different failure modes of equipment is done as follows:
  i) Assign a probability of occurrence for each type of failure mode using information on how common a failure mode is
  ii) At the simulated failure time of the equipment, the type of failure mode that actually occurred is sampled in accordance with the failure modes' probability of occurrence.
- Ranking of repairs: Equipments to be repaired are ranked according to the consequences of failures: 1 is emergent and 5 is affordable to go unrepaired. The maintenance of equipments with higher rank (higher priority) takes precedence over the lower ranked ones (Table 17.2).

Probability of Subsequent Catastrophic Failure	Consequence of Failure		
	High	Medium	Low
High	1	2	3
Medium	2	3	4
Low	3	4	5

TABLE 17.2 Ranking of Equipments for Maintenance (following Tischuk, 2002)

The objective value is the total maintenance cost plus economic loss (to be minimized). The economic loss is the loss caused by equipment failures that lead to reduced production rate or downtime. It is the economic indicator of maintenance performance. Thus by minimizing the maintenance cost plus the economic loss, one simultaneously optimizes the cost and the performance of maintenance.

The cost term includes four types of cost: the *PM cost* and *CM cost*, which are the material costs associated with PM and CM actions, respectively (e.g, the cost of parts replacement, lubricating oils, cleaning agents), the *labor cost* (labor salary) and the *inventory cost* (the cost associated with storing spare parts).

The economic loss term includes two types of losses: (i) economic loss associated with failed equipments that have not been repaired (e.g., a fouled heat exchanger can continue operating but at reduced heat transfer rate), (ii) economic loss due to unavailability of equipment during repair time.

The assumptions are

- Failure follows exponential distribution (i.e., failure rate is independent of time).
- CM is perfect "as good as new" (AGAN), and PM is imperfect.
- There is no interaction between units in multiunit systems.
- Equipment failures are detected immediately and failed equipment is repaired instantaneously at failure.

Maintenance Policy and Decision Variables

The standard periodic PM is used (which is the most suitable for failures with constant failure rate). Three decision variables are considered in the model: (i) the PM time schedule that involves two parameters: the time to perform the first PM (called PM starting time) and the PM time interval, (ii) the spare part inventory level, (iii) the number of maintenance employees.

Interfering/Noninterfering Units

When preventive maintenance (PM) is performed on a specific equipment, there are two possibilities as follows:

(i) PM action on that equipment does not affect production and the economic loss during the maintenance time is negligible; the equipment whose PM does not interfere with production is termed *noninterfering unit*; example of such equipment is valve, pump (with spare unit on line).

(ii) PM action on that equipment significantly affects production, for example, it causes production loss or even downtime, which leads to economic loss; the equipment whose PM interferes with production is termed *interfering unit*.

The distinction between interfering and noninterfering units is incorporated into the simulation-based model; it also has implications in the optimization procedure.

Input Data

For each piece of equipment, the following data are needed:

(i) Reliability data like the mean time between failures (MTBF).

(ii) Information on the failure modes and the associated probability of occurrence for each type of failure mode.

(iii) The time and the associated material cost of CM (for each type of failure mode) and PM.

(iv) The economic loss associated to each type of failure mode.

(v) The inventory cost rate for each type of spare parts.

(vi) Other input data are the waiting time for an purchasing order of spare parts to arrive, the labor paid rate, the available labor hours per employee per week (default value = 40), the ranking (for repair), and the classification (interfering or noninterfering) of the equipments.

Spare Parts Inventory

It is usually the case that an equipment shares common spare parts with the others because they all belong to the same type/group of equipments (like pumps, valves). In such case, the common spare parts can be kept altogether at one storage place and the number of a specific common spare part to keep can be less than the number of equipments it services; hence the problem of determining optimal inventory level arises.

We assume that this optimal inventory level (determined by the maintenance optimization model or user-specified) is well maintained: if the inventory level falls below the prespecified minimal level, then a purchasing order for the spare part (to be stocked) is made immediately to replenish the stocking level.

Labor Assignment

A worker has necessary skills for PM and CM of only a specific group of equipments (e.g., he/she can take care of only rotating machines like pumps, compressors); each group of equipments is assigned to a group of maintenance employees, whose size is to be optimized.

Imperfect Maintenance

Imperfect maintenance is the type of maintenance that leaves equipment in a state somewhere between "as good as new" and "as bad as old" (which is more realistic than the assumption of perfect maintenance). Our imperfect maintenance modeling follows the improvement factor method proposed by Malik (1979), which is the cumulative failure curve of equipment after maintenance lies between the cumulative failure curves corresponding to the two cases: "as good as new" and "as bad as old," as shown in Fig. 17.16.

Referring to Fig. 17.16, the degree of imperfectness is indicated by the improvement factor $\gamma(0 \leq \gamma \leq 1)$ defined by $(T^* - t_0) = \gamma(t_f - t_0)$. If $\gamma = 0$ we have minimal maintenance and if $\gamma = 1$ we have perfect maintenance.

Maintenance Rules

1. Failed equipment are scheduled for repair from highest to lowest priority.

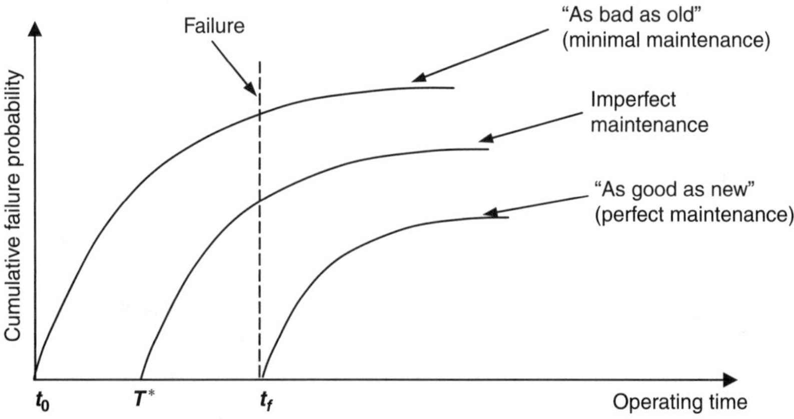

FIGURE 17.16 Imperfect maintenance cumulative probabilities.

2. Maintenance requests that have been delayed due to resources unavailability will be fulfilled as soon as the resources are available again.
3. If equipment has undergone corrective maintenance a predetermined period of time prior to the scheduled action (current value is 7 days), such action is suspended so that resources can be used elsewhere.
4. If, due to resources unavailability, repair of an equipment has been delayed more than a predetermined threshold value (current value is 21 days), the priority for repair of that equipment is upgraded one level.

Monte Carlo Simulation Procedure

The Monte Carlo simulation procedure is similar to the sampling procedure to evaluate stochastic accuracy described in Chap. 10. It is briefly described as a five-step procedure as follows. More details can be found in Nguyen et al. (2008).

- Failure times of equipment are sampled using the reliability function.
- At failure times of equipment, the type of failure modes that caused equipment failure is sampled in accordance with the probability of occurrence.
- The cost of corrective maintenance, the repair time, and the economic losses are determined corresponding to the type of failure modes identified.
- Preventive maintenance requests for equipment are generated in accordance with the predetermined preventive maintenance schedule (predetermined PM policy).
- The planning time horizon is divided into time intervals of weeks. In each week, all the maintenance requests are identified and waited to be fulfilled according to the maintenance rules (e.g., starting first with the tasks with highest priority and resuming the delayed ones as soon as resources are available again). Resources consumptions in maintenance activities are computed and resources availability is kept track of and compared against the minimal value to decide whether there are enough resources to fulfill a specific maintenance task.

Example 17.9: We now compare the three modeling tools: renewal processes, Markov processes, and Monte Carlo simulation. Consider the following problem:

- A system consisting of 10 identical equipment is studied.
- Equipment failure follows exponential distribution with failure rate $\lambda = 0.01$ (1/day).

- The mean time to repair is 2 business days (MTTR $= t_{ER} = 2$).
- Number of available maintenance employee is 1.
- A time horizon of 2 years (730 days) is considered.
- The cost parameters (for each equipment) are $C_c = C^l_{i,cm} + C^p_{i,cm} = 10{,}000$, $L_{UA} = 5000$ (\$/day).

Renewal Process: Because the interaction between factors in maintenance operations cannot be modeled by renewal processes (i.e., the constraint of labor resource availability on maintenance activities is not considered), the average availability $A^{av}_i(t)$ for each equipment is calculated separately, and is equal to 0.9804 [Eq. (17.71)].

Markov Process: Like in the case of the renewal process, the availability for each equipment is calculated separately because the interactions with labor are not modeled. The time-continuous homogenous Markov model is used to evaluate the average availability [Eq. (17.1 and 17.85)] with $\lambda_i = 0.01$ and repair rate $= \mu_i = 1/\text{MTTR} = 0.5$. The average availability $A^{av}_i(t)$ is found to be 0.9804.

Monte Carlo Simulation: The Monte Carlo simulation was used to simulate failure events, the downtimes (i.e., repair times) within the time horizon [0, 730] for each equipment are counted, upon which the average availability (for each equipment) is calculated. The average availabilities of the equipment range from 0.9801 (for equipment with highest priority in repair) to only 0.8739 (for equipment with lowest priority). The mean value of availabilities of the 10 equipment is 0.9447.

The results (average values per equipment) are summarized in Table 17.3.

Methods	$NR_i(t)$	Availability	Cost Rate	Lost Rate	Obj. Value
Renewal processes	7.3	0.9804	100	98	198
Markov processes	7.3	0.9804	100	98	198
Monte Carlo	6.9	0.9447	94.5	276.5	371

Table 17.3 Comparison of the Three Modeling Methods for Example 17.8

Thus, it can be seen that the three modeling methods surprisingly give almost the same result: the availability of equipment with highest priority in repair (so there is no delay in repair for this equipment) obtained by Monte Carlo is the same as the availability obtained by renewal processes and Markov process (= 0.9804). However, the Monte Carlo simulation gives smaller availabilities for equipment with lower priority, which is attributed to the fact that the Monte Carlo simulation has the ability to capture the constraint of labor resource availability on maintenance activities. This constraint leads to the situation that a repair has to be delayed because the employee is occupied with another ongoing repair; hence total downtime is larger than the case where there is no limitation on labor resource, which is the underlying assumption in renewal processes and Markov processes (so that the repair time is constant). The result of this is that the Monte Carlo simulation gives a smaller availability, which leads to a greater value of economic loss.

Example 17.10: The well-known Tennessee Eastman (TE) plant is used. The process flow diagram of the plant is given in Fig. 11.9. The corresponding list of equipment is given in Table 17.4. The information (MTBF, mean maintenance time) were obtained from Bloch and Geltner (2006) and Mannan (2005) or estimated if the information is not available.

Equipment	Quantity	MTBF (days)	Time needed for CM (h)	Time needed for PM (h)	Priority
Valves	11	1,000	2–5	2	3
Compressors	1	381	12–18	6	1
Pumps	2	381	4–12	4	4
Heat exchangers	2	1,193	12–14	8	2
Flash drum	1	2,208	24–72	12	1
Stripper	1	2,582	48–96	12	1
Reactor	1	1,660	12–72	12	1

TABLE 17.4 List of Equipment of the TE Process

In order to predict failures in a plant, the equipment in the plant have to be studied individually and in detail. Table 17.5 summarizes the failure analysis for the TE process.

It is assumed that there are spare pumps in place ready to replace the failed ones so the associated economic losses of pump failures are negligible.

Equipment	Failure	Part Failure	PM Action	CM Action
Heat exchanger	Leak Increased pressure drop Increased wear and tear	Bolts and nuts Excess fouling of tube Control valves	Tighten/ lubricate bolts Cleaning yearly	Replace bolts and nuts Replace tube
Recycle compressor	Overheating Reduced capacity	Bearings/ rotating parts Condenser coils/filters	Lubrication Clean	Replace parts
Exothermic two phase reactor	Stress corrosion	Structural member Filter gauges	Impinge water jet Change (6 months) Calibration (1 yr)	Replace parts
	Suction pressure too low Seal leak	Worn impeller Broken/bent seal	Lubrication Impeller back	Replace pump
Pump	Excessive power Fails to open Fails to close	Clogged impeller Threads/lever Valve seat	Flush Lubricate Clean	
Valves	Leakage through valve	Corrosion	Paint	Replace part
Product condenser	Leakage Reduction in water quality	Fuel corrosion layer	Change filter Change water	Replace part

TABLE 17.5 Equipment Failure Analysis for the Tennessee Eastman Process

Moreover, bypass valves are also installed ready to operate manually once the valves (usually control valves) fail so the associated economic losses of valve failures are small. The main economic losses are therefore attributed to failures of main process equipments such as compressors, heat exchangers, reactors, etc., where the spare ones are not available. A part of the full data, the data for the valve V1 (the valve located in stream 1 in Fig. 10.9), is shown in Table 17.6.

Because the TE plant has no specific chemical and no specific chemical reaction associated to it, we assigned tentatively, a low economic loss associated to a failure of all equipment that have a spare (only $10 per day for pumps and $1000 per day for valves). For major equipment that do not have a spare (reactor, column, compressor, heat exchangers), we assigned a larger value associated to the plant shutdowns. The value is $60,000 per day, which is about one-tenth of the 0.5 to 0.7 millions a day of estimated economic loss in a typical refinery (Tan and Kramer, 1997).

The economic losses were obtained by performing an analysis of the economic impact of equipment failures on the plant production. The maintenance cost is the cost of materials (e.g., the cost of parts replacement or equipment replacement, the cost of lubrication oil or cleaning agents, etc.) needed for maintenance activities. The equipment was priced using the information provided in Peters et al. (2003). Valves, pumps, and compressors are assumed to be noninterfering units (i.e., their PM does not interfere with production) while the main process equipment (heat exchangers, flash drum, stripper, reactor) are interfering units.

The results shown below are obtained under the following assumptions:

- PM is considered to be perfect (improvement factor = 1).
- A labor can take care of any kind of equipment.

The planning time horizon is 2 years (730 days). The average objective value, including the two terms, the cost term and the economic loss term, is shown as a function of decision variables. The average value is calculated as the mean value obtained after N simulation runs $(= \frac{1}{N}\sum_{i=1}^{N} Obj.value_i)$. The number of simulations $N = 10{,}000$.

Effect of Labor Resource Availability: The effect of available labor resource on maintenance performance is investigated when both CM and PM are used. The results shown below are obtained under two conditions:

(i) The spare parts are always available

(ii) PM is applied for noninterfering units only.

Figure 17.17 shows the maintenance labor cost, the economic losses, and the total costs plus losses as functions of the number of maintenance workers in the case PM time interval = 0.1 × MTBF. The objective value, total costs plus losses, for the case no PM is used is also shown for comparison with this case.

If there are too few available maintenance workers, the maintenance requests would rarely be fulfilled on time or may never be fulfilled, which in turn results in increased economic losses. In fact, when only CM is used, the average value of the total number of CM orders that have to be delayed (this number is calculated over all equipments in the whole time horizon), is 0.5158, 0.2027, 0.1049, and 0.1042 corresponding to 1, 2, 3, and 4 workers, respectively. Thus, it can be seen that increasing the number of workers from 1 to 3 will decrease the number of delayed CM requests significantly. On the other hand, if there are more maintenance workers than needed, the maintenance requests are always fulfilled on time (economic loss is minimized) but the labor costs become unnecessarily high. Figure 17.7 also shows that: (i) the use of PM reduces the objective value;

Failure Mode	Failure Mode Description	Probability of Occurrence	Time Needed for CM (h)	CM Cost ($)	Inventory Cost ($/y)	Economic Loss Due to Equipment Failure ($/day)	Economic Loss Due to Unavailability of Equipment ($/day)
1	Severe corrosion	0.1	2	250	35	1,000	1,000
2	Moderate corrosion	0.1	2	250	31	1,000	1,000
3	Slight corrosion	0.1	0	0	0	100	100
4	Severe wear	0.1	2	250	35	1,000	1,000
5	Moderate wear	0.1	2	250	32	1,000	1,000
6	Slight wear	0.1	2	250	28	400	400
7	Severe fatigue	0.1	5	90	6	1,000	1,000
8	Slight fatigue	0.1	4	90	5	100	100
9	Overload	0.1	4	60	4	1,000	1,000
10	Misalignment	0.1	4	60	4	1,000	1,000

Table 17.6 Data for the Valve V1

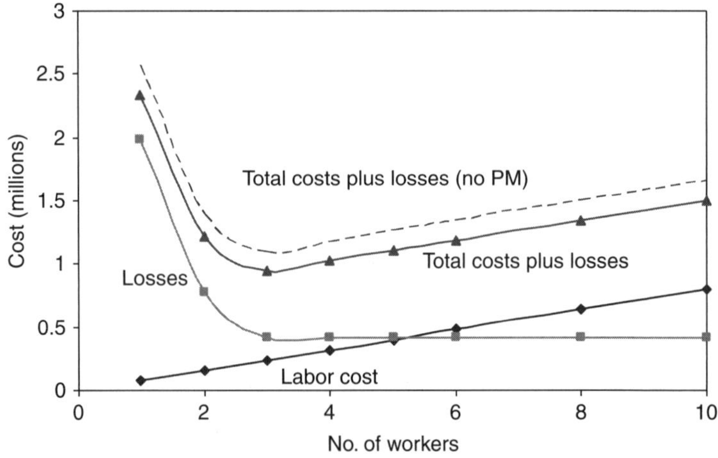

FIGURE 17.17 Effect of labor resource (PM for noninterfering units, PM cycle = 0.1 × MTBF).

(ii) the optimum labor force size for this example is found to be three (the same result was obtained when the PM frequency was changed within the range 0.1–0.9 × MTBF.

Effect of Spare Parts Availability: Assuming that only CM is used and 10 maintenance workers are available. Also for simplicity, assuming that if a specific spare part associated with a group of equipment (e.g., group of 11 valves) is stockpiled, then the number of that spare part = the number of equipments belonging to that group (thus, the decision is whether to keep inventory for a specific spare part, not the inventory level). We compare results of the following three cases:

(a) All spare parts are available.

(b) Only spare parts for repairing noninterfering units (11 valves, 2 pumps, and 1 compressor) are available.

(c) No spare parts are available.

The results are shown in Table 17.7.

Option	(a) All Spare Parts Are Available	(b) Some Spare Parts Are Available	(c) Spare Parts Are Not Available
Inventory cost	197,210	36,776	0
Economic losses	543,173	655,189	770,404
All costs (including inventory)	1,115,089	954,157	916,727
Total costs + losses	1,658,262	1,609,346	1,687,131

TABLE 17.7 Results for the Case Only CM is Used with Consideration of Spare Parts Policy

The maintenance requests can be fulfilled only when the necessary tools/equipment parts are available. Hence, if the spare parts are not stockpiled, the inventory cost is reduced or eliminated at the expense of increased economic losses. Because the equipment's repair has to be delayed until the needed parts are available through emergent purchasing, extra economic loss incured during the time that the equipment' repair is delayed due to spare parts unavailability (besides the loss associated with equipment failure), or eliminating/reducing inventory leads to increased economic loss, a well-known phenomenon. The spare parts necessary for repairing (or replacing) the main equipment in the process such as reactor, stripper, heat exchanger, flash drum are expensive and their associated inventory costs are high. If these spare parts are not stockpiled (the option b), we save a significant amount of money in the inventory cost ($160,434) at the expense of increase in economic losses ($112,016). Although the totals seem to be similar, one does gain benefit by choosing option (b) because this option renders lower objective value. On the other hand, if one decides not to stockpile any spare part, the inventory cost is eliminated but the economic losses increase by $227,231, which results in increased total costs plus losses. While the differences between the different values are low for this small example, the results confirm the common practice in the industry: some degree of inventory is kept but not the extreme cases (do not keep inventory at all or keep inventory at maximum level).

Effect of Preventive Maintenance: The effect of preventive maintenance (PM) under the condition of no resource limitation (there are 10 maintenance workers and spare parts are always available) is investigated. The results are shown in Figs. 17.18 (for noninterfering units) and 17.19 (for interfering units). Note that to illustrate the effect of PM in each group, when the effect of PM done in one group of equipment is being investigated, the other group is subjected to CM only.

It is expected that when PM is performed on equipment, the equipment's condition is preserved and failure is prevented, which results in less equipment failure, less repair cost, increased uptime, less production loss, and less economic losses. In fact, the average values of the total number of CM (i.e., equipment repair) over all equipment in the whole time horizon in the following three cases:

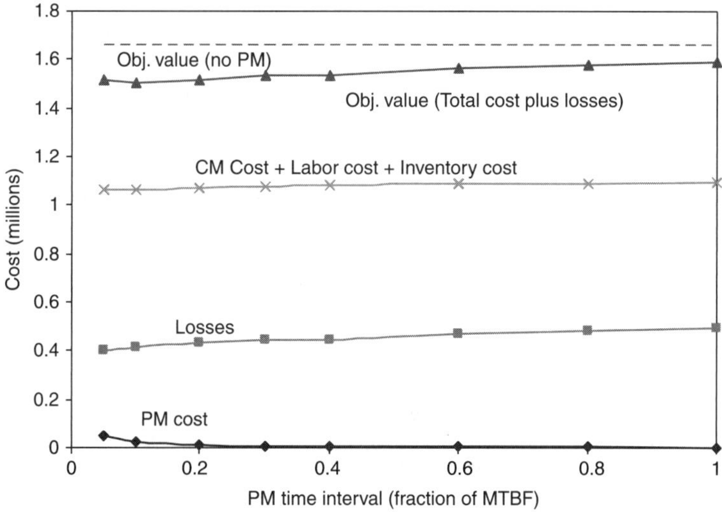

FIGURE 17.18 Effect of PM for noninterfering units (planning time horizon = 2 years).

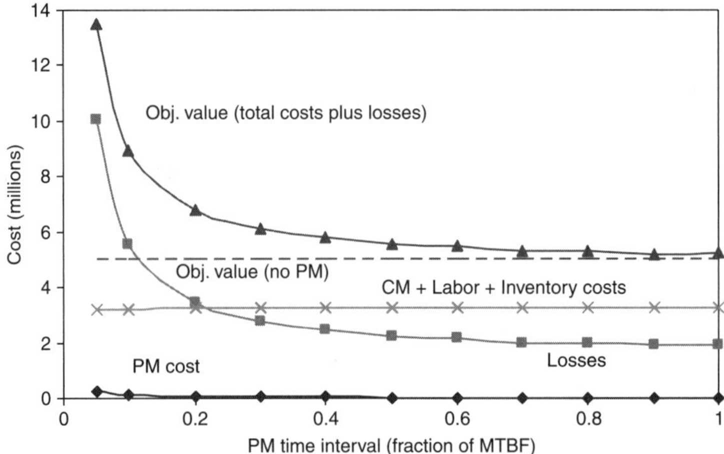

FIGURE 17.19 Effect of PM frequency for Interfering units (planning time = 6 years).

(i) no PM, (ii) PM time interval (noninterfering units) = 0.9 × MTBF, (iii) PM time interval (noninterfering units) = 0.1 × MTBF; are calculated to be 16.02, 10.5, and 5.12, respectively. Thus, applying PM reduces the number of equipment failures. Figure 17.18 confirms the expected results: it clearly shows that the losses decrease and the CM cost also decreases (that makes the curve "CM cost + Labor + Inventory costs" go down; recall that the labor cost and inventory cost are fixed costs) as one increases PM frequency. The benefit of PM is also confirmed: the objective value when PM is used is lower than the objective value when only CM is used. Nonetheless, even the interference of PM activities on production is negligible; doing PM too frequently is not a good idea either because

(i) It leads to high PM cost that may be greater than the benefit it provides as is the case where PM time interval = 0.05 × MTBF (shown in Fig. 17.18).

(ii) It increases the chance for human errors in maintenance to occur, which causes the equipment (after PM) to be impaired and induces extra economic loss. The optimal PM time interval for the noninterfering units is shown to be 0.1 × MTBF.

On the other hand, if the PM activities do interfere with production, then significant economic loss takes place while the equipment are being preventively maintained (e.g., due to unavailability of equipment during maintenance time). This is illustrated in Fig. 17.19. In such situation, there are two competing effects of PM on the equipment availability and the associated economic loss as follows:

(i) The equipment unavailability due to failure is reduced.

(ii) Equipment unavailability during PM activities increases, the plant needs to shut down, and the corresponding economic losses increase.

As a result, the economic loss may increase or decrease. Therefore, for interfering units, if PM frequency is increased,

- The repair cost is reduced but the PM cost increases.
- The economic loss may increase or decrease.

	Perfect PM	Imperfect PM
Economic loss (millions, $)	0.617	0.649
Objective value (millions, $)	0.971	1.009
Increase in obj. value by imperfectness of PM (%)		3.9%
Computational time (10,000 simulations), sec	56	59

Table 17.8 Effect of Imperfectness of PM, Optimal PM Policy Obtained by Inspection, TE Example

These competing effects lead to increased or decreased objective value (total costs plus losses) when PM is applied. An increased cost plus losses discourages the application of PM to equipment whose PM interrupts production, which is the case shown in Fig. 17.19, which shows that when frequency of PM on the main process equipments is increased, the PM cost, the economic losses and the total costs plus losses increase. The minimum objective value is found at PM time interval being $0.9 \times$ MTBF but the change of objective value in the proximity region around the minimum point (PM time interval from $0.7 \times$ MTBF to $1.0 \times$ MTBF) is small (about 0.5% of the objective value).

To investigate the effect of the assumption of perfect/imperfect PM, the objective value corresponding to the optimal PM policy obtained by inspection is evaluated under the assumption of imperfect PM (with improvement factor $\gamma = 0.5$). The results are shown in Table 17.8.

As expected, the assumption of imperfect PM leads to higher number of failures; hence higher economic loss and objective value than the case perfect PM is assumed; the computational time in the former case is longer than in the latter case because more failures are sampled and more maintenance requests are processed, but the difference in computational time is small.

Advantages and Limitations

Renewal Process

- *Advantages:* The renewal processes theory is probably the simplest maintenance modeling tool. It can model different types of failure distributions and the resulting models are analytical models that can be solved to optimality. Imperfect maintenance is also usually modeled by using renewal processes.
- *Limitations:* Complicated situations cannot be handled. For example, constraints on resources availability or opportunistic preventive maintenance policies cannot be added. Situations such as gradual deterioration of equipment or equipment with finite states (failed, functioning, or in between) cannot be modeled.

Markov Process

- *Advantages:* Allow the consideration of multiple states.
- *Limitations:* Complicated maintenance problems require significant effort in modeling. Time-independent failure rate and repair rates are required. Otherwise the analytical results cannot be obtained. Dynamic interactions between factors in maintenance operations cannot be considered.

Monte Carlo Simulation

- *Advantages:* It can virtually model any kind of maintenance optimization problems.
- *Limitations:* These are stochastic models that can only be optimized using stochastic search methods such as genetic algorithms, which do not guarantee optimality. Finally, they generally require significant computational time.

References

Barlow, R. E. and F. Proshan, *Mathematical Theory of Reliability*, John Wiley & Sons, New York, USA (1965).

Birolini, A., *Reliability Engineering: Theory and Practice*, 5th ed., Springer-Verlag Berlin, Germany (2007).

Bloch, H. P. and F. K. Geltner, "Maximizing Machinery Uptime," Elsevier, MA, USA (2006).

Center for Chemical Process Safety, *Guidelines for Hazard Evaluation Procedures*, 3d ed., Wiley-AIChE (2002).

Chan, G. K. and S. Asgarpoor, "Optimum Maintenance Policy with Markov Processes," *Electric Power Systems Research*, **76**(6–7):452–456 (2006).

Charles, A. S., I. R. Floru, C. Azzaro-Pantel, L. Pibouleau, and S. Domenech, "Optimization of Preventive Maintenance Strategies in a Multipurpose Batch Plant: Application to Semiconductor Manufacturing," *Comput. Chem. Eng.* **27**(4):449–467 (2003).

Dhillon, B. S., *Engineering Maintenance*, CRC Press, Boca Raton, USA (2002).

Gertsbakh, I. B., *Models of Preventive Maintenance*, North-Holland Pub, Co. Amsterdam, Holland:355 (1977).

Gertsbakh, I., *Reliability Theory: With Applications to Preventive Maintenance*, Springer-Verlag Berlin, Germany (2000).

Lavaja, J. and M. Bagajewicz, "A New MILP Model for the Planning of Heat Exchanger Network Cleaning," *Industrial and Engineering Chemistry Research*, Special issue honoring Arthur Westerberg, **43**(14): 924–3938 (2004).

Lavaja, J., and M. Bagajewicz, "A New MILP Model for the Planning of Heat Exchanger Network Cleaning. Part II: Throughput Loss Considerations," *Ind. Eng. Chem. Res.* **44**(21):8045–8056 (2005a).

Lavaja, J. and M. Bagajewicz, "A New MILP Model for the Planning of Heat Exchanger Network Cleaning. Part III: Multiperiod Cleaning under Uncertainty with Financial Risk Management," *Ind. Eng. Chem. Res.* **44**(21):8136–8146 (2005b).

Malik, M. A. K., "Reliable Preventive Maintenance Policy," *AIIE Transactions* **11**(3):221–228 (1979).

Mannan S. (editor), *Lees' Loss Prevention in the Process Industries*, 3d ed., Elsevier Butterworth-Heinemann, Oxford, UK (2005).

Márquez, A. C, *The Maintenance Management Framework: Models and Methods for Complex Systems Maintenance*, Springer Series in Reliability Engineering, Springer-Verlag London, UK:355 (2007).

Medhi, J., *Stochastic Processes*, 2d ed., New Age International Publishers, India (2000).

Murthy, D. N. P, A. Atrens, and J. A. Eccleston, "Strategic Maintenance Management," *Journal of Quality in Maintenance Engineering*, **8**(4):287–305 (2002).

Nakagawa, T., *Maintenance Theory of Reliability*, Springer Series in Reliability Engineering, Springer-Verlag London, UK (2005).

Nguyen, D. Q, C. Brammer, and M. Bagajewicz, "New Tool for the Evaluation of the Scheduling of Preventive Maintenance for Chemical Process Plants," *Ind. Eng. Chem. Res.* **47**(6):1910–1924 (2008).

Peters, M. S., Timmerhaus K. D., and West R. E., *Plant Design and Economics for Chemical Engineers*, 5th ed. McGraw-Hill, New York, USA (2003).

Roup, J. P., "Use Equipment Failure Statistics Properly," *Hydrocarbon Processing* **78**(1):55–60 (1999).

Smith, A. M. and G. R. Hinchcliffe, *RCM Gateway to World Class Maintenance*, Elsevier, MA, USA (2004).

Tan, J. S. and M. A. Kramer, "A General Framework for Preventive Maintenance Optimization in Chemical Process Operations," *Comput. Chem. Eng.* **21**(12): 1451–1469 (1997).

Tischuk, J. L., *The Application of Risk Based Approaches to Inspection Planning*, Tischuk Enterprises, UK (2002).

Wang, H. and H. Pham, "A Quasi Renewal Process and its Application in the Imperfect Maintenance," *International Journal of Systems Science* **27**(10):1055–1062 (1996).

Wang, H. and H. Pham, *Reliability and Optimal Maintenance*, Springer Series in Reliability Engineering, Springer-Verlag London, UK (2006).

CHAPTER 18

Maintenance Optimization*

After reviewing the existing maintenance representation methods in Chap. 17, we now attempt to optimize the parameters of the models using tools that are more powerful than simple inspection.

Maintenance optimization research focuses on maintenance planning, the heart of a maintenance program, where several decisions need to be made: maintenance time plan over a time horizon of months or years, inventory level, replacement strategy, maintenance work force size, etc. Maintenance scheduling, in turn, organizes maintenance activities on a daily or a weekly basis. It allocates necessary and existing labor and material resource to fulfill maintenance tasks considering several factors: priority of maintenance tasks, availability of current labor, and material resources, etc.

The relationship between long-term planning and short-term scheduling has been considered hierarchical, that is, once planning set picks its corresponding variables taking into account the limitations of scheduling in a simplified manner, scheduling follows with the details. This hierarchy is a product of the historical inability of maintenance models to deal with the complexity of the task, hence the decomposition. The price often paid is that the scheduling model cannot accommodate well to the mandates of planning. However, as we saw in the last chapter, the ability of simulation-based models to address the complexity removes the need for the hierarchy and allows the consideration of all aspects of the problem simultaneously using the real interactions, not simplified ones.

The problem of optimization of preventive maintenance has been extensively studied, mainly by people in the field of industrial engineering and operation research. An extensive review of maintenance optimization models was given in Wang and Pham (2006). A few works on maintenance optimization in chemical process plants have

*This chapter was written by DuyQuang Nguyen and Miguel Bagajewicz.

also been published. In chemical process industry, maintenance optimization is either studied separately or in conjunction with the problems of optimal process design and operation planning to achieve maximum profit or minimum environmental risk. For example, Tan and Kramer (1997) presented a general framework for maintenance optimization in chemical process plants using Monte Carlo simulation as modeling tool and Genetic Algorithm. Vassiliadis and Pistikopoulos (2000) combinatorially optimized process configuration and operation as well as preventive maintenance (PM) strategies that accomplish the conflicting environmental and profitability targets using a nonlinear programming (NLP) problem formulation. As discussed in Chap. 17, periodic PM is the most common method while age-dependent PM or imperfect maintenance is still the object of maintenance optimization research (Wang and Pham, 2006). Its application in practice is being hindered by the difficulty to obtain data on failure rates as function of time/age.

Components of Maintenance Optimization

A maintenance planning and scheduling optimization model is a mathematical model used to assess the effect of maintenance actions on reliability and economic performance as well as maintenance costs and to choose the optimal path. The components of the model are as follows:

- *Objective:* The most common objective is minimizing cost, composed of parts cost, labor cost, and spare parts inventory cost. Another common objective is maximizing reliability (reliability, availability, average uptime) usually used for certain types of systems such as nuclear power plants, aerospace systems, and power-generating systems where reliability and/or safety are much more important than cost.
- *Decision variables:* The decision variables depend on the maintenance policy. The most common decision variable is PM time schedule (i.e., PM frequency, PM time intervals). Maintenance resources (labor, spare parts) are also considered, as we saw in Chap. 17.
- *Constraints:* These are the requirements needed to achieve of a maintenance program and/or the constraints imposed on the maintenance activities. Desired levels of reliability, average uptime can be imposed in cost minimization models, and cost/budget constraints can be used when reliability is maximized.
- *Models:* Precedence order of actions, IF-THEN-ELSE decisions, needs to be somehow incorporated. These relationships are

in fact embedded in the different assessment models that were presented in Chap. 17.

- *Integration with models of other activities:* Usually, the PM policy alone is optimized; there are few research works on the problem of simultaneous optimization of PM policy and maintenance resources (e.g., Shum and Gong, 2006). Above and beyond these simple integration, maintenance optimization is in reality a subset of plant activities, and should start to be integrated with other models that maximize the plant efficiency as a whole, not just the maintenance effects.

Renewal Process-Based Models

The framework to optimize PM using renewal processes is as follows:

- The system within a PM cycle (the time in between two consecutive PM times) is considered a renewal process, and the maintenance cost and availability in this period of time can be evaluated.
- The general renewal processes–based PM problem formulation presented in the literatures (e.g., Wang and Pham, 2006) is to minimize maintenance cost rate subjected to constraint in availability.

$$\left.\begin{aligned} &\min_{0<T<\infty}\{\mathrm{MC}_i(T)\} = \frac{C_c \mathrm{NR}_i(T) + C_p}{T} \\ &\text{s.t.} \\ &A_i^{\mathrm{av}}(T) = \frac{\mathrm{TU}(T)}{\mathrm{TU}(T)+\mathrm{TD}(T)} = \frac{\mathrm{TU}(T)}{T} \geq A_0 \end{aligned}\right\} \tag{18.1}$$

where TU and TD are accumulative uptime and downtime, respectively; C_c and C_p are the cost per maintenance action of CM and PM $C_c = C^l_{i,\mathrm{cm}} + C^p_{i,\mathrm{cm}}$; $C_p = C^l_{i,\mathrm{pm}} + C^p_{i,\mathrm{pm}}$.

- The input parameters of the model are the distributions of failure times and repair times, the unit cost (cost per maintenance action) of CM and PM, and the parameters characterizing the imperfectness of maintenance, etc.
- The variable T, the PM time interval, is the only decision variable.
- The model formulation given by Eq. (18.1) is the most commonly used one, one can of course formulate a problem that maximizes availability subjected to budget constraint, etc.
- The approach recommended is to translate downtime/unavailability to economic loss; the summation of loss and

maintenance cost is then simultaneously minimized, as we have shown in Chap. 17. More specifically, we propose to solve the following problem:

$$\operatorname*{Min}_{0<T<\infty}\left(\frac{C_c \mathrm{NR}_i(T)+C_p}{T}+L_{\mathrm{UA}} * [1-A_i^{\mathrm{av}}(T)]\right) \tag{18.2}$$

which is an unconstrained problem that can be solved analytically. In the case of budget limitations, a lower bound on T needs to be added. Thus, if the solution of the unconstrained problem is a value of T lower than the lower bound, the lower bound is adopted.

The cost and availability (which is directly related to loss) contain two terms associated to repair $[\mathrm{NR}_i(T)/T]$ and preventive maintenance (t_{pm}/T). The dependence of $[\mathrm{NR}_i(T)/T]$ and (t_{pm}/T) on T are usually monotonic and in opposite direction: when T decreases, $[\mathrm{NR}_i(T)/T]$ decreases while (t_{pm}/T) increases. Thus, in overall the cost and loss usually have their own unique minimum point and objective value; the summation of cost and loss may have multiple local minimums but the global minimum can be easily identified by graphical method. Exception to this generalization occurs when failure events are characterized by exponential distributions: in such case the repair term $[\mathrm{NR}_i(T)/T)]$ is constant while the objective value is monotonic function of T. Thus, in such case, the objective value is minimum when $T \rightarrow$ infinity (i.e., not to use PM at all).

We now obtain the expressions that render the optimum taking the derivative of Eq. (18.2) with respect to the only decision variable (T) and equating it to zero as follows:

$$\frac{\partial}{\partial T}\left[\frac{C_c \mathrm{NR}_i(T)+C_p}{T}\right]-L_{\mathrm{UA}}\frac{\partial A_i^{\mathrm{av}}(T)}{\partial T}=0 \tag{18.3}$$

For the average availability we use Eq. (17.71), where the downtime is assumed to be very small as compared to T (the usual case). Thus, we get

$$\frac{\partial}{\partial T}\left[\frac{C_c \mathrm{NR}_i(T)+C_p}{T}\right]-L_{\mathrm{UA}}\frac{\partial}{\partial T}\left[\frac{T}{T+\mathrm{NR}_i(T)t_{\mathrm{ER}}+t_{\mathrm{PM}}}\right]=0 \tag{18.4}$$

Equation (18.4) contains the number of repairs, which comes from an expression such as Eq. (17.36) or Eq. (17.49). Thus to obtain the answer, it is appropriate to consider the particular distribution. We do that in the next example.

Example 18.1: Consider an alternating renewal processes with random failures described by Erlang distribution with parameters $\lambda_i = 0.01$ (1/day), $n = 3$, that is, $f_{i,n}(t) = 0.5t^2\lambda_i^3 e^{-\lambda_i t}$. Repair time t_{ER} and PM time are constant: $t_{\text{ER}} = 2$ (days) and $t_{\text{PM}} = 1$ (day) (they are considered to be negligibly small when compared with the PM time interval (T). The unit costs of CM and PM are $C_c = \$10{,}000$ and $C_p = \$500$, the loss rate $L_{\text{UA}} = 1000$ (\$/day). Note that this failure distribution has mean time to failure, $\text{MTTF} = n/\lambda_i = 3/0.01 = 300$ (days).

The Laplace transform of the number of failures $\text{NR}_i(s)$ is given by Eq. (17.37) with $f_i(s) = \lambda^3/(s+\lambda)^3$, then the expected number of failures within a PM cycle $\text{NR}_i(\text{T})$ can be found from inverse Laplace transform to be

$$\text{NR}_i(T) = \frac{1}{3}\left[\lambda_i T - 1 + \frac{2}{\sqrt{3}} e^{-3\lambda_i T/2} \sin\left(\frac{\sqrt{3}\lambda_i T}{2} + \frac{\pi}{3}\right)\right] \tag{18.5}$$

The expected cost rate is $\text{MC}_i(T) = \dfrac{C_c \text{NR}_i(T) + C_p}{T}$; Eq. (17.71) is used to evaluate the availability.

The optimal PM time interval T that simultaneously minimizes the cost rate $\text{MC}_i(T)$ and loss rate $L_{\text{UA}} * [1 - A_i^s(T)]$ can be found by solving Eq. (18.4) together with Eq. (18.5). Figure 18.1 illustrates the trade-offs. The optimal PM time interval T and the optimal objective value are given in Table 18.1.

Note that because the time to do PM introduces downtime, decreasing PM time interval T (use PM more often) has two competing effects on availability: (i) reduced downtime due to equipment failures [the failure rate $\text{NR}_i(T)/T$

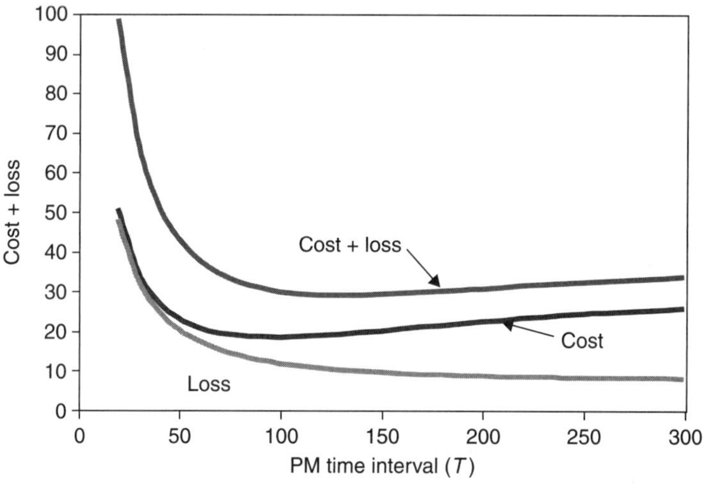

FIGURE 18.1 Optimal PM time interval for Example 18.1.

Objective	Optimal *T* (Days)	Availability	Cost Rate (\$/Day)	Cost Plus Loss Rate (\$/Day)
Minimal cost rate	100	0.9885	18.09	29.57
Minimal (cost + loss)	130	0.9902	18.856	28.683

TABLE 18.1 Optimal PM Policy for Example 18.1

decreases when T decreases], (ii) increased downtime due to the PM actions. The same thing can be said for the effect of PM on repair cost rate and PM cost rate: the repair cost rate $C_c\mathrm{NR}_i(T)/T$ decreases while PM cost rate (C_p/T) increases when T decreases. However, for this example, the overall effects of PM on availability and on maintenance cost rate are different: while the curve of maintenance cost rate vs. T exhibits a minimum point, the dependence of availability (and hence the loss rate) on T is a monotonic function: availability decreases (hence loss rate increases) when T decreases. The optimal PM time interval T is one-third of the MTTF.

Remark: When many equipment are considered, a total cost (repair cost + losses) can be calculated for each equipment as a function of the value of T for each one. When budget limitations are included, then some expressions of the labor and spare parts are needed for all equipment as a function of all the PM time intervals. Here is where we start noticing that renewal processes are insufficient.

Markov-Based Model

We present here optimization strategy for models based on time-continuous homogenous Markov processes. This Markov model is based on the following arguments:

- In the Markov model, λ_i, μ_i, μ_p are parameters. They are the reciprocals of the mean time to failure, mean time to repair, and mean time to do PM, respectively. The only decision variable is λ_p, which is the reciprocal of the mean time to take the unit offline for PM (i.e., the PM time interval T).
- PM action is done only when the unit is taken offline for PM, that is, the PM cost rate is given by $C_p\lambda_p$.

Using $\lambda_p = 1/T$ in the calculation of $\mathrm{NR}_i(T)$, the expected cost + loss rate is given by

$$\min_{0<T<\infty}\left[\frac{C_c\mathrm{NR}_i(T)}{T} + C_p\lambda_p + L_{\mathrm{UA}} * [1 - A_i^{\mathrm{av}}(T)]\right] \tag{18.6}$$

This simple optimization problem can be solved using numerical methods. The difference between this model [Eq. (18.6)] and the one based on renewal processes [Eq. (18.2)] is that $\mathrm{NR}_i(T)$, the expected number of failures, and the average availability $A_i^{\mathrm{av}}(T)$ are evaluated differently.

Example 18.2: Consider a simple Markov PM model with the following parameters: mean time to failure = 200 (days), mean time to repair = 2 (days), mean time to do PM = 1 day, $\lambda_i = 0.01$, $\mu_i = 0.5$, $\mu_p = 1$. The cost parameters are: C_c = 15,000; $C_p = 100$; unit loss $L_{\mathrm{UA}} = 1000$. The optimal PM time interval, T, is obtained numerically. The results are shown in Fig. 18.2 where the cost rate and the loss rate are given by $C_c\mathrm{NR}_i(T)/T + C_p\lambda_p$ and $L_{\mathrm{UA}} * [1 - A_i^{\mathrm{av}}(T)]$, respectively.

When T decreases (use PM more often): (i) the repair cost rate loss $C_cNR_i(T)/T$ decreases (and the maintenance cost rate decreases), (ii) the availability decreases

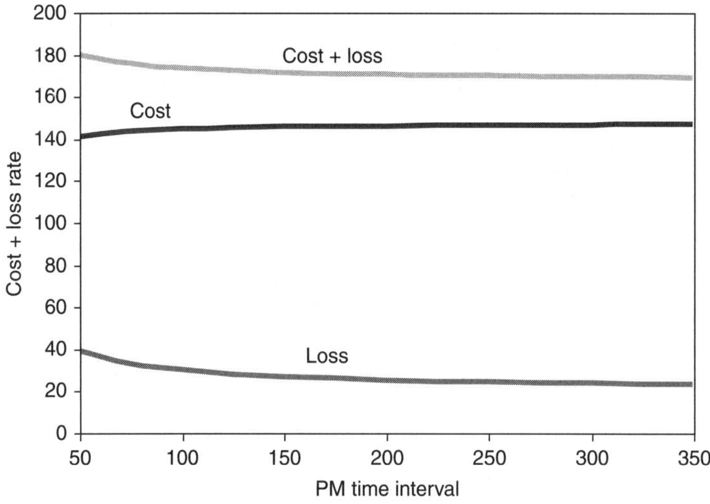

Figure 18.2 Optimal PM time interval for Example 18.2.

(downtime due to PM increases) hence loss rate increases. However, the curve of the objective function, total cost plus loss, does not exhibit any minimum point. In this example, the objective value is a monotonic function of T and the best solution is not to use preventive maintenance ($T \rightarrow \infty$). The same result is obtained (it is not recommended to use PM or $T \rightarrow \infty$) if renewal processes model is used in this example (exponentially distributed failures, the underlying assumption in the Markov model, are used in the renewal processes model).

Monte Carlo–Based Models Using Genetic Algorithms

We used a genetic algorithm (GA) because this method is well-established and was shown to have good performance (although it does not guarantee optimality). In brief, genetic algorithm method is based on the principles of genetics, natural selection, and evolution; it "allows a population composed of many individuals to evolve under specified selection rules to a state that maximizes the 'fitness,' that is minimizes the cost function" (Haupt and Haupt, 2004). The algorithmic procedure and detailed description of the well-known genetic algorithm method can be found in various textbooks such as the book by Haupt and Haupt (2004). The GA is briefly described as a seven-step procedure as follows:

- *Variable encoding and decoding*: This step involves the conversion (i.e., encoding) of the values of decision variables into an appropriate representation (a chromosome). Decoding is the reverse process of encoding.
- *Initialization of population*: This step involves randomly generating a population of N chromosomes.

- *Natural selection*: This step involves three operations: (i) evaluating the cost function corresponding to each chromosome/individual in the population, (ii) sorting the population in descending order of "fitness" (e.g., if the cost function/objective value is to be minimized, then lower cost = larger "fitness" value), (iii) selecting a portion of population with good fitness value to keep and discarding the rest.
- *Selection*: Selecting and pairing the retained (survived) chromosomes to produce offspring for the next generation. Usually two chromosomes are paired to produce two offsprings.
- *Mating*: Offspring of the paired chromosomes (parent) are produced through the crossover process whereby the parent's genetic codes are passed on to the offspring.
- *Mutation*: Random mutations alter a certain percentage of the bits in the list of chromosomes.
- *Convergence*: At this step, next–generation population is generated which contains new chromosomes. The cycle of evaluating cost functions—selecting, pairing, producing offspring—is repeated unless convergence criterion is met, which is to terminate the GA procedure if the best objective value obtained in each iteration does not change after a predetermined number of iterations.

The parameters involved in the GA method are the size of population, the portion of population to keep, the mutation rate, and the selection and crossover methods. Details of calculation steps/operators in the genetic algorithm we used are described below.

Variable Encoding and Decoding

All the decision variables in the model, the PM starting time (P_{ini}) and PM time interval (PMI), the spare part inventory level (SPIL) for each spare part and the number of employees (NE) in each group, are integer variables. The P_{ini} and PMI are expressed as fractions of the MTBF (PMI = a*MTBF and P_{ini} = b*MTBF). The inventory level SPIL is also "normalized": the inventory level for a "common" spare part is expressed as a fraction c of the number of pieces of equipment it services. There are two justifications for the normalization of the variables:

(i) The fractions a, b, c give a better understanding of the magnitude of the variables than their absolute values,

(ii) Because the values of the variables PMI, P_{ini}, and SPIL can vary greatly, normalizing the variables reduces their range of variability, which ultimately helps the GA converge faster.

The reasonable range of a and b is [0, 2] while the reasonable range of c is [0, 1]. The fractions a, b (for each equipment), and c (for each spare part) together with the number of employees NE (for each group) are to be optimized. We decided to use the standard binary GA, that is, the variables *a, b, c,* and NE are coded using a binary representation. The variable NE, the number of employees, is represented by a string of binaries using a decimal-to-binary transformation. The reasonable range of number of employees is [1, 8], thus a string of 3 binaries is used to represent NE (recall that NE is the number of employees in a group, not the total labor workforce size). For practical reasons, we postulate that the variables *a, b,* and *c* can take only discrete values (like 0.1, 0.2, 0.3, etc.). Indeed, it is a common industrial practice that the preventive maintenance time schedules P_{ini} and PMI take only discrete values like 30 days, 60 days, etc. We confine the possible values of *a, b* (representing P_{ini} and PMI) to be one of the following 16 discrete values: /0.1, 0.15, 0.2, 0.25, 0.3, 0.35, 0.4, 0.45, 0.5, 0.55, 0.6, 0.7, 0.8, 0.9, 1.0, 1.1/ for interfering units (vector U) and /0.5, 0.6, 0.7, 0.8, 0.85, 0.9, 0.95, 1.0, 1.05, 1.1, 1.15, 1.2, 1.3, 1.4, 1.5, 1.6/ for noninterfering units (vector V). The value of *c* is confined to take one of the following 8 discrete values: /0.1, 0.2, 0.3, 0.4, 0.5, 0.6, 0.7, 0.8/ (vector W); the possible values of *a, b, c* can be easily changed by the user if desirable.

Thus, the problem of determining the optimal values of *a, b, c* turns into the problem of selecting values for *a, b, c* from the pools of discrete values. This is done in two steps, which are illustrated for the variable *a* for interfering units as follows:

(i) The value of *a* is indicated by its location (the index *i*) in the corresponding vector containing possible values of *a* (vector U), e.g., if $i = 2$ then $a = U[2] = 0.15$.

(ii) A gene consisting of 4 binaries is used to represent the index *i* whose value ranges from 1 to 16. The variables *b* and *c* are treated in the same way (4 binaries are needed to represent *b* and 3 binaries for *c*).

GA Operators and GA Parameters

The methods for the GA operators and the values are intuitively chosen in accordance with the scale of the problem using the guidelines provided in the literature (Haupt and Haupt, 2004). They are as follows:

- Selection: Roulette wheel rank.
- Crossover: Uniform crossover method.
- Population size = 40 for small-scale problems such as the Tennessee Eastman example; population size = 60 for large-scale problems such as the FCC unit example.
- Fraction of population to keep = 0.5.
- Mutation rate = 0.3.

Example 18.3: We use the Tennessee Eastman (TE) process of Example 17.10. All the assumptions stated above are kept. The results are shown in Tables 18.2, 18.3, and 18.4. One of the GA runs (run 1) gave a similar result to the one obtained by inspection but with a better objective value, as can be seen in column 3 of Tables 18.2, 18.3, and 18.4. When PM is not used, it is because the scheduled PM time is outside the time horizon. The result obtained by the second GA run (run 2 shown in column 4 of the three tables) is the best one among the three. The best result has one fewer labor and PM is not used for the valves, which leads to smaller labor cost and PM cost and an increase in expected loss. However, there is an opposite effect: the economic loss decreases due to the fact that more spare parts inventory is used (hence larger inventory cost). Overall, the economic loss and the objective value decrease. The convergence of the GA is shown in Fig. 18.3. The computation time in each GA run is about 2 h, 5 min.

		By GA Optimization	
Equipment	**By Inspection**	**Run 1**	**Run 2**
11 Valves	0.1	0.1 (6 valves) and 0.2 (5 valves)	PM not used
1 Compressor	0.1	0.4	0.1
2 Pumps	0.1	0.1	0.15
2 Heat exchanger	0.9	PM not used	PM not used
1 Flash drum	0.9		
1 Stripper	0.9		
1 Reactor	0.9		

Table 18.2 Optimal PM Time Frequency (Fraction of MTBF) for TE Example

		By GA Optimization	
Equipment	**By inspection**	**Run 1**	**Run 2**
11 Valves	Yes	Inventory for only 5 valves	Inventory for only 5 valves
1 Compressor	Yes	Yes	Yes
2 Pumps	Yes	Yes	No
2 Heat exchanger	No	Yes	Yes
1 Flash drum	No	No	No
1 Stripper	No	No	No
1 Reactor	No	No	No

Table 18.3 Optimal Spare Parts Inventory Policy for TE Example

	By Inspection	By GA Optimization	
		Run 1	Run 2
No. of labor	3	3	2
Objective value (millions)	0.971	0.856	0.823

Table 18.4 Optimal Number of Labor and Objective Value for TE Example

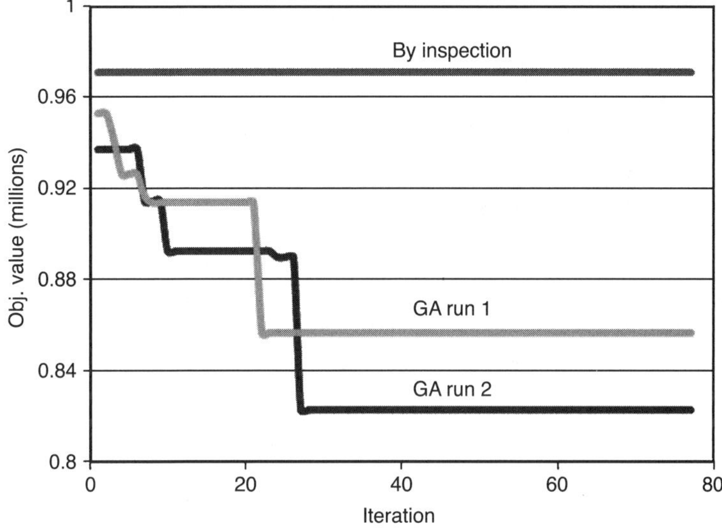

Figure 18.3 GA convergence, TE example.

Remark: If renewal processes and homogeneous Markov processes are used to solve this example (under the same assumptions of perfect maintenance and exponentially distributed failures), they both give the same result, which is not to use PM (PM interval $\rightarrow \infty$) for all equipment. Moreover, the problem of determining optimal sizes of maintenance resources (labor and spare parts) cannot be addressed by using the two analytical methods because these analytical methods cannot capture the effect of resources availability on maintenance activities. Advanced analytical methods such as nonhomogeneous Markov or semi-Markov processes could be used for this example, but significant effort is required to develop such an analytical model. Simulation is the preferred option for complicated maintenance problems like this example.

Example 18.4: A large-scale problem, the fluid catalytic cracker (FCC) unit in a refinery, is considered. A large west coast refinery volunteered equipment and volume specifications for its FCC unit. This unit, which processes roughly 50,000 barrels a day (bbl/day) of feed, comprises 61 pumps (31 primary, 30 spare), 2 compressors, 4 heaters, 87 heat exchangers, 15 vessels, 1 catalytic reactors and its associated catalyst regenerator, and 12 columns and strippers. The valves are not considered in this study. The main process equipments (process vessels, the catalytic reactor and its associated catalyst regenerator, columns, and strippers) are not included in this study because (i) the failure of main process equipment is very rare, the failure rate is in the magnitude of $10^{-4} - 10^{-2}$ (failure/year) (from data listed in Mannan, 2005), (ii) practically, the main process equipments are preventively maintained only at turnaround (i.e., when an entire processing unit or the refinery is shut down for overhaul). Thus, only rotating equipment (pumps, compressors) and heaters, exchangers are included in this study. These types of equipment are indeed the ones subjected to preventive maintenance program in refineries.

The MTBF and the MTBR of the equipment considered in this study are listed in Table 18.5. These values are estimated (corresponding to the operating condition in a refinery) based on the values provided in Mannan (2005). The mean time to perform preventive maintenance is estimated.

The following assumptions were made in estimating the economic losses:

- Economic loss of product is assumed to be $10/bbl. This results in an economic loss of $500,000/day if the process unit is fully shut down.
- For the pumps in the process:
  1. Spare pumps always work. If a spared pump fails, the spare instantaneously comes online.
  2. If a spare is insufficient to maintain a stream at its normal operating rate, economic loss is proportional to the loss in throughput.

With these assumptions, the economic loss corresponding to failure of pumps with spare is essentially zero (it takes nominal value of $10/day in the model).

- For the heat exchangers and heaters in the process:
  1. Failed exchangers transfer heat, but at a reduced rate (20–30% heat transfer loss).
  2. Any exchanger located in series with other exchangers may be bypassed while being serviced without interrupting the process.
  3. Economic loss is proportional to the portion of heat-duty lost due to the failure.
- For the compressors:
  1. If the component fails, the process goes offline.
  2. The result is a maximum economic loss per day ($500,000/day).

A sample of economic data is given in Table 18.6, which shows the cost of corrective maintenance (CM cost), the economic loss due to unrepaired failure of equipment (type 1), the economic loss due to unavailability of equipment during repair time (type 2), and the probability of occurrence for each type of failure

Units	Quantity	MTBF (Days)	Time Needed for CM (h)	Time Needed for PM (h)	Priority
Pumps	61	694	6–8	4	1, 4, 5
Compressors	2	381	30	8	1
Heaters	4	1344	25–36	8	2
Heat exchangers	87	1344	25–36	8	1, 2, 3, 4

TABLE 18.5 List of Equipment of the FCC Unit

Unit	Failure Mode	Failure Mode Description	Prob. of Occurrence	CM Cost ($/CM Action)	Econ. Loss, Type 1 ($/Day)	Econ. Loss, Type 2 ($/Day)	Invent. Cost ($/Part/Year)
Pump	1	Seals failure	0.4	6,900	10	10	98
	2	Leak	0.05	6,900	10	10	88
	3	Motor failure	0.05	6,900	10	10	79
	4	Couplings	0.05	6,900	10	10	101
	5	Bearings	0.05	6,900	10	10	91
	6	Corrosion	0.2	6,900	10	10	0
	7	Wear/tear	0.2	6,900	10	10	0
Compressor	1	Lubrication breakdown	0.15	37,400	200,000	500,000	1,927
	2	Seal failure	0.2	37,400	200,000	500,000	1,730
	3	Excessive vibration	0.15	37,400	200,000	500,000	1,554
	4	Fatigue/rupture	0.2	37,400	200,000	500,000	1,927
	5	Corrosion	0.15	37,400	200,000	500,000	1,730
	6	Erosion/wear	0.15	37,400	200,000	500,000	1,554

Table 18.6 Sample of Economic Data in the FCC Example

Unit	Failure Mode	Failure Mode Description	Prob. of Occurrence	CM Cost ($/CM Action)	Econ. Loss, Type 1 ($/Day)	Econ. Loss, Type 2 ($/Day)	Invent. Cost ($/Part/Year)
Heater	1	Fouling	0.5	69,000	50,000	100,000	930
	2	Fatigue/crack	0.1	69,000	50,000	100,000	831
	3	Tube rupture	0.1	69,000	50,000	100,000	747
	4	Corrosion	0.2	69,000	50,000	100,000	1,002
	5	Others	0.1	69,000	50,000	100,000	897
Process heat exchanger	1	Fouling	0.5	12,600	14,940	49,800	465
	2	Fatigue/crack	0.1	12,600	14,940	49,800	416
	3	Tube rupture	0.1	12,600	14,940	49,800	374
	4	Corrosion	0.2	12,600	14,940	49,800	501
	5	Others	0.1	12,600	14,940	49,800	449
Heat exchanger	1	Fouling	0.5	6,700	1,500	3,000	310
	2	Fatigue/crack	0.1	6,700	1,500	3,000	277
	3	Tube rupture	0.1	6,700	1,500	3,000	249
	4	Corrosion	0.2	6,700	1,500	3,000	334
	5	Others	0.1	6,700	1,500	3,000	299

Table 18.6 Sample of Economic Data in the FCC Example (*Continued*)

modes for some equipment. Full data for all the equipment, which also include other types of data such as waiting time for an emergent purchasing order to arrive, can be obtained by contacting the authors.

The exact maintenance model described in the maintenance assessment section is used for this example, that is, the PM is assumed to be imperfect (improvement factor, $\gamma = 0.5$); there are different groups of employees assigned to appropriate groups of equipments. Regarding the spare parts policy, the spare parts inventory levels are optimized instead of the much simpler inventory policy used in the TE example, which is whether to keep inventory for a specific spare part.

To save computational time, only 100 simulation runs are used to evaluate the total cost plus loss of a candidate PM policy (i.e., a chromosome). The objective value is the total cost plus losses for a 10-year span. The solutions (optimal and near-optimal) obtained by GA are re-evaluated by using a higher number of simulations (1000) to confirm the optimality of the solutions (it is the difference in objective values of the obtained solutions that matters, not their absolute values). Note that, since GA does not guarantee global optimality, the term "optimal" is meant to be the best possible obtained by GA. The computational time is 6 h 20 min. Results for the FCC unit example are shown in Fig. 18.4 and Table 18.7.

The inventory level for a group of equipment is calculated as the average inventory level of all types of spare parts serving that group of equipment. The inventory level for a specific spare part is in turn calculated as the number of stored items divided by the number of pieces of equipment it services (assuming one spare—one equipment relationship). For a specific group of equipment, the inventory levels for each type of spare part are optimized separately (they are generally different from each other but the difference is small) but only the average inventory level is reported. The result shows that, in general, a reasonable inventory level of 50% is recommended.

The results show that PM is used for all types of equipment under consideration in this FCC unit but too frequent PM is not recommended (the PM frequency is generally in the magnitude of 1.0*MTBF). There are two reasons for this.

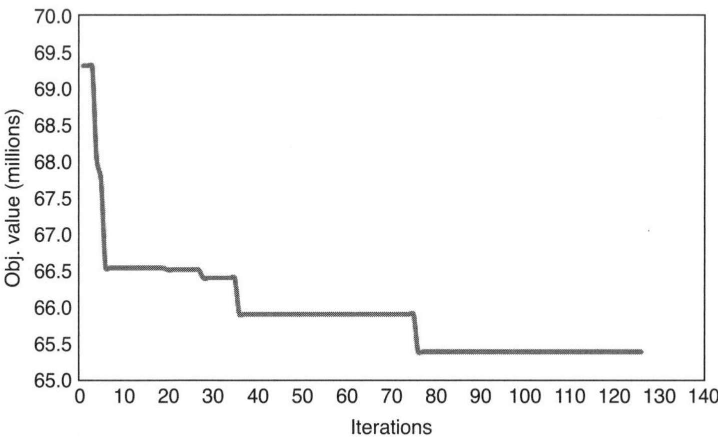

Figure 18.4 Optimal objective value by GA, FCC unit.

Equipment Group	Description	Group Size	PM Starting Time	PM Frequency	Inventory Level	Range of Economic Loss, Type 2 (Thousands/Day)
1	Pumps	14	1.3	0.9	0.29	0.01
2	Pumps	14	0.5	0.1	0.43	0.01
3	Pumps	3	1	0.35	0.8	10–73
4	Compressors	2	1.3	1.6	0.63	500
5	Heaters	4	1.3	1.05	0.5	100
6	Exchangers	13	1.1	0.85	0.54	7–9
7	Exchangers	12	0.5	0.9	0.47	15–33
8	Exchangers	6	0.7	1.05	0.39	45–58
9	Exchangers	6	1	1.3	0.33	115–450
10	Exchangers	13	0.6	1.2	0.56	5–7
11	Exchangers	9	0.8	0.7	0.3	8–15
12	Exchangers	10	0.7	0.9	0.47	3
13	Exchangers	9	0.7	1.6	0.56	22
14	Exchangers	9	1	1.2	0.37	3

Table 18.7 Optimal Maintenance Policy by GA Optimization for FCC Unit

Labor Group	Equipment Covered	Number of Employees
1	Rotating equipments (pumps, compressors)	2
2	Heaters, heat exchangers	5
	Total	7

TABLE 18.8 Optimal Labor Workforce Size, FCC Example

- For equipment with spare online, such as pumps, their PM does not interfere with production (favorable condition to perform PM) but the economic loss incurred by their failures is also small (the reduction in loss, which is the benefit or the incentive to perform PM, is small). Moderate PM frequencies (0.9, 0.1, and 0.35) are used for this type of equipment.
- For equipment without spare copy such as heat exchangers and compressors, it is assumed that their PM interferes with production (the extent of interference is quantified by the economic loss incurred during the maintenance time of the equipment). As Nguyen et al. (2008) pointed out, there are two competing effects of PM on economic loss for this type of equipment: as PM frequency increases (doing PM more often), the economic loss may decrease because PM reduces failure-induced downtime but the economic loss may also increase because PM increases PM-induced downtime. It may be beneficial to apply PM for this kind of equipment, but the PM should not be done so frequently. The result shows that, for exchangers with minor impact of PM activities on production (groups 6, 7, 10, 11, 12, 14), PM is used with moderate frequencies (from 0.7 to 1.2). For main process equipment (compressors, heaters, and exchangers group 8, 9, and 13) whose PM activities cause significant PM-induced downtime and economic loss, PM is generally not recommended (PM frequency ranges from 1.05 to 1.6) because the gain (reducing failures-related cost and lost) is shadowed by the undesirable side effect of PM (economic loss increases).

The optimal labor workforce size is shown in Table 18.8.

Contributions of different terms in the 65.39 million optimal objective value are as follows: 3.79 million of (PM + CM) cost, 2.8 million of labor cost, 0.56 million of inventory cost, and 58.25 million of expected economic losses. Thus, roughly 11% of the objective value is the total cost while the rest (89%) is the expected economic losses, that is, the economic losses account for a large part in the objective function. Thus, to reduce the total costs plus losses, it is necessary to reduce economic losses by maintaining sufficient resources (labor and spare parts) for punctuality of maintenance actions. The obtained optimal number of employees (7) is larger than the actual number in the actual FCC plant used in this example (5), and the optimal inventory level may also be greater than the standard level in industrial practice where minimal inventory level is desired.

References

Haupt, R. L. and S. E. Haupt, *Practical Genetic Algorithms*, 2d ed., Wiley-Interscience, New Jersey, USA (2004).

Mannan, S. (editor), *Lees' Loss Prevention in the Process Industries*, 3d ed., Elsevier Butterworth-Heinemann, Oxford, UK (2005).

Nguyen, D. Q, C. Brammer and M. Bagajewicz, "New Tool for the Evaluation of the Scheduling of Preventive Maintenance for Chemical Process Plants," *Ind. Eng. Chem. Res.* **47**(6):1910–1924 (2008).

Shum, Y. S. and D. C. Gong, "The Application of Genetic Algorithm in the Development of Preventive Maintenance Analytic Model," *The Int. J. Adv. Manuf. Technol.* **32**:169–183 (2006).

Tan, J. S. and M. A. Kramer, "A General Framework for Preventive Maintenance Optimization in Chemical Process Operations," *Comp. Chem. Eng.* **21**(12): 1451–1469 (1997).

Vassiliadis, C. G. and E. N. Pistikopoulos, "Maintenance-Based Strategies for Environmental Risk Minimization in the Process Industries," *J. Haz. Mat.* **71**: 481–501 (2000).

Wang, H. and H. Pham, *Reliability and Optimal Maintenance*, Springer Series in Reliability Engineering, Springer-Verlag, London, UK (2006).

CHAPTER 19

Value and Optimization of Instrument Maintenance*

It was mentioned in Chap. 17 that activities like the scheduling of the cleaning of heat exchanger preheating trains and instrumentation maintenance (calibration mostly) are run by crews that are fully devoted to the task and are not usually part of the overall maintenance unit. They are run fairly in separate. In this chapter, we discuss the case of instrumentation maintenance.

Despite the existence of data reconciliation and gross/bias detection techniques, and even when these techniques are being used in plant, very often instrument maintenance is performed using schedules that do not make use of any external information.

The right thing to do is to establish instrument maintenance schedules that will follow the bias detections that stem from using data reconciliation. Therefore, we discuss the merits of exploiting hardware redundancy for instrumentation maintenance.

Prior research work on the effect of maintenance on plant instrumentation performance is scarce. Lai et al. (2003) optimized a hardware-redundant sensor network used in a corrective maintenance (CM) program. They used Markov processes to evaluate the availability and the expected number of repairs and replacements under different sensor network configurations. The objective function, which includes sensor cost and maintenance cost, was minimized by using genetic algorithm. Sanchez and Bagajewicz (2000) investigated the impact of corrective maintenance in the design of sensor network where system reliability/availability, which is dependent on a corrective maintenance

*This chapter was written by DuyQuang Nguyen and Miguel Bagajewicz.

program, is used as an objective function or a constraint in the sensor network design problem. In both cases, the target is to optimally design sensor networks with consideration of the impact of corrective maintenance from a technical/cost perspective rather than a benefit/profit viewpoint. Liang and Chang (2008) simultaneously generated the design specifications and the corresponding preventive maintenance (PM) policies for protective systems (the systems of sensors and valves that perform two basic functions: alarm and shutdown). They used Markov processes to model maintenance activities and used integer programming to minimize the total expected expenditure, which is the sum of the capital investments, the expected maintenance costs, and the expected losses due to system failures. There is only one study of economic impact (benefit) of sensor maintenance activities and resources on an existing system (Nguyen and Bagajewicz, 2009). Because accuracy has been tied to economics (Chap. 12), one can now study the effect of reducing sensor failure through preventive maintenance.

Quite clearly, as is the case of general maintenance, the more one performs preventive maintenance, the less often the sensors fail, and therefore, the smaller the downside expected financial loss is. Because the maintenance has a cost, there is, as in the general case, a trade-off.

The economic benefit of performing a certain schedule of PM on sensors is given by the difference of the financial loss with PM and without PM (i.e., usually with CM only). To make this assessment in a very general conceptual fashion, we introduce a general optimization model to find the best maintenance scheme that renders maximum economical benefit.

Let ψ be the set of selected sensors subjected to PM, $\mathbf{T}$ be the vector of scheduled maintenance times for the selected sensors. The first element of this vector gives the maintenance time of the first sensor in ψ, and so on. We assume that the maintenance is cyclical. This assumption is a practical one.

Example 19.1: To clarify the issue, we use the three streams and one unit example presented in Chap. 11 to illustrate accuracy concepts (see Example 11.1). The flow sheet of this example is shown again in Fig. 19.1.

In this case, the set of sensors is $\psi = \{FM_1, FM_2, FM_3\}$ and the vector of scheduled maintenance time intervals $\mathbf{T}$ is $\{T_1, T_2, T_3\}$ (we assume that the maintenance is

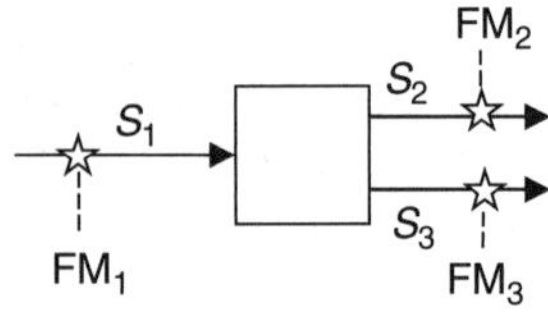

Figure 19.1 Flowsheet for Example 19.1.

cyclical), then the maintenance times of all sensors are $\{T_1, 2T_1, 3T_1, \ldots, kT_1\}$ for sensor FM_1, $\{T_2, 2T_2, 3T_2, \ldots, kT_2\}$ for sensor FM_2, $\{T_3, 2T_3, 3T_3, \ldots, kT_3\}$ for sensor FM_3.

To obtain the most appropriate schedule, one needs to construct the economic model. We first consider the benefit, which we compute as the difference between the downside expected financial loss DEFL(ψ, T^0) when the current level of preventive maintenance is performed (given by T^0), and the downside expected financial loss when the preventive maintenance with a schedule T is implemented [DEFL(ψ, T)]. Thus the benefit is given by

$$\text{Benefit} = \text{DEFL}(\psi, T^0) - \text{DEFL}(\psi, T) \tag{19.1}$$

The downside expected financial loss is, as we saw in Chap. 12, a function of the frequency at which only one bias shows, the frequency of two simultaneous biases, and so on. When PM has increased frequency, these frequencies are smaller.

Preventive and corrective maintenance have a certain cost given by a certain function, Cost(ψ, T). Thus the benefit will have to outweigh the change in cost.

$$\text{Cost change} = \text{Cost}(\psi, T) - \text{Cost}(\psi, T^0) \tag{19.2}$$

There are, however, some constraints. For example, performing preventive maintenance to the set of sensors ψ with a schedule T has to be implemented by a certain number of available personnel AP. Now the needed number of personnel is a function of the chosen schedule NP(ψ, T), which needs to be smaller than AP. In addition, or as an alternative, the number of labor hours needed for the preventive maintenance HN(ψ, T) needs to be smaller than the available hours AH. These are very difficult to express analytically and will not be attempted here.

If one wants to find the optimal schedule, the following optimization model is needed:

$$\left.\begin{aligned}
&\underset{T}{\text{Max}}\left\{[\text{DEFL}(\psi,T^0)-\text{DEFL}(\psi,T)]-[\text{Cost}(\psi,T)-\text{Cost}(\psi,T^0)]\right\}\\
&\text{s.t.}\\
&\qquad \text{NP}(\psi,T)\le \text{AP}\\
&\qquad \text{NH}(\psi,T)\le \text{AH}
\end{aligned}\right\} \tag{19.3}$$

The objective maximizes the benefit minus the cost. The optimization is done searching over all possible PM schedules T, but can also be done searching for the best set of installed sensors. In such case, the cost function should compute the cost of the new sensors in addition to the cost of the PM. We omit other constraints such as spare parts inventory decisions.

The above optimization model is very difficult to implement. One possible solution procedure would be using genetic algorithms, but to the date this book is written this has not been implemented. In addition, the model has shortcomings that are similar to those that were pointed out in general maintenance models, that is, it is very difficult to include all aspects in detail. We already mention that resource constrains (number of personnel available for PM, or for both PM and CM, number of total hours, inventory policies, delays stemming from unavailable spare parts, etc.) are difficult to introduce in analytic models. Thus we will resort to concentrating on using a Monte Carlo simulation model built using some realistic plant maintenance practices.

The following alternatives are considered:

1. All instruments are considered equally important, are subjected to PM, and are inspected at the same time. This is usually the case if PM is performed periodically by a third-party contractor. We call this strategy "periodic-all instruments at once."
2. All instruments are subjected to PM program, which is performed by available maintenance personnel at the plant. However, due to limited maintenance human resource, it is required to appropriately schedule maintenance activities so that all instruments are preventively maintained at least once during the time horizon. This is achieved by associating sensors to different groups, which are to be maintained at different scheduled times. We call this strategy "periodic-all instruments in sequential groups." The size of each group would be commensurate with the number of available employees.
3. Only some sensors are subjected to PM, because the human resource in the plant is limited. Usually, sensors that measure important variables for production accounting, critical control loops are selected for PM. We call this strategy "periodic-some instruments at once" or "periodic-some instruments in sequential groups," depending on the resource limitations.

To complement the above choices, the following decision variables are needed:

- Selection of important sensors that are subjected to preventive maintenance
- Cycle (periodic) time interval (CT) to perform PM
- Scheduling (group choices and sequential order) of PM of important sensors

To simplify the scheduling task we propose to

1. Choose the cycle (periodic) time interval CT and use it as a parameter.
2. Group important sensors into n groups. The number of groups of sensors (n) is chosen a priori and all groups have, if possible, the same number of sensors. In addition,
   - The maximum number of selected sensors is calculated as the available maintenance labor hours within the duration of the time horizon divided by the average labor hours for inspecting a sensor.
   - The selection of sensors to be included in each group is done in such a way that sensors with high failure rate or operating under harsh conditions (i.e., they are more likely to fail than the others) should be inspected first. Thus sensors are sorted in descending order of degree of urgent need for PM (e.g., descending order of failure rate), and added to the different groups until all candidates are chosen.
   - In the extreme cases where observability is lost (e.g., when sensor fails and data reconciliation is not used), we assume that the estimation of the variable during sensor's repair time is possible by using historical data. We assume a precision of 3 times the historical precision observed.
   - We first decide on the time interval between PM actions (TI) of two groups of sensors (illustrated in Fig. 19.2).
   - We assume that at the beginning ($t = 0$), all sensors are as good as new. The first time at which PM is performed (Tinitial) is a parameter in the model. We illustrate the scheduling in Fig. 19.2, where the allocation of different times for PM of two different sensor groups (1 and 2) is

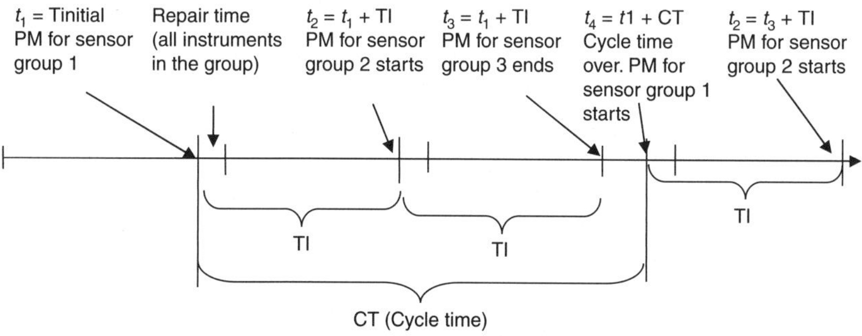

FIGURE 19.2 Allocation of preventive maintenance times for different sensor groups.

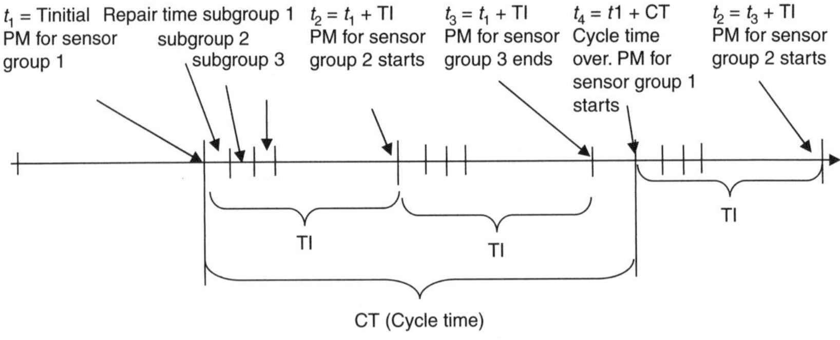

FIGURE 19.3 Allocation of preventive maintenance times with resource limitations.

shown for the "periodic-all instruments at once" option. Note that the cycle time need not be the same as the sum of the time intervals, that is, n TI ≤ CT. In turn, Fig. 19.3 shows the "periodic-all instruments in sequential groups" option. When only some instruments are maintained, the figures are conceptually equivalent.

3. In calculating the maintenance cost, we consider only the labor cost and the material costs (e.g., cost of spare parts).

4. Sensors can be inspected, calibrated online without interfering with production.

Financial Loss Evaluation

The downside expected financial loss is given by:

$$\text{DEFL} = \Psi^0 \text{DEFL}^0 + \sum_i \Psi_i^1 \text{DEFL}^1\big|_i + \cdots \sum_{i1,i2,\ldots,iN} \Psi^n_{i1,i2,\ldots,iN} \text{DEFL}^N\big|_{i1,i2,\ldots,iN} \tag{19.4}$$

The same Monte Carlo sampling procedure, as described in Chap. 11, is used (but only one type of bias is considered, which is the sudden fixed value bias). From the information obtained from the sampling procedure, the fraction of time that the system is in a specific state $\Psi^n_{i1,i2,\ldots,iN}$ can be obtained. Following the procedure described in Nguyen et al. (2006), the financial loss in a specific state $\text{DEFL}^N\big|_{i1,i2,\ldots,iN}$ (which are integrals with discontinuous integrand functions) can also be calculated based on information about (undetected) bias sizes provided by Monte Carlo simulation. The final financial loss is then calculated as summation of the product of $\Psi^n_{i1,i2,\ldots,iN}$ and $\text{DEFL}^N\big|_{i1,i2,\ldots,iN}$

Example 19.2: We now focus on the three streams, one unit case of Fig. 19.1. We highlight the effect of the different decision making, especially PM-related ones. We focus on accuracy, leaving financial losses and optimality to be discussed in the next example.

The sensor data is as follows (Bagajewicz, 2005): sensor precision $\sigma_i^2 = 1, 2,$ and 3; failure rate: 0.025, 0.015, 0.005 (1/day); and repair time: 0.5, 2, and 1 (day), respectively. It is also assumed that biases follow a normal distribution with zero mean and standard deviations $\rho_k = 2, 4,$ and 6, respectively. The accuracy of the estimator for product stream flowrate S_3 is calculated.

When any two out of the three sensors fail at the time and are removed for repair, the measurements associated with those fault sensors become unobservable. If historical data is recorded, it can be used to estimate the variable during repair time. The residual precision (during the repair time) is assumed to be 3 times the precision.

The following situations are considered:

- *Case* **1:** Neither data reconciliation (DR) nor any kind of maintenance is used; the bias in measurements is therefore not detected.
- *Case* **2:** There is no data reconciliation but PM is used; hence, sensor failures can be detected by sensor inspection only, which takes place every 365 days.
- *Case* **3:** Data reconciliation is used without any kind of maintenance. It is assumed that when a bias is detected due to sensor failure, the measurement is simply ignored.
- *Case* **4:** Data reconciliation is used together with CM; hence, when a bias is detected through software-detection techniques, the sensor is repaired and resumes service afterward. No PM is used.
- *Case* **5:** Data reconciliation is used together with CM and PM; hence, sensor failures can be detected either by data-treatment techniques or by sensor inspection under a PM program. For this case, consider the three PM schemes as follows:
  - All three sensors are inspected every 180 days.
  - All three sensors are inspected every 365 days.
  - Only the sensor measuring the product stream is inspected every 180 days.

The number of Monte Carlo sampling is 10^5. The results are shown in Table 19.1. The computation time varies from less than 5 s when neither data reconciliation nor maintenance is used to about 8 min when data reconciliation and PM are used. This computation time reduces if fewer samples are taken.

Although cases 1 and 3 are unrealistic (at least some basic types of maintenance policies are implemented in the plant), they are included for comparison. The results show how much the use of maintenance and data reconciliation improves accuracy and how much more the use of preventive maintenance improves accuracy. Moreover, they also assess the impact of frequency. Data reconciliation (together with gross error detection) improves accuracy by reducing the effect of random noises and detecting biases above threshold values. The effect of the maintenance policy is better explained when one looks at the behavior of accuracy value with time. We show in Fig. 19.4 the accuracy value as function of time for the cases 1, 2, 3, and 5b.

Cases	Data Management	Maintenance Management	Accuracy
1	No data reconciliation	No maintenance	5.9905
2		PM every 365 days	4.2879
3	Using data reconciliation	No maintenance	3.2089
4		Only CM	3.0752
5a		CM and PM every 180 days for all sensors, all at once	2.0668
5b		CM and PM every 365 days for all sensors, all at once	2.4868
5c		CM and PM every 180 days only for sensor S_3	2.2694

TABLE 19.1 Calculation Results for Example

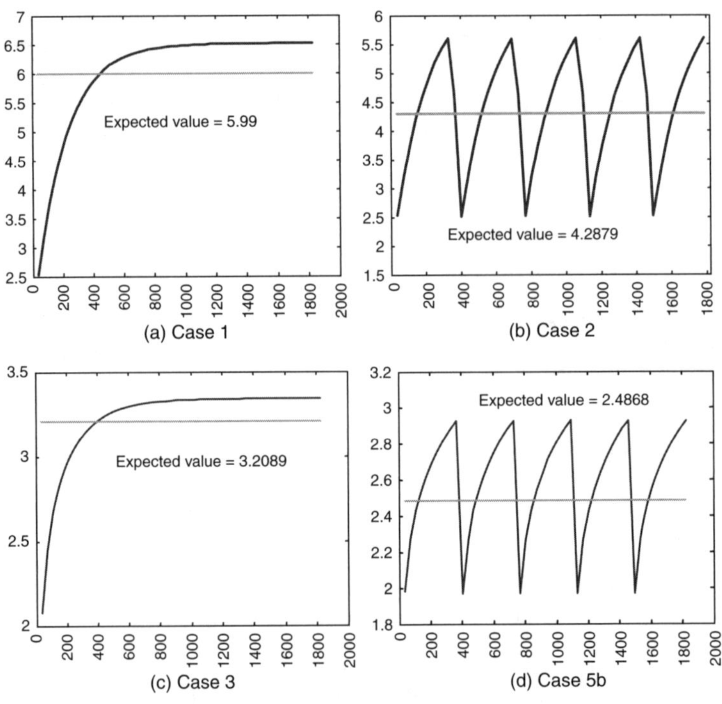

FIGURE 19.4 Accuracy at specific points in time when preventive maintenance is used.

Case 4 (data reconciliation and CM are used) was already shown by Nguyen and Bagajewicz (2009) and it has the same trend as the one shown for case 1 and case 3, that is, the accuracy value increases progressively with time. When PM is used (case 2 and case 5b), at the time sensors are inspected, undetected biases (when no data reconciliation is used or they are too small to be detected) are eliminated because the sensors are repaired. The results of using PM are

1. Accuracy improves (e.g., accuracy in case 2 is less than in case 1).
2. Accuracy increases with time, but it will be improved (back to the level of "as good as new" if the maintenance is perfect) at the scheduled PM times. Obviously, the larger the frequency of maintenance, the shorter the cycles, and therefore the accuracy is smaller.

We now focus on the effect of maintenance polices on financial loss by considering another example, a larger-scale process. This example is used to demonstrate the use of expected financial loss in the planning/scheduling of preventive maintenance.

Example 19.3: Consider the process of Example 11.7. We repeat its flow sheet in Fig. 19.5. The total flowrates are variables of interest. Assume that all streams are measured. The flowrates are the same as those given in Example 11.7 (Table 11.10).

The parameters for this example are

- Sensor precision = 2.5% (for all sensors).
- Sensor failure rate: $\lambda_i = 0.01$ (1/day), $i = 1, 3, 5, ..., 23$ and $\lambda_i = 0.02$ (1/day), $i = 2, 4, 6, \ldots, 24$. Sensor repair time $R_i = 1$ day, $i = 1, 3, 5, \ldots, 23$ and $R_i = 2$ days, $i = 2, 4, 6, \ldots, 24$.
- A time horizon of 2 years is used.
- The first PM time (Tinitial) is half of the cycle time (CT).
- The economical benefit of using PM is calculated using the difference of downside financial losses: DEFL (with PM) – DEFL (without PM).
- The average hours for inspecting/calibrating a sensor is 2 h.

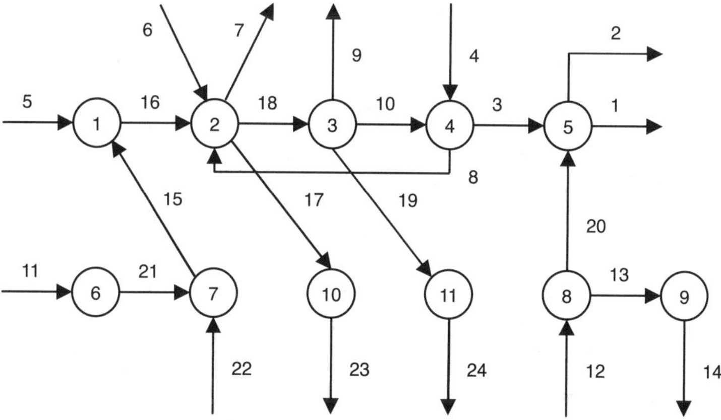

FIGURE 19.5 Example 19.3.

- Cost of PM and CM includes labor cost and material cost.
- If PM is performed by a contractor, the PM labor cost (including traveling cost to the site, other costs) is \$50/h. The PM cost (per month basis) is calculated as follows: 50 (\$/labor hours) × 2 (labor hours/sensor) × 24 sensors × (number of PM cycles in a year)/12 (months).
- Maintenance labor cost (for CM and PM) in the plant is calculated as the number of employees × 4000 (\$/person/month).
- The materials cost for CM (e.g., cost of part replacement) is estimated to be 75% of the price of the brand new sensor, the materials cost for PM (e.g., cost of lubricating oil, calibrating agents) is estimated to be 5% of the price of the brand new sensor. Prices of sensors are 1000 for sensors 1, 2, 5, 6, 7, 11, 12, 22, 23, 24; 800 for sensors 3, 4, 8, 9, 10, 13, 14; and 600 for sensors 15, 16, 17, 18, 19, 20, 21.
- The time window of analysis T in the calculation of financial loss is 30 days (this is based on the argument that by mean of production accounting calculation every month, one can detect the loss in production that has been covered by biased measurement).
- The cost of product (or cost of inventory) (Ks) per day is \$5000.
- Cost of license for data reconciliation (per month basis) is \$2000.

The product stream for which the financial loss is calculated is stream S_1. The following assumptions are made:

1. No resource limitation for corrective maintenance is included, that is, *all* recognized fault sensors will be repaired in time, right after the failures are identified.
2. Resource limitations on PM are taken care of by choosing the different options of grouping the sensors. For example, when the sensors are arranged in groups of three, it is assumed that there are resources to inspect three sensors at a time.

The number of Monte Carlo samplings used is 1.10^4. We used a 2.8-GHz Intel Pentium, 1028-MB-RAM PC for this example. The computation time varies from about 45 min when neither data reconciliation nor maintenance is used to about 4 h when data reconciliation and PM are used.

For completion and comparison, we will discuss all PM strategies and compare them with the use of CM only or no maintenance at all.

Periodic maintenance of all instruments at once: Table 19.2 summarizes all gross benefits as compared to the case of no maintenance and no data reconciliation (first row). It is assumed that, if maintenance (CM and PM) is used, there is one employee providing fast-response CM in the plant while PM is provided by third-party contract. The total cost includes material cost of CM (determined from simulation), labor cost for CM (one employee), material and labor cost for PM (dependent on PM schedule), and the license fee for data reconciliation. All the cost results are shown on per month basis. All the assumptions stated above are also applied in this example.

Table 19.2 shows clearly the benefit of using data reconciliation (to detect biases) and preventive maintenance (to detect hidden failures and preserve sensor condition): both of them significantly improve accuracy and reduce financial loss. The best result (lowest financial loss) is obtained when both data reconciliation and PM are used.

Data Management	Maintenance Management	Accuracy S_1	Financial Loss (DEFL)	Gross Benefit	Total Cost	Net Benefit
No data reconciliation	No maintenance	4.46	$150,750	0	0	0
	CM and PM every 180 days, all sensors	3.59	$130,935	$19,815	$6,890	$12,925
	CM and PM every 90 days, all sensors	3.21	$122,235	$28,515	$8,506	$20,009
With data reconciliation	No maintenance	3.86	$123,030	$27,720	$2,000	$25,720
	CM only	2.55	$90,495	$60,255	$6,927	$53,328
	CM plus PM every 180 days, all sensors	2.14	$43,455	$107,295	$9,938	$97,357
	CM plus PM every 90 days, all sensors	1.97	$32,715	$118,035	$11,476	$106,559

Table 19.2 Results for Example 19.3—Strategy: "Periodic-All instruments at Once"

Periodic maintenance of all instruments in sequential groups: For this maintenance strategy, CM and PM are provided by available maintenance employees and the cost is calculated correspondingly. We consider two cases related to the number of maintenance employees available: case a, 3 people and case b, 1 person. Grouping of sensors (case a) is done in numerical order (from 1 to 24): three sensors are included at a time. For case b, sensors are sorted according to their failure rate. The sorted sensor list is: sensors 1, 3, 5, 7, 9, 11, 13, 15, 17, 19, 21, 23 (failure rate = 0.01) then sensors 2, 4, 6, 8, 10, 12, 14, 16, 18, 20, 22, 24. The PM schedule described by the cycle time (CT), PM starting time (Tinitial), and time interval (TI) between groups of sensors (or between individual sensor as in case b) together with calculated results are shown in Table 19.3. The gross benefit is calculated against the case of no data reconciliation, no maintenance.

Case b renders much better benefit than case a because case b achieves the same job (preventively maintaining all sensors in a 90-days cycle); hence essentially the same financial loss at a much lower labor cost (1 employee vs. 3 employees in case a). The results show that the financial loss depends mainly on the cycle time (i.e., PM frequency) and is relatively insensitive to the sequence at which the sensors are preventively maintained. The results also suggest that one employee is enough for this 24-sensor system.

Periodic maintenance of some instruments in sequential groups: We now explore the case where only some sensors are of concern and need PM and the rest is subjected to CM only. This is the case when maintenance resources (labor, tools, budget, etc.) are limited and PM for all instruments is impossible. Assuming that there is one employee responsible for both the process and the utilities system (hence labor resource is limited), then prioritization of PM duties is needed in this case. We assume that only 12 out of 24 sensors are preventively maintained and choose a cycle time of 6 months. We consider two cases: c (sensors 1–12 are selected) and d (sensors 13–14 are selected). Description of the two cases and the calculated results are shown in Table 19.4.

For case c, the financial loss is lower than in case d. This is expected because the product stream is stream S_1 and therefore, if sensors measuring product stream flowrate and the streams that are redundant with it (directly connected to stream S_1 through data reconciliation) are selected for PM, the accuracy value and also the financial loss is lower than the case that are not measured (case d).

Case	Maintenance Management	Accuracy S_1	Financial Loss (DEFL)	Gross Benefit	Cost	Net Benefit
a	CM plus PM, 3 sensors in a group, CT = 90 days, TI = 10 days, Tinitial = 15 days	2.02	\$33,948	\$116,802	\$31,691	\$86,111
b	CM plus PM, one group, CT = 90 days, TI = 3 days, Tinitial = 15 days	2.01	\$33,255	\$1174,95	\$14,719	\$102,776

Table 19.3 Results for Example 19.3—"Periodic Maintenance of All Instruments in Sequential Groups"

Case	Maintenance Management	Accuracy S_1	Financial loss (DEFL)	Gross Benefit	Cost	Net benefit
c	CM plus PM, sensors 1–12 are selected, CT =180 days, TI = 15 days, Tinitial = 10 days	2.25	$43,575	$107,175	$12,531	$94,644
d	CM plus PM, sensors 13–24 are selected, CT =180 days, TI = 15 days, Tinitial = 10 days	2.44	$72,645	$78,105	$12,118	$65,987

Table 19.4 Results for Example 19.3—Strategy: "Periodic Maintenance of Some Instruments Sequentially"

References

Bagajewicz, M., "On a New Definition of a Stochastic-based Accuracy Concept of Data Reconciliation-Based Estimators," *Proc. 15th European Symposium on Computer-Aided Process Engineering* (2005).

Lai, C.-A., C.-T. Chang, C.-L. Ko, and C.-L. Chen, "Optimal Sensor Placement and Maintenance Strategies for Mass-Flow Networks," *Ind. Eng. Chem. Res.* **42**(19):4366–4375 (2003).

Liang, K.-H. and C.-T. Chang, "A Simultaneous Optimization Approach to Generate Design Specifications and Maintenance Policies for the Multilayer Protective Systems in Chemical Processes," *Ind. Eng. Chem. Res.* **47**(15):5543–5555 (2008).

Nguyen, D. Q. T., Kitipat Siemanond, and M. Bagajewicz, "Downside Financial Loss of Sensor Networks in the Presence of Gross Errors," *AIChE J.* **52**(2):3825–3841 (2006).

Nguyen, D. Q. T. and M. J. Bagajewicz, "On the Impact of Sensor Maintenance Policies on Stochastic-Based Accuracy," *Comp. Chem. Eng.* submitted (2009).

Sanchez, M. C. and M. J. Bagajewicz, "On the Impact of Corrective Maintenance in the Design of Sensor Networks," *Ind. Eng. Chem. Res.* **39**(4):977–981(2000).

Index

Note: Italicized "*f*" and "*t*" indicate figures and tables, respectively.